Ultra-Wideband Short-Pulse Electromagnetics 4

Ultra-Wideband Short-Pulse Electromagnetics 4

Edited by

Ehud Heyman

Tel Aviv University
Tel Aviv, Israel

Benjamin Mandelbaum

Rafael
Haifa, Israel

and

Joseph Shiloh

Rafael
Haifa, Israel

Kluwer Academic / Plenum Publishers
New York, Boston, Dordrecht, London, Moscow

Proceedings of the 4th Ultra-Wideband Short-Pulse Electromagnetics Conference, held June 14–19, 1998, in Tel Aviv, Israel

ISBN: 0-306-46206-0

233 Spring Street, New York, N.Y. 10013

10 9 8 7 6 5 4 3 2 1

A C.I.P. record for this book is available from the Library of Congress

Printed in the United States of America

PREFACE

The first and the second international conferences on Ultra-Wideband (UWB), Short-Pulse (SP) Electromagnetics were held at Polytechnic University, Brooklyn, New York in 1992 and 1994, while the third conference was held in 1996 in Albuquerque, New Mexico. The first two conferences were dedicated only to UWB-SP, but the last one has been organized under the auspices of AMEREM'96, which included, in addition, the 10th Conference on High Power Electromagnetics (HPEM-10) and the third conference on Unexploded Ordnance Detection and Range Remediation (UXO-3). The same format has been maintained in the fourth UWB-SP Conference, which was held in June 14–19, 1998 in Tel-Aviv, Israel, under the auspices of EUROEM'98, together with HPEM-11 and UXO-4. Joseph Shiloh and Benjamin Mandelbaum chaired the conference, while Ehud Heyman coordinated the UWB-SP part. The fifth conference in the UWB-SP series will be held as a part of the EUROEM'2000 in May 20–June 2 in Edinburgh, Scotland.

The purpose of the first conference was to focus attention on new advanced technologies for generating, radiating, and detecting UWB-SP signals, on mathematical modeling of short pulse propagation and scattering phenomena, and on potential future applications. Since then we have witnessed the emergence of new systems, applications, and theoretical models. The papers which appear in this volume, the fourth in the UWB-SP series, update subject areas from the earlier UWB-SP conferences. These topics include: Pulse generation and detection, Broadband electronic systems, Antennas—theory, design, experiments and systems, Pulse propagation, Scattering theory, Signal processing, and Buried targets—detection and identification.

The editors of this volume wish to thank many individuals and institutions that contributed to the success of this conference including those involved in the two parallel conferences (HPEM/NEM and UXO), for their efforts and participation. Special thanks are devoted to Carl E. Baum who was the driving force behind these conferences.

The editors and the Organizing Committee thank the Summa Foundation, the Permanent HPEM Committee, and Yuma Proving Ground for their sponsorship and assistance in organizing the conference. The sponsorship of the following Institutes and organizations is appreciated: IEEE, URSI, Los-Alamos National Laboratory (USA), United States Air Force, European Office of Aerospace Research and Development, Maxwell Technologies and Maxwell Physics International (USA), Ministry of Science (Israel), Ministry of Defense (Israel), Ministry of Tourism (Israel), and RAFAEL (Israel). Euroem'98 was hosted by RAFAEL (Israel).

Ehud Heyman
Benjamin Mandelbaum
Joseph Shiloh
December 1998

CONTENTS

PULSE GENERATION AND DETECTION

ANTENNAS

BROADBAND SYSTEMS AND COMPONENTS

SCATTERING: MODELING, MEASUREMENTS, AND PROCESSING

BURIED OBJECTS

GENERATION OF HIGH-POWER SUBNANOSECOND PULSES

G.A.Mesyats, S.N.Rukin, V.G.Shpak, M.I.Yalandin

Institute of Electrophysics Russian Academy of Sciences
34 Komsomolskaya St., Yekaterinburg, Russia

INTRODUCTION

The formers of short (0.1-1.0 ns) pulses based on spark gaps remained for long the only type of low-impedance (<100 Ohm) subnanosecond devices capable of producing an output voltage of 100 kV and more. For 22 years elapsed since the appearance of the article [1] that described the pioneering experiments on the production of high-power electron beams of 200-500-ps pulses of voltage peaking at 200-300 kV, much progress has been achieved in research and development of high-current, high-voltage systems operating on the subnanosecond scale. First, high-voltage modulators have been improved to miniaturize them [2], to increase the pulse repetition rate [3], and to simplify the retuning of their parameters (duration, pulse risetime and waveform) [4]. Second, the progress in research and technology of high-voltage semiconductor switches has led to the appearance of conceptually new high-current generators operating on the nanosecond and subnanosecond scales [5,6]. Finally, in the last decade, high-current subnanosecond devices have found new applications in vacuum and gas breakdown studies [7], for the production and investigation of the dynamics of short high-density electron beams [8], for the production of high-power short microwave pulses [9], and for the formation of superwideband radiopulses [10].

GENERATORS WITH SPARK GAPS

Any subnanosecond generator consists of a series of energy compression units each containing an energy store, a switch, and energy input/output systems. Any attempt to peak and/or shorten the original pulse to less than 1 ns bypassing the nanosecond compression stage at the preceding stage generally fails because of the difficulties related to the need to have a sufficiently high level of prebreakdown overvoltage across fast switches. Several design concepts for subnanosecond generators are known which are based on

- shortening the voltage pulse risetime immediately at the output of the nanosecond driver (Fig. 1a);
- energy compression in the output, subnanosecond, energy storage unit (Fig. 1b), and
- cutting a subnanosecond pulse from a nanosecond one (Fig. 1c,d).

In any version of subnanosecond generator, the output stage is an additional unit of the nanosecond driver which is largely responsible for the operation of the generator as a whole. We have developed a series of compact high-voltage repetitive pulse generators of the Radan type [2,3] (25-50 Ohm, 100-200 kV, 2-5 ns, 1000-100 Hz) that are currently widely used as drivers in subnanosecond modulators. Among various types of fast switch used in high-repetitive-rate pulsed systems, high-pressure gas gaps are most popular. From the viewpoint of safe operation of compact devices, nitrogen is preferable to hydrogen. With that, the operating pressure in these gas switches is 40 atm and more.

A generator with a short energy store (see Fig.1b) is capable of producing a pulse of highly reproducible duration, but it calls for additional high-voltage insulators which reduce reliability. Since all switches shown in the circuit diagram operate in the self- breakdown mode, the instability of their operation

Ultra-Wideband, Short-Pulse Electromagnetics 4
Edited by Heyman *et al.*, Kluwer Academic / Plenum Publishers, New York, 1999.

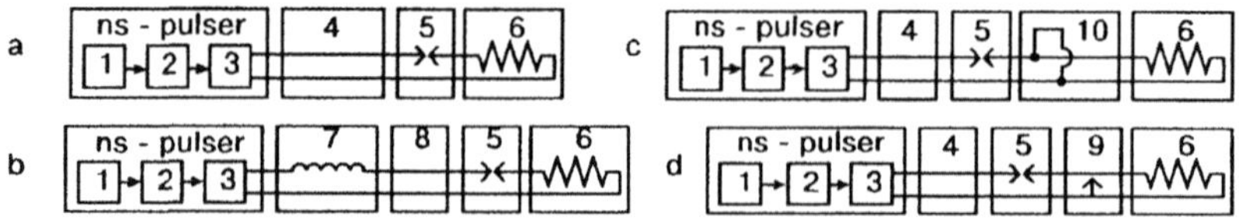

Fig.1. Design principles of subnanosecond generators. a- formation a nanosecond pulse with a short leading edge; b- energy compression in a subnanosecond pulse-forming unit; c,d- cutting the subnanosecond pulse out of a nanosecond pulse. 1 - charging device, 2 - high-voltage energy store, 3 - nanosecond switch, 4 - transmitting line, 5 - peaking switch, 6 - load, 7 - inductance, 8 - subnanosecond energy store, 9 - cutoff switch, 10 - short circuited stab.

makes the energy gain of the generator version presented in Fig. 1b insignificant compared to that of the versions illustrated by Fig.1c,d.

In out experiments with subnanosecond generators we conventionally used the circuit design (Fig.1d) with a peaking gap and a cutoff gap which make it possible to rather easily vary the output pulse duration over a rather wide range [4]. The output pulse duration for such a slicer generator is determined by the subnanosecond time lag between the operations of the cutoff gap and the peaker. Each gap is equipped with a gear for varying the gap spacing during the operation of the generator, which allows a smooth variation of the output pulse peak voltage and duration. Slicer versions with a varied number of serial peaking spark gaps have been tested (Fig.2). For all versions the voltage rise rate at the first peaking gap was the same and equal to $\sim 0.5 \times 10^{14}$ V/s, which was sufficient for multichannel switching.

In a slicer with a single-gap peaker, a 160-kV pulse with a minimum FWHM of ~300 ps has been obtained. Shorter pulses could be produced only on decreasing peak voltage. The voltage rise rate dV/dt of this subnanosecond pulse was $\sim 5 \times 10^{14}$ V/s. The voltage fall rate for the so produced pulse was 1.5-2 times greater than the rise rate, since the cutoff gap operates with a subnanosecond overvoltage. This makes it possible to reduce the inductance of the cutoff gap. Owing to the fact that the voltage rise rate and the electric strength increase at every next spark gap, a pulse of ~150 kV with a FWHM of 150 ps (dV/dt = 10^{15} V/s) could be produced by a three-gap peaker. The pulse fall time (50 ps) corresponded to the transient response of the oscilloscope used.

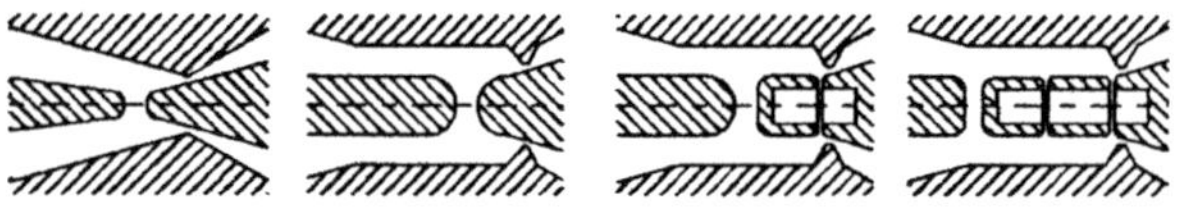

Fig.2. Design performance of electrodes of the peaking and cutoff gas spark gaps.

Subnanosecond slicers with a single-gap peaker were used to generate subnanosecond pulses with a repetition rate of 100 and 1000 Hz by reducing the output peak voltage to 100 and 70 kV, respectively. Used as a nanosecond driver was a multipurpose double-pulse-forming-line (DPFL) generator of the RADAN series for the first case and a single-pulse-forming-line modulator [3] for the second one. For systems operating with a high pulse repetition rate, low-jitter capability is of particular importance. It has been demonstrated [10] that the slicer spark gaps operated with jitters low enough for a digital stroboscopic oscilloscope (e.g., Tek TDS-820) could be used. In this case, signals are digitized with the total number of readings equal to 500. It follows

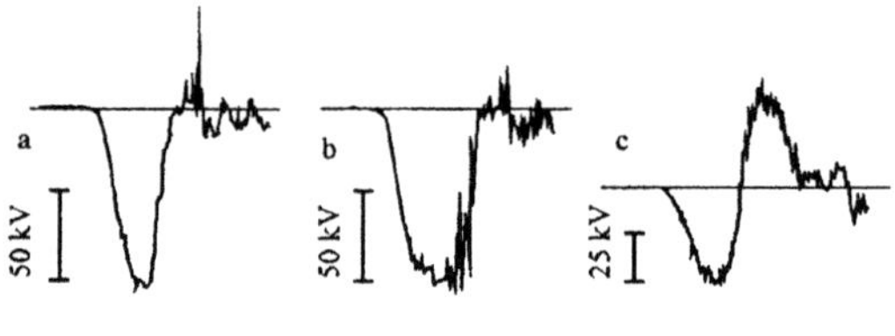

Fig.3. a,b-unipolar pulses formed by the slicer; c-pulse (a) as it transformed by the passive waveforme converter

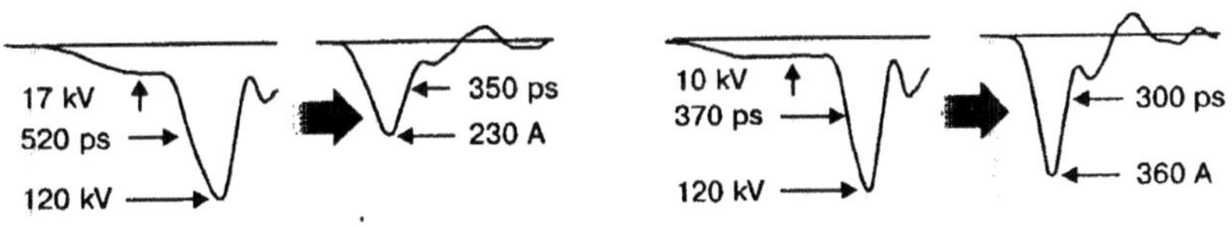

Fig.4. Reduced-width accelerating pulse accompanied with extended pre-pulse provides highest peak current and the charge of subnanosecond electron bunch

from the signal waveform (Fig.3a) that the 250-ps pulse is stable by amplitude to ±2-3%; with that, the pulse duration is stable at a level of 50 ps. Increasing the pulse duration to 500 ps (Fig.3b) by varying the gap spacing of the cutoff gap doubles jitter. This is due to a decrease in the prebreakdown voltage of the cutoff gap.

Stability of the pulse duration is important for subnanosecond systems designed for producing superwideband probing signals. It is this stability that is responsible for the reproducibility of the radiation spectral characteristics from pulse to pulse. Replacement of the cutoff gap by a short-circuited loop (see Fig. 1c) substantially improves stability; however, the energy efficiency of such a system is conceptually not above 50 %. Moreover, the main pulse is followed by an inverted-polarity stray pulse. Nevertheless, the system based on a short-circuited loop works well when it is required to transform an already formed subnanosecond unipolar pulse of a modulator into a bipolar pulse not having a constant component in its spectrum. The stability of the waveform of such a pulse (see Fig.3c) is entirely determined by the stability of the original unipolar pulse of the modulator.

APPLICATION OF SUBNANOSECOND PULSES

Historically, vacuum diodes with simple cold cathodes were used first as loads for subnanosecond high-voltage generators [1]. They were used to produce high-density electron beams (200-300 keV, 2-3 kA, 10^5-10^6 A/cm^2). Further development of this work has made it possible to create repetitive subnanosecond pulsed accelerators [7,8] that have become a basis for a new experimental lead for further research in pulsed power: the studies on the generation of subnanosecond wideband pulses of superradiation (SR) [9].

Experiments [7] have been performed to study the features of the formation of a tubular electron beam in a coaxial magnetically insulated coaxial diode. It has been discovered that the nanosecond prepulse plays a dominant role in the initiation of emission centers at the cathode. It has been established that for pulses with $t_{0.5}$~200 ps at a voltage of ~75 kV the high-current stage of electron emission was not attained at all, notwithstanding that the macroscopic electric field at the cathode surface was as high as 2-3 MV/cm. In the same conditions for a peak voltage of ~120 kV, increasing the time for which the prepulse was effective made it possible to obtain a higher current and a greater total charge of the beam electrons even at a shorter accelerating voltage pulse (Fig.4). The stability of operation of the subnanosecond modulator combined with the capabilities of the wideband pickups and digital instrumentation used have allowed real-time investigations of the behavior of the cross structure of a magnetized subnanosecond electron bunch and of the dynamic processes accompanying its transportation [8]. The use of a subnanosecond high-current electron beam has made it possible to provide evidence in support of the hypothesis that the Dicke superradiation effect known in quantum electronics has its classical analog in the microwave range.

Theoretically it has been shown that the superradiation in the microwave range may occur only in a single blob of electrons oscillating in an undulator and rotating in a uniform magnetic field (cyclotron SR). Similar effects take place at Cherenkov interaction of a short electron bunch with slow waves generated in corrugated metal waveguides or in dielectric slowdown systems (see, e.g. [11]). In the experimental

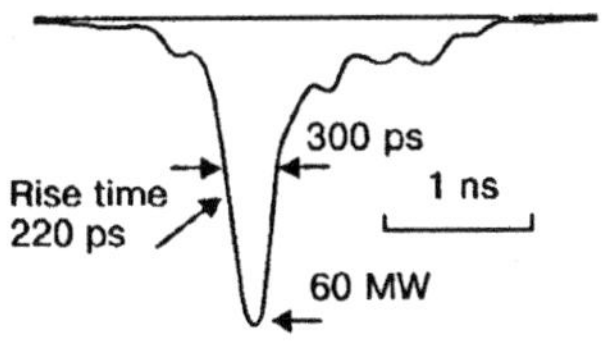

Fig.5. 38-GHz superradiation pulse probuced by subnanosecond electron beam

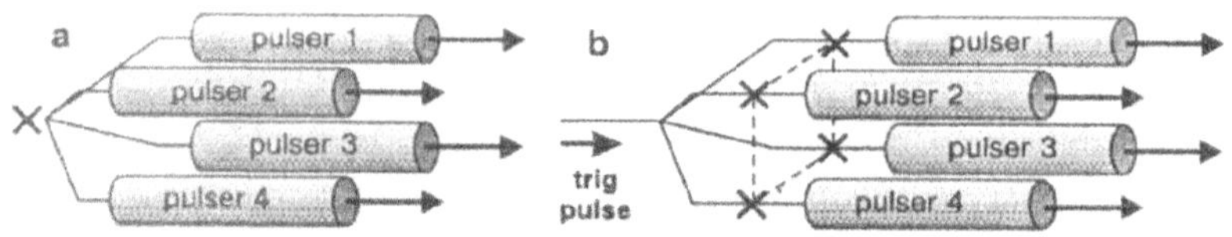

Fig.6. Basic design performance of multisection subnanosecond modulators

verification of the SR effect, the initial stages of transient processes in relativistic microwave devices such as FEL, CARM, BWO, and Cherenkov masers, could be observed in real time. The observations were performed under conditions where there were no feedback since the spatial length of the electron bunch (5-10 cm) was equal to or much less than the length of the interaction space (10-30 cm). As for now, for a beam of duration < 1 ns interacting with the reverse harmonics of the TM_{01} wave of a corrugated cylindrical waveguide, pulses with a central frequency of 38 GHz, a duration of 300 ps, and a peak power of 50-60 MW have been produced. It should be noted that a pulse with these parameters (Fig.5) includes a little more than ten periods of microwaves and has an unique bandwidth of several percents. It is of importance that this device realized by joint efforts of the Institute of Electrophysics (Ekaterinburg), the Institute of Applied Physics (Nizhny Novgorod), and Strathclyde University (UK) operates with a pulse repetition rate of several tens of hertz and has a beam-to-radiation energy conversion efficiency at a level of 5%.

The field of uses of the subnanosecond generators is extended substantially on enhancement of the amplitude stability of the output pulses and on provision of the possibility of their rigorous synchronization. Of importance is the possibility to start synchronously several microwave generators or amplifiers, superwideband antennas with pulsed excitation, and the like. For these cases, it is proposed to create synchronous or phased microwave radiators with superhigh-power unit components (Fig.6). These prospects call for a highly low-jitter operation of the gas switches of the high-voltage modulator output stage. In the experiment described in Ref. 12, the 200-kV triggered three-electrode gap of a RADAN-303 generator was synchronized to 0.5-ns jitter when started by a pulse of duration 300 ps. To attain this, for the gap voltage control range up to 5%, the

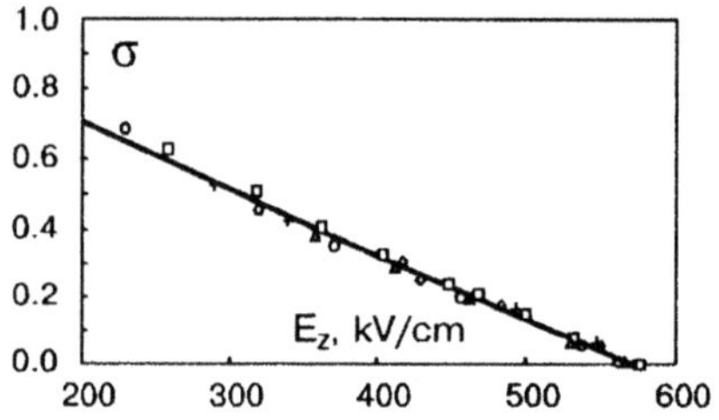

Fig.7. Control band of the spark gap measured at subnanosecond triggering mode as a function of electric field strength at the positive electrode

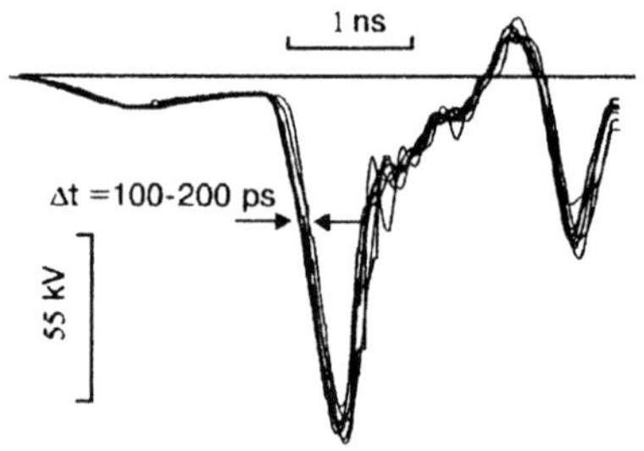

Fig.8. Output pulses of two-section subnanosecond modulator accumulated at the oscilloscope monitor. Envelope performance includes triggering jitter (up to 100 ps) of transient digitizing oscilloscope Tek-7250

starting pulse energy was no more than 10^{-3} of the switched energy and the peak voltage was no more than 10% of the charge voltage of the terminated energy store. Increasing the energy of the starting pulse extends substantially the control range for the switch (Fig.7). It has been shown experimentally that the main prerequisite for low-jitter operation of the gas gap switch is the advancing subnanosecond breakdown between the trigger electrode and the opposite potential electrode, stimulated by the starting pulse.

The RADAN nanosecond DPFL generators, being the base for subnanosecond modulators, may be connected in parallel (be sectioned) with the use of a special common spark gap and thereby be rigorously synchronized (see Fig.6a). As the spark gap operates, each DPFL connected in parallel discharges into its own load. With that, the pulses formed have equal amplitudes. A double-channel generator [12] designed based on this concept was used to produce synchronous subnanosecond pulses. To do this, a slicer was connected at the output of each generator. The root-mean-square jitter of the risetimes of the subnanosecond pulses formed by these slicers was 70-80 ps, which is quite acceptable for summing signals of duration 0.5-1 ns (Fig.8).

Superwideband RF or microwave radiation can be produced without an electron beam by direct conversion of a short pulse into electromagnetic energy with the use of a TEM antenna. In view of the concern with the electric strength of output circuits and antennas, the systems producing a short pulse consisting only of leading and trailing edges and having no plateau are most compact. The subnanosecond modulators with the input impedance equal to 50 om based on the RADAN drivers and slicers comply with this requirement and are used to create radiators with a characteristic frequency of a few gigahertz at ~200 MW of power delivered to the antenna [10]. To excite a superwideband antenna, it is beneficial to use bipolar pulses whose spectral function turns to zero in the low-frequency limit. To do this, the subnanosecond unipolar pulse of the modulator was transformed into a bipolar pulse ($\pm$ 50-100 kV) using either a system based on a short-circuited loop or a more efficient but less stable "active" pulse former with two independent spark gaps operating with less than 100-ps jitter [10].

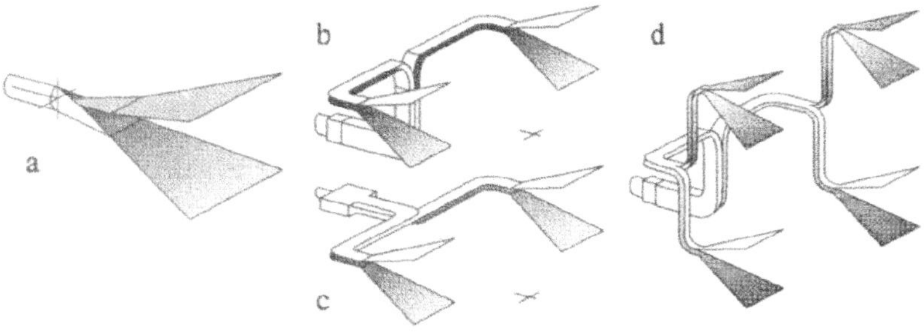

Fig.9. TEM-horn antennae: a-single; b,c-dual (synphased and opposite phase, respectivelly); d- two dimensional, four antennae array

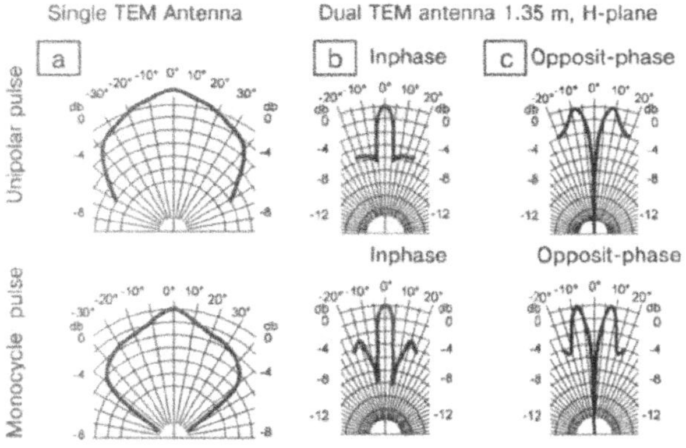

Fig.10. Directional diagrams of the TEM antennae

Based on the repetitive pulse modulators, a series of superwideband generators with single TEM antennas (Fig.9a) or antenna systems consisting of two (Fig.9b, c) or four (Fig.9d) elements have been created [10,13]. The sectioned high-voltage TEM antennas were powered by a 300-600 ps pulse produced by one or two rigorously synchronized modulators. With the linear dimension of the "output aperture" being ~2 m, they are able to synthesize a directional diagram with a 10° main lobe at a level of -6 dB in the H plane or in both H and E planes (Fig.10). When used in experiments on location of closely situated objects, a generator with a single TEM antenna has demonstrated ~20-30 cm resolution [13]. The electric field at the reception site being ~30 m off the generator was at a level of some tens of volts per centimeter.

The studies performed have provided comprehensive experimental data on the electric strength of the air insulation on the nanosecond and subnanosecond scales under the action of high voltage pulses and pulses with a microwave carrier (Fig.11). All measurements were carried out in the atmospheric air at an ordinary ambient pressure. Used as test gaps were elements of cylindrical and strip waveguides and coaxial lines. These data testify to the limiting capabilities of the air insulation to be used in further work aimed at increasing both the peak and the average power of the sources of wideband microwave pulses and superwideband radiopulses.

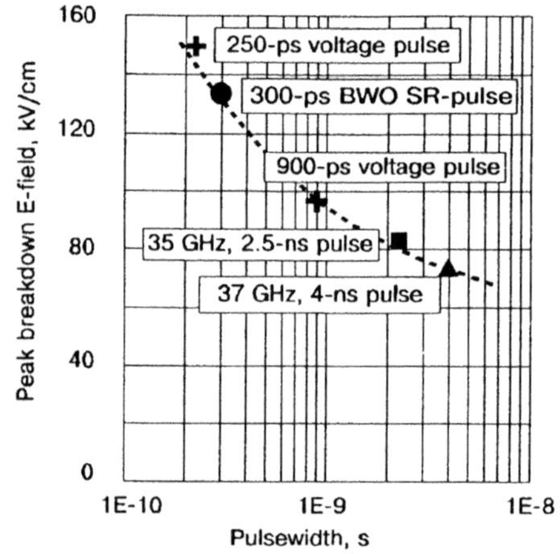

Fig.11. Breakdown strength of ambient air with affect of K-band microwave pulses and subnanosecond unipolar voltage pulses

GENERATORS BASED ON SEMICONDUCTOR DEVICES

In the range of pulse durations of a few nanoseconds, semiconductor devices are used as opening switches and operate as parts of generators whose power-amplifying output stage incorporates an inductive energy store. The method was founded in 1983 by Professor I.V. Grekhov with co-workers as they developed drift step-recovery diodes (DSRD) [14]. The process of current cutoff in a DSRD is the removal of equilibrium carriers from the structure base with a saturated rate of ~100 μm/ns. For a base thickness of 100-200 μm, this ensures the switching time about 1-2 ns. The absence of an extra plasma in the structure at the stage of current cutoff reduces the highest possible current density to 160-200 A/cm^2. However, the opening switch performance by current and voltage can be improved by increasing the structure area and by assembling several structures in series. Record parameters of a DSRD have been obtained in the experiment [5] where several DSRDs connected in series were used to produce pulses with a peak voltage of 80 kV, a current of 800 A, an FWHM of 1.5 ns, and a repetition rate of 1 kHz.

A substantial increase in pulse power and in voltage for semiconductor opening switches was achieved after the discovery, in 1991, of the SOS effect - a nanosecond cutoff of superdense currents in semiconductors [15]. The SOS effect is basically different from other types of current switching in semiconductor devices in that the process of current cutoff occurs not throughout the structure base, but only in its narrow highly doped regions. As this takes place, the structure base remains filled with an excessive dense plasma. This provides simultaneously a high density of the current being cut off (reaching 60 kA/cm^2) and a subnanosecond cutoff time. Another remarkable feature of the SOS effect is that a mechanism for even partition of voltage over the

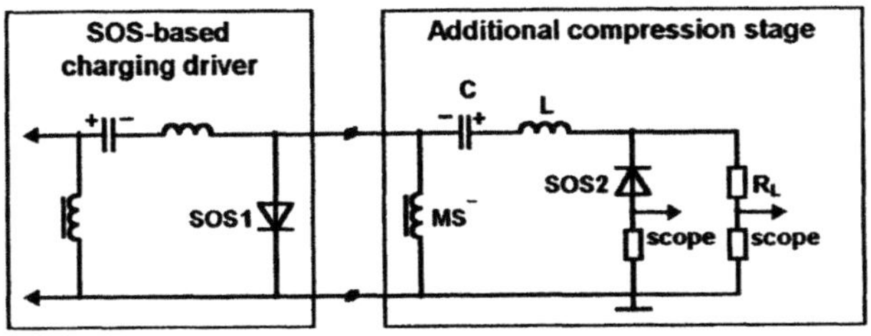

Fig. 12. Experimental circuit of SOS generator

structures connected in series is self-acting at the stage of current cutoff. This offers the possibility to create opening switches with megavolt voltages by connecting semiconductor structures in series not using external voltage dividers. Based on this effect, SOS diodes have been developed which are nanosecond solid-state switches intended for interruption of high-density currents. They have an operating voltage of some hundreds of kilovolts and are capable of switching several gigawatts of power and interrupting a current of a few or some tens of kiloamps at kilohertz pulse repetition rates [16]. With the structure pumping time being of the order of 100 ns, the minimum duration of the pulse formed, which is determined by the current cutoff time, is about 10 ns.

The situation changed crucially as a subnanosecond current cutoff had been detected in a SOS diode [6]. The experimental arrangement containing a charge generator and an additional energy compression stage for the test SOS diode is shown in Fig.12. Used as a charge generator was a system operating into an active load and capable of producing a 100-140-kV pulse with a current of 200-400 A, a duration of 30-40 ns, and a pulse repetition rate of up to 5 kHz in the mode of 30-s pulse packets. The additional energy compression stage intended for pumping of the test SOS diode (SOS2) consists of a capacitance energy store, C, a magnetic switch, MS-, an inductive energy store, L, and a resistive load, R_L.

When a direct pumping current passes through the diode SOS1 placed in the charge generator, part of this current branches out and flows through the magnetic switch MS- and remagnetize its core in the needed direction. As the charge generator operates, the capacitor C is charged to 30-60 kV in 25-40 ns. The current of this process peaking at 100-400 A flows through SOS2 and pumps it in the straightforward direction. After saturation of the core of the switch MS-, the capacitor C discharges through SOS2 and switches the energy into the energy store L. As this takes place, a backward current, whose rise rate and flow time can be controlled by varying the parameters L and C, is introduced into the diode SOS2. As the backward current is cut off by the diode SOS2, this current is switched into the load RL, and a short high voltage pulse is formed across the load. The current cutoff time as a function of the duration of the backward pumping is shown in Fig.13. The current cutoff time was fixed at a level of 0.1-0.9 of the peak current. It can be seen that for the backward pumping time ranging between 9 and 15 ns, the current cutoff time is in the range 500-700 ps which is bounded below by the transient response of the oscilloscope used. Figure 14,a illustrates the stage of subnanosecond cutoff of a 750-A current. The cutoff time is 600 ps. Figure 14,b gives an oscillogram of the voltage pulse across a 290-ohm load.

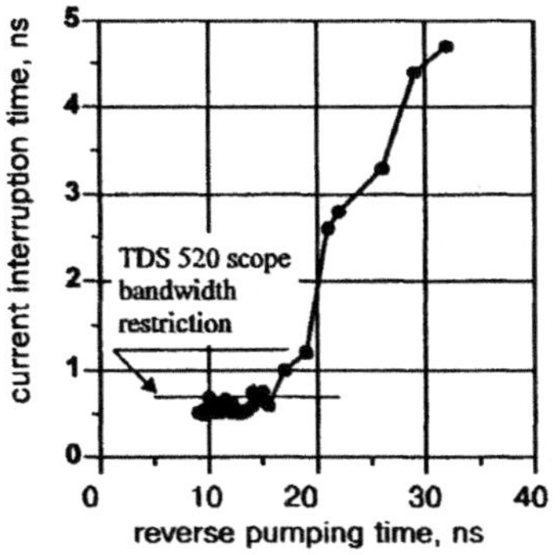

Fig.13. Current interruption time vs.reverse pumping time

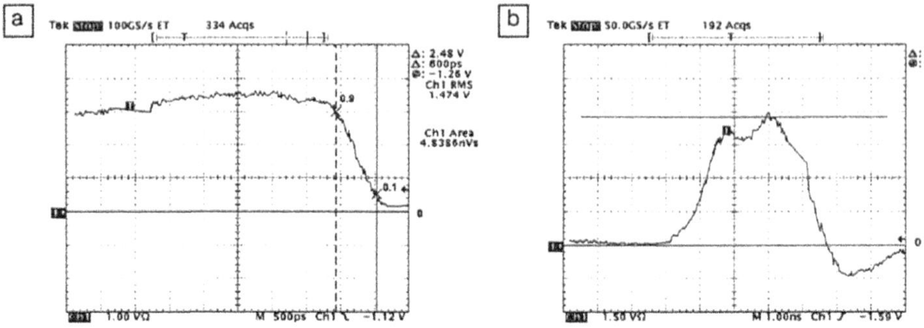

Fig. 14. a- Current interruption stage with 600 ps cut-off time: vertical scale - 235 A/div; b- Output voltage across 290-Ω resistive load: vertical scale - 30 kV/div.

In general, in experiments with the cutoff current ranging from 0.5 to 1 kA, the cutoff time 500-700 ps, and the load resistance from 200 to 2000 ohm, the output pulse peak voltage across the load R varied from 60 to 150 kV, the pulse risetime was about 1 ns, and the pulse FWHM was in the range from 2 to 4 ns. The operation of the system features high reproducibility of the output pulse parameters, which is due to the absence of gas-discharge switches. Thus, the waveforms given in Fig.14a,b have been obtained in the build-up mode and represent a superposition of 334 and 192 pulse waveforms, respectively.

From the practical viewpoint, the experiment has demonstrated the feasibility of a simple method for the formation of high-current pulses of amplitude some hundreds of kilovolts and duration a few nanoseconds. With this approach, two pulse peaking stages connected in series are used, each containing its own SOS diode. The first stage forms a pumping pulse of duration several tens of nanoseconds for pumping the second-stage SOS diode. As the current is cut off by the second-stage SOS diode, a pulse of duration several nanoseconds is formed across the load. The output pulse peak parameters are determined by the energy stored in the pumping generator and by the generator output voltage. These circumstances offer promise for further increasing pulsed power and voltage in producing pulses of duration a few nanoseconds.

The principal advantages of the generators with semiconductor opening switches are the practically unlimited (solid-state) service life, no conceptual limitation for the pulse repetition rate (the recovery time for DSRD and SOS diodes is not over 1 μs), and the high average specific power at kilohertz pulse repetition rates. However, the now available semiconductor generators are still unacceptable of producing high-power pulses of peak voltage much over 100 kV and subnanosecond risetime. Therefore, the high-pressure gas gap peakers still remain out of competition in producing pulses of duration 0.1-1 ns.

In view of this, the approach combining the advantages of semiconductor generators and gas-discharge peakers is a promising line in the development of the subnanosecond pulse power technology. With this approach, the semiconductor generator provides fast (a few nanoseconds) charging of the pulse-forming unit of the gas peaker that forms a subnanosecond output pulse. The use of superhigh-pressure hydrogen gaps makes it possible, providing the pulse repetition rate of the overall system at a level of a few kilohertz, to retain the repetitive operation capabilities of the semiconductor generator.

SUMMARY

The results of fundamental importance obtained in a series of experiments on studying the operating modes of subnanosecond high-pressure gas gaps are as follows: repetitive pulse modulators with highly reproducible and tunable parameters have been realized, subnanosecond jitters have been achieved for the starting of electrically triggered gas switches, and compact module systems have been created.

New experimental data on the electric strength of various types of insulation exposed to pulsed electric and microwave fields have made it possible to create rather compact and high-power superwideband generators.

The studies on the generation of subnanosecond high-current beams have led to the realization of new high-power sources of millimeter microwaves operating in the superradiation mode.

High-voltage semiconductor opening switches have been used as a basis for a new class of short pulse generators showing promise for a wide spectrum of uses in subnanosecond high-voltage and microwave technology.

REFERENCES

1. B.M.Kovalchuk, G.A.Mesyats, and V.G.Shpak, Generators of Powerful Subnanosecond Pulses, *IEEE Proc. Pulsed Power Conf., Lubbok, TX, USA*. [1D5]. (1976).
2. V.G.Shpak, S.A.Shunailov, M.I.Yalandin, and A.N.Dyad'kov, The RADAN SEF-303A, a small high-current pulsed power supply, *IET*.V.36, №1:106 (1993)
3. V.G.Shpak, M.I.Yalandin, M.R.Oulmascoulov, and S.A.Shunailov, 1000-pps Subnanosecond High-Voltage Generator, *Proc.of the 11th IEEE International Pulsed Power Conference.Baltimore, ML,USA.*: 1575 (1997)
4. G.A.Mesyats, V.G.Shpak, S.A.Shunailov, and M.I.Yalandin, Desk-top subnanosecond pulser research, development and applications, *Proc. of SPIE International Symposium: Intense Microwave Pulses.-Los Angeles, CA, USA*. Vol.2154: 262(1994)
5. V.M.Efanov, A.F.Kardo-Sysoev, M.A.Larionov, et al., Powerful semiconductor 80 kV nanosecond pulser, *Abstracts of 11th IEEE Int. Pulsed Power Conf. Baltimore, Maryland, USA.*: 273 (1997)
6. S.K.Lyubutin , G.A.Mesyats, S.N.Rukin, and B.G.Slovikovskii, Subnanosecond high-density current interruption in SOS diodes, *Abstracts of 11th IEEE Int. Pulsed Power Conf. Baltimore, Maryland, USA.*: 162 (1997)
7. M.I.Yalandin, V.G.Shpak, S.A.Shunailov, and M.R.Ulmaskulov, An Experimental Study of the Formation and the Dynamics of Transportation of a Magnetically Insulated High-Current Subnanosecond Tubular Electron Beam, *Proc. XVII th International Symposium on Discharges and Electrical Insulation in Vacuum, Berkeley, CA, USA*. Vol.2: 635(1996)
8. V.G.Shpak, S.A.Shunailov, M.R.Ulmaskulov, M.I.Yalandin, and I.V.Pegel, Experimental Investigation of Dynamics of High-Current Subnanosecond Electron Bunch, *Sov. Tech. Phys. Lett.* V.22, Iss.7: 65 (1996)
9. N.S.Ginzburg, A.S.Sergeev, I.V.Zotova, I.V.Konoplev, A.D.R.Phelps, A.W.Cross, S.J.Cook, P.Aitken, V.G.Shpak, M.I.Yalandin, S.A.Shunailov, and M.R.Ulmaskulov, Experimental Observation of Cyclotron Superradiance under Group Synchronism Conditions, *Phys. Rev. Lett.* Vol.78, No.12: 2365 (1997)
10. V.G.Shpak, S.A.Shunailov, M.R.Ulmaskulov, and M.I.Yalandin, Generation of high-power broadband electromagnetic pulses with PRF of 100 pps, *Proc.of the 10th IEEE International Pulsed Power Conference, Albuquerque, NM, USA.*: 539 (1995)
11. N. S. Ginzburg, Yu.V. Novozhilova and A.S. Sergeev, Generation of Short Electromagnetic Pulses by Electron Banch in the BWO-type Slow-Wave System, *Sov. Tech. Phys. Lett.* V.22: 359 (1996)
12. V.G.Shpak, S.A.Shunailov, and M.I.Yalandin, Investigations of compact high-current accelerators RADAN 303 synchronization with nanosecond accuracy, *Proc.of the 10th IEEE International Pulsed Power Conference, Albuquerque, NM, USA.*: 666 (1995).
13. V.P.Gubanov, S.D.Korovin, I.V.Pegel, V.V.Rostov,A.S.Stepchenko, M.R.Ulmaskulov, V.G.Shpak, S.A.Shunailov, and M.I.Yalandin, Generation of Powerful Electromagnetic Pulses, *Sov. Tech. Phys. Lett.* V.20, Iss.14: 89 (1994)
14. I.V. Grekhov, V.M. Efanov, A.F. Kardo-Sysoev, and S.V. Shenderei, Formation of high-voltage nanosecond drops across semisonductor diodes with the drift mechanism for voltage recovery *Sov. Tech. Phys. Lett.* 1983, V.9: 435 (1983)
15. S.A. Darznek, Yu.A. Kotov, G.A. Mesyats, and S.N. Rukin, The SOS effect: a nanosecond cutoff of superdense currents in semiconductors, *Doklady AN.* V.334: 304 (1994)
16. S.K.Lyubutin, G.A.Mesyats, S.N.Rukin, B.G.Slovikovskii, and A.M.Turov, New solid-state opening switches for repetitive pulsed power technology, *Proc.: XI Int. Conf. on High Power Particle Beams., Prague, Czech Republic.* V.1: 135 (1996)

FUNDAMENTAL PHYSICAL CONSIDERATIONS FOR ULTRAFAST SPARK GAP SWITCHING

Jane M. Lehr, Carl E. Baum, William D. Prather and Robert J. Torres

Air Force Research Laboratory, Phillips Site
High Power Systems Branch
Kirtland AFB, NM 87117-5776 USA

ABSTRACT

A problem of interest concerns compact, high power ultra-wideband systems which minimize the volume of high pressure gas. Thus, single channel spark gaps are considered because the volume of gas under pressure is much less than that needed for sources containing ring gap switches. However, the intrinsic inductance of single channel spark gaps is prohibitively high to achieve large rates of voltage rise.

For future applications, the limit of spark gap technology for ultrafast switching is explored. Specifically, an estimate of the fastest risetime achievable with a single channel spark gap has been investigated using three approaches. The first examines the growth of the electron avalanche in gases to estimate its growth rate. The avalanche growth rate determines the fastest possible risetime of the resultant pulse. The second approach uses the components of the velocity of electromagnetic propagation to estimate the achievable risetime. The third uses an equivalent circuit of a single channel spark gap to calculate the maximum achievable rate of voltage rise. The first two estimates indicate that risetimes on the order of 1-10 ps are achievable. The last treatment, however, illustrates the dependence of the pulse risetime on the peak voltage and calculates the maximum rate of voltage rise to be on the order of 10^{16} V/s.

To reduce the effect of the intrinsic inductance of the channel, a simple geometrical alteration to the spark gap geometry has been devised which effectively reduces the inductance per unit length of the spark gap to that of its transmission line feed. This simple change alleviates the constraint imposes by the maximum rate of voltage rise and is anticipated to permit the realization of picosecond risetime high power electromagnetic sources.

Ultra-Wideband, Short-Pulse Electromagnetics 4
Edited by Heyman *et al.*, Kluwer Academic / Plenum Publishers, New York, 1999.

INTRODUCTION

The general limitation on the achievable risetime for gaseous ultra-wideband (UWB) high power electromagnetic sources is the spark gap. It is of interest, for a myriad of applications, to produce, control and accurately measure ultrafast breakdown in spark gaps. Thus, knowledge of the fundamental limitations of spark gap technology is desirable. Three treatments are used to examine the limits. The first examines the rate of carrier generation in the gaseous insulating media under uniform field conditions to estimate the minimum time to achieve the critical charge carrier density. Secondly, the limitation imposed by the finiteness of the electromagnetic propagation. Finally, a circuit model for the post-breakdown circuit is invoked to derive the limitation on the rate of voltage rise. These treatments provide insight into ultrafast breakdown and guidelines for its generation.

Pulse conditioning systems are being used to generate fast rising electromagnetic fields in the 10s of gigawatt power range [1,2]. In general, this is accomplished by successively manipulating the timescale at which energy is delivered to a load. A block diagram of such a power conditioning system, is shown in Figure 1. Pulsewidths in the sub-nanosecond regime are produced by charging transmission lines of successively shorter electrical length along with an accompanying increase in voltage.

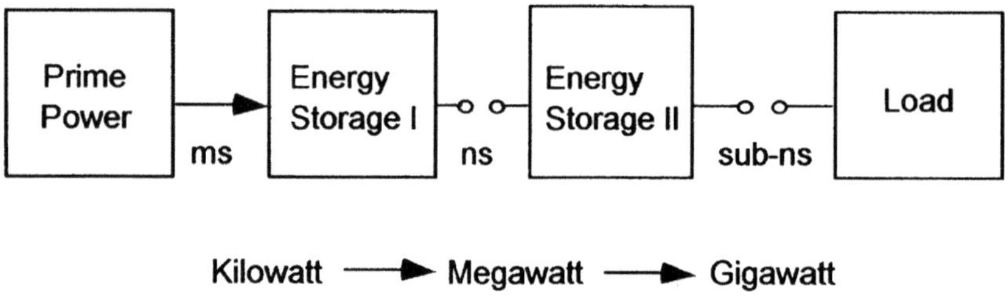

Figure 1. UWB production. Sub-nanosecond duration pulses are generated by successively reducing the timescale at which electrical energy is stored.

The switching element is a major component of any power conditioning system and, for UWB high power electromagnetic field generation, ultrafast closing capability, along with fast voltage recovery are desired. A fast pulse risetime is critical because the risetime contains the high frequency components of the resulting spectrum. To sharpen the risetime on a pulse, a spark gap configuration, called a peaking gap is used. The crux of the peaking gap is the establishment of very high electric fields in the interelectrode spacing. The velocity of propagation of the electron avalanche is proportional to the electric field applied to electrodes, and thus, gap closure is dominated by the applied electric field. To produce ultrafast switching, the spark gap is dramatically overvolted; that is, the spark gap is charged far in excess of its self-breakdown voltage. Peaking gaps typically operate at gas pressures in the range of 100 atm and electric fields in the MV/cm range. The self breakdown curve for gases is known to saturate in the vicinity of 100 MV/m for pressures to 50 atm[3]. To achieve overvolting without switching at the self breakdown voltage, the spark gap is pulse charged very quickly. This allows a large overvoltage to be achieved, and overvoltages of over 300% are achievable.

Interelectrode gap distance, δ, is chosen as small as possible to minimize the intrinsic inductance of the spark channel, since the inductance is known to limit the achievable risetime of the resultant pulse. Spark channel inductances of less than 1 nH have been achieved with gap lengths of 1 mm and less. Because of the short gap lengths, even charge voltages of less than 100 kV can produce interelectrode electric fields in the MV/cm range with voltage doubling at the spark gap. These small interelectrode distances, however, yield high spark gap capacitances, even for relatively small diameter electrodes. Moreover, this high spark gap capacitance, and the fast charging times lead to a strong displacement current which manifests as an undesirable prepulse on the load voltage. The prepulse is observed on the output pulse with a magnitude, V_{pp}, which is related to the capacitance of the spark gap, C, by

$$V_{pp} = Z_L C \frac{dV_C}{dt}$$

where Z_L is the output impedance and V_C is the voltage which is applied to the spark gap. Since a fast charge is critical to peaking gap operation, small diameter electrodes are desirable. Moreover, minimizing the electrode diameter of the peaking gap may lead to enhanced performance. The generation of 50 ps risetime pulse with a 60 kV charge in a single channel switch of very small dimensions has been reported [4].

PREDICTIVE RELATIONS

The spark gap is said to have closed when the potential difference between the electrodes is sharply reduced and current becomes circuit limited by the source and load impedances. The predictions for the limitation of spark gap technology are broken into two parts: the prebreakdown limitation and the post breakdown limitation. The development of the prebreakdown phase of the spark development determines the minimum closure time of the spark gap, which determines the risetime. Once the voltage across the channel has collapsed, the electromagnetic wave must re-establish in the interradial distance. The post breakdown phase consists of the inhibition of the voltage rise due to the inductance of the spark channel itself.

Avalanche Propagation Limit

The breakdown of a spark gap initiates with an electron avalanche, even when a streamer mechanism is operative. An avalanche initiates from an initial electron distribution which may be resident in the gas, or from an external source. If the initial electrons are thermally distributed, the electron population is small. The electrons accelerate under the influence of an external field. After some time, an electron undergoes a collision with the background gas, with a resultant energy exchange. If this energy exchange is sufficient, an ionizing collision occurs and an electron/ion pair is created. The growth of the current density in this manner results in an electron avalanche. The growth of the electron avalanche is described by the evolution of the electron number density. The electron number density, N(t), grows from its initial value, N_0, under the influence of an applied electric field, according to

$$N(t) = N_0 \exp^{\int \alpha(v) v dt} ,$$

where v is the velocity of the electron, t is time, and $\alpha(v)$ is the electron ionization rate coefficient. In general, the rate coefficient, α, is generally a function of the relative velocity of the collision pair and is gas specific. The above equation leads to the definition of the time constant related to the above equation, the avalanche growth time, given by

$$t_a = \frac{1}{\alpha v} \quad .$$

For gas pressures of 100 atm, avalanche transit times on the order of 1 ps has been calculated for reduced electric field, E/p of 90 V/cm-torr[5]. If a single avalanche results in breakdown, the associated risetime for avalanche growth is the risetime of the pulse. This treatment however assumes the absence of the streamer mechanism and, thus, inherently assumes uniform field conditions apply.

It should be noted that this estimation is still valid when a streamer mechanism is operative. The streamer mechanism initiates with an electron avalanche. The electron avalanche grows until the electric field produced by the cumulative space charge in the avalanche head is on the order of the applied field. The critical electron number for the avalanche to streamer transition is on the order of 10^8. When this criteria is met, the discharge is said to have entered the streamer phase. The advent of the streamer mechanism serves to greatly enhance the propagation speed of the discharge. Thus, the major component of the discharge transit time is the electron avalanche development time.

Electromagnetic Propagation Limit

The electromagnetic propagation limit explores the timescale of the propagating wave in the insulating media. Prior to breakdown, a spark gap is a capacitor. This capacitor, however, does not necessarily charge fully before breakdown occurs. The spark gap cannot close in less time than the transit time of the electromagnetic wave though the gap, as shown in Figure 2.

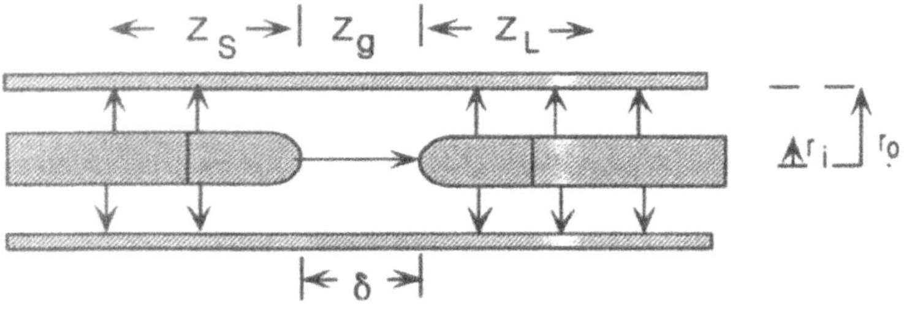

Figure 2. The electric field lines as a pulse propagates through a spark gap, with an interelectrode gap distance, δ, prior to breakdown. The field lines on the right hand side are the transmitted wave after the spark gap switches.

To examine the propagation time in the spark gap spacing, the common scaling parameter Pδ is used, where P is the gas pressure, and δ is the interelectrode gap distance.

Thus, the transit time of an electromagnetic pulse through a spark gap filled with an insulating media with electric permittivity, ϵ, and magnetic permeability, μ, is

$$P\delta \approx \frac{Pt_r}{\sqrt{\epsilon\mu}} \ .$$

Thus for a high pressure gaseous spark gap, with an interelectrode gap spacing of 1 mm, the minimum risetime is approximately 3 ps. However, for a spark gap inserted into a coaxial transmission line, the risetime is determined not only by the gap closure but also by the re-establishment of the propagating electric field in the inter-radial distance. The achievable risetime is longer since the electromagnetic wave must be re-established between the inner and outer conductors. Thus, the above limit must be modified by the interradial transit time, yielding,

$$P[\delta + \Delta r] \approx \frac{Pt_r}{\sqrt{\epsilon\mu}} \ ,$$

where Δr is the difference between the outer and inner radii of the transmission line. The resolution time, then, for measurements is given by $\Delta r\ (\epsilon\mu)^{1/2}$. For most high power systems, the interradial distance, Δr, is made large to insure voltage holdoff in the coaxial systems. Of course, the breakdown voltage is inversely proportional to both the inner radius and the logarithm of the radius ratio. Thus, for ultrafast switching, for a given impedance, the inner conductor radius should be chosen to be small.

Maximum Rate of Voltage Rise

Spark gaps are very good candidates for transferring high peak powers. However, the small radius of the spark channel makes these switches intrinsically inductive and limits the achievable risetime by inhibiting the rate of voltage rise. The spark gap is represented by the circuit of Figure 3, which is valid once the discharge channel current is circuit limited. From this simple equivalent circuit, the maximum rate of voltage rise, dV/dt, in the spark gap can be calculated.

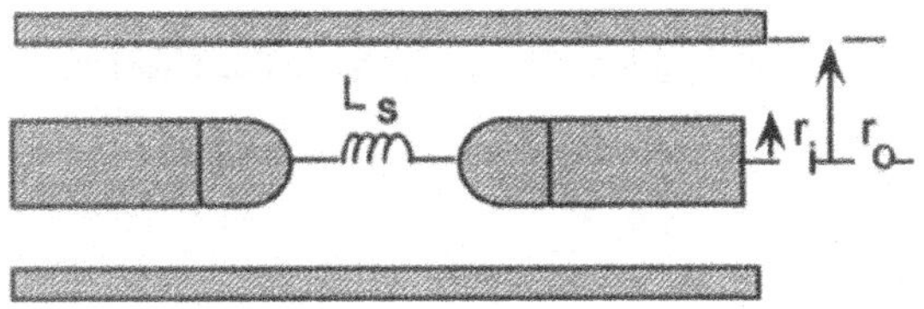

Figure 3. A model of the discharge channel of a spark gap.

When breakdown has progressed so that it is self sustaining, the single channel of the spark gap is modeled as an inductor, L_S. The coaxial transmission line feed of the equivalent circuit has an inductance per unit length, L′, given by

$$L' = \frac{\mu_0}{2\pi} \ln \left[\frac{r_o}{r_i}\right] ,$$

where r_o and r_I are the outer and inner radii of the transmission line. The value of the inductance of the spark channel, L_S, can be difficult to calculate exactly. However, it has been shown[6] that a very good approximation can be obtained by treating the current carrying channel as a transmission line of inner radius r_c and outer radius r_o . Thus, the intrinsic inductance, L_S, of the current carrying breakdown channel is

$$L_S = \delta \frac{\mu_0}{2\pi} \ln \left[\frac{r_o}{r_c} \right] ,$$

where δ is the interelectrode spacing of the spark gap. The inductance per unit length of the spark gap, L_S', is

$$L_S' = \frac{L_S}{\delta} .$$

For a system with electrodes which are large compared to the channel radius, the value of the inductance per unit length, L_S', of the spark channel can be approximated. For a system with large electrodes, and reasonably low impedance,

$$\frac{r_o}{r_c} \sim 30,\ 50,\ 100 .$$

Since the natural logarithm function is slowly varying,

$$\ln \left[\frac{r_o}{r_c} \right] \sim 4 .$$

Thus, the inductance per unit length of the spark channel can be approximated for practical purposes as

$$L_S' \simeq \frac{2\mu_0}{\pi} .$$

The voltage across this equivalent inductor of the spark channel, V_c, is given by

$$V_C = L_S \frac{dI}{dt}$$

This is related to the source voltage, V_S, and load voltage, V_L, by

$$V_C = V_S - V_L$$

where the source voltage, V_S, is the voltage on the left electrode of the spark gap. The rate of voltage rise into the load is expressed as

$$\frac{dV_L}{dt} = Z_L \frac{dI}{dt}$$

where Z_0 is the load impedance. Rearranging terms,

$$\frac{dV_L}{dt} = \frac{Z_L V_C}{L_S} = \frac{Z_L}{L_S} [V_S - V_L]$$

If the electrodes of the spark gap are large, and spaced a distance, δ, apart, the electric field is uniform and given by

$$E = \frac{V_S}{\delta} \ .$$

The rate of voltage rise is initially related to the electric field by

$$\frac{dV_L}{dt} = \frac{Z_L E}{L_S'} \ .$$

To estimate the maximum rate of voltage rise, let the electric field, E, be 100 MV/m, and the load impedance be 50 Ω. Thus, with these representative values, the the rate of voltage rise calculates to be

$$\frac{dV}{dt} \approx 6 \cdot 10^{15} \frac{V}{s} \ .$$

This estimate implies that for a switching voltage of 100 kV, the achievable risetime is on the order of 10 ps. However, for a switching voltage of 1 MV, 100 ps is possible.

This estimate of the maximum rate of voltage rise in a spark gap is the order of magnitude which is currently being produced through the use of peaking gaps[1,4]. The above equation suggests that the even higher rates of voltage rise can be achieved by increasing the electric field on the spark gap and reducing the inductance per unit length of the spark channel.

THE COMPENSATED SPARK GAP: A MINIMUM INDUCTANCE DESIGN

In ultrafast switching, the interelectrode spacing in the spark gap is chosen to be very short to minimize the inductance of the spark channel. It can be shown, in general, the inductance of the spark channel in a spark gap dominates switch performance. This implies that by removing the additional inductance caused by the reduced crossectional area of the spark channel in relation to the electrode area, the switch characteristics will be completely determined by the inductance of the switch hardware. For long gap spark channels this can be accomplished by the addition of a "taper" in the outer conductor as shown in Figure 4. The purpose of this taper is to match the impedance of the spark channel to the driving system impedance and hence, matches the inductance per unit length of the hardware to the inductance per unit length of the interelectrode gap region.

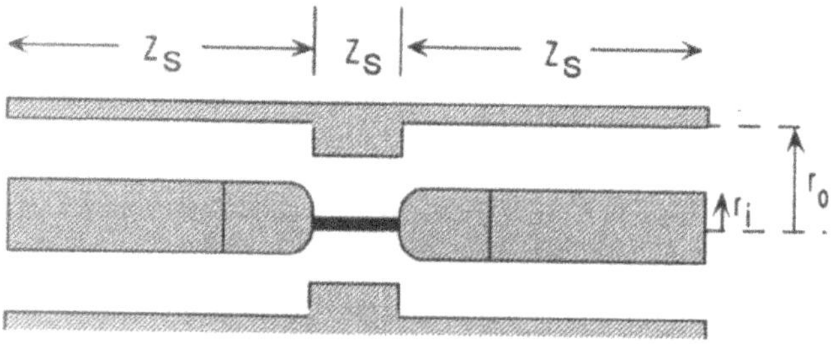

Figure 4. Taper in the outer conductor. The added dimension of the outer conductor compensates for the reduced impedance between the electrodes of the spark gap due to the reduced radius of the conducting path through the spark channel. The taper dimension is chosen so that $\ell n(r_o/r_c)$ is approximately equal to $\ell n(r/r)$ and the impedance of the interelectrode region of the spark gap is close to that of the driving system.

The sharp edges, as shown in Figure 4, introduce field enhancement to the spark gap design which are detrimental to the spark gap performance. A practical design, as illustrated in Figure 5, contours both the inner conductor and the outer conductor to maintain a constant impedance as well as the voltage holdoff throughout the switch.

The design of Figure 5, called a biconical spark gap, satisfies these objectives. The bicone geometry is congenial to the natural coordinate system of the electromagnetic wave emitted from the spark gap. That is, the electromagnetic field follows a smooth transition to rejoin the transmission line. This principal is sometimes referred to as an optical path and is important in inhibiting the generation of higher order modes in the coaxial transmission line. Moreover, this design minimizes the interradial term in the electromagnetic propagation limit. This term may be considerably larger than the spark gap contribution to the electromagnetic limit. The biconical spark gap geometry is superior for ultrafast switching because it minimizes the intrinsic inductance of the spark channel, reduces the capacitance of the spark gap and resists the generation of higher order modes. Additionally, since the impedance of the system is maintained throughout the switch region, the power loss associated with the switch, is minimized.

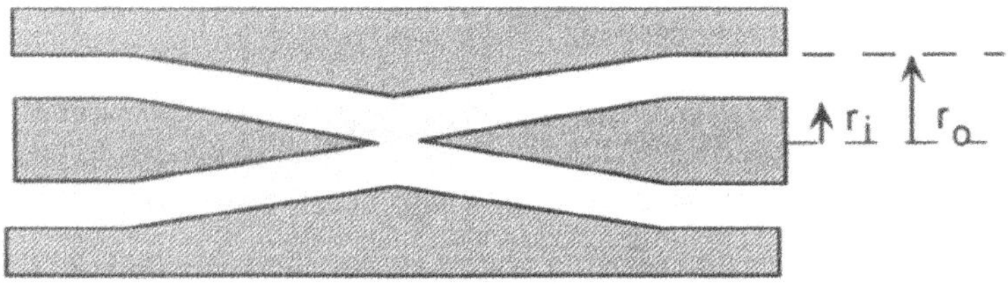

Figure 5. A biconical spark gap design. This design maintains a quasi-constant impedance throughout the electrode section which minimizes the intrinsic inductance of the spark channel. The smooth transition resists the generation of higher order modes. Since the impedance of the system is maintained throughout the switch region, the power loss associated with the switch is minimized.

CONCLUSIONS

The peak achievable risetime has been examined in three ways: investigation of the critical carrier production rate to produce an electron avalanche which results in closure of a gaseous spark gap; an examination of the limits imposed by the propagation of an electromagnetic wave in a transmission line encased spark gap; the third method results from a circuit model of the spark gap after breakdown. These calculations are in remarkably good agreement: risetimes not less than a few picoseconds can be expected. Maximum achievable rates of voltage rise for single channel spark gaps is calculated to be on the order of 10^{16} V/s. This switch parameter is of paramount importance in predicting the acheivable peak radiated electric field levels in transient systems.

The above calculations provide guidelines for peaking switch design. Foremost, the electric field which is applied to the switch is critical in determining the achievable risetime. The electromagnetic wave calculations indicate the difference in the inner conductor dimensions of coaxial spark gaps should be minimized for a given impedance. It has been shown that different expectations are valid for achieving peak voltages of different magnitudes. Moreover, the notched spark gap design may alleviate the conclusions regarding the maximum rate of rise.

REFERENCES

1. W.D. Prather, C.E. Baum, F.J. Agee, J.P. O'Laughlin, D.W. Scholfield, J.W. Burger, J.Hull, J.S.H. Schoenberg, and R. Copeland, "Ultrawide Band Sources and Antennas: Present Technology, Limitations, Future Challenges," pp. 381-389 in *Ultra Wideband/ Short Pulse Electromagnetics 3*, C.E. Baum, et.al (eds.), Plenum Press, New York, 1996.

2. J.P. O'Laughlin and R.P. Copeland, "Subnanosecond Power Conditioning Technique Using Transmission Line to Transmission Line Charging," Proc. 20th Power Modulator Symposium, pp. 351-354, 1992.

3. E.A. Avilov and N.V. Belkin, “Electrical Strength of Nitrogen and Hydrogen at High Pressures,” Sov. Phys. Tech. Phys., Vol. 19, No. 12, June 1975.

4. C.A. Frost, T.H. Martin, P.E. Patterson, L.F. Rinehart, G.J. Rohwein, L.D. Roose, J.F.Aurand, M.T.Buttram, "Ultrafast Gas Switching Experiments," Proc. 9th IEEE Pulsed Power Conference, K. Prestwich and W. Baker, Eds., pp. 491-494, 1993.

5. R.E. Cassell and F. Villa,"High Speed Switching in Gases," SLAC Publication 1858, February, 1989.

6. S. Levinson, E.E. Kunhardt and M. Kristiansen, “Simulation of Inductive and Electromagnetic Effects Associated with Single and Multichannel Triggered Spark Gaps,” Proc. 2nd IEEE Pulsed Power Conference, A.H. Guenther and M. Kristiansen, Eds., pp. 433-436, 1979.

NOVEL SOURCE OF POWERFUL SUBNANOSECOND MICROWAVE PULSES BASED ON SUPERRADIANCE

N.S.Ginzburg[1], I.V.Zotova[1], N.Yu.Novozhilova[1], A.S.Sergeev[1], A.D.R.Phelps[2], A.W.Cross[2], W.He[2], K.Ronald[2], S.M.Wiggins[2], V.G.Shpak[3], M.I.Yalandin[3], S.A.Shunailov[3], M.R.Ulmaskulov[3], V.P.Tarakanov[4]

[1]Institute of Applied Physics, RAS, 603600, Nizhny Novgorod, Russia
[2]University of Strathclyde, Glasgow, G4 ONG, UK
[3]Institute of Electrophysics, RAS, 620049 Ekaterinburg, Russia
[4]High Energy Research Center, RAS, 127412, Moscow, Russia

INTRODUCTION

Coherent emission from short electron bunches with length much smaller than the wavelength is well known and was first observed in the Smith-Purcell experiments. Another type of conventional coherent emission, which is used extensively in microwave electronics arises from continuous or quasi-continuous electrons beams with length much greater than the wavelength. In this case coherent emission is related with stimulated processes resulting in selfbunching of the electrons. For the intermediate case, i.e. for electron bunches with a size of several wavelengths it was traditionally assumed that radiation can be related only with density fluctuations inside the electron bunch (incoherent component) or with coherent emission from the sharp edges of the bunch. Only recently it was recognized that much more intense coherent emission can occur from the entire volume of such bunches due to selfbunching arising from the mutual influence of different fractions of the electron bunch. Based on a similar mechanism first suggested by Dicke for ensembles of inverted quantum atoms [1] coherent emission of such bunches can be regarded as superradiance. According to theoretical considerations of superradiance from non-equilibrium ensembles of classical electrons undertaken in recent year [2-9], this phenomenon involves features present in stimulated processes (selfbunching and coherence) as well as spontaneous processes (absence of threshold) and can be utilized to generate intense ultrashort electromagnetic pulses.

Present paper is devoted to description of novel source of powerful subnanosecond microwave pulses based on superradiance. This source exploits coherent stimulated emission of intense subnanosecond electron bunches moving through periodic waveguide and interacting with backward propagating wave. The 300 ps /60 MW microwave Ka band pulses were generated with repetition frequencies of up to 25 pps. Observation of RF breakdown of ambient air as well as the illumination of a panel of neon bulbs with a finely structured pattern corresponding to the excitation of the TM_{01} mode confirms high level of the absolute power. The experimental results are compare with results of the numerical simulation based on the particle-in-cell (PIC) code KARAT. The simulations shows that mechanism of microwave pulse generation is associated with self-bunching and the mutual influence of different part of the electrons pulse due to slippage of the wave with respect to the electrons and thus can be interpreted as superradiance.

EXPERIMENTAL SET-UP

A RADAN 303 compact accelerator with a subnanosecond slicer was used to inject 0.6-2.5 ns, 1.5-2 kA, 250 keV electron pulses [10]. These electron pulses were generated from a magnetically insulated coaxial diode, which utilised a cold explosive emission cathode. A new design of cathode unit has been developed to provide the possibility of the smooth, precise adjustment of the accelerating gap to vary the electron current. The fast rising e-beam current and accelerating voltage pulses were measured using a Faraday cage strip line current probe and an inline capacitive voltage probe respectively, with both signals recorded using a 7 GHz Tektronix 7250 transient digitizing oscilloscope. High current electron pulses were transported through the interaction space in the form of corrugated waveguide of 6 cm of a total length in a longitudinal guiding magnetic field of up to 10 T created by a superconducting magnet. The period of corrugation of the slow wave structure was approximately 3.5 mm, the corrugation depth was 0.75 mm and had a mean radius of 3.75.mm.

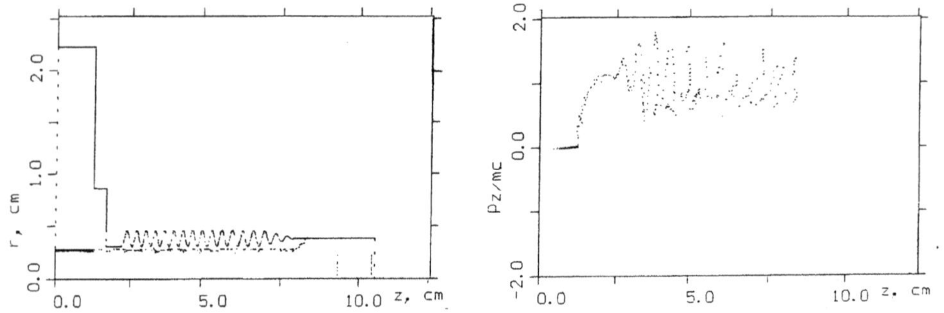

Fig.1 (a) Geometry of interaction region and position of electrons at the time 1 ns; (b) phase plane.

SIMULATION OF SUPERRADIATIVE EMISSION OF ELECTRON BUNCH MOVING IN PERIODICAL STRUCTURE

Simulation of radiation of subnanosecond electron bunches passing through the periodic waveguide structures was carried out using the particle in cell code KARAT. In Fig 1 a is shown the system geometry (all sizes in centimeters) and electron bunch 1 ns after a 240 kV driving voltage pulse was imposed in coaxial line. Selfconsistent current pulse through cross-section z=2.5 cm is presented in Fig. 2. The pulse duration about 0.8 nc and peak power ~1.5 kA corresponds to experimental measurements of current pulse. In Fig.1b we see modulation of bunch density. The phase plane (p_z,z) (Fig.1b) shows strong modulation of the longitudinal momentum developing by this time. The dependence of RF output power on time is presented in Fig.3. The spectrum of radiation (see Fig.4) with central frequency of 38.6 GHz corresponds to the experimental data and the frequency which can be found from the resonance conditions. The average peak power reached 75 MW (after averaging over the high frequency it should be half of the magnitude presented in Fig.3) and the electronic efficiency was about 7%.

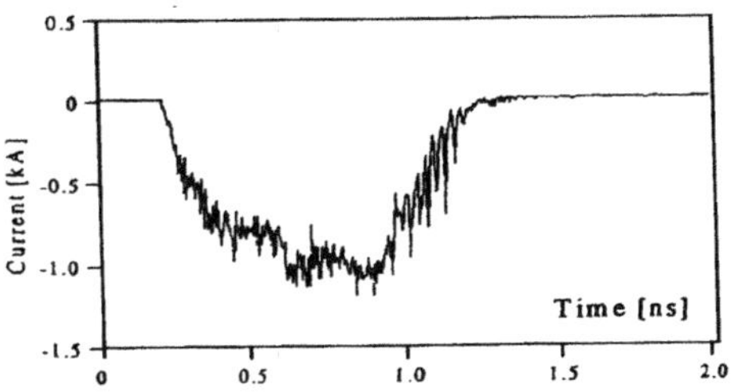

Fig.2 Current pulse

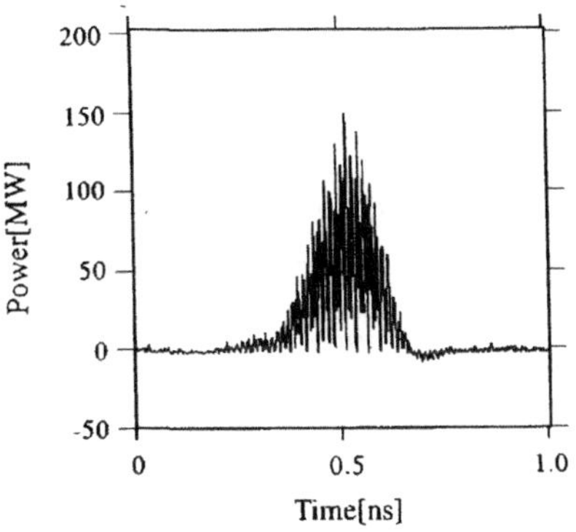

Fig.3 Poynting vector integrated over the transverse cross section as function of time

EXPERIMENTAL RESULTS

Typical oscilloscope trace of microwave signal from an electron pulse passing through a periodical waveguide of total length 6 cm is presented in Fig.5. The observed microwave spikes have duration about 300 ps and the rise time of 200 ps. Frequency measurements have been made using a set of cut-off waveguides filters showed that the main peak had a central oscillating frequency of approximately 38 GHz. The relative radiation spectrum bandwidth found from these measurements was about 5 %. The radiation had a polarization corresponding to the TM_{01} mode. The measured radiation pattern also corresponded with good accuracy to the excitation of the TM_{01} mode. This measurement allowed the absolute peak power to be estimated by integrating the detected signal from the detector over its radial position. The peak power estimated by this method was about 60 MW. A rather high level of radiation power was indicated also by the illumination of a neon bulb panel when the radiation signal irradiated the panel at a distance of 30 cm from the output horn. The radiation pattern accurately corresponded to the excitation of the designed TM_{01} mode. One more important evidence of the high peak power was obtained from the observation of RF breakdown of ambient air for subnanosecond pulses in the focus of parabolic reflector as well

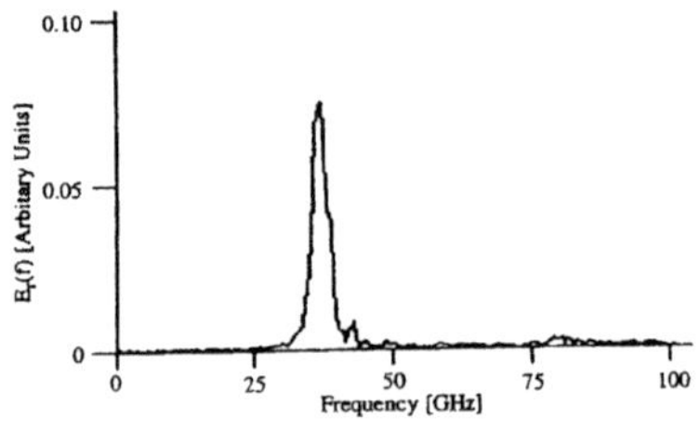

Fig.4 Frequency spectrum of output signal

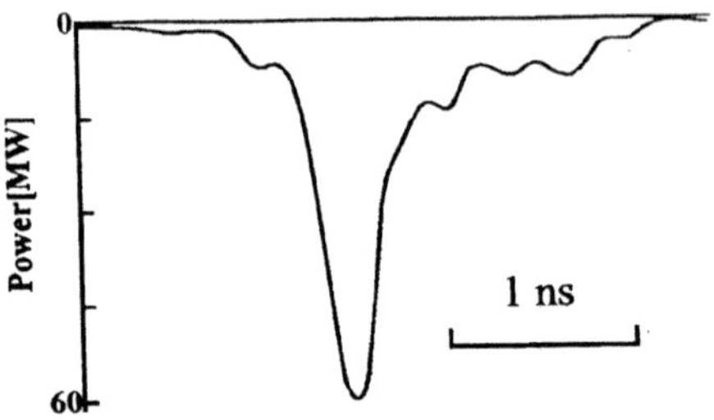

Fig.5 Superradiance microwave pulse recorded by detector

as inside receiving concentrating conical horn. Estimation of the strength of electrical fields inside horn and comparison with the results of previous observations of breakdown from long nanosecond microwave pulses [11-15] demonstrates that the absolute power was not less then 60 MW. The calculations have shown that above peak power corresponds to 5% efficiency of the energy transformation from the e-bunch into microwave pulse. Besides it takes to note that the rate of microwave power rise time represents rather high value of 0.3-0.4 GW/ns.

Fig.6 The dependence of peak power on electron pulse duration

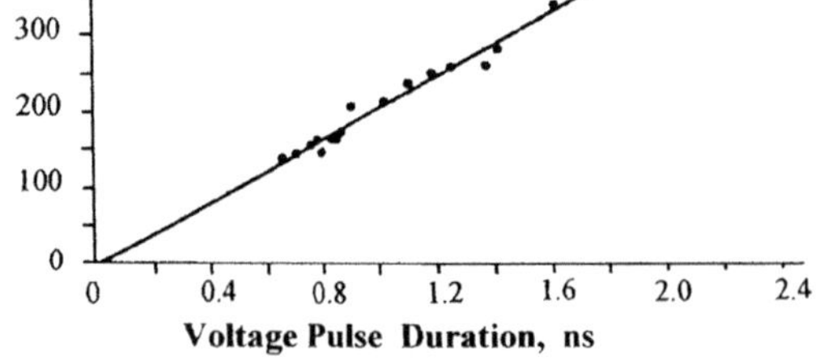

Fig.7 The dependence of total electron charge on pulse duration

To study the dependence of peak power on the total bunch charge the accelerating pulse duration was varied while keeping the accelerating voltage amplitude constant. The dependence of peak power on pulse duration is presented in Fig 6. This dependence is rather close to square law in some region of pulse duration. This mean that electrons radiate coherently from the entire volume of the bunch. Note that for small pulse duration given interaction length 6 cm is not sufficient for formation of SR spikes. With increasing electron pulse duration saturation of grows of peak power obviously related with the fact that pulse became too long to provide coherent radiation from all over pulse length. Actually, subsequent microwave spikes emitted with increasing pulse duration. Note that experimental dependence is rather close to results of simulations. Thus, microwave pulse shape, pulse width as well as absolute power found from simulations are in good agreements with experimental data.

In Fig.7 dependence of total electron charge on pulse duration is presented. This dependence is found by integrating of electron current pulse. Based on curves shown in Fig.6,7 the dependence of peak power on square of total charge is drawn in Fig.8. This dependence is very close to linear. Some shift of fitted linear function from origin can be explained that at leading and trailing edges of accelerating pulse there are low energy electrons. These electrons can't participate in resonance interaction of e.m pulse. With decreasing accelerating pulse duration relative number of such electrons increases. That is why the radiation power tends to zero faster then total charge.

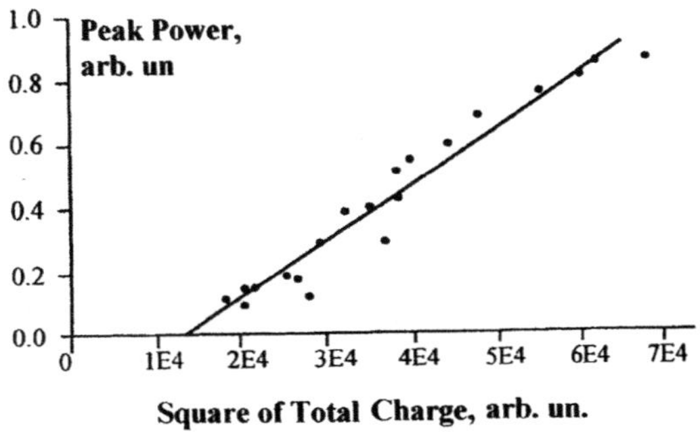

Fig. 8 The dependence of peak power on square of total charge

CONCLUSION

The generated microwave pulses possess high stability and reproducibility. Alongside with capability of operation of RADAN 303 modulator in repetition rate mode and using for bunch transportation dc magnetic fields it gives possibility to generate microwave pulse with repetition frequency of 25 Hz. Observation of RF breakdown of ambient air as well as the illumination of a panel of neon bulbs with a finely structured pattern corresponding to the excitation of the TM_{01} mode confirms high level of the absolute power. The unique characteristic of microwave pulses such as duration (0.3 ns) in combination with 60 MW power levels and repetition rate mode is encouraging for future applications in areas such as novel diagnostics and the studying of nonlinear phenomena in plasmas and solids.

Another advantage of the RADAN subnanosecond accelerator and consequently any associated experiments is that the whole system is in the form of a table-top system.

This work was supported by the Russian Fund of Fundamental Research, grant 98-02-17308 and by the United Kingdom DRA and EPSRC.

REFERENCES

[1] R.H.Dicke Phys.Rev. **93**, 99, (1954)
[2] R.H.Bonifacio, B.W.J. McNeil and P.Pierini, Phys. Rev. **A 40**, 4467 (1989).
[3] R.Bonifachio, C.Maroli, N.Piovella, Opt.Comm., **68**, 364 (1988).
[4] V.V.Zhelyeznyakov, V.V.Kocharovsky and V.Vl.Kocharovsky, Usp.Fiz.Nauk, **159**, 194 (1989)
[5] N.S.Ginzburg, Sov. Tech. Phys. Lett. **14**, 197 (1988)
[6] N.S.Ginzburg, Yu.V.Novozhilova and A.S.Sergeev, Sov. Tech. Phys. Lett. **22**, 359 (1996).
[7] D.A.Jaroszynski, P.Chaix, N.Piovella et al., Phys. Rev. Lett, **78**, 1699 (1997)
[8] N.S.Ginzburg et al, Phys. Rev. Lett, **78**, 2365 (1997)
[9] N.S.Ginzburg *et al.* Nucl. Inst. and Meth. in Phys. Res. **A393**, 352 (1997).
[10] M.I.Yalandin, V.G.Shpak, S.A.Shunailov, M.R.Ulmaskulov Proc. XVII th International Symposium on Discharges and Electrical Insulation in Vacuum.- Berkeley, CA, USA.- 1996.- Vol.2.- P.635-639.
[11] N.F.Kovalev et al, Pis'ma v ZhETF, **18**, 232 (1973)
[12] S.D.Korovin *et al.* Pis'ma v ZhTF, **11**, 1972 (1985).
[13] M.I.Yalandin, G.T.Smirnov, V.G.Shpak, S.A.Shunailov, Proc. 9th IEEE International Pulsed Power Conference- Albuquerque, NM, USA, 1993.- P.388-391.
[14] A.S.Elchaninov, S.D.Korovin, G.A.Mesyats., V.V.Rostov, V.G.Shpak, M.I.Yalandin, Proc. VI Int. Conf. High-Power Part. Beams: BEAMS-86.- Kobe, Japan,1986.- P.552-555.
[15] M.I.Yalandin, V.G.Shpak, S.A.Shunailov, M.R.Ulmaskulov, Proc. of SPIE International Symposium:Intense Microwave Pulses III.- San Diego, CA, USA, 1995.-Vol. 2557.-P.289-297.

DEMONSTRATION OF SUBMILLIMETER RADIATION GENERATION FROM STATIC FIELD BY SUPERLUMINOUS IONIZATION FRONT IN SEMICONDUCTOR CAPACITOR ARRAY

D.Hashimshony, C.Cohen, A.Zigler
Hebrew University, Jerusalem , Israel.
and
K.Papadopoulos
University of Maryland, College Park MD, USA

Abstract

Generation of frequency upshifted monochromatic radiation in the 100 GHz to FIR range by the interaction of a superluminous photoconducting front with an electrostatic "frozen wave" configuration in a semiconductor is reported. The interaction converts the energy contained in the "frozen wave" into monochromatic radiation, whose frequency depends on the energy in the laser pulse creating the superluminous front and the wavelength of the static wave. Monochromatic free space propagating waves with frequency between 100 GHz and 1.6 THz were generated using a 100 fsec Ti:Sapphire laser pulse incident obliquely on a ZnSe photoconducting slab containing a frozen wave field. A superluminous photoconducting front was formed by two photon absorption. Tunability was achieved by varying the laser pulse energy from .1 to 1 mJ. The electric field waveform of the emitted radiation pulse was measured by an optically gated miniature dipole antenna. The concept can be scaled to develop narrowband, tunable and powerful THz and FIR sources for spectroscopic imaging and other applications.

In recent years the concept of frequency up shift of electromagnetic radiation propagating in the plasma medium by a rapid spatial and temporal modification of plasma parameters has received a considerable attention due to its potential use as tunable high power radiation source[1]. There is a special interest in a compact source of high power far-infrared radiation which may play an important role in diverse applications such as advanced radar systems, ultrafast chemical and biological imaging and etc. Several alternate schemes have been proposed and some of them were demonstrated[2,3]. One of the original schemes is up shift of radiation propagating through the gaseous medium by reflection out of fast ionization front. The front can be generated by laser beam that propagates inside the medium or by sweeping the medium with a laser beam from the side[4]. Frequency upshift is achieved due to Doppler shift from the overdense front [5]. The frequency shift can be achieved even when the font is underdense [6]. It was shown theoretically that very large frequency up-shifts are possible by superluminous ionization front. Many of the schemes of frequency up-shift of an initially propagating wave share the same setback. Their output power is limited by the power of the initial wave , therefore requires an initial high power radiation sources.

Recently it was suggested to use a laser produced ionization front to generate tunable radiation directly from a static electric field[7] , thereby eliminating the need for the initial high power radiation source. In this scheme - a dc to ac radiation converter(ARC), radiation is up shifted from zero frequency to a tunable value. In the proof of principle experiment of that scheme, an alternately biased capacitors produce a static filed inside working gas. An ionization front created by a short laser pulse move between the capacitors plate produced a 40 GHz when the plasma density was about $10^{12}cm^{-3}$.

In this work, we present the first measurements of tunable Far IR radiation obtained using a reflecting regime in semiconductor plasma. The plasma of electrons and holes was induced in ZnSe crystal by two photon ionization using 100fsec Ti Sapphire laser. The crystal was placed in an alternatively DC biased capacitor array. The superluminous ionization front velocity kept fixed on 1.6 times the light speed, while the plasma frequency was changed monotonically by varying the laser energy. The emitted radiation frequency was observed in a wide range span from 100Ghz to a several THz . The measurement set up was designed to measure the full temporal shape of the emitted radiation electrical field amplitude. This feature was used for the detection of both amplitude and phase of the frequency spectrum up to a few THz.

There are a several advantages in using a semiconductor as the capacitor substrate.
a) The breakdown threshold is much larger in semiconductor, thus a very large electrical fields can be used.b). A low ionization energy, only ~2ev per electron-hole pair production c.) The relatively short recombination time allows to obtain a high repetition rates and avoids the need for removal of the used plasma. d) The radiated frequency can be easily tuned by the varying laser energy and angle of incident.

In interpreting and scaling the experimental results it is instructive to refer to the traditional frozen wave generator concept[8]. Such a device consists of segments of transmission lines sections with one way transit times equal to the desired microwave half period arranged in series or parallel and connected with optically activated switches. The sections are charged alternatively positive and negative to form a dc “frozen-wave”. Radiation is produced when the optically controlled switches are activated simultaneously. The critical differences in the physics controlling our device and the traditional frozen wave generator, is that the shorting of the capacitor array is sequential and the radiation produced by shorting it propagates inside the plasma which is generated at superluminous speeds. This, however has profound consequences since the frequency is upshifted to satisfy the appropriate dispersion relation and phase matching conditions. Following Lampe et al.[10] we work in the laboratory frame since the front is superluminous. There are three modes in the system. The first is the frozen wave, with zero frequency and wavelength $k_o=\pi/d$, where d is the distance between the capacitors (Fig. 1). The other two are the rigthward(+) and leftward (-) propagating waves in the semiconductor plasma. They must satisfy the usual dispersion relation

$$k_{1,2}c/w = (+,-)\sqrt{e}\,\{1-(w_e^2/w_{1,2}^2)\,[w_{1,2}/(w_{1,2}-ig)]\}^{1/2} \qquad (1)$$

where e is the dielectric constant of the semiconductor, w_e the plasma frequency and the g phenomenological dephasing rate. The continuity conditions require that the waves are in phase at the front, z= -Ut, where U= c/sinq and q is the incidence of the laser on the semiconductor. This gives

$$k_oc/\sin q = w_{1,2} + k_{1,2}c/\sin q \qquad (2)$$

Neglecting the leftward wave since it never catches up with the front[7] and droping the subscript from the frequency and wavenumber of the forward wave, we find from (1) and (2) the relationship between the emitted frequency w and the other parameters as

$$(k_oc\beta-\omega)^2-\omega^2\varepsilon\beta^2+\omega_e^2\varepsilon\beta^2\frac{\omega}{\omega-i\gamma}=0 \qquad (3)$$

where b= 1/sinq. It is easy to see that in the absence of plasma, i.e. w_e->0 (or w>> w_e), equation (3) reduces to the frozen generator one. However, for k_ocb <<w equation (3) becomes

$$\omega(\omega - i\gamma) = \omega_e^2 \frac{\varepsilon\beta^2}{\varepsilon\beta^2 - 1} \qquad (4)$$

It is clear that for w_e>>g and eb^2>>1, conditions easily fulfilled in our experiment the emitted frequency w is near the plasma frequency and proportional to √n where n is the carrier density. The value of n is of course a function of the laser energy per pulse W. It is easy to see that for two photon absorption, since the absorption length is $(aI)^{-1}$, where I is the laser intensity and a the two photon absorption coefficient, the average density is given by

$$n = a W^2 / hn S^2 t \qquad (5)$$

where S is the illuminated area of the crystal and t the laser pulse length. From eqs (4) and (5) we find that the emitted frequency scales linearly with the laser pulse energy consistent with the results shown in Figure 3.

Experimental and Results

The experimental set up is described on fig,1. Semiconductor ZnSe crystal was cut and polished to the shape as shown in figure 1. The crystal was chosen so its energy gap will be larger then the energy of the single photon of the ionizing laser. Therefore, the generation process of carriers involved two photon and the absorption depth was intensity-dependent. This allowed a depeer penetration of laser radiation and creation of large volume of plasma. Thus, FIR radiation is generated by all volume of plasma.

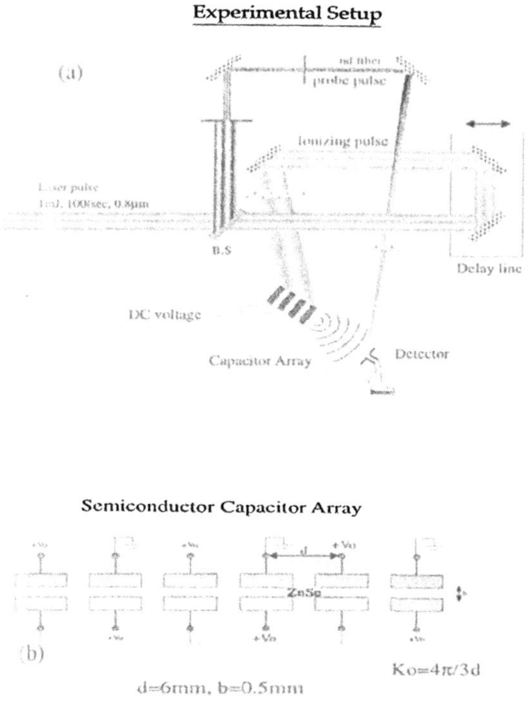

FIG. 1 .(a) Experimental set-up. (b) Capacitor Array.

The crystal was placed between the two thin glass plates with a multiple electrode structure . This capacitor array made by an evaporation of Al structure on a thin glass plate. The capacitors were alternately biased with voltage of 50 volts to form an alternately electric dc field inside the crystal. The capacitor array was charged by a fast rise time pulser that produced a relatively long voltage pulse. The 3mmx10mm capacitor plate were separated by 3mm. Thus, the initial static wave wavelength was centered at 9mm. The capacitor plates were extended over the crystal to create an uniform field inside the crystal . The overall length of the structure was 20mm.

The emitted pulsed electric field was measured following the standard pump and probe experiments. The pump laser source was a 100fsec Ti:Sapphire laser working at 0.8 micron wavelength. The laser energy was limited to 2mj and its repetition rate was 10 Hz . The amplitude of electrical field of the generated radiation was monitored by means of gated planar dipole antenna. The laser beam was split into two beams, the main beam carrying more then 90% of the energy served as the pump beam and the second beam served as the probe beam. The main beam was focused by a cylindrical lens on the side of the crystal to form the propagating ionizing front. The crystal was placed at 20 degrees angle to the pump laser beam front The optical path of both beams is made that of both of beams will reach the detector simultaneously. The adjustment of the optical paths made by two photon correlation using BBO crystal that was placed at detector position. The optical path of the beams was adjusted to cause the probe beam to reach the dipole antenna detector at the same time with the emitted radiation.

The detector was a dipole antenna build on a heavily ion implanted silicon layer on a sapphire substrate. The implantation was performed using a 10^{15} cm^{-2} dose of 100 KeV O^+ ions and resulted in a subpicosecond temporal response[12] due to reduction of the recombination time. The detector was gated at different delay times by varying the optical delay between the pump and probe laser pulses. The antenna was driven by the voltage induced by the incoming parallel polarized radiation from the capacitor array . The induced, time dependent voltage across the gap was determined by measuring the average current produced in the antenna circuit. This current flows through the detector only when the gap is irradiated and shorted by the delayed 100 fs laser probe beam. A SRS250 boxcar integrator is used to collect the short current pulses. The current was averaged over 10000 pulses for each delay stop. For the 100 fsec laser pulse the gap in the antenna detector remains closed for a time period of the order the carrier life time < 0.5 psec, resulting in a maximum frequency response of about 2 THz .The profile of electric field of the emitted radiation was measured by monitoring the average current vs. the time delay between the pump laser beam and the probe laser beam.

Figure 2 shows a typical measured temporal scan of the emitted radiation electric field. The scan represents the measured signal amplitude vs. the delay between the pump and the probe laser pulse. The signal was monitored by means of the boxcar integrator. The delay stop was controlled mechanically by the movement of the microstage. The duration of the signal spans several psecs, and is controlled only by the laser power. The scan length and step size were adjusted so the full waveform could be recovered. Each scan was initiated long before the emitted radiation reached[the detector. At that point the boxcar reading was close to zero. Then mechanically varying the time delay between the probe beam and the signal, the point where the boxcar reading is close to zero again was reached. Starting from this delay, the boxcar output remained zero when the delay was increased. This was done to insure that the full emitted signal was scanned and recorded. Figure 3 shows the cumulative results of the frequency vs. the laser pulse energy in the range between .1 and 1 mJ. In each case the frequency was measured using the procedure outlined above. It is clear that the frequency of the radiation scales approximately as the energy in the laser pulse. As will be discussed later the pulse energy is proportional to the carrier density generated or equivalently to the plasma frequency.

Disscussion:

Before closing we comment on the scaling properties of the concept presented with respect to power, frequency and bandwidth. As noted in Lampe et al.[10] for superluminous fronts the transmission coefficient is close to unity. As a result the output radiation has a field amplitude close to the static field. During the sweep time the electrostatic energy stored in the photoconductor is transformed into electromagnetic radiation. The energy stored in a photoconductor of length l, area S composed of N capacitors occupying a fraction f of the volume with a dc field E_o is given by W

$$K = 1/2\, f\, e\, e_o\, E_o^2\, (S\, l)$$

The discharge time $t = l/V_g$, where V_g is the group velocity of the radiation. Approximating 1/2 f e by unity, we find that the maximum power density P will be given by

$$P = 10^5\, (V_g/c)\, (E_o/10\ \mathrm{kV/cm})\ \mathrm{W/cm^2} \qquad (6)$$

The power scaling clearly the potential for producing high power radiation. Frequency tunability scaling is determined by equation (5). It is a linear function of W. Using 10 mJ pulses frequencies exceeding 10 THz can be generated.

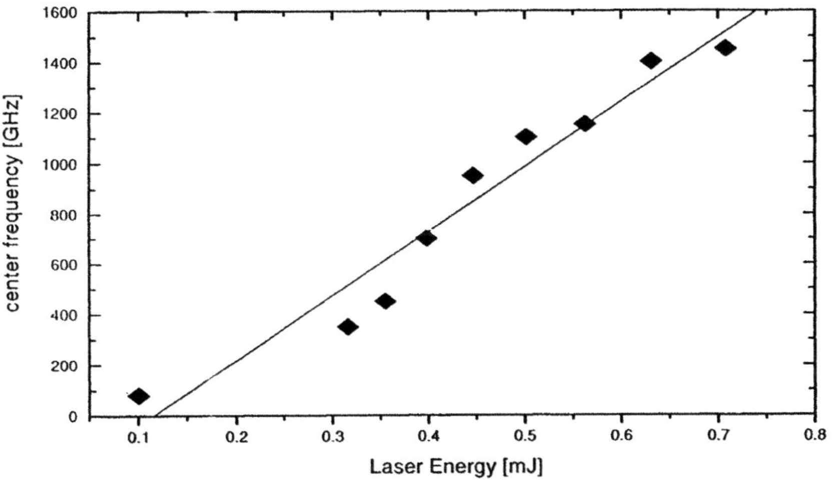

FIG. 3. Center frequency of the emitted radiation vs. laser energy.

Finally the bandwidth is controlled by the number of capacitors N so that Dw/w = 1/N. For example using d = 1 mm the radiation bandwidth will be 0.2/cm. Besides using longer lengths, the bandwidth can be significantly decreased by implementing phased array concepts. The partial evidence of this effect can be seen from Fig2 . The first half cycle of the emitted waveform is much narrower and sharper then the following peak. The possible explanation can befound by considering the time domain. In the time domain scan each half cycle can be considered as the footprint of one of the capacitors, where the first one to measured in the free space is the last one to be generated. The radiation created by the last capacitor is less affected by the plasma and by destructive interference effects, since it radiated almost directly into the free space. The following half cycles of the radiation were originally generated deeper inside the plasma and effected more by the plasma presence and the interference between sources originated by other capacitors. Their contribution to the radiation measured outside, cannot longer be considered as purely generated by one capacitor alone but is a product of the interference between the sources. At frequency domain observed the spectrum of the first half cycle looked like a spectrum of the single biased photoconductive THz source, it radiates wider frequency and with center frequency as twice as larger then the full array.

In summary we have presented a proof-principle of a novel radiation concept that can generate tunable, narrowband generation spanning the range of 100 GHz to the FIR. The concept is a hybrid using concepts of frozen wave generation, large aperture photoconducting antennas and frequency upshifting by superluminous plasma fronts. The resultant is fully scalable and has the potential for providing powerful, tunable sources in a frequency range valuable for many spectroscopic applications.

References.

[1] Savage, Jr., C. Joshi, and W. B. Mori, Phys. Rev. Lett. **68**, 946 (1992).
[2] S. C. Wilks, J. M. Dawson, and W. B. Mori, Phys. Rev. Lett. **61**, 337 (1988).
[3] E.Esarey, P. Sprangle, B. Hafizy, and P. Serafim, Phys. Rev. E. **53**, 6419 (1996).
[4] D. Hashimshony, C. Cohen, A. Zigler, K. Papadopoulos,Optics. Comm. **124**, 443 (1996).
[5] B. I. Greene, P. N. Saeta, Douglas R. Dykaar, IEEE, S. Schmitt-Rink, and Shun Lien Chuang,IEEE, J. Quantum Electron. **28**, 2302 (1992).
[6] C. Yeh, J. Appl. Phys. **38**, 2871 (1967).
[7] X. Xu, N. Yugami, and Y. Nishida, Phys. Rev. E. **55**, 3328 (1997)
[8] C. H. Lai, R. Liou, and T.C. Katsouleas, Phys. Rev. Lett. **23**, 4764 (1996).
[9].W. B. Mori, T. Katsouleas, J. M. Dawson, and C. H. Lai, Phys. Rev. Lett. **74**, 542 (1995).
[10]] Martin Lampe, and Edward Ott, Jerome H Walker, Phs. fluids, **21**, 42 (1987)

ABOUT MECHANISM OF WIDEBAND MICROWAVE RADIATION AT EXPLOSION OF CONDENSED HIGH EXPLOSIVES

V.A.Cherepenin[1], V.P.Shumilin[2]

[1]Institute of Radioengineering and Electronics, Mohovaya Str., 11, GSP-3, 103907, Moscow, Russia.
[2]High Energy Density Research Center, Russian Academy of Sciences, IVTAN, Izhorskaja Str. 13/19, Moscow, 127412, Russia.

INTRODUCTION

Electromagnetic disturbances at explosions of charges of condensed high explosives (HE), apparently, for the first time were found out by A.G.Ivanov in summer of 1938[1]. In the subsequent years these effects were repeatedly investigated in many laboratories of the world[2-8], and the frequency range, in which it was possible to find out electromagnetic disturbances, continuously extended toward the high frequencies. According to the results of these experiments, it has appeared, in particular, that at frequencies up to megahertz ($f \leq 10^6$Hz) these disturbances represent themselves quasi-stationary electromagnetic fields, which quickly decrease depending on distance from a point of explosion ($\sim R^{-3}$). At higher frequencies the microwave radiation itself was found out, when the field decreases in inverse proportion to distance ($\sim R^{-1}$). We will be interested only in this high-frequency radiation, the presence of which is not explained till now. Concerning low-frequency disturbances see review[7].

Purely speaking, the set of the experimental facts, concerning the generation of microwave radiation at explosions of condensed HE, is not big. In 1955 T.Takakura[2], working with small charges of lead azide (M=0.1÷0.4 g), has shown, that the radiation is registered on frequencies up to 10^8Hz (more precisely, up to 90MHz). At f=190MHz and 3.3GHz the radiation was not observed. At frequency 14MHz by direct measurements it was shown, that the field decreases $\sim R^{-1}$. In 1959 B.Koch[3] has published results of measurements of radiation at frequency f=24.1MHz from the charge of tolite-hexogene (50/50) with mass M=7.5·10^2g. From these data it follows, that brightness temperature of radiation at this frequency is $\sim 10^{15}$°K. In 1965 W.H.Andersen and C.L.Long[4] measured a spectrum of radiation from the charge of tetryl M=20÷10^3 g in mass over frequency range

Ultra-Wideband, Short-Pulse Electromagnetics 4
Edited by Heyman *et al.*, Kluwer Academic / Plenum Publishers, New York, 1999.

$4\cdot10^8 \div 5\cdot10^8$Hz. It appears to be close to line spectrum. In 1970 V.I.Pechkovskii and G.S.Kalchik[8] have registered radiation over frequency range f=$7.8\cdot10^9 \div 1.12\cdot10^{10}$Hz at a distance 300÷350 m apart from a place of explosion M=10^4g of ammonium nitrate. The experiments were carried out under conditions of brightly expressed channel effect. At last in 1993-1998 A.B.Prishchepenko and co-authors has published a large cycle of work (see, for example,[9-11]), devoted to creation of electroexplosive sources of wide-band microwave radiation. It is not excluded, that the registered radiation is connected not to presence of radio-technical circuits, but is caused directly by HE. In such case it is possible presumably to assert, that at explosions of HE charges (M=$10^2 \div 10^3$g) radiation with frequency up to $1.5\cdot10^{11}$Hz is registered.

Thus, there is the some of the experimental facts, showing that at explosions of condensed HE charges (as with a shell or without it) of small weight (M=$10^2 \div 10^4$g) wide-band microwave radiation (f=$10^8 \div 10^{11}$Hz) is registered. The intensity of this radiation much surpasses intensity of thermal radiation.

The present work is devoted to attempt to explain this phenomenon. The essence of the offered mechanism consists in following: during expansion of products of detonation a highly non-equilibrium medium is generated, oscillatory temperature of which can essentially exceed rotational one. Such medium is basically active from the point of view of generation and amplification of microwave radiation. As against laser active media, the frequency of collisions of molecules is about frequency of radiation, therefore the radiation will be wide-band. As an air behind a shock wave is strongly ionized, the radiation can leave the explosion zone boundaries only after the decrease of this ionization because of its expansion and instability of border of detonation products. As presence of a shell as a way of initiation of the charge can accelerate process of destruction of a conducting layer behind a shock wave.

SPHERICAL EXPLOSION IN AIR

For definition and by virtue of obvious advantages of symmetry we shall consider the explosion of a spherical condensed HE charge without a shell in air under normal conditions. We shall also assume, that the initiation comes true in the center of charge.

At the moment of an exit of detonation wave through a surface of charge, a shock wave is formed in air. This wave, gradually reducing the speed, is expanded into environment. The products of detonation during the expansion on the initial stage behave like a piston, which push a shock wave before itself. In later moment of time this piston ever more lags behind a shock wave and finally should stop. Estimations show, that in case of spherical explosion in air under normal conditions the size of area, occupied by the products of detonation, does not surpass approximately ten initial sizes of charge.

Experimental study of the initial stages of condensed spherical HE charges explosions[12-14] has shown, that the speed of detonation products is well described by a relationship

$$\frac{dr}{dt} = \frac{D_0}{R_0}\frac{1}{r^n}, \tag{1}$$

where R_0 - initial radius of charge, D_0 - initial speed of a shock wave, $r=R/R_0=1\div10$, n=const ($0<n<1$). The quantities of D_0 and n are experimentally determined for various HE.

Assuming detonation to be normal, we shall consider, that the substance behind the detonation wave is characterized by Jouguet parameters. Thus, the distribution of state variables over the charge in initial moment of time is extremely non-uniform, though it does not affect the external dimensions of a charge. For further evolution of the detonation products the rather important circumstance is that the substance was exposed to heating up to Jouguet temperature (T_H) and is compressed up to density ρ_H.

As estimations show, the moment of non-equilibrium appearance at expansion of detonation products rather poorly depends on the specific law of cooling $T(r)$, the only important factor is that this cooling obeys the power law. Therefore, for many practically important cases it is possible to consider, that this process is an adiabatic one with a constant adiabatic index. At $\gamma=5/3$

$$T = T_H(\rho/\rho_H)^{2/3} \propto \frac{T_H}{r^2}. \tag{2}$$

The initial speed of a shock wave in air can reach $D_0 \cong 10^6$ cm/sec, that corresponds to Mach number $\cong 30$. According to shock adiabat for air[15], equilibrium temperature in this case reaches 15 000°K and the electron concentration - $n_e \cong 10^{19}\,cm^{-3}$. The maximum electron concentration is near to a contact surface which separates the detonation products and heated air behind a shock wave, while the effective thickness of a conducting layer is much less than thickness of an air between a shock wave and a contact surface[16]. To make an estimation let us accept, that the concentration is equilibrium one. Then from[15] it is possible to derive n_e as a function of a Mach number, or dependence of critical frequency on r. For $D_0 \cong 10^6$ cm/sec, $n=2/3$

$$f_{cr} \cong 5 \cdot 10^{13} e^{-r}, \tag{3}$$

where $r \geq 1.3$, and the frequency is expressed in Hz. One can see, that at $r>5$ the microwave output of radiation from the zone of explosion is possible. Actually, the process of destruction of a conducting layer behind a shock wave is more complicated. The fact is, that the contact surface appears to be unstable, especially in relation to small-scale disturbances[17]. This instability is well visible on numerous photos of an initial stage of the condensed spherical HE charges explosion at $r>5$ (see, for example,[14]). Thus, the destruction of a conducting layer will occur faster and is non-uniform over the surface of a shock wave.

NON-EQUILIBRIUM IN DETONATION PRODUCTS

Since the temperature is exactly the parameter determining the rates of various relaxation processes , we shall characterize explosion process by time

$$\tau_T = \left(\frac{1}{T}\left|\frac{dT}{dt}\right|\right)^{-1} \propto \frac{R_0}{D_0} r^{n+1}, \tag{4}$$

while the initial charge radius, obviously, is proportional to weight of HE to power 1/3.

Though the characteristic time between collisions of molecules (τ_{col}) at expansion grows proportionally with r^4, as translational degrees of freedom, as rotational ones will be in equilibrium, since the initial frequency of collisions is rather high (~10^{13}Hz). The situation with oscillatory degrees of freedom is different. As the resonant exchange of oscillatory quantums between molecules (VV-relaxation) proceeds rather quickly, so Boltzmann's distribution in energy of oscillatory will not be disturb[18]. However, characteristic time of energy exchange between oscillatory and translational (and rotational) degrees of freedom (VT- relaxation) grows exponentially with reduction of temperature

$$\tau_{VT} = \tau_{col} \cdot A \cdot \exp((T_{VT} / T)^{1/3}), \tag{5}$$

where A and T_{VT} - constants depending on a type of molecules. Therefore, a time moment inevitably comes, when τ_{VT} will exceed τ_T. At further expansion the oscillatory temperature (T_V) already will not decrease. As estimations show, for typical HE it occurs at r=1.5÷2. Both weight and initial HE density influence a moment of oscillatory temperature freezing only as a logarithmic function, while the reduction of T_H value (i.e. the use low-calorific HE) can essentially speed up this process.

It is necessary to note also, that in case of strong gas dissociation, an inverse process of recombination is a source of exited molecules, that can in turn increase the time of VT-relaxation more than by a factor of 100[19]. Generally speaking, it concerns any chemical reaction, because during expansion the endothermic channels of reactions "are blocked", while the remained exothermic channels are sources of exited molecules.

One more important process, which equilibrium state is broken at expansion of detonation products, is ionization. Since the temperature T_H is not great, the equilibrium degree of ionization in an initial moment is small (~10^{-8}), that is, the basic process, in difference from[20,21], is ionization by heavy particles. In connection with strong temperature dependence of ionization speed (~exp(-I/T), where I- the energy of ionization) the non-equilibrium state is reached rather quickly (r~1) and at further gas expansion the degree of ionization falls down only by recombination processes, that is following the power law. So, for hexogene with initial density ρ_0 =1.66 g/cm^3 and M=1gramm the degree of ionization at r=5 will be ~$2\cdot10^{-11}$, which corresponds to electron concentration ~10^{10}cm^{-3} (f_{cr}~10^9Hz). The residual electron concentration grows up with increase of T_H and falls down at growth of both M and ρ_0.

$$n_e \propto T_H^3 M^{-1/3} \rho_0^{-2/3}. \tag{6}$$

Thus, the detonation products during the process of expanding represent themselves the medium, in which oscillatory temperature exceed rotational one by an order of magnitude, and the electron concentration is unusually high, despite of strong cooling of gas.

RADIATIONAL INSTABILITY IN DETONATION PRODUCTS

Non-equilibrium of medium is necessary, but not sufficient condition of amplification and generation of electromagnetic radiation. Twiss[22] has shown, that for realization of negative absorption over some frequency range it is necessary to comply simultaneously with two conditions:

$$\frac{\partial F}{\partial E} > 0 \quad (7)$$

and

$$\frac{\partial Q_{\omega}}{\partial E} < 0, \quad (8)$$

where F(E) - function of energy (E) distribution, and $Q_{\omega}(E)$ - effective probability of stimulated radiation.

As over a microwave range of frequencies the energy of quantum of radiation is much less than temperature, so the effect of negative absorption should be described in classical approach. Such consideration was spent for one-dimension oscillators[23-26]. It has appeared, that the ensemble of harmonic oscillators, as well as in a quantum case, always absorbs energy, and for negative absorption to take place in the monoenergetic oscillators medium they must necessarily be unharmonic ones

$$\frac{d\omega}{dE} \neq 0. \quad (9)$$

For unharmonic oscillators with distribution function F(E) the negative absorption will be observed, if

$$\frac{\partial}{\partial E}\left[F(E)\cdot\omega(E)\right] > 0 \quad (10)$$

across the energy interval

$$\Delta E \geq \nu\left(\partial\omega/\partial E\right)^{-1}, \quad (11)$$

where ν - frequency of collisions.

Actually the condition (10) is analogous to requirement of population inversion in a quantum case. Usually two types of inversion are distinguished when considering the molecular systems[27]: oscillatory one, or complete, and oscillatory- rotational, or partial. In our case, the complete inversion is not realized, since the function of distribution in oscillatory degrees of freedom is Boltzmann's. Therefore, only the partial inversion is possible. The classical analogue of partial inversion can be received by considering vibrating rotator, when function of distribution in oscillatory (E_V) and rotational (E_R) energies are Boltzmann's ones:

$$F(E_V, E_R) \propto \exp\left(-\frac{E_V}{T_V} - \frac{E_R}{T}\right). \quad (12)$$

The oscillations of such rotator are non-harmonic ones, and difference of energies $E = E_V - E_R$ plays the role of energy in relations (9) - (11). In this case

$$\frac{\partial F}{\partial E} \propto \left(\frac{1}{T} - \frac{1}{T_V}\right) > 0 \qquad (13)$$

at $T_V > T$. As against effects, in which the basic role plays oscillators unharmonism[23-26], this effect has a kinetic character and at large frequencies of collisions is realized over a wide range of radiation frequencies. Over the optical range of frequencies this effect was used for creation of lasers[28,29].

Thus, in a considered case, the detonation products represent themselves an active medium, capable to amplify and generate the microwave radiation. Certainly, it is necessary, that this amplification exceeds the absorption in plasma of detonation products themselves and conducting layer behind a shock wave.

CONCLUSION

Within the framework of offered model a number of observable effects find its natural explanation, in particular, the absence of high-frequency radiation at explosions of lead azide charges[2]. The fact is, that the detonation products can represent themselves an active media only if the composition of detonation products is rather rich in molecules with exited oscillatory degrees of freedom, and a great life time. In a case of lead azide (PbN_6) there is an obvious contradiction between the low heat of explosive transformation (and, hence, the low temperature T_H) and large energy of oscillatory excitation of molecules N_2 ($\cong$3340°K). For obtaining of wide-band microwave radiation the most optimum for composition are, apparently, the secondary HE such as $C_\alpha H_\beta N_\gamma O_\delta$.

The well-known fact of radiation intensity increase at explosions of charges in shells can be explained by acceleration of destruction process of a conducting layer behind a shock wave. An essential moment is also the influence of both a shell and a way of charge initiation on the radiation pattern.

Thus, the considered mechanism of radiation will take effect after some delay from the moment of HE initiation, and this delay time is proportional to weight of HE to power 1/3. In absence of nonlinear effects the average frequency of radiation should decrease during a pulse. The important parameters, influencing the considered mechanism of radiation are: a type of HE and geometry of charge, existence of a shell, the way of initiation, external pressure.

In summary we would like to thank Dr. E.V.Chernykh, Dr. E.V.Nesterov and Prof. S.L.Ziglin for fruitful discussion of work.

REFERENCES

1. A.G.Ivanov, Seismic-electrical effect of second kind, *Transaction of USSR Academy of Sciences. Geography and Geophysics Series.* (Russian). 5:699 (1940).
2. T.Takakura, Radio noise radiated on the detonation of explosive, *Publications of the Astronomical Society of Japan.* 7:210 (1955).
3. B.Koch, Émission d'ondes radioélectriques par des détonations, *Académie des Sciences. comp. rend.* 248:2173 (1959).
4. W.H.Andersen and C.L.Long, Electromagnetic radiation from detonating solid explosives, *J. Appl. Phys.* 36:1494 (1965).
5. A.P.Boronin, V..A.Velmin, Yu.A.Medvedev, B.M.Stepanov, Experimental study of electromagnetic field in an adjacent zone at explosions of condensed HE, *Journ. of Appl. Mech. and Tech. Phys.*

(Russian). 9:99 (1968).
6. A.P.Boronin, V.N.Kapinos, S.A.Krenev, About physical mechanism of electromagnetic field generation by explosion of condensed HE charge. Experimental research data, *The Phys. of Combus. and Explos.* (Russian). 26:117 (1990).
7. A.P.Boronin, V.N.Kapinos, S.A.Krenev, V.N.Mineev, About physical mechanism of electromagnetic field generation by explosion of condensed HE charge. Review, *The Phys. of Combus. and Explos.* (Russian). 26:110 (1990).
8. V.I.Pechkovskii, G.S.Kalchik, Attenuation of detonation under action of a magnetic field, arising at explosion HE charges in shells, *The Phys. of Combus. and Explos.* (Russian). 6:123 (1970).
9 .A.B.Prishchepenko, V.V.Kiseljov and I.S.Kudimov, Radio frequency weapon at the future battlefield, *in:*"Electromagnetic Environments and Consequences. Proc. of the EUROEM 94 Int. Symp.", D.J.Serafin, J.Ch.Bolomey, D.Dupouy ed., Gramat, France, I:266 (1995).
10.A.B.Prishchepenko, V.P.Zhitnikov, Microwave ammunitions: suum cuique, *in:*"Ultra Wideband, Short-Pulse Electromagnetics 3. Proc. of the AMEREM 96 Int. Symp." A.Stone, C.Baum and L.Carin ed., Plenum, New York, (In Press).
11.A.B.Prishchepenko, Invisible death of electronics, *Soldier of Fortune.* (Russian). 3:45 (1996).
12.Yu.N.Ryabinin, I.I.Tamm, About similarity of air shock waves, formed by HE charges, *in:*"Physics of Explosion", M.A.Sadovskii and A.F.Belyaev ed., USSR Academy of Sciences, Moscow, Russia, 5:71 (1956).
13.B.D.Ckristoforov, Air shock wave front parameters at explosion of PETN and lead azide charges of different density, *Journ. of Appl. Mech. and Tech. Phys.* (Russian). 2:175 (1961).
14.V.V.Adushkin, About formation of a shock wave and explosion products flight in air., *Journ. of Appl. Mech. and Tech. Phys.* (Russian). 4:107 (1963).
15.N.M.Kuznetsov."Air Thermodynamic Functions and Shock Adiabats at High Temperatures." Mashinostroenie, Moscow (1965).
16.B.S.Punkevich, B.M.Stepanov, Air electroconductivity behind shock wave front at explosion of condenced HE, *The Phys. of Combus. and Explos.* (Russian). 16:109 (1980).
17.S.I.Anisimov, Ia.B.Zeldovich, Rayleigh-Taylor instability of boundary between detonation products and gas at spherical explosion, *J. Tech. Phys. Lett.* (Russian). 3:1081 (1977).
18.E.V.Stupochenko, S.A.Losev, A.I.Osipov."Relaxation Processes in Shock Waves", Nauka, Moscow (1965).
19. N.M.Kuznetsov, Oscillatory relaxation in recombinating extending gas, *High Temperature Thermophysics* (Russian). 4:282 (1966).
20.Yu.P.Raizer, About residual ionization of gas, expanding in emptiness, *J. Exper. And Theor Phys.* (Russian). 37:580 (1959).
21. N.M.Kuznetsov, Yu.P.Raizer, About electron recombination in plasma, extending in emptiness, *Journ. of Appl. Mech. and Tech. Phys.* (Russian). 6:10 (1965).
22.R.Q.Twiss, Radiation transfer and the possibility of negative absorption in radio astronomy, *Aust. J. Phys.* 11:564 (1958).
23.A.V.Gaponov, Instability of a system of excited oscillators with respect to electromagnetic perturbations, *J. Exper. and Theor Phys.* (Russian). 39:326 (1960).
24.I.I.Sobelman, I.V.Tyutin, Stimulated radiating processes in quantum and classical theories, *Usp. Fiz. Nauk.* (Russian). 79:595 (1963).
25.A.V.Gaponov, M.I.Petelin, V.K.Yulpatov, Stimulated emission by exited classical oscillators and its use in high-frequency electronics, *Izv. Vuzov. Radiophysics.* (Russian). 10:1414 (1967).
26.V.M.Fain."Quantum Radiophysics. v.1. Fotons and Nonlinear Media". Sovetskoe radio, Moscow, (1972).
27.M.S.Jijoev, V.G.Platonenko, R.V.Khokhlov, Chemical lasers, *Usp. Fiz. Nauk.* (Russian). 100:641 (1970).
28.M.S.Jijoev, V.V.Korolev, V.N.Markov, V.G.Platonenko, R.V.Khokhlov, Detonating gasdynamic laser, *J. Exper. and Theor Phys. Lett.* (Russian). 14:73 (1971).
29.V.M.Marchenko, A.M.Prochorov, About an opportunity of creation of inverse medium for lasers by means of explosion, *J. Exper. and Theor Phys. Lett.* (Russian). 14:116 (1971).

CALORIMETRIC SPECTROMETER FOR MEASURING SINGLE MICROWAVE PULSES IN RELATIVISTIC MICROWAVE ELECTRONICS DEVICES

Irina L. Bogdankevich[1], Yannick Caillez[3], Anatoly G. Shkvarunets[1], Pavel. S. Strelkov[1], Vladimir P. Tarakanov[2], Denis K. Ul'yanov[1], Raimond Wioland[3]

[1]General Physics Institute, Russian Academy of Sciences, Vavilova str. 38, Moscow, 117942 Russia
[2]Institute of High Temperatures, Russian Academy of Sciences, Izhorskaya ul. 13/19, Moscow, 127412 Russia
[3]Centre d'Etudes de Gramat, France

INTRODUCTION

Broadband plasma relativistic microwave oscillators fabricated at the General Physics Institute of the Russian Academy of Sciences have the following output parameters: microwaves are generated in a frequency band on the order of the mean frequency, the output power is ~100 MW, the mean microwave frequency is 10÷20 GHz, the duration of a single pulse is 20÷800 ns, and the diameter of the emitting horn window is no less than 15 cm (because of the possible microwave breakdown) [1–3].

The spectrum of high-power broadband emission from the plasma in the course of a REB-plasma interaction was studied in experiments [4] with a 15-channel spectrometer in the frequency band 6÷40 GHz, the bandwidth of each channel being 2.3 GHz. The power collected at the entrance of the spectrum analyzer was ~1 W, the total radiation power being several tens of MW. Since, in those experiments, it was necessary to significantly attenuate the signal from the receiving horn to the detection assembly, the error in measuring the absolute spectral power density was very large (±50%).

A similar five-channel spectrometer with the bandwidth 18–40 GHz was used in [5]. In [1], the power collected at the entrance of a five-channel spectrometer was < 1 MW, but the accuracy in measuring the spectral power density was also very low because of the difficulties in performing the absolute calibration of the channels at high powers.

Here, we describe the calorimetric spectrometer that has already been used in experiments [6] and provides the possibility for measuring the spectral power density with a

Ultra-Wideband, Short-Pulse Electromagnetics 4
Edited by Heyman *et al.*, Kluwer Academic / Plenum Publishers, New York, 1999.

much higher accuracy (<20%). The main advantage of our spectrometer is that it can measure the spectrum of the total microwave energy flux in absolute units, while the spectrometers used in [1,4,5] were capable of measuring only a small fraction of the total energy flux. Since the emission spectrum from the sources studied in those papers may be different at different points over the cross section of the microwave beam and the measurement results depend on the polarization of the receiving antenna, the data presented in [1,4,5] are insufficient to reconstruct the spectrum of the total microwave energy flux. The spectrum of the total energy flux from oscillators similar to those described in [1-3] is the fundamental characteristic calculated theoretically and used to estimate the possible applications.

DESCRIPTION OF THE DEVICE

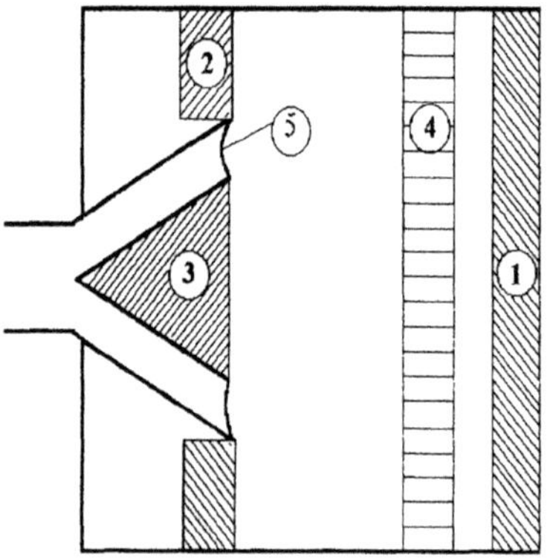

Figure 1. Schematic of the calorimetric spectrometer: (1) calorimeter, (2) calorimeter recording reflected microwaves, (3) horn of the microwave oscillator, (4) low-frequency filter, and (5) window.

Figure 1 shows a schematic of the proposed calorimetric spectrometer aimed at measuring the total power of a microwave pulse in the given frequency band over the pulse duration. Radiation from the output window (5) of the horn (3) of a microwave oscillator comes at the low-frequency filter (4), which is a metal plate of thickness l with apertures of diameter d. Since $l > 2.5d$, each of the apertures is a waveguide. So filter (4) transmits only those microwaves whose frequencies are $f > f_{cr} \cong c/(1{,}7 \cdot d)$ with c the speed of light, and d the aperture diameter. The energy Q_I of these microwaves is measured by a calorimeter (1) [7], and the energy Q_2 of the reflected microwaves, by a calorimeter (2). The geometric sizes of the horn (3) and calorimeter (2) and the distance between the horn and filter (4) are chosen so that the microwaves reflected from the filter do not pass through the horn. Thus, the quantity Q_I+Q_2 is equal to the total power of the microwave pulse.

CALCULATION AND MEASUREMENT OF THE FREQUENCY CHARACTERISTICS OF THE FILTERS

Each of the filters is a diffractive grid with apertures of different diameters and with different periods (Fig. 2). The geometry of the grid is such that $h \sim d$. The transparency coefficient (K) of the grid depends on the relation between λ_{cr} ($\lambda_{cr} \cong 1{,}7 \cdot d$) and the wavelength λ of the incident wave. For $\lambda_{cr} >> \lambda$, the K is governed exclusively by the grid geometry and is equal to the ratio of the area of the apertures to the total area of the filter. For $\lambda < \lambda_{cr}$, the transparency coefficient is equal to zero, which is typical of cut-off waveguides. In the resonance region $\lambda \sim \lambda_{cr}$, the transparency coefficient depends on frequency and can vary from values smaller than the geometric transparency coefficient to values close to unity. For each grid, the dependence of the transparency coefficient on the incident wave frequency has been computed with a 3D version of the KARAT code [8], which solves the boundary-value problem for time-dependent Maxwell's equations using the difference scheme with shifted meshes. The results obtained were tested against the exact analytic solution to the problem of the diffraction of a planar wave by a circular aperture of radius a in an ideally conducting plane [9]. Figure 3 shows numerical and analytical profiles of P/P_{geom} vs the parameter ka with $k = 2\pi/\lambda$, P the power transmitted through the aperture, and P_{geom} the fraction of the incident energy flux within an area of πa^2. The transparency coefficient of the aperture is seen to substantially increase in the wavelength range close to λ_{cr}. Good agreement between numerical and analytical profiles allowed us to employ the KARAT code to calculate the parameters of the filters to be used in experiments.

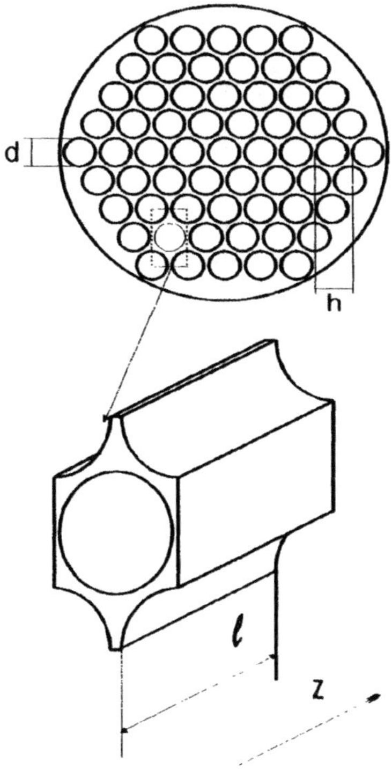

Figure 2. Diffractive grid used as a low-frequency filter and an element of an infinite periodic structure that models the diffractive grid.

In calculations, a finite-size filter was modeled by an infinite structure periodic in the (x, y) plane. As an element of the structure, we chose a cell shown in Fig. 2. A wave packet with a prescribed envelope is incident upon this cell along the z-axis. The boundary conditions hold: the tangential component of the electric field is equal to zero at the grid surface, which corresponds to the condition $[E\ n] = 0$ at a perfectly conducting surface. We also used the periodicity condition for the structure to be invariant with respect to translations along the x- and y-axes by distances h_x and h_y, respectively.

Figure 4 depicts the computed temporal profiles of the Pointing vector flux along the z-axis over the beam cross section before (Fig. 4a) and after (Fig. 4b) the beam passes through the grid. Figure 4a shows that, by the time $t = 0.8$ ns, the energy flux that comes from the left becomes maximum and, by the time $t = 1.1$ ns, reduces (the nonresonant case) due to reflection from the grid. Figure 4b shows that the energy flux that has passed

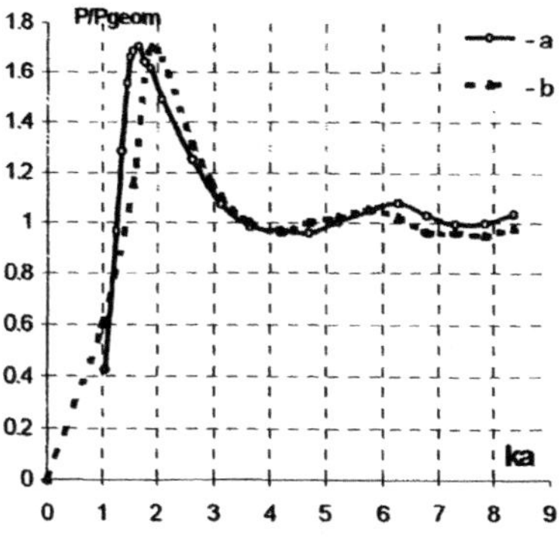

Figure 3. Comparison between the computation results (curve a) and the exact solution (curve b).

through the grid becomes steady by the time $t = 1.3$ ns. The transparency coefficient is calculated as a ratio of the microwave power

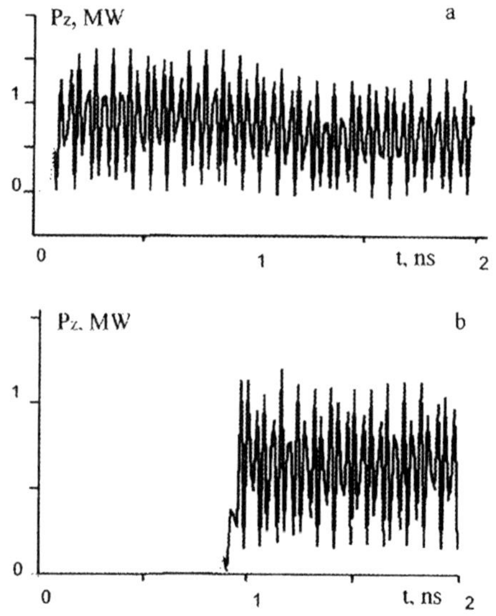

Figure 4. Temporal profiles of the Poynting vector flux along the *z*-axis (a) before and (b) after the beam passes through the grid.

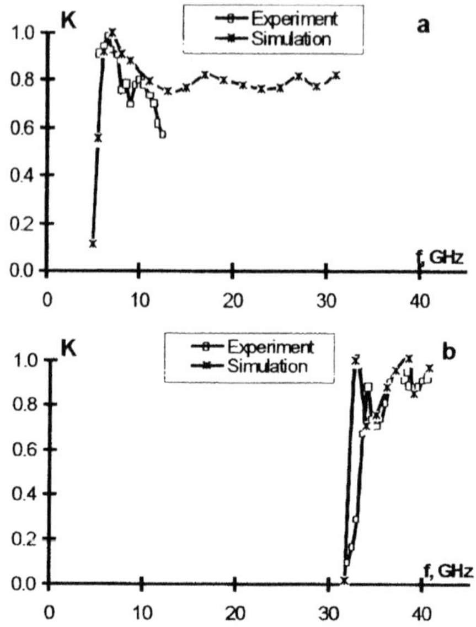

Figure 5. Measured and calculated profiles of the transparency coefficient vs the incident wave frequency.

passing through the filter after the point in time 1.3 ns to the total power irradiating the filter. These results allows us to determine the frequency characteristics (two of which are presented in Fig. 5) of the transparency coefficients for seven filters in the frequency band from 5 to 42 GHz.

The same characteristics were measured in experiments with the device elaborated by D.H. Ghodgaonkar, V.V. Varadan, and V.K. Varadan (Gramat, France) and consisting of two conical transmitting and receiving antennas between which the filter was installed. The measurements were carried out in the frequency band 5.85÷110 GHz. The diameter of the Gaussian microwave beam at a power level of 3 dB in the plane of the filters was $\sim 3\lambda$.

The calculated and measured profiles of the transparency coefficient vs the incident wave frequency for the same filter assembly are plotted in Fig. 5. Figure 5a shows that, in the wavenumber range near λ_{cr}, the calculated and measured transparency coefficients are close to unity. For $\lambda > \lambda_{cr}$, the calculated transparency coefficient is ~0.8 and coincides with the geometric transparency coefficient . The transparency coefficient measured in the range $\lambda \approx 2\lambda_{cr}$ is small (~0.6, Fig. 5a) and differs from the computed one. This might stem from the fact that, in experiments, the beam diameter was comparable to the aperture diameter, whereas the simulations were performed for a larger diameter beam. Agreement between the calculated and experimental profiles in Fig 5b is quite satisfactory. In measuring the emission spectra from high-power relativistic microwave oscillators, we used the computed frequency profiles of the transparency coefficient.

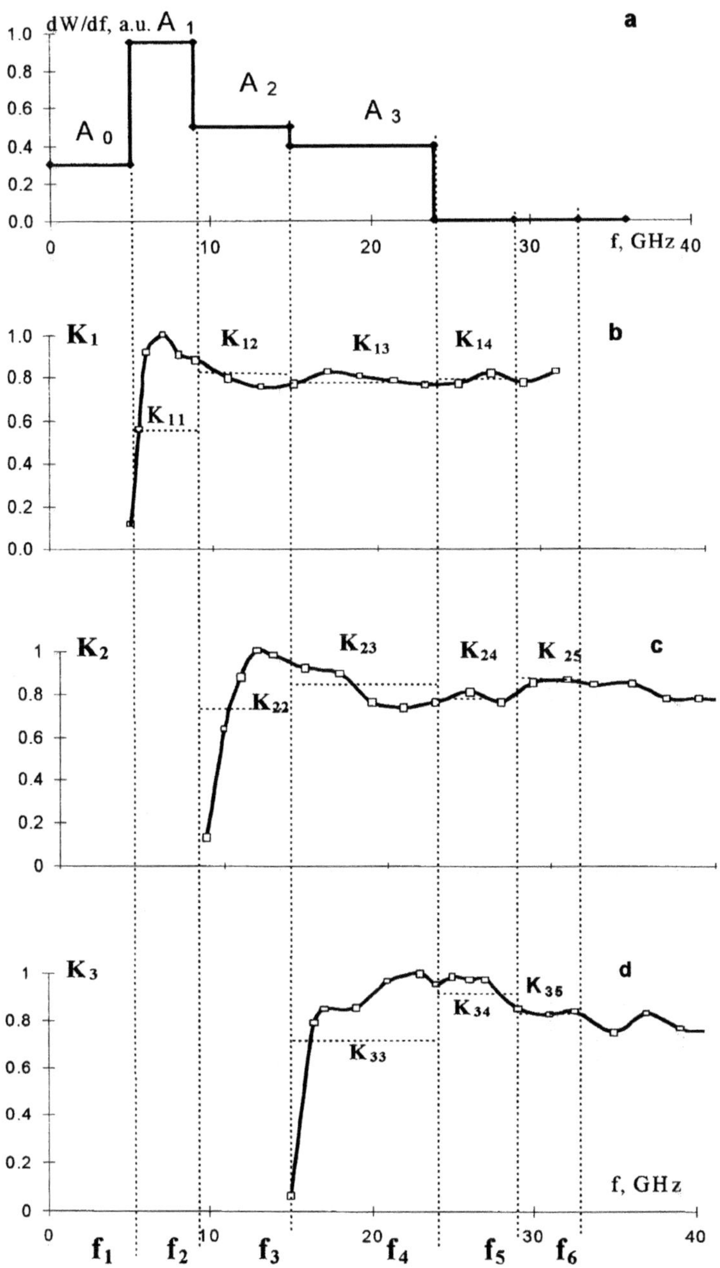

Figure 6. Illustration of the spectral measurement technique: (a) histograms of signal spectrum, (b) the transparency coefficient: K_1 of filter 1 with critical frequency f_1, (c) the K_2 of filter 2 with critical frequency f_2 and (d) K_3 filter 3 with critical frequency f_3.

Our method can be applied when the emission spectrum does not change from pulse to pulse. This means that the ratio $Q_1/(Q_1+Q_2)$ measured for filters with different f_{cr} should be the same in each microwave pulse under identical experimental conditions. This method provides the possibility of obtaining the spectra in the form of histograms (Fig. 6a) with amplitudes A_0, A_1, A_2, ..., A_n. Below, we present an example of how the results of spectral measurements are processed when the maximum emission frequency is low than $f_4 = 24$ HGz, i.e., when the recorded value $Q_1 = 0$, with filter 4 being installed. The frequency profiles of the transparency coefficient for the filters are shown in Figs. 6b–6d, in which K_n is the transparency coefficient of the n-th filter with the critical frequency f_n. We denote by K_{nk} the mean value of the transparency coefficient for the n-th filter in the frequency band from f_k to f_{k+1} to arrive at the following equations:

$$A_0+A_1+A_2+A_3 = W_0 \quad \text{without filter;}$$

$$K_{11}A_1+K_{12}A_2+K_{13}A_3 = W_1 \quad \text{for filter 1;}$$

$$K_{22}A_2+K_{23}A_3 = W_2 \quad \text{for filter 2;}$$

$$K_{33}A_3 = W_3 \quad \text{for filter 3}$$

Here, W_0 is the total microwave energy flux averaged over all the pulses. The product of the average value of the ratio $Q_1/(Q_1+Q_2)$ for filter 1 with W_0 is equal to W_1. The quantities W_2 for filter 2, W_3 for filter 3, etc. are determined in a similar way. Ordinarily, the quantity W_0 is averaged over ≈20 pulses, and the ratio $Q_1/(Q_1+Q_2)$ for each filter, over ≈5 pulses.

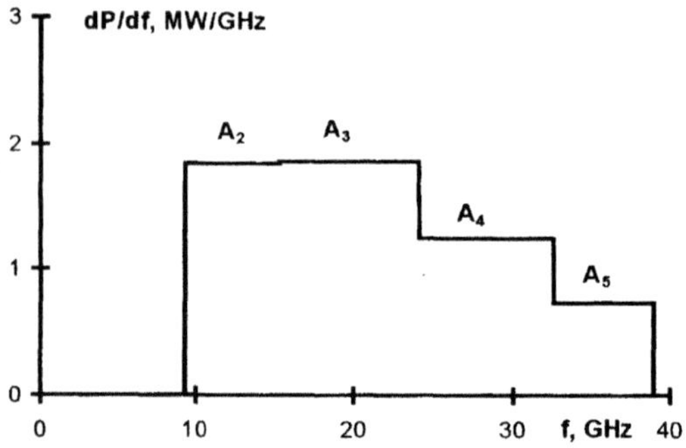

Figure 7. An example of the spectrum of the broadband relativistic microwave oscillator.

As an example, Fig. 7 displays the spectrum from a broadband plasma relativistic microwave oscillator [6]. On the ordinate, the spectral power density expressed in MW/GHz is plotted.

The power was determined from the measured microwave pulse energy and pulse duration (~20ns).

This work was supported in part by the Ministry of Science of the Russian Federation and the Russian Foundation for Basic Research (project no. 97-02-16948) and was carried out with the Relativistic Microwave Plasma Oscillator (registration number 01-04).

REFERENCES

1. A.G. Shkvarunets, A.A. Rukhadze and P.S. Strelkov Wide-band relativistic plasma microwave oscillator, *Plasma Phys. Rep.*, 20;613-616(1994).
2. M.V. Kuzelev, O.T. Loza, *et al.*, Spectral characteristics of a relativistic plasma microwave generator. *Sov. Phys. JETP,* 82;pp.1102-1111(1996).
3. O.T. Loza, P.S. Strelkov, and I.E. Ivanov, "Relativistic Cherenkov plasma maser of microsecond pulse duration," *IEEE Trans. Plasma Sci.,* 7th Special Issue on High Power Microwave Generation, (1998)
4. K.G. Kato, G. Benford, and D. Tzach, Detailed spectra of electron-beam plasma interactions, *Phys. Rev. Lett.,* 51;p.1587(1983)
5. R. Ando, M. Masuzaki, *et al.,* Microwave power emission from a beam-plasma system *J. Phys. Soc. Jpn.,* , 65;p. 2518(1996)
6. M.V. Kuzelev, O.T. Loza, *et al.,* High-power broad-band tunable microwave oscillator, driven by REB in plasma.. *11th International Conference on High Power Particle Beams,* Prague, Czech. Republic, 1; p. 225(1996)
7. A.G. Shkvarunets, A broadband microwave calorimeter of large cross section, *Instruments and Experimental Techniques*, 39;pp.535-538(1996).
8. V.P. Tarakanov, User's Manual for Code KARAT, Springfield, VA: Berkeley Research Associates, Inc., 1992.
9. L.A. Vainshtein, "Electromagnetic Waves", Moscow: Sov. Radio, 1957.

UNIVERSAL SENSOR USING ELECTRO-OPTIC SENSING PRINCIPLES

Marco Mailand, Bertrand Daout

montena emc, CH-1723 Rossens, Switzerland
Tel.: ++41 (0)26 411 31 51, Fax: ++41 (0)26 411 31 80
Internet: www.montena.com

1 ABSTRACT

Techniques like high power microwaves (HPM), pulse scattering, EMP, ESD and transient radar still lack reliable electric field monitors. One of the main obstacles to develop such field sensors are the enormous difficulties to miniaturize the sensor head providing simultaneously sufficient sensitivity, dynamic range and electrical isolation. This paper will present a recent sensor development of a universal photonic field probe for transient and continuous wave electric field measurements. The probe head incorporates three electrooptic Lithium Niobate ($LiNbO_3$) modulators packaged in purely dielectric material. The modulators cover a frequency range from dc to 3.5 GHz, where in the range dc to 1 GHz an amplitude flatness of ± 0.5 dB can be specified. The modulator input impedance has been optimized to provide maximum electric field sensitivity in conjunction with a broad band dipole antenna.

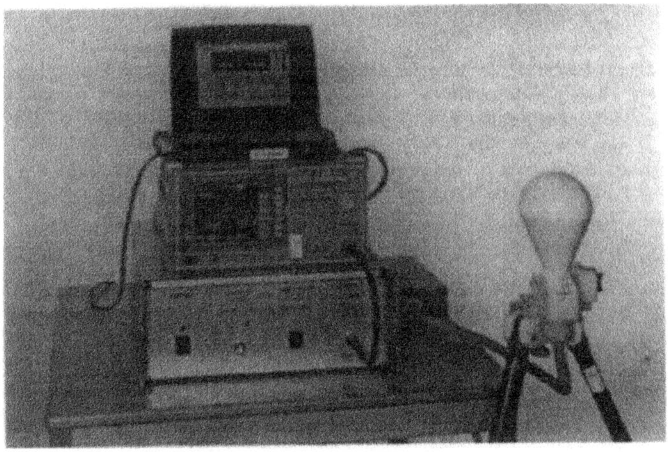

Fig. 1: Sensor system

The sensor system (see example fig. 1) including an 240 mW low noise, single frequency laser and 18 GHz bandwidth GaInP photo detectors reaches a pulse sensitivity of 4.5 V/m. For narrow band measurements a sensitivity limit of about 100 µV/m/√Hz can be extrapolated. The 1 dB compression point has been set to 12 kV/m peak to peak and could be increased to about 50 kV/m loosing sensitivity correspondingly. The sensor operates passively

Ultra-Wideband, Short-Pulse Electromagnetics 4
Edited by Heyman *et al.*, Kluwer Academic / Plenum Publishers, New York, 1999.

and is inherently potential-free by means of optical fibers with a length of up to 2 km. The modulators have been hermetically sealed and allow measurements in liquids in a depth of up to 4 meters.

2 INTRODUCTION

The recent sensor design is based on the experiences, gained during preceding projects [1, 2 and 3]. The new universal sensor allows to reconfigure the sensor head and broadband antennas, providing maximum flexibility of the whole sensor system. By extending the length of the broadband dipole rods one can increase the sensitivity of a single interferometer while loosing bandwidth.

The universal sensor allows now to dismount the electro-optical Mach-Zehnder modulators from the beam splitter. Therefore high precision optical connectors[1] with incorporated protection cap have been employed for all optical connections. The protection cap will automatically cover the highly sensitive tip of the optical connector during unplugging. Another advantage of the new connectors is the eye protection against the invisible laser beam as well as its compatibility to the FC/PC connector standard. Input and output fiber of the modulator will be plugged together into the beam splitter, using twin-type connectors.

The size of the dielectric beam splitter case has been drastically decreased to 150 x 100 x 60 mm^3 and 50 m polarization maintaining fiber (PMF) have been mounted directly to the beam splitter input. This measure reduces the number of optical connectors in the PMF path and increases thus the incident light power and reliability of the whole system.

The modulator input impedance has been optimized to provide maximum electric field sensitivity in conjunction with a broad band dipole antenna.

3 A FLEXIBLE MODULATOR SYSTEM

Electrooptic crystals like Lithium Niobate change their refracting index when an electric field is applied. The electrooptic coefficient of $LiNbO_3$ is in the range of $30 \cdot 10^{-12}$ m/V. Thus very high electric fields are necessary to provide acceptable sensitivity for a bulk crystal modulator. Titanium diffused or annealed proton exchange wave guides and integrated electrode structures with a gap width of only 6 μm are one possibility of transducing relatively small voltages to high field strengths. The change of the refractive index in an electrooptic crystal is defined by

$$\Delta n = \frac{n^3 r}{2} E \qquad (1)$$

where n is the refractive index and r is the electrooptic coefficient corresponding to the polarization of the electric field E. The phase shift of the light wave is given by

$$\Delta\varphi = \frac{2\pi}{\lambda} L \Delta n, \qquad (2)$$

where L is the length over which the electric field is present (= electrode length). Expressing the electric field as a voltage per gap width d one gets

$$\Delta\varphi = \frac{2\pi}{\lambda} \frac{n^3 r}{2} \frac{\Gamma L}{d} V \qquad (3)$$

1 DIAMOND Inc., Via dei patrizi 5, CH-6616 Losone-Locarno TI, Switzerland
Tel.: ++41/(0)91/785 45 45, Fax: ++41/(0)91/785 45 00

where Γ (0.5...1) is the overlap of the electric field with the light wave in the crystal. To shift the phase by π, it needs a voltage

$$V_\pi = \frac{\lambda d}{n^3 r \Gamma L} \tag{4}$$

which is a standard indicator for the sensitivity of an electrooptic modulator. Typical V_π of integrated electrooptic modulators are in the range of 5 to 15 Volts. Increasing the electrode length means reducing V_π.

3.1 The Electric Field Sensor

Since $LiNbO_3$ has a high permittivity ($\varepsilon_r = 35$) the electrode capacitance has significant influence on the overall sensitivity when a broad band dipole antenna is connected. An electrically short cylindrical dipole has an effective length

$$h_e = \frac{h(\Omega - 1)}{2(\Omega - 2 + \ln 4)} \tag{5}$$

with

$$\Omega = 2\ln\left(\frac{2h}{a}\right) \tag{6}$$

where $2h$ is the dipole length and a the wire radius. Applying an electric field E, the open circuit dipole voltage is

$$V_a = h_e E. \tag{7}$$

The dipole terminal capacitance is given by

$$C_a = \frac{4\pi\varepsilon_0 h}{\Omega - 2 - \ln 4} \tag{8}$$

Consequently the dipole output voltage when loaded with the interferometer capacitance C_m is

$$V_m = \frac{C_a}{C_a + C_m} h_e E \tag{9}$$

The electrode capacitance per electrode length L is given by

$$C_m / L = \varepsilon_{eff} \frac{K\left(\sqrt{1 - r_s^2}\right)}{K\left(r_s\right)} \text{ (F/m).} \tag{10}$$

K is the elliptical integral of the first kind with

$$r_s = \frac{1}{\frac{2w}{d} + 1} \tag{11}$$

where w = electrode width and d = electrode gap. For the chosen electrode dimensions we get approximately 1 pF per Millimeter electrode length.

A dipole length 2h = 3 cm and a wire radius a = 0.5 mm results in a dipole terminal capacitance of only 0.35 pF. Despite of the relatively low Vπ, typical interferometers with 5 to 10 pF input capacitance therefore reduce significantly the sensitivity of a connected dipole.

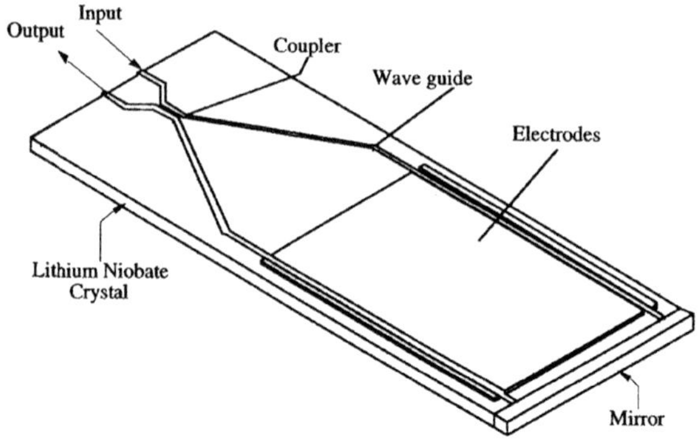

Fig. 2: Schematic of the reflective modulator

The electrode length was set to 2.5 mm yielding a Vπ = 10 V and a 3 dB bandwidth of 3.5 GHz. To reach maximum sensitivity a reflective type modulator (figure 2) was chosen. Since the light travels twice along the electrodes, the half wave voltage Vπ is doubled and the electrode capacitance is minimal. The reflective modulator was designed with a passive bias laser trimming introducing an optical path difference of $\Delta\varphi = \pi/2$.

The overall sensitivity of the sensor system depends on the antenna size, sensor loads, modulator Vπ and laser/detector noise performance. The average current on the detector i_m is given by:

$$i_m = \frac{1}{s}\frac{rP_{laser}}{2}l_{tot} \qquad (12)$$

where $s = 3$ is the splitter ratio $r = 0.68$ the photodetector responsivity, P_{Laser} the laser power (100 mW) and $l_{tot} = 0.126$ (9 dB) represents the total losses in the system (including modulator, splitter access, and fiber connector losses).

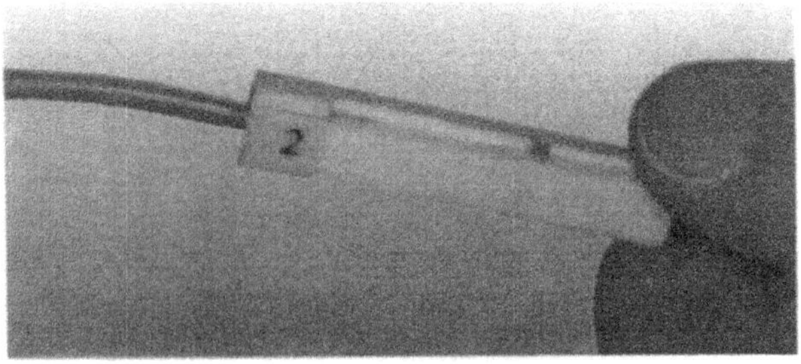

Fig. 3: Packaged interferometer

The average noise current $\langle i_{noise} \rangle$ is given by [6]:

$$\langle i_{noise} \rangle = \sqrt{B_n\left[\frac{4k_BT}{R_t} + 2ei_m + n_{rin}(i_m)^2\right]} \qquad (13)$$

where B_n is the detection bandwidth, $k_B = 1.38054 \cdot 10^{-23}$ J/K, $R_t = 50\ \Omega$ (photo detector termination impedance), $e = 1.6\ 10^{-19}$, and $n_{rin} = 3.16\ 10^{-17}$ (which corresponds to a relative intensity noise $N_{rin} = -165$ dB/Hz). The increase of the photo detector current due to an electric field component E is

$$\Delta i_s = \frac{\pi V_m}{V_\pi} i_m \tag{14}$$

where $V_\pi = 10$ V. From equation 12 through 14 we finally get the sensitivity limit

$$E_{\min} = 75 \cdot 10^{-6} \sqrt{B_n} \quad \left(\frac{V}{m}\sqrt{Hz}\right) \tag{15}$$

Considering the interferometer bandwidth of about 3.5 GHz the theoretical broad band sensitivity is 4.5 V/m. The 1 dB compression of the interferometer is reached at about 36 % of the halve wave modulator voltage $V\pi$. With a dipole length of 3 cm the compression voltage of 3.6 V corresponds to an electric field of 4.1 kV/m.

In the recent development the modulators have dipole connectors incorporated, allowing the exchange of the dipole rods and thus extending the flexibility and sensitivity range of the modulator.

3.2 The Magnetic Field Sensor

Since the modulators can be unplugged from beam splitter is it now possible to carry out magnetic field measurements too with the same sensor system setup. A magnetic field sensor consists of a wire loop and two interferometers. The magnetic field sensor interferometers have the same electrode and optical structure like those used for the electric field sensor. Additionally the electrodes are loaded with a precision microwave 4 Ω (0.01%) resistor. Connecting two interferometers to the loop (at 0° and 180°) we get a symmetrically double loaded loop (Fig.4).

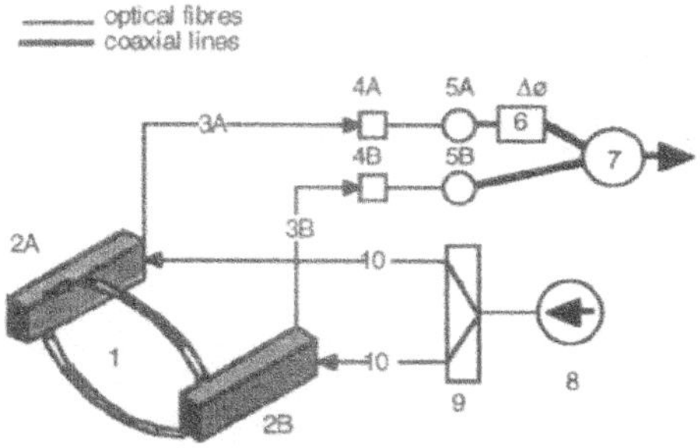

1) Loop
2) Modulators
3) Single mode output fibers (SMF)
4) Optical variable attenuators
5) Photo detectors
6) Coaxial line stretcher
7) Power combiner
8) Laser
9) Fiber splitter
10) Polarization maintaining input fibers (PMF)

Fig 4: Magnetic field sensor

The sum of the load voltages is directly proportional to the magnetic current mode on the loop. More details and references for the double loaded loop are found in [4]. Despite of having very good loading symmetry, the modulator (fig 4: 2A and 2B) characteristics and fiber lengths (fig 4: 3A and 3B) are not exactly the same. Thus an amplitude and phase calibration is necessary. The amplitude is equalized in the optical path using two continuously variable optical attenuators (fig 4: 4A and 4B). The optical attenuator has the advantage of not having an influence on the phase difference between the optical paths (fig 4: 3A and 3B). After the conversion to electrical signals with the photo detectors 5A and 5B the phase difference is corrected by tuning the 50 Ω cable length (Δø) by means of a coaxial line stretcher (fig 4: 6) before summing (fig 4: 7).

To perform the amplitude and phase calibration a pulse generator was connected simultaneously to the two magnetic sensor interferometers (fig. 4: 2A and 2B) with 180° opposite phase. This simulates electric field without magnetic component. The uncalibrated output signal at the power combiner output (fig 4: 7) is depicted in figure 5a.

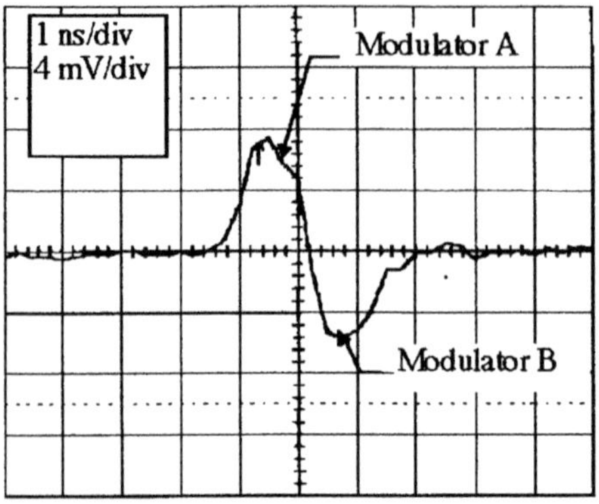

Fig. 5a: Pulse response of magnetic field modulators without phase and amplitude calibration

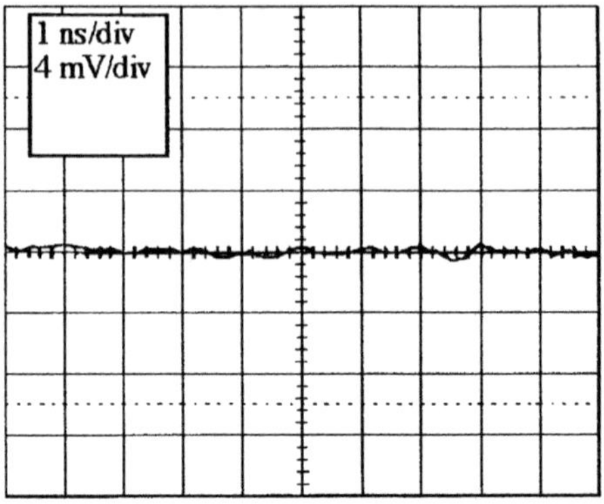

Fig. 5b: Adjusted phase difference and amplitude

After equalizing of the amplitudes the phase difference was eliminated by prolonging the electrical path of the coaxial line stretcher (fig. 4: 6), yielding in zero response to the simulated pure electric field (see fig. 5b). Subsequently to the calibration procedure in the time domain the sum to difference ratio was maximized to 38 dB in the frequency domain at 1 GHz.

4 MEASUREMENT RESULTS

Test measurements for the electric field sensors were carried out in a small TEM cell with 7.5 cm septum height yielding a ±2 dB flat frequency response up to 1 GHz. The cell frequency response was calibrated by a reference monopole.

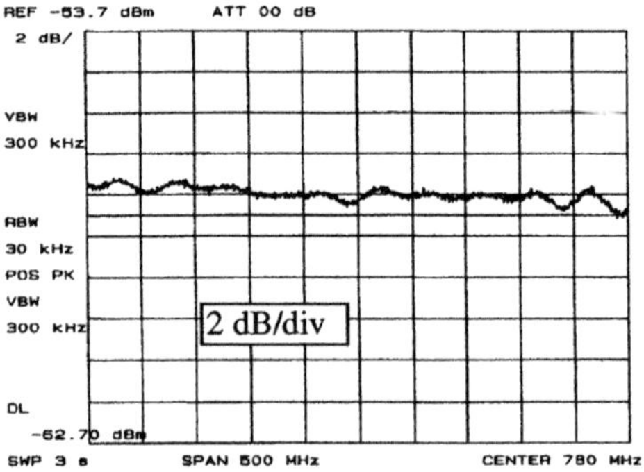

Fig. 6: Frequency response of the electric field sensor

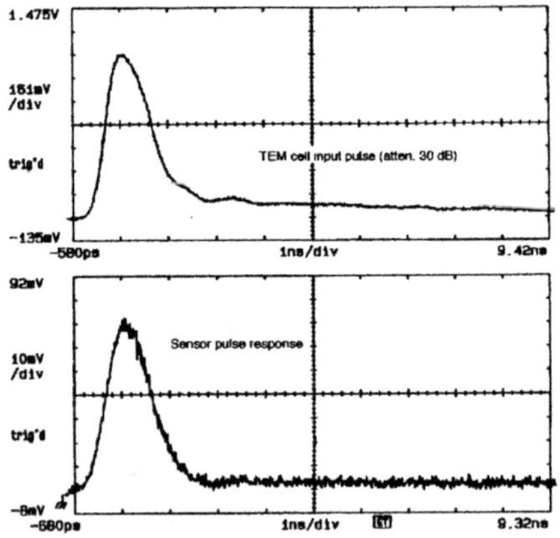

Fig. 7: Comparison of TEM cell input pulse (upper) and sensor pulse response (1 ns/div, rise time ≈ 300 ps).

Figure 6 shows the frequency response of an electric field sensor in the TEM cell in the range 530 to 1030 MHz. Measurements were carried out down to a few kHz indicating a very flat response (< ±0.5 dB). Figure 7 compares the TEM cell input pulse and sensor response. The rise time is in the range of 300 ps.

Because of limited space in the 7.5 cm TEM cell, the magnetic field sensor was tested in a GTEM 500 cell with 500 mm septum height. As expected the frequency characteristic shows first order high pass performance caused by the loop inductivity and the 4 Ω loads (see Figure 8 and [3]). The high pass characteristic of the magnetic field pulse response has to be corrected by a corresponding low pass function.

Figure 9 shows an ESD pulsed field measurement. The field was produced by discharging a 5 nF capacitor at 3.2 kV over a pressurized spark gap and leading the discharge current through a 50 Ω coaxial cable into the GTEM cell, containing one modulator.

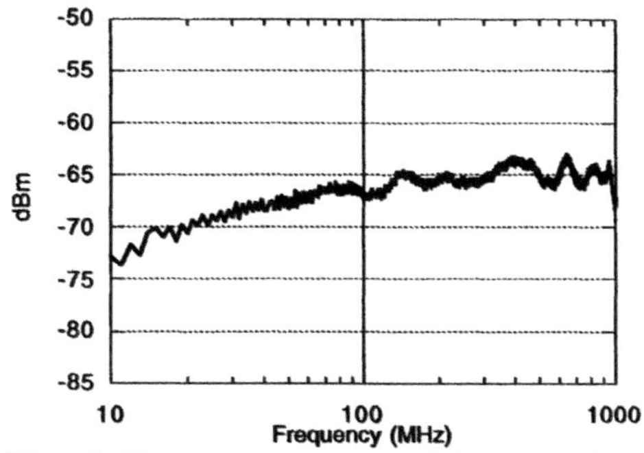

Fig. 8: Frequency response of the magnetic field sensor in a GTEM 500

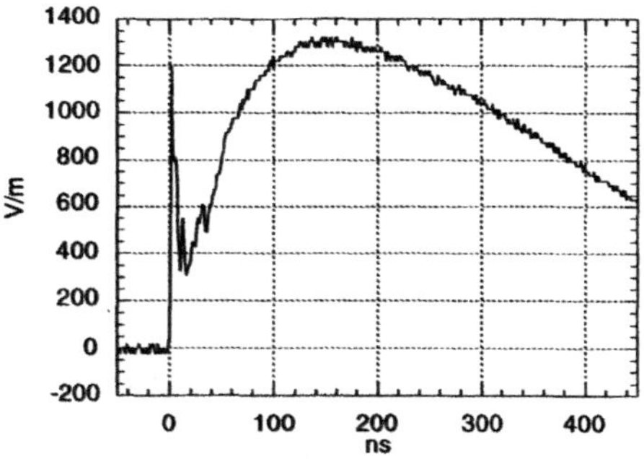

Fig. 9: Electro static discharge (ESD) field measurement.

Figure 9 gives an impression of the large dynamic range combined with fast pulse response and large bandwidth. This makes the electric field sensor also useful for direct high power microwave (HPM) measurements and related military applications. Due to the small size of the modulators is it possible to examine pulsed electric and magnetic fields inside of small volumina with minimum cross sections in the range of 50 x 50 mm².

For isotropy measurements the GTEM 500 cell was used too. Three modulators were mounted orthogonal to each other on a dielectric support. The support was then rotated by 15° steps in the GTEM EUT volume at a frequency of 1 GHz and a field level E = 10 V/m.

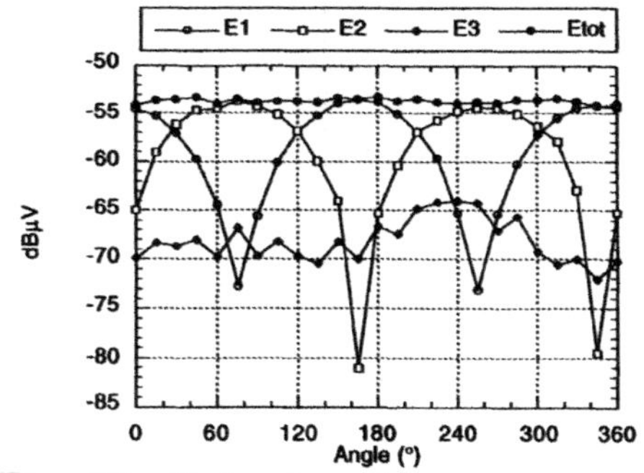

Fig. 10: Probe channel response versus probe rotation at 1 GHz

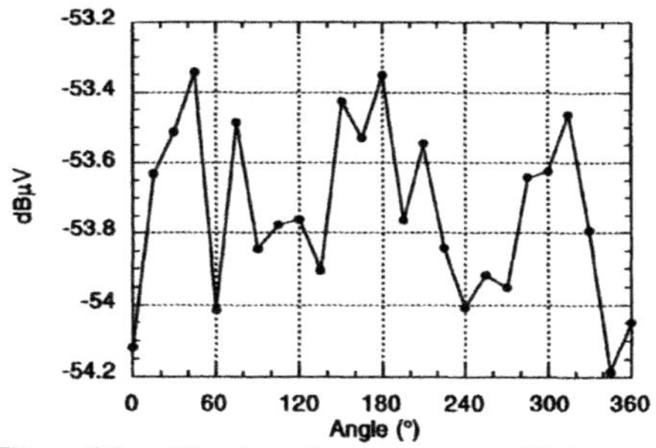

Fig. 11: Total probe response (E_{tot}) versus probe rotation at 1 GHz

All electric field channels E1 to E3 were recorded using a spectrum analyzer. The results of the isotropy measurements are depicted in figures 10 and 11. From figure 11 one concludes that an isotropy performance better than ± 0.5 dB has been reached.

5 CONCLUSION

The theory, practical realization and measurement results of a photonic field sensor system for simultaneous electric and magnetic field measurements has been discussed. A pulsed electromagnetic field with a 300 ps rise time was detected with minimal distortion. The 3.5 GHz modulator bandwidth theoretically allows rise times below 100 ps which has not been shown yet due to lack of fast pulse generators. The small size, high dynamic range and high sensitivity open a wide application range covering ESD, high power microwaves (HPM), EMC and radiation hazards. Since the electro-optic modulator works fully passive, long term surveys are possible. Also a minimum distortion of the electromagnetic field is assured because the modulators consist mainly of dielectric material. The hermetically sealing of the modulator allows measurements in liquids in a depth of up to 4 meters. Typical single mode fibers have losses < 1 dB/km at 1319 nm wavelength. Thus the distance between modulator/beam splitter and the laser/detector unit can be several kilometers. The new optical connectors incorporate eye protection, high reliability as well as best performance and compatibility with existing connector standards. The overall flexibility of the sensor system could be improved by optical connections at the beam splitter output and exchangeable dipole antennas.

6 REFERENCES

1. Felix Gassmann, Marco Mailand: "Isotropic photonic electric field sensor for sub-nanosecond pulse measurements", AMEREM'98, Albuquerque, New Mexiko, USA, May 27-31, 1996

2. Felix Gassmann, Marco Mailand: "A 9 channel photonic isotropic electric and magnetic field sensor with sub-nanosecond rise time", EMC Zürich '97, February 18-20, 1997, Zürich, Switzerland, pp 217-221 / 41G8

3. Felix Gassmann, Bernhard Eicher, Urs Knafl: „Isotrope Feldstärkemessung im Frequenz- und Zeitbereich mit einem interferometrischen Sensor“, EMV ’98, 10. - 12. Februar 1998, Düsseldorf, Germany, pp 631-639

4. Felix Gassmann: „Elektrooptischer Sensor für die isotrope Messung des elektrischen und magnetischen Feldes“, Ph. D. No. 11’190, ETH Zürich, Switzerland

WAVE EQUATIONS AND TRANSMISSION FORMULAS FOR THE OUTPUT OF A RECEIVING ANTENNA

Arthur D. Yaghjian[1] and Thorkild B. Hansen[2]

[1]AFRL/SNHE, Hanscom AFB, MA 01731, USA
[2]Schlumberger-Doll Research, Ridgefield, CT 06877, USA

INTRODUCTION

The vector output of a receiving antenna that translates without rotation with respect to a transmitting antenna is shown to satisfy the homogeneous wave equation outside the region of intersection of the receiving and transmitting antennas. This result holds for antennas operating in either the frequency domain (narrow-band antennas) or the time domain (wide-band antennas), provided multiple interactions between the transmitting and receiving antennas are negligible. In the frequency domain, if $\mathbf{b}_{p\omega}(\mathbf{r})$ is the vector sum of the scalar outputs of two mutually perpendicular orientations of the receiving antenna at the position $\mathbf{r}$, plane-wave representations for the antennas can be used to prove that $\mathbf{b}_{p\omega}(\mathbf{r})$ satisfies the homogeneous vector Helmholtz equation (wave equation in the frequency domain)

$$\nabla^2 \mathbf{b}_{p\omega}(\mathbf{r}) + k^2 \mathbf{b}_{p\omega}(\mathbf{r}) = 0 \tag{1}$$

for the receiving antenna outside the transmitting antenna. The propagation constant k equals ω/c, where ω is the frequency and c is the speed of light. The time-domain vector output $\mathbf{b}_p(\mathbf{r}, t)$ of the receiving antenna, which is related to the frequency-domain vector output by the inverse Fourier transform

$$\mathbf{b}_p(\mathbf{r}, t) = \int_{-\infty}^{+\infty} \mathbf{b}_{p\omega}(\mathbf{r}) e^{-i\omega t} d\omega \tag{2}$$

satisfies the time-domain homogeneous wave equation

$$\nabla^2 \mathbf{b}_p(\mathbf{r}, t) - \frac{1}{c^2} \frac{\partial^2}{\partial t^2} \mathbf{b}_p(\mathbf{r}, t) = 0 \tag{3}$$

for the receiving antenna outside the transmitting antenna. The frequency-domain vector Helmholtz equation (1) has been used to efficiently compute the near-field coupling

Ultra-Wideband, Short-Pulse Electromagnetics 4
Edited by Heyman *et al.*, Kluwer Academic / Plenum Publishers, New York, 1999.

between co-sited antennas,[1] to suggest a novel method of spherical near-field scanning with probe correction,[2] and to derive probe-corrected transmission formulas for plane-polar, near-field antenna measurements.[3] In this paper we rigorously derive (1) and (3), and use these wave equations to facilitate the derivation of probe-corrected, plane-wave transmission formulas needed for the direct processing of planar near-field scanning data taken in either the frequency domain or the time domain.[4]

DERIVATION OF THE HELMHOLTZ EQUATION

To simplify the derivations, we shall consider an acoustic receiving "antenna" (referred to as the probe) whose output voltage at a reference plane S_p in the matched, single-mode waveguide of the probe is given in the frequency domain by $b_{p\omega}(\mathbf{r})$. The position $\mathbf{r}$ of a reference point fixed to the probe is given in a coordinate system fixed in the acoustic transmitting "antenna," and it is assumed that the orientation of the probe does not change with respect to the transmitting antenna. That is, the probe translates without rotation with respect to the transmitting antenna. (See Figure 1.)

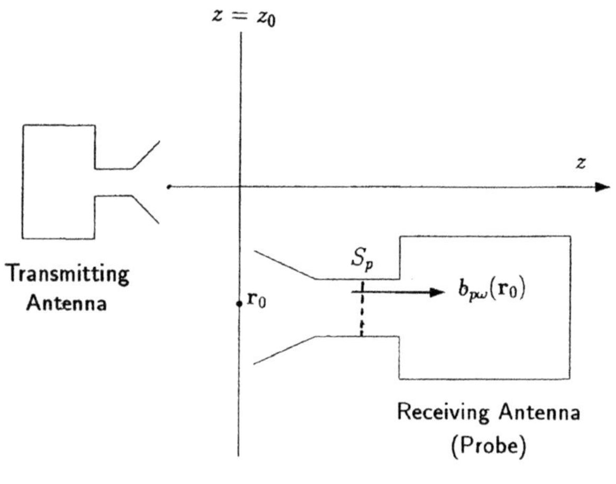

Figure 1: Schematic of transmitting and receiving antennas.

If multiple interactions between the probe and transmitting antenna are negligible, the probe output $b_{p\omega}(\mathbf{r})$ can be expressed in terms of the plane-wave transmission formula[4,5]

$$b_{p\omega}(\mathbf{r}) = \int_{-\infty}^{+\infty}\int_{-\infty}^{+\infty} R_{p\omega}(k_x, k_y) T_\omega(k_x, k_y) e^{i(k_x x + k_y y + \gamma z)} dk_x dk_y \qquad (4)$$

where $e^{-i\omega t}$ $(\omega > 0)$ harmonic time dependence is suppressed, $\mathbf{r} = x\hat{\mathbf{x}} + y\hat{\mathbf{y}} + z\hat{\mathbf{z}}$, and $\gamma = (k^2 - k_x^2 - k_y^2)^{\frac{1}{2}}$ is chosen positive real for the propagating spectrum $(k_x^2 + k_y^2 < k^2)$ and positive imaginary for the evanescent spectrum $(k_x^2 + k_y^2 > k^2)$. The function

$T_\omega(k_x, k_y)$ is the transmitting plane-wave spectrum of the transmitting antenna, and $R_{p\omega}(k_x, k_y)$ is the plane-wave receiving characteristic of the probe.

Each of the plane waves in the integrand of (4) satisfies the homogeneous Helmholtz equation $(\nabla^2 + k^2)e^{i(k_x x + k_y y + \gamma z)} = 0$. Thus, the probe output satisfies the homogeneous Helmholtz equation $(\nabla^2 + k^2)b_{p\omega}(\mathbf{r}) = 0$ at points in space where the operator $\nabla^2 + k^2$ can be moved under the integral signs in (4). Moreover, the theorems of Section 6.23 of Kaplan[6] show that the operator $\nabla^2 + k^2$ can be brought under the integral signs in (4) if the integrand is exponentially decaying at infinity in the (k_x, k_y) plane. Hence, the probe output satisfies the homogeneous Helmholtz equation at points in space where the integrand in (4) decays exponentially as $|\gamma| \to \infty$. We shall therefore examine the asymptotic behavior of the integrand in (4).

Let z_s be the maximum value of z for which the source function of the transmitting antenna is nonzero, that is, the physical body of the transmitting antenna is located in the region $z < z_s$. In Reference 7 we show that the spectrum of the transmitting antenna can be written as

$$T_\omega(k_x, k_y) = C_\omega(k_x, k_y)\, e^{|\gamma| z_s}, \quad k_x^2 + k_y^2 > k^2 \tag{5}$$

where $C_\omega(k_x, k_y) \to 0$ as $|\gamma| \to \infty$.

A corresponding asymptotic formula for the receiving characteristic $R_{p\omega}(k_x, k_y)$ of the probe can also be found, namely

$$R_{p\omega}(k_x, k_y) = C_{p\omega}(k_x, k_y) e^{-|\gamma| z_p}, \quad k_x^2 + k_y^2 > k^2 \tag{6}$$

where $C_{p\omega}(k_x, k_y)$ is bounded as $|\gamma| \to \infty$, and the probe is located in the region $z > z_p$ when the reference point of the probe is located at the origin of the coordinate system fixed in the transmitting antenna.

The asymptotic relations (5) and (6) show that the integrand of the integral representation (4) for the probe output satisfies

$$R_{p\omega}(k_x, k_y) T_\omega(k_x, k_y) e^{i(k_x x + k_y y + \gamma z)} = C'_\omega(k_x, k_y) e^{|\gamma|(z_s - z_p - z)}, \quad k_x^2 + k_y^2 > k^2 \tag{7}$$

where $C'_\omega(k_x, k_y) \to 0$ as $|\gamma| \to \infty$. Consequently, when $z_s < z + z_p$ the integrand of (4) decays exponentially as $|\gamma| \to \infty$, and the theorems of Section 6.23 of Kaplan[6] show that the operator $\nabla^2 + k^2$ can be brought inside the integral sign to prove that

$$\nabla^2 b_{p\omega}(\mathbf{r}) + k^2 b_{p\omega}(\mathbf{r}) = 0, \quad z > z_s - z_p. \tag{8}$$

The condition $z > z_s - z_p$ in (8) (where $\mathbf{r} = x\hat{\mathbf{x}} + y\hat{\mathbf{y}} + z\hat{\mathbf{z}}$ denotes the position of the reference point of the probe in the coordinate system fixed in the transmitting antenna) simply states that the probe cannot touch the transmitting antenna during a scan of the reference point of the probe in the z plane. If, however, the probe gets too close to the transmitting antenna during a scan, the formula (4) is no longer accurate because it neglects multiple interactions, and (8) no longer holds.

The probe output $b_{p\omega}(\mathbf{r})$ (for a fixed orientation of the probe with respect to the transmitting antenna) satisfies the homogeneous Helmholtz equation (8) not only for the probe in a half space in front of the transmitting antenna, but also for the probe in the entire region outside a closed surface formed by the locus of points $(z_s - z_p)$ for all orientations of the z axis; that is, in the region in which the minimum convex surface enclosing the transmitting antenna, and the minimum convex surface enclosing the probe, do not intersect. This can be shown by repeating the analysis that led to (7) and (8) for different orientations of the z axis.

DERIVATION OF THE FREQUENCY-DOMAIN TRANSMISSION FORMULA FROM THE HELMHOLTZ EQUATION

The inverse of the frequency-domain probe-corrected transmission formula (4) can now be derived from the result that the output of the probe satisfies the homogeneous Helmholtz equation (8).

The far-field acoustic pressure radiated by the transmitting antenna (in the absence of the probe) is related to its transmitting plane-wave spectrum by the formula[5]

$$p_\omega(\mathbf{r}) \sim -ik\cos\theta\, T_\omega(k\cos\phi\sin\theta, k\sin\phi\sin\theta)\frac{e^{ikr}}{r} \tag{9}$$

where (r, θ, ϕ) are the usual spherical coordinates.

Since the far field is locally a plane wave, the definition of the receiving characteristic of the linear probe shows that the probe output is given by

$$b_{p\omega}(\mathbf{r}) \sim -2\pi ik\cos\theta R_{p\omega}(k\cos\phi\sin\theta, k\sin\phi\sin\theta) \cdot T_\omega(k\cos\phi\sin\theta, k\sin\phi\sin\theta)\frac{e^{ikr}}{r} \tag{10}$$

when the reference point of the probe is at the far-field point (r, θ, ϕ). Equation (10) is the acoustic analog of the generalized Friis transmission formula.[8] Hence, the probe output $b_{p\omega}(\mathbf{r})$ satisfies the homogeneous Helmholtz equation (8) as well as the radiation condition in the region $z > z_s - z_p$. Consequently, it has the Green's function far-field representation (see eq. (15) of Reference 9)

$$b_{p\omega}(\mathbf{r}) \sim -\frac{e^{ikr}}{r}\frac{ik\cos\theta}{2\pi}\int_{-\infty}^{+\infty}\int_{-\infty}^{+\infty} b_{p\omega}(\mathbf{r}_0)\, e^{-ik\hat{\mathbf{r}}\cdot\mathbf{r}_0}dx_0dy_0, \quad z_0 > z_s - z_p. \tag{11}$$

A comparison of the far-field expressions (10) and (11) for the probe output shows that

$$T_\omega(k_x, k_y)R_{p\omega}(k_x, k_y) = \frac{e^{-i\gamma z_0}}{(2\pi)^2}\int_{-\infty}^{+\infty}\int_{-\infty}^{+\infty} b_{p\omega}(\mathbf{r}_0)\, e^{-i(k_x x_0 + k_y y_0)}dx_0dy_0,$$

$$k_x^2 + k_y^2 < k^2, \quad z_0 > z_s - z_p \tag{12}$$

which is equivalent to the inverse Fourier transform of (4) for propagating modes.

Although (12) has been derived for the propagating spectrum only, it can be argued to hold for all (k_x, k_y) as follows. Assume the probe were a point-source receiver that measured the pressure ($b_{p\omega}(\mathbf{r}_0) = p_\omega(\mathbf{r}_0)$) at that point. Such a probe would have a constant receiving characteristic, and (12) would reduce to simply the definition of the plane-wave spectrum for the acoustic pressure of the transmitting antenna for all (k_x, k_y). Since an arbitrary probe can be represented by a superposition of point-source receivers,[10] it follows that (12) is also valid for all (k_x, k_y) and an arbitrary probe.

The formula (12) allows one to compute the frequency-domain transmitting spectrum $T_\omega(k_x, k_y)$ of the transmitting antenna from the measured output $b_{p\omega}(\mathbf{r}_0)$ of the probe on the near-field scan plane $z = z_0$ (assuming the receiving characteristic $R_{p\omega}(k_x, k_y)$ of the probe is known). Once $T_\omega(k_x, k_y)$ is determined, the far field of the transmitting antenna at the frequency ω can be found from (9).

DERIVATION OF THE TIME-DOMAIN TRANSMISSION FORMULA FROM THE WAVE EQUATION

The inverse of the time-domain probe-corrected transmission formula for antennas excited by wide-band pulses can be derived from the result that the probe output satisfies the time-domain wave equation. Taking the inverse Fourier transform of the frequency-domain Helmholtz equation (8), we obtain the homogeneous time-domain wave equation for the time-domain output of the acoustic probe

$$\nabla^2 b_p(\mathbf{r},t) - \frac{1}{c^2}\frac{\partial^2}{\partial t^2} b_p(\mathbf{r},t) = 0, \qquad z > z_s - z_p. \tag{13}$$

Moreover, as explained previously, for a probe that translates without rotation with respect to the transmitting antenna, the probe output $b_p(\mathbf{r},t)$ satisfies the homogeneous wave equation (13) everywhere except where the minimum convex surface enclosing the probe intersects with the minimum convex surface enclosing the transmitting antenna.

Like the corresponding frequency-domain equation (8), the time-domain wave equation (13) holds only if multiple interactions between the probe and transmitting antenna are negligible. However, in the time domain, unlike the frequency domain, the multiple interactions can often be eliminated by time gating the output of the probe.

The inverse Fourier transform of (9) shows that the time-domain acoustic far-field pressure can be expressed in terms of its time-domain spectrum as

$$p(\mathbf{r},t) \sim \frac{\cos\theta}{cr}\frac{\partial}{\partial t} T(c^{-1}\cos\phi\sin\theta, c^{-1}\sin\phi\sin\theta, t - r/c) \tag{14}$$

where the time-domain transmitting spectrum $T(\xi,\eta,t)$ of the transmitting antenna is equal to the inverse Fourier transform of the frequency-domain spectrum, $T(\xi,\eta,t) = \int_{-\infty}^{+\infty} T_\omega(\omega\xi,\omega\eta)e^{-i\omega t}d\omega$, and, of course, $p(\mathbf{r},t) = \int_{-\infty}^{+\infty} p_\omega(\mathbf{r})e^{-i\omega t}d\omega$.

Since the far-field is locally a plane wave, the output of the probe, when the probe is located at the far-field point (r,θ,ϕ) and is illuminated by the pressure field (14), is determined from the definition of the time-domain receiving characteristic $R_p(\xi,\eta,t)$ of the probe as

$$\begin{aligned} b_p(\mathbf{r},t) &\sim \frac{\cos\theta}{cr}\int_{-\infty}^{+\infty} R_p(c^{-1}\cos\phi\sin\theta, c^{-1}\sin\phi\sin\theta, t - t' - r/c) \\ &\qquad \cdot\frac{\partial}{\partial t'}T(c^{-1}\cos\phi\sin\theta, c^{-1}\sin\phi\sin\theta, t')dt' \\ &= \frac{\cos\theta}{cr}\frac{\partial}{\partial t}\int_{-\infty}^{+\infty} R_p(c^{-1}\cos\phi\sin\theta, c^{-1}\sin\phi\sin\theta, t - t' - r/c) \\ &\qquad \cdot T(c^{-1}\cos\phi\sin\theta, c^{-1}\sin\phi\sin\theta, t')dt'. \end{aligned} \tag{15}$$

The receiving characteristics in the time and frequency domains are related through the Fourier transform; specifically $R_p(\xi,\eta,t) = \int_{-\infty}^{+\infty} R_{p\omega}(\omega\xi,\omega\eta)e^{-i\omega t}d\omega$. The far-field formula (15) can also be obtained by taking the inverse Fourier transform of (10) and applying the convolution theorem. The far-field output of the probe in (15), which is the time-domain acoustic analog of the generalized Friis transmission formula,[8] has the form $\mathcal{F}_b(\theta,\phi,t - r/c)/r$, so that the probe output satisfies the outgoing radiation condition at infinity. Since the probe output satisfies the wave equation (13) and the outgoing radiation condition, it has the Green's function far-field representation

$$b_p(\mathbf{r},t) \sim \frac{\cos\theta}{2\pi\, cr}\frac{\partial}{\partial t}\int_{-\infty}^{+\infty}\int_{-\infty}^{+\infty} b_p(\mathbf{r}_0, t - r/c + \mathbf{r}_0\cdot\hat{\mathbf{r}})\, dx_0 dy_0, \quad z_0 > z_s - z_p \tag{16}$$

which can be obtained by taking the inverse Fourier transform of (11).

Comparing the two far-field expressions (15) and (16) yields the relation

$$\int_{-\infty}^{+\infty} R_p(\xi,\eta,t-t')T(\xi,\eta,t')dt' = \frac{1}{2\pi}\int_{-\infty}^{+\infty}\int_{-\infty}^{+\infty} b_p(\mathbf{r}_0, t+\xi x_0+\eta y_0)\, dx_0 dy_0,$$
$$\xi^2+\eta^2 < c^{-2}, \quad z_s - z_p < 0 \tag{17}$$

where, for the sake of simplicity, we have let the z coordinate of the scan plane equal zero, that is, $z_0 = 0$. The equation (17) is an inverse time-domain plane-wave transmission formula for the propagating modes. To isolate $T(\xi,\eta,t)$ from this equation, multiply both sides by the inverse receiving characteristic $R_p^{-1}(\xi,\eta,t''-t)$ defined by

$$\int_{-\infty}^{+\infty} R_p^{-1}(\xi,\eta,t')R_p(\xi,\eta,t-t')dt' = 4\pi^2\delta(t) \tag{18}$$

and integrate the resulting equation with respect to t from $-\infty$ to $+\infty$ to get

$$T(\xi,\eta,t'') = \frac{1}{(2\pi)^3}\int_{-\infty}^{+\infty} R_p^{-1}(\xi,\eta,t-t'')\int_{-\infty}^{+\infty}\int_{-\infty}^{+\infty} b_p(\mathbf{r}_0, t+\xi x_0+\eta y_0)\, dx_0 dy_0 dt,$$
$$\xi^2+\eta^2 < c^{-2}, \quad z_s - z_p < 0. \tag{19}$$

The wave-equation derivation has thus led to an expression (19) for the time-domain spectrum $T(\xi,\eta,t)$ that is valid for propagating modes. Nonetheless, one can argue the validity of (19) for all (ξ,η) by returning to (17). Assume that the probe is a point-source receiver at $\mathbf{r}_0$ that measures the pressure ($b_p(\mathbf{r}_0,t) = p(\mathbf{r}_0,t)$). Such a probe would have $R_p(\xi,\eta,t) = \delta(t)$ and (17) would reduce to the definition of the time-domain plane-wave spectrum for the acoustic pressure of the transmitting antenna at $z_0 = 0$ for all (ξ,η).[9] Since an arbitrary probe can be reproduced by a superposition of point-source receivers,[10,11] equations (17) and (19) are valid for all (ξ,η) and an arbitrary probe.

One can compute the time-domain plane-wave spectrum $T(\xi,\eta,t)$ of the transmitting antenna from the measured time-domain output $b_p(\mathbf{r}_0,t)$ of the probe in the near-field scan plane $z = z_0 = 0$. The time-domain far field of the transmitting antenna can then be found from $T(\xi,\eta,t)$ through equation (14).

SUMMARY

In summary, the output of a linear receiving antenna satisfies the homogeneous wave equation and the outgoing wave condition in both the frequency and time domains. The homogeneous wave equations are satisfied throughout all space except where the minimum convex surface enclosing the receiving antenna (probe) intersects the minimum convex surface enclosing the transmitting antenna. It is assumed that the probe translates without rotation with respect to the transmitting antenna, and that multiple interactions between the probe and transmitting antenna are negligible (or gated out in the time domain). The frequency-domain wave equation (Helmholtz equation) has been used in the past to efficiently compute the coupling between co-sited antennas,[1] and to suggest a simplified probe-corrected spherical near-field measurement method.[2] In this paper we first gave a rigorous derivation of these wave equations and their region of validity. We then used the wave equations to derive both the frequency-domain and time-domain probe-corrected plane-wave transmission formulas that determine the plane-wave spectrum (and thus the near and far fields) of the transmitting antenna in terms of the measured output and the plane-wave receiving characteristic of the probe.

ACKNOWLEDGMENTS

The research presented in this paper was supported by the United States Air Force Office of Scientific Research, Bolling AFB, DC.

REFERENCES

1. A.D. Yaghjian, "Efficient computation of antenna coupling and fields within the near-field region," *IEEE Trans. Antenna Propagat.*, vol. AP-30, pp. 113-128, January 1982.

2. F.H. Larsen, *Probe-Corrected Spherical Near-Field Antenna Measurements*, Ph.D. Thesis, Technical University of Denmark, pp. 63-65, December 1980.

3. A.D. Yaghjian, "Antenna coupling and near-field sampling in plane-polar coordinates," and "Sampling in plane-polar coordinates," *IEEE Trans. Antenna Propagat.*, vols. 40 and 44, pp. 304-312 and pp. 696-700, March 1992 and May 1996.

4. T.B. Hansen and A.D. Yaghjian, "Probe-corrected planar near-field scanning in the time domain," *IEEE Trans. Antenna Propagat.*, vol. AP-43, pp. 569-584, June 1995.

5. D.M. Kerns, "Scattering-matrix description and near-field measurements of electroacoustic transducers," *J. Acoust. Soc. Am.*, vol. 57, pp. 497-507, February 1975.

6. W. Kaplan, *Advanced Calculus, 4th Edition*, Reading MA: Addison-Wesley, 1991.

7. A.D. Yaghjian, T.B. Hansen, and A.J. Devaney, "Minimum source region for a given far-field pattern," *IEEE Trans. Antenna Propagat.*, vol. 45, pp. 911-912, May 1997.

8. P.K. Park and C.T. Tai, "Receiving antennas," ch. 6, p. 6-8 in *Antenna Handbook*, Eds. Y.T. Lo and S.W. Lee, New York: Van Nostrand Reinhold, 1988.

9. T.B. Hansen and A.D. Yaghjian, "Planar near-field scanning in the time domain, Part 1: Formulation," *IEEE Trans. Antenna Propagat.*, vol. AP-42, pp. 1280-1291, September 1994.

10. A.D. Yaghjian and R.C. Wittmann, "The receiving antenna as a linear differential operator: application to spherical near-field measurements," *IEEE Trans. Antenna Propagat.*, vol. AP-33, pp. 1175-1185, November 1985.

11. T.B. Hansen, "Spherical expansions of time-domain acoustic fields: Application to near-field scanning," *J. Acoust. Soc. Am.*, vol. 98, pp. 1204-1215, August 1995.

ENERGY CONSIDERATION IN SPACE–TIME SYNTHESIS OF COLLIMATED PULSED APERTURES

Amir Shlivinski and Ehud Heyman

Faculty of Engineering, Tel-Aviv University,
Tel Aviv 69978, Israel

ABSTRACT

We consider measures for an effective realization of short pulse antennas. We discuss such new concepts as TD multipole decomposition, time-dependent reactive and radiative energies and introduce the ratio between them, termed the time-domain Q as a measure of the radiation effectiveness. These near-zone characteristics are examined vis a vis the far-zone characteristics, the time dependent radiation pattern and the directivity. The theory is used to evaluate and compare several different realizations of highly collimated, short-pulse aperture distributions.

INTRODUCTION

In recent papers [1, 2] we introduced measures for the effective realization of short pulse antennas. These measures involve such new concepts as the time domain (TD) reactive energy, the TD reactive power flow and the TD Q. They are based on the TD spherical waves expansion (also termed multipole expansion) of the antenna field. These concepts have been motivated by the well known frequency domain (FD) concepts. Yet, the FD concept of reactive energy that expresses the energy stored in the near zone of antenna, has to modified in the TD (for short pulse fields) where no stored energy is involved. The reactive energy had therefore been redefined for short pulse fields using TD multipole

Ultra-Wideband, Short-Pulse Electromagnetics 4
Edited by Heyman *et al.*, Kluwer Academic / Plenum Publishers, New York, 1999.

expansion. It has been shown that the TD reactive energy is is a relatively strong energy pulse concentrated in the near zone of the antenna. Furthermore, unlike the radiative energy pulse, the reactive energy pulse discharges back into the source once the radiated pulse has ended. Thus the effectiveness of the radiation can be quantified by the TD Q that measures the ratio between the total TD reactive energy outside the sphere enclosing the antenna and the radiative energy. Large Q implies large TD reactive energies and thereby a less effective source realization. These near and far zone characteristics are, of course, related. Large directivity require high order TD spherical modes with large values of Q. Furthermore, smaller antennas with the same directivity have larger reactive energy and larger Q. These considerations pose a tradeoff between directivity, size and radiation effectiveness which is similar to the one in conventional frequency domain (FD) analysis. Yet time domain system have the additional degree of freedom of both space *and* time synthesis.

The considerations in the preceding paragraph are explored in the present paper for several realizations of collimated short-pulse antennas, and in particular for the class of space-time synthesized apertures considered in [3] which has not been studied in [1, 2]. The time-dependent radiation patterns are calculated via the slant stack transform (SST) formulation of [4] which furnishes a direct geometrical projection relation between the current distribution and the radiation pattern. Next, the near zone properties and the TD Q are calculated via the TD multipole expansion formulation of [2]. Finally, these measures are used to quantify and contrast the effectiveness of the different realizations.

TD MULTIPOLE EXPANSION OF THE FIELD

We start with TD multipole expansion of the radiated field, which is used here to analyze the near-to-far-zone field transition. This formulation can be applied in fact in any spherically stratified configuration $\epsilon(r), \mu(t)$ bounded by a general conical surface $\mathcal{C}(\theta, \phi)$ whereon the boundary conditions are imposed (Fig. 1(a)), but here we shall only be interested in radiation in free space. The field in the configuration of Fig. 1(a) may be decomposed into E and H type modes denoted, respectively, by superscript e and h, and expressed as a sum of all these modes, with ν denoting the mode index and $\alpha = e$ or h

$$\mathbf{E}_t(\mathbf{r}, t) = \frac{1}{r} \sum_{\alpha,\nu} V_\nu^\alpha(r, t)\, \mathbf{e}_\nu^\alpha(\theta, \phi), \quad E_r(\mathbf{r}, t) = \frac{\eta c}{r^2} \sum_{\nu} k_{t\nu}^e \Phi_\nu^e(\theta, \phi)\, \partial_t^{-1} I_\nu^e(r, t) \tag{1}$$

with a similar expression for the $\mathbf{H}$ field. Here $\mathbf{r} = (r, \theta, \phi)$ denotes the conventional spherical coordinate system, $c(r) = 1/\sqrt{\epsilon\mu}$ and $\eta(r) = \sqrt{\mu/\epsilon}$. The functions $\Phi_\nu^\alpha(\theta, \phi)$ and $\mathbf{e}_\nu^\alpha(\theta, \phi)$ are the transversal scalar and vector mode functions for the E and H modes. These functions depend only on the conical cross section and are independent of (r, t) and of (ϵ, μ). The amplitude functions V_ν^α and I_ν^e depend only on (r, t) (independent of (θ, ϕ) and account for the source excitation and the medium inhomogeneity via a solution of the time-dependent radial transmission-line equation. A detailed discussion of this general expansion is given in [1].

For the present problem of radiation in free space we have $\Phi_\nu^\alpha = Y_{nm}^{(s)}(\theta, \phi)$ and $\mathbf{e}_\nu^e = \frac{y_{nm}}{\sqrt{n(n+1)}} \left[\hat{\boldsymbol{\theta}} \sin\theta P_n'^m(\cos\theta) \left\{ {\cos \atop \sin} \right\} m\phi + \hat{\boldsymbol{\phi}} \frac{m}{\sin\theta} P_n^m(\cos\theta) \left\{ {\sin \atop -\cos} \right\} m\phi \right]$ while $\mathbf{e}_\nu^h$ are obtained by

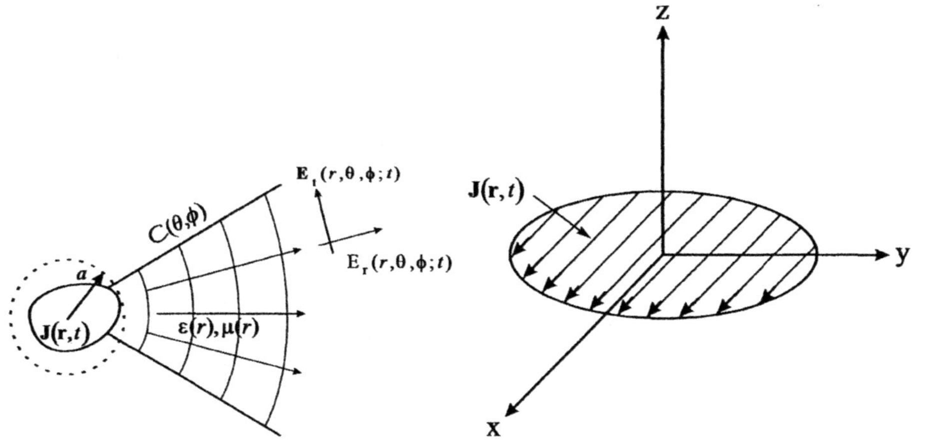

(a) (b)

Figure 1: (a) TD radiation in general spherically stratified conical configuration. (b) A disk of x-polarized pulsed current distribution.

replacing $\widehat{\boldsymbol{\theta}} \to -\widehat{\boldsymbol{\phi}}$ and $\widehat{\boldsymbol{\phi}} \to \widehat{\boldsymbol{\theta}}$. Here ν is a triple index $\nu = (n, m, s)$, $s = 1, 2$, $n = 0, 1, \ldots$ and $0 \le m \le n$, $Y_{nm}^{(s)}$ are the spherical harmonics, $P_n^m(x)$ are the associate Legendre functions [8, Chap. 8] and the upper and lower terms correspond to $s = 1, 2$, respectively. $y_{nm} = \sqrt{\varepsilon_m \frac{2n+1}{4\pi} \frac{(n-m)!}{(n+m)!}}$ are normalization constants with $\varepsilon_m = 1$ or 2 for $m = 0$ or $m \ge 1$, respectively. Finally, these functions satisfy spherical orthogonality and completeness conditions.

In free space, the time-dependent radial transmission-line equation can be solved analytically. Thus, outside the source region ($r > a$) (1) yields

$$\mathbf{E}_t(\mathbf{r}, t) = \frac{\eta}{r} \sum_{\nu} \mathbf{e}_\nu^e(\theta, \phi) \mathcal{L}_n^{(2)} M_\nu^e(\tau) + \frac{1}{r} \sum_{\nu} \mathbf{e}_\nu^h(\theta, \phi) \mathcal{L}_n^{(1)} M_\nu^h(\tau) \tag{2a}$$

$$E_r(\mathbf{r}, t) = \frac{\eta c}{r^2} \sum_{\nu} \sqrt{n(n+1)} \Phi_\nu(\theta, \phi) \partial_t^{-1} \mathcal{L}_n^{(1)} M_\nu^e(\tau), \quad \text{with } \tau = t - r/c \tag{2b}$$

where the TD operators $\mathcal{L}_n^{(1,2)}$ are given by $\mathcal{L}_n^{(1)} = \sum_{l=0}^{n} a_{n,l} \left(\frac{c}{r}\right)^l \partial_t^{-l}$ and $\mathcal{L}_n^{(2)} = \sum_{l=0}^{n+1} b_{n,l} \left(\frac{c}{r}\right)^l \partial_t^{-l}$ with $a_{n,l} = 2^{-l} \frac{(n+l)!}{l!(n-l)!}$ and $b_{n,l} = a_{n,l} + (l-1) a_{n,l-1}$ for $0 \le l \le n+1$ (note that $b_{n,0} = a_{n,0} = 1$ and $b_{n,n+1} = n a_{n,n}$). It follows that the field is completely described by the "TD multipole moments" M_ν^α in (2). They are calculated by a weighted projection of $\mathbf{J}(\mathbf{r}, t)$ onto the spherical basis functions via

$$M_\nu^e(t) = \frac{-1}{2n!} \partial_t^n \iiint_V dV \left(\frac{r}{2c}\right)^n \mathbf{h}_\nu^e \cdot \int_{-1}^{1} d\xi (1 - \xi^2)^n (\nabla \times \mathbf{J}(\mathbf{r}, t + \xi r/c) \tag{3a}$$

$$M_\nu^h(t) = \frac{\eta/c}{2n!} \partial_t^{n+1} \iiint_V dV \left(\frac{r}{2c}\right)^n \mathbf{e}_\nu^h \cdot \int_{-1}^{1} d\xi (1 - \xi^2)^n \mathbf{J}(\mathbf{r}, t + \xi r/c) \tag{3b}$$

(note, the $\nabla\times$ in (3a) operates only on the $\mathbf{r}$ part in $\mathbf{J}$). The weighted integration in (3) should be understood as follows: First, for a given spherical shell of radius r, the ξ-integration represents a local and temporal average of the current in the time window $t \pm r/c$ and thus it serves as a weighting factor for the sources in that spherical shell. Next, the weighted source distribution at r is projected onto the angular basis functions $\mathbf{h}_\nu^e$ and $\mathbf{e}_\nu^h$ via the angular (Ω) integration, and finally the result is integrated as a function of r. Note that for the large order modes (higher n's) the main contribution comes from sources close to the external boundary of the source distribution ($r \simeq a$). A thorough large n analysis of the operation in (3) and thereby of the convergence rate of the series in the near and far zones have been carried out in [1].

Example for the TD multipoles

The multipole structure is demonstrated for a circular disk of radius a carrying a uniform distribution of x-polarized pulse current $\mathbf{J}(\mathbf{r}, t) = \widehat{\mathbf{x}}\delta(z)f(t)$ for $r < a$ where $f(t)$ is an arbitrary pulse (cf. Fig. 1(b)). We use a twice differentiated Gaussian $f(t) = Ae^{-(t/\gamma)^2} H_2(t/\gamma)$ where $H_2(x) = 4x^2 - 2$ is the Hermite polynomial of order 2. The normalization constant A is taken so that the $\mathcal{L}_2$ norm of $f(t)$ is a unity (i.e. $\int dt f^2(t) = 1$) and γ is taken to be $\gamma = (12/7)^{1/2} T$ where $T = \int dt\, t^2 f^2(t)$ is the pulse length. Results of the numerically calculated multipole moments are given in Fig. 2 for a disk with radius $a = 9cT$. The figure depicts low order ($n \ll a/cT$ and intermediate order ($n \simeq a/cT$) multipole moments. The high order multipole moments decay fast as a function of n [1].

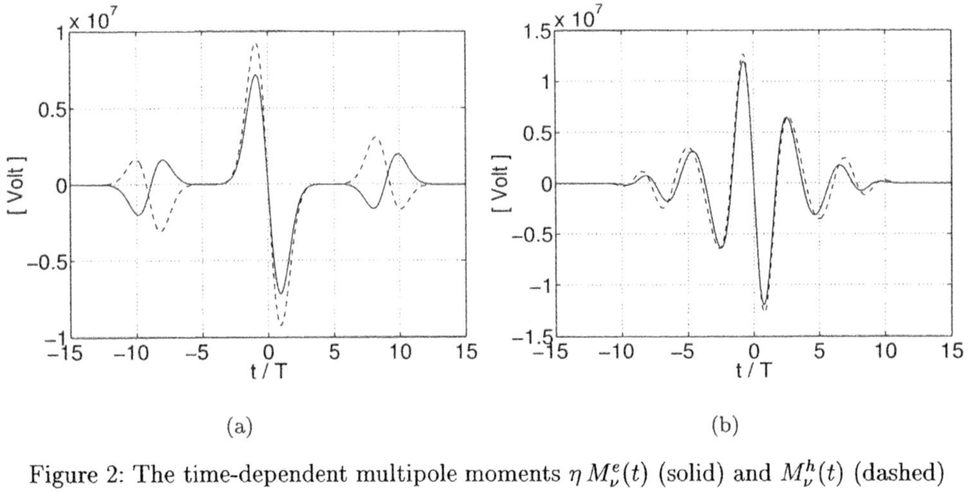

Figure 2: The time-dependent multipole moments $\eta\, M_\nu^e(t)$ (solid) and $M_\nu^h(t)$ (dashed) for a disk of radius $a = 9cT$ vs. the normalized time t/T. (a) Low order moments: $n = 1$ and 2 for the E and H mode, respectively. (b) High order moments: $n = 10$ and 11 for the H and E mode, respectively.

THE TIME DOMAIN RADIATION PATTERN

The field representation in (2) is valid for the near and far zones. Noting that for large r $\mathcal{L}_n^{(1,2)}$ are dominated by the $l = 0$ term, giving $\mathcal{L}_n^{(1,2)} \sim 1$, we obtain

$$\mathbf{E}(\mathbf{r}, t) \sim \mathbf{E}_t(\mathbf{r}, t) \sim \frac{1}{r}\mathbf{F}(\widehat{\mathbf{r}}, \tau), \tag{4a}$$

where $\mathbf{F}(\hat{\mathbf{r}},\tau)$ is the "TD radiation pattern" given by

$$\mathbf{F}(\hat{\mathbf{r}},\tau)=\eta\sum_{\nu}\mathbf{e}_{\nu}^{e}(\theta,\phi)M_{\nu}^{e}(\tau)+\sum_{\nu}\mathbf{e}_{\nu}^{h}(\theta,\phi)M_{\nu}^{h}(\tau) \tag{4b}$$

The TD radiation pattern may be expressed, instead, via a TD plane wave representation. One finds that it is given by the *slant stack transform* (SST) of the current distribution $\mathbf{J}(\mathbf{r},t)$ (see [4]),

$$\mathbf{F}(\hat{\mathbf{r}},\tau)=\frac{-\mu}{4\pi}\int_{V}dV'\,\partial_{\tau}\mathbf{J}_{\parallel}(\mathbf{r}',\tau+\hat{\mathbf{r}}\cdot\mathbf{r}'/c)=\frac{-\mu}{4\pi}\partial_{\tau}(\mathbf{1}-\hat{\mathbf{r}}\hat{\mathbf{r}})\cdot\tilde{\mathbf{J}}(\hat{\mathbf{r}},\tau) \tag{5}$$

where $\mathbf{J}_{\parallel}$ is the transverse component of $\mathbf{J}$ relative to the observation direction $\hat{\mathbf{r}}$. The function $\tilde{\mathbf{J}}(\hat{\mathbf{r}},\tau)$ is the SST of $\mathbf{J}(\mathbf{r},t)$: It consists of projections of $\mathbf{J}$ onto planes perpendicular to $\hat{\mathbf{r}}$, and stacked with a delay corresponding to a plane wave propagating in the $\hat{\mathbf{r}}$ direction, thus extracting from $\mathbf{J}$ the TD plane wave information that propagates in that direction. Further details of this transform can be found in [4].

THE TD POWER FLOW: Radiative and reactive power

The radial power flow with respect to a sphere of radius r is defined as

$$P(r,t)=\oiint_{S}r^{2}d\Omega\,\hat{\mathbf{r}}\cdot\mathbf{S}(\mathbf{r},t),\qquad \mathbf{S}(\mathbf{r},t)=\mathbf{E}(\mathbf{r},t)\times\mathbf{H}(\mathbf{r},t) \tag{6}$$

where $\mathbf{S}(\mathbf{r},t)$ is the time dependent Poynting vector. Using the modal representation (2) and the mode orthogonality P may be expressed as a sum of modal powers $P(r,t)=\sum_{\alpha,\nu}P_{\nu}^{\alpha}(r,t)$ ($\alpha=e,h$ for E modes and H modes respectively) where $P_{\nu}^{\alpha}=\eta^{\pm1}\sum_{\ell=0}^{n}\sum_{q=0}^{n+1}a_{n,l}b_{n,q}\left(\frac{c}{r}\right)^{q+\ell}\left[\partial_t^{-l}M_{\nu}^{\alpha}(\tau)\right]\left[\partial_t^{-q}M_{\nu}^{\alpha}(\tau)\right]$. P_{ν}^{α} can be reordered as an ascending series of powers of $\frac{1}{r}$ and recast in the form

$$P_{\nu}^{\alpha}=P_{\nu_{\mathrm{rad}}}^{\alpha}(\tau)+P_{\nu_{\mathrm{react}}}^{\alpha}(r,t), \tag{7a}$$

where

$$P_{\nu_{\mathrm{rad}}}^{\alpha}(\tau)=\eta^{\pm1}\left[M_{\nu}^{\alpha}(\tau)\right]^{2} \tag{7b}$$

$$P_{\nu_{\mathrm{react}}}^{\alpha}(r,t)=\eta^{\pm1}\sum_{p=1}^{2n+1}\left(\frac{c}{r}\right)^{p}\sum_{q=q_{\min}}^{q_{\max}}a_{n,p-q}b_{n,q}\left[\partial_t^{q-p}M_{\nu}^{\alpha}(\tau)\right]\left[\partial_t^{-q}M_{\nu}^{\alpha}(\tau)\right] \tag{7c}$$

with $q_{\min}=\max\{0,p-n\}$ and $q_{\max}=\min\{n+1,p\}$ (i.e., for $p=1,\ldots n$ the q summation goes from 0 to p, while for $p=n+1,\ldots 2n+1$, it goes from $p-n$ to $n+1$).

The leading order term $P_{\nu_{\mathrm{rad}}}^{\alpha}(\tau)$ (associated with the $l=0$ term in $\mathcal{L}_{n}^{(1,2)}$) is the modal "TD radiation power": It is always positive and, being a function of τ only, it propagates without distortion all the way to infinity. The remainder term $P_{\nu_{\mathrm{react}}}^{\alpha}(r,t)$ (associated with the higher order term in $\mathcal{L}_{n}^{(1,2)}$) is denoted as the "TD reactive power-flow": It is a series of the form r^{-p}, $p=1,\ldots 2n+1$, and thus vanishes at infinity.

Furthermore, by analyzing this expression it can be shown that this pulse has a zero mean, i.e.,

$$\int_{-\infty}^{\infty} dt\ P^{\alpha}_{\nu_{\text{react}}}(r,t) = 0 \tag{8a}$$

implying that the total transmitted energy is described only by $P^{\alpha}_{\nu_{\text{rad}}}$:

$$\mathcal{E}^{\alpha}_{\nu} = \int_{-\infty}^{\infty} dt\ P^{\alpha}_{\nu}(r,t) = \int_{-\infty}^{\infty} dt\ P^{\alpha}_{\nu_{\text{rad}}}(\tau) \tag{8b}$$

and is thus independent of r. It follows that the pulsed reactive power-flow is initially positive and then exhibits several oscillations so that finally Eq. (8a) is satisfied. The positive part in $P^{\alpha}_{\nu_{\text{react}}}(r,t)$ indicates power transmitted from the source to build up the local pulse reactive energy while the negative part indicate power transmitting back to the source as the local energy of the field decays. Note, though, that $P^{\alpha}_{\nu_{\text{react}}}$ is not a function of τ but rather of t and r so that the backward energy transmission is *not* a backward traveling wave.

The functional form of $P^{\alpha}_{\nu_{\text{react}}}$ also implies that in the "near zone" this term is stronger than $P^{\alpha}_{\nu_{\text{rad}}}$. Thus the total pulsed power-flow in the near zone is strong but only a fraction of it is used effectively for the radiation power, while the rest is transmitted back to the source. In the "far zone," on the other hand, $P^{\alpha}_{\nu_{\text{react}}} \ll P^{\alpha}_{\nu_{\text{rad}}}$ and in particular it vanishes as $r \to \infty$, as can be seen in Fig. 3 where the radiative and reactive parts of P are depict for H-mode multipole number $n = 10$ of Fig. 2(b). A detailed analytic analysis where this near to far zone transition takes place can be found in [1, 2]: In particular it has been shown there that the modal reactive energies are growing for $r < \nu c T$.

THE TD ENERGY DENSITY: Radiative and reactive energy

The electromagnetic pulsed energy densities in a spherical shell of radius r (i.e., densities per unit of radius) is defined as

$$w(r,t) = \oiint d\Omega\, r^2\, W(\mathbf{r},t), \quad W(\mathbf{r},t) = \frac{1}{2}\mu \mathbf{H}(\mathbf{r},t)\cdot\mathbf{H}(\mathbf{r},t) + \frac{1}{2}\epsilon\mathbf{E}(\mathbf{r},t)\cdot\mathbf{E}(\mathbf{r},t) \tag{9}$$

Performing the surface integration in (9) and using mode orthogonality, the pulsed energy densities can be rearrange into two parts

$$w^{\alpha}_{\nu}(r,t) = c^{-1} P^{\alpha}_{\nu_{\text{rad}}}(\tau) + w^{\alpha}_{\nu_{\text{react}}}(r,t) \tag{10a}$$

where $P^{\alpha}_{\nu_{\text{rad}}}$ is the TD modal radiation power defined in (7b) and the reminder is termed the "TD modal reactive energy",

$$w^{\alpha}_{\nu_{\text{react}}}(r,t) = c^{-1} P^{\alpha}_{\nu_{\text{react}}}(r,t) + \frac{\eta^{\pm 1}}{c}\sum_{p=1}^{2n+1}\left(\frac{c}{r}\right)^{p+1}\sum_{q=q_{\min}}^{q_{\max}} p a_{n,p-q} b_{n,q} \partial_t^{-1}\left\{\left[\partial_t^{q-p} M^{\alpha}_{\nu}(\tau)\right]\left[\partial_t^{-q} M^{\alpha}_{\nu}(\tau)\right]\right\} \tag{10b}$$

where $P^{\alpha}_{\nu_{\text{react}}}$ is defined in (7c). Equation (10a) indicates that only $c^{-1}P^{\alpha}_{\nu_{\text{rad}}}$ is associated with the radiation power flow. The other part, $w^{\alpha}_{\nu_{\text{react}}}$, describes a local build up of the

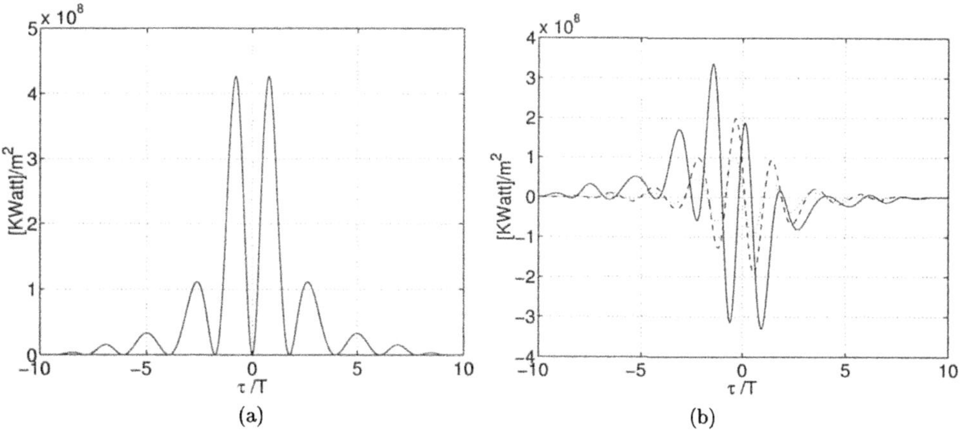

Figure 3: The radiative (a) and the reactive (b) pulsed power-flow for a H mode multipole of order $n = 10$ of Fig. 2(b) shown as a function of normalized time $\tau/T = (t - r/c)/T$. The radiative power is unchanged as a function of r. The reactive power is shown for several view points: $\frac{r}{cT} = 1.5a$, $9a$ and $18a$ (full, dashed and dotted lines, respectively).

pulsed energy needed to support the radiation power. This pulsed energy decays as the pulse is transmitted and part of it is transmitted back to the source via $P^{\alpha}_{\nu_{\text{react}}}$ (see discussion following (8)). Finally, the reactive energy is dominant in the near zone, i.e. $w^{\alpha}_{\nu_{\text{react}}} \gg c^{-1} P^{\alpha}_{\nu_{\text{rad}}}$, while in the far zone $w^{\alpha}_{\nu_{\text{react}}}$ is weaker and in fact vanishes as $r \to \infty$ (recall the discussion following (8)).

TIME DOMAIN Q FACTOR

In this section we shall introduce a parameter Q that quantifies the energy properties of the field by comparing the total radiation energy with a measure of the total reactive energy needed to generated the radiated field. We start by defining the total TD radiating energy $\mathcal{E}$ and the total radiated power $P_{T_{\text{rad}}(t)}$ (cf. (8b))

$$\mathcal{E} = \int_{-\infty}^{\infty} dt\, P_{\text{rad}}(t), \quad P_{\text{rad}}(t) = \sum_{\nu,\alpha} P^{\alpha}_{\nu_{\text{rad}}}(t) \tag{11}$$

The measure for the total reactive energy has a more complicated definition since this energy is time dependent. We therefore define the *time-averaged* reactive energy at $r > a$ as

$$\mathcal{W}(a) = \frac{1}{T} \int_{a}^{\infty} dr \int_{-\infty}^{\infty} dt\, w_{\text{react}}(r, t) \tag{12}$$

where $w_{\text{react}}(r,t) = \sum_{\nu,\alpha} w^{\alpha}_{\nu_{\text{react}}}(t)$ is the total TD reactive energy density and T is the RMS pulse-length of the radiated field, defined by

$$T = \sqrt{\mathcal{E}^{-1} \int_{-\infty}^{\infty} dt\, t^2\, P_{\text{rad}}(t)}. \tag{13}$$

Without loss of generality it is assumed in this definition, that $P_{T_{rad}}(t)$ is centered at $t = 0$ i.e., $\int_{-\infty}^{\infty} dt\, t\, P_{rad}(t) = 0$.

A TD Q factor of the field with respect to the sphere a enclosing the antenna may now be defined as

$$Q(a) = 2\pi \frac{\mathcal{W}(a)}{\mathcal{E}}. \tag{14}$$

This general definition applies to *any* pulse shape with a finite energy. It reduces to the conventional definitions of Q for time harmonic sources [6] given by $Q(a) = \omega W_{react}(a)/P_R$, where ω is the radian frequency, P_R is the average radiation power and $W_{react}(a)$ is the average reactive energy stored in $r > a$. This monochromatic expression is obtained from (14) if T, $\mathcal{E}$ and $\mathcal{W}(a)$ are replaced by $2\pi/\omega$, $2\pi P_R/\omega$, and $W_{react}(a)$, respectively.

In order to calculate the general definition in (14) for the *total* $Q(a)$, it may be rephrased in terms of *modal* quantities via [2]

$$Q(a) = \frac{\sum_{\alpha,\nu} \mathcal{E}_\nu^\alpha T_\nu^\alpha Q_\nu^\alpha(a)}{\sqrt{\mathcal{E} \sum_{\nu,\alpha} \mathcal{E}_\nu^\alpha T_\nu^{\alpha 2}}} \tag{15}$$

where the modal quantities $\mathcal{E}_\nu^\alpha$, T_ν^α and $Q_\nu^\alpha(a)$ are calculated via Eqs. (11), (13) and (14) respectively, where instead of the *total* properties we place the *modal* properties (e.g., $\mathcal{E} \leftrightarrow \mathcal{E}_\nu^\alpha$). The modal expressions for $\mathcal{E}_\nu^\alpha$, T_ν^α and $Q_\nu^\alpha(a)$ may readily be calculated via the modal analysis as discussed in the two preceding sections. In particular it can be shown that $Q_\nu^\alpha(a)$ start to grow rapidly with n for $n > a/cT$ where a and T are the antenna size and pulselength, respectively.

Thus Eq. (15) readily clarifies the tradeoff between directivity, antenna size, pulse length and effective realization. Large directivity, for example, requires larger $\mathcal{E}_\nu^\alpha$ at the higher order modes while, on the other hand, $Q_\nu^\alpha(a)$ increase rapidly for $n > a/cT$. This tradeoff will be demonstrated in the simulations below.

EXAMPLE REALIZATION OF COLLIMATED PULSE FIELD

We consider a class of x-polarized pulsed current distributions on a circular disk of radius a located in the $z = 0$ plane (Fig. 1(b)). The current for $r < a$ is given by

$$\mathbf{J}(\mathbf{r}, t) = \hat{\mathbf{x}}\delta(z)(b^2 - r^2)^{-\frac{1}{2}} \mathrm{Re} f_+(t - ir^2/2bc), \quad b > 0, \tag{16}$$

where $\hat{\mathbf{x}}$ is a unit vector in the x direction, b is a parameter to be discussed below and $f_+(t)$ is any analytic signal, as defined by the positive frequency spectrum $f_+(t) = \frac{1}{\pi}\int_0^\infty d\omega\, \hat{f}(\omega) e^{i\omega t}$. Specifically we shall use here a twice differentiated analytic δ pulse

$$f_+(t) = A\delta_+^{(2)}(t - i\sqrt{3}T) = 2A/\pi i(t - i\sqrt{3}T)^3, \quad T > 0 \tag{17}$$

Where T is a parameter. It can be shown that T defines the pulselength as defined as in (13). Also the normalization constant A is taken so that the $\mathcal{L}_2$ norm of $\mathbf{J}(\mathbf{r}, t)$ in the space-time domain is normalized to a constant (πa^2 in this case). One observes now that the current's decay away from the center of the aperture is affected by the imaginary part in the argument of f_+ in (16): Specifically using (17), one finds that the

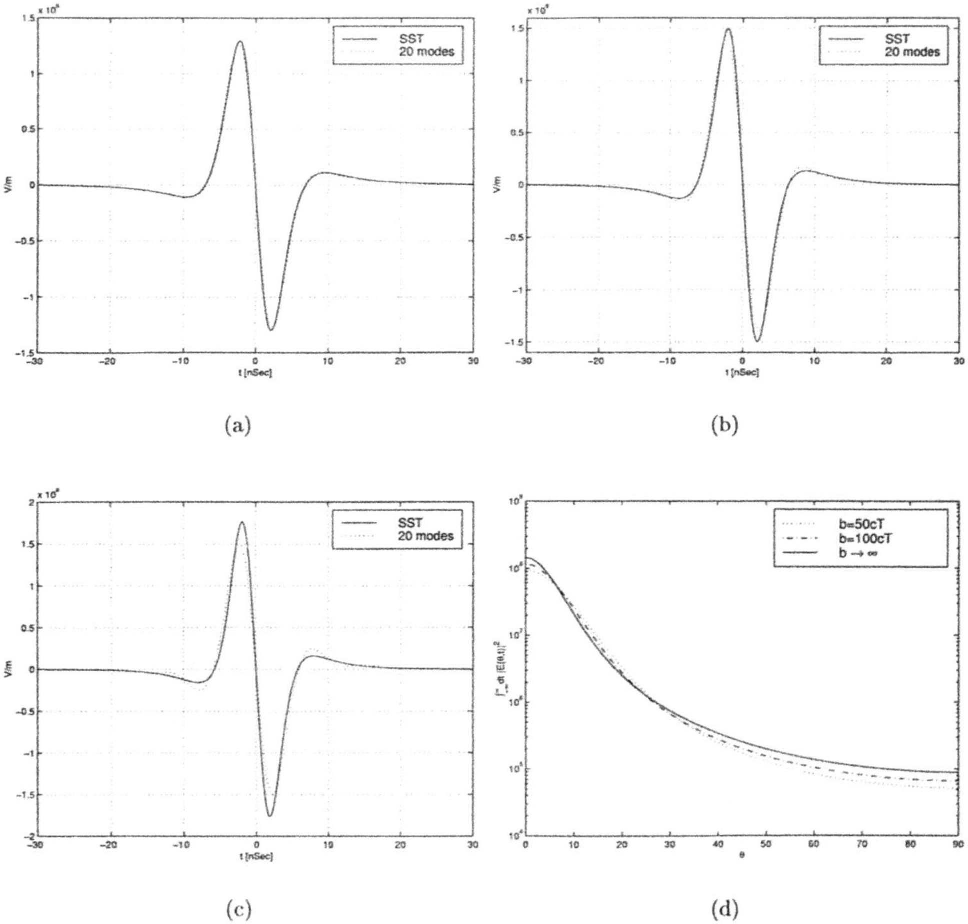

Figure 4: Axial far zone pulse for: (a) $b = 50cT$. (b) $b = 100cT$. (c) $b \to \infty$. (d) Energy radiation pattern.

peak of the pulse $\mathrm{Re} f_+(t - ir^2/2bc)$ at $t = 0$ decays like $-2A/\pi(\sqrt{3}T + r^2/2bc)^3$. The effective aperture width of this distribution, defined as the diameter where the pulse energy decays to one half of its value on the axis, is found to be $W \simeq \sqrt{2cTb}$.

For large a such that $2a \gg W$, the truncation $r < a$ in the current distribution (16) may be neglected. This current distribution gives rise to an iso-diffracting pulsed beam field (PB) of the type considered for example in [3], with b being the collimation distance. If however, $2a \lesssim W$ than the truncation effect is significant and the radiating field is not an iso-diffracting PB.

Below we shall consider three different distributions, all having the same T and $a = 9cT$, but with different values for b: $b = 50cT$, $b = 100cT$ and $b \to \infty$. In the first and second cases, the effective aperture widths as determined above are $W = 10cT$ and $14cT$, respectively. Since they are smaller than the disk diameter, they approximately generate iso-diffracting PBs. In the third case (where $W >> 2a$) the aperture

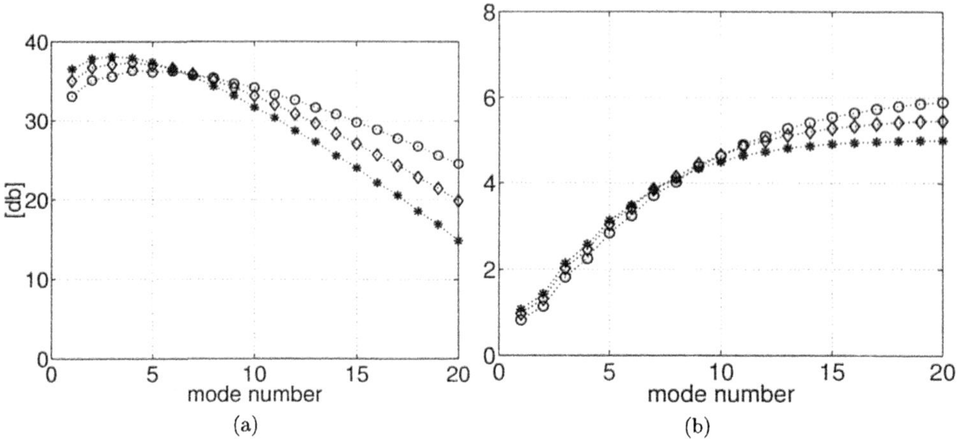

Figure 5: (a) The modal energy distribution $\mathcal{E}_\nu^\alpha$. (b) Cumulative TD Q. Legend: $*\ b = 50cT$, $\diamond\ b = 100cT$ and $\circ\ b \to \infty$.

distribution is practically uniform over the entire disk.

Fig. 4 shows the far zone electric field pulse on the z-axis as computed via the TD multipole expansion (4b). Only the first 20 multipole modes have been included in the summation (dotted lines). In order to validate the TD multipole expansion results and establish the number of terms needed, we compare the results with the the SST formulation (5) of the far field radiation pattern (full lines), which is exact in the far zone. In the large b case the effective aperture is wider and thus more multipoles are needed in order to recover the correct SST result (indeed it has been established in [1] that the number of relevant modes for uniform distributions is approximately $n \sim 3a/cT$). The H-plane energy radiation pattern as a function of the angle from the z axis (the bore sight) is depicted in Fig. 4(d) (cases (a), (b) and (c) are described by the dotted the dashed-dotted and the full lines, respectively). One may readily observe that the directivity increases and beam width decreases as b and, thereby the effective aperture width, increase.

Finally, in Fig. 5 we explore the near field effectiveness of the three source realizations, by plotting the modal energies $\mathcal{E}_\nu^\alpha$ as a function of the mode number (Fig. 5(a)), and the total TD Q as a function of highest mode number in the summation (15) (Fig. 5(b)). From the modal energy distributions we note that the dominant modes are those with $n \simeq \frac{1}{2}W/cT$. The results for the total quality factor Q indicate that the number of relevant modes is $n \simeq 2W/cT$: Beyond this value, the modal energies are small and their contributions to the total Q are negligible. One also observe that the PB-type distributions in (a) and (b) are characterized by lower TD Q: This can be explained by noting that as the effective source diameter increases the total Q increases too, indicating that more higher order modes are dominant in the complete description of the field.

Acknowledgement

This work is supported in part by the U.S. Air Force Office of Scientific Research under Grant No. 96-1-0039 and in part by the Israel Science Foundation under Grant No. 404/98.

References

[1] A. Shlivinski and E. Heyman, "Time Domain Near Field Analysis of Short Pulse Antennas. Part I: Spherical wave (multipole) expansion," *IEEE Trans. Antennas Propagat.*, (submitted) 1997.

[2] —, "—. Part II: Reactive Energy and the Antenna Q" *IEEE Trans. Antennas Propagat.*, (submitted) 1997.

[3] E. Heyman and T. Melamed, "Certain consideration in aperture synthesis for ultra-wideband/short-pulsed fields," *IEEE Trans. Antennas Propagat.*, **AP-42**, 518–525, 1994.

[4] A. Shlivinski, E. Heyman and R. Kastner, "A unified antenna parameterization in the time and frequency domains," *IEEE Trans. Antennas and Propagat.*, **AP-45**, 1140–1149, 1997.

[5] A.J. Devaney and E. Wolf, "Multipole expansions and plane wave representations of the electromagnetic field," *J. Math. Phys.*, Vol. 15, pp. 234–244, 1974.

[6] R.E. Collin and S. Rothschild, "Evaluation of antenna Q," *IRE Trans. Antennas Propagat.*, **AP-12**, 23–27, 1964.

[7] R.F. Harrington, "Effect of antenna size on gain, bandwidth and efficiency," *J. Res. NBS*, **64D**, 1–12, 1960.

[8] M. Abramowitz and I. A. Stegun, *Handbook of Mathematical Functions*, Dover Publications, 1972,

INTERMEDIATE FIELD OF AN IMPULSE-RADIATING ANTENNA

Carl E. Baum
Air Force Research Laboratory
AFRL/DEHP
3550 Aberdeen Ave., SE
Kirtland AFB, NM 87117-5776

1. INTRODUCTION

Much has been done to characterize an impulse radiating antenna (IRA) and has been summarized.[14, 17, 18, 20] In a previous discussion[3] which began this odyssey, the impulsive part of the waveform (on boresight) is not a true impulse. It has a peak which is just the aperture field, but its width decreases like r^{-1}, so that its time integral decreases like r^{-1}, appropriate to a far field. Extending the model based on an aperture integral one can look in more detail at the behavior for large distances from the aperture, so as to obtain a higher-order approximation and obtain the shape of the approximate delta function,[5] called $\delta_a(t)$.

Some development in this regard has already been made.[16, 19, 21] For a given retarded time there is a circle on the aperture plane, the radius of which expands as time increases. The field is represented as an integral of the aperture field on this circle. For a uniform, single-polarization aperture field closed-form expressions for the field are developed. There are also other approaches to this problem.[12, 13] For the special case of a uniform aperture field on a circular aperture, the fields can be evaluated exactly on the z axis perpendicular to the aperture plane and centered in the circular aperture.[6, 9–11, 15] While, in most cases, the aperture integration is performed in the frequency domain, with the resulting field transformed to the time domain, the integration can also be performed directly in the time domain,[6] simplifying the derivation. Furthermore, it is also shown[6] that this result applies to any TEM wave (uniform phase) on the aperture provided that one use the field at the aperture center ($\vec{r} = \vec{0}$) in the expressions.

Our starting point is the general expressions for the fields[2, 3, 6] from a specified tangential electric field on the aperture S_a as

$$E_x'(x',y';t)\ \vec{1}_x + E_y'(x',y';t)\ \vec{1}_y \tag{1.1}$$

Referring to fig. 1.1, note that a prime designates a parameter on the aperture S_a which lies on the $z = 0$ plane. Parameters at the observation point $\vec{r}$ are unprimed. Our coordinates are as usual

$$\begin{gathered}\vec{r} = (x,y,z) = x\ \vec{1}_x + y\ \vec{1}_y + z\ \vec{1}_z = r\ \vec{1}_r \\ r = |\vec{r}|\ ,\quad R = |\vec{r} - \vec{r}'|\ ,\quad x = \Psi\cos(\phi)\ ,\quad y = \Psi\sin(\phi)\end{gathered} \tag{1.2}$$

Referring to fig. 1.1, the aperture boundary C_a encloses the aperture S_a consisting of a set of points $(x',y') \in S_a$. If one extends the points $(x',y') \in C_a$ as the points (x, y) with $x = x'$ and $y = y'$ for all positive z, there is formed a cylinder with generators parallel to the z axis. The same extension of $(x',y') \in S_a$ to equal (x, y) gives the interior of this cylinder which we call the aperture extension V_a.

Ultra-Wideband, Short-Pulse Electromagnetics 4
Edited by Heyman *et al.*, Kluwer Academic / Plenum Publishers, New York, 1999.

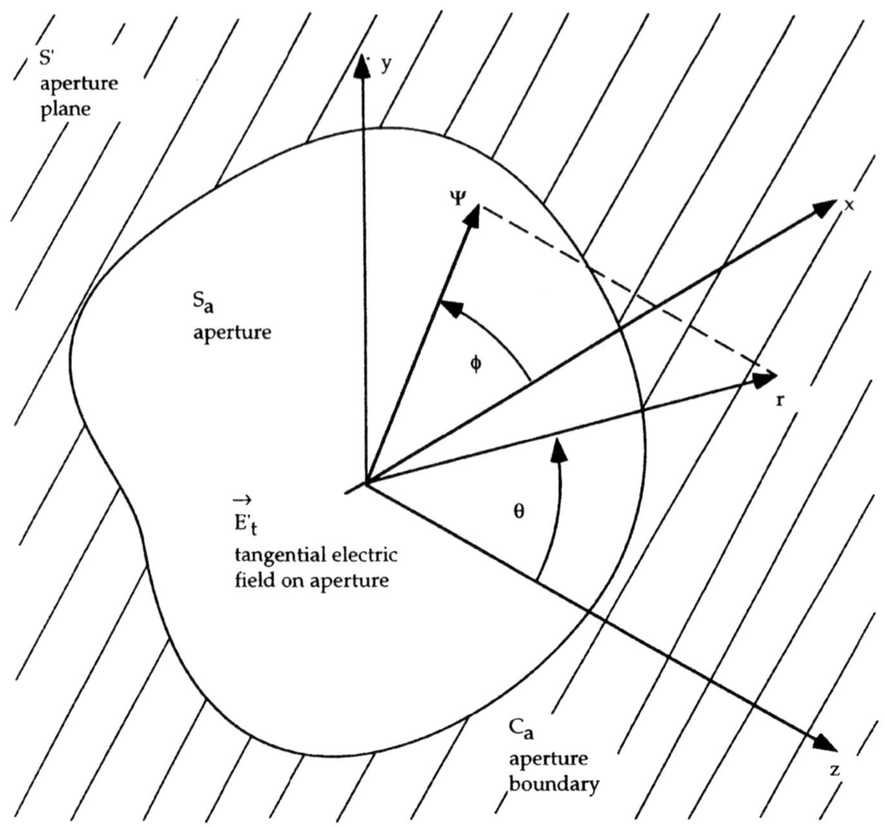

Fig. 1.1 Aperture Antenna with Specified Tangential Electric Field

2. FORMULATION OF THE INTERMEDIATE FIELD

For the intermediate-field formulation of the impulsive part of the waveform (becoming an impulse in the limit of large r on the beam axis) let us first consider only the R^{-1} terms. This defines what is sometimes called the "radiating near field."[23] Furthermore, let us consider the asymptotic form for $z \to \infty$ with fixed x and y. First we have

$$R = z\left[1 + \left[\frac{x-x'}{z}\right]^2 + \left[\frac{y-y'}{z}\right]^2\right]^{\frac{1}{2}} = z\left[1 + \frac{1}{2z^2}\left[[x-x']^2 + [y-y']^2\right] + O\left(z^{-4}\right)\right] \tag{2.1}$$

$$= z + O\left(z^{-1}\right) \quad \text{as } z \to \infty$$

Note that x' and y' are bounded by the assumed finite dimensions of the aperture. Now in (1.2) retain only the leading z^{-1} terms in the coefficients of the aperture fields giving

$$\vec{E}(\vec{r},t) = \frac{1}{2\pi cz}\frac{\partial}{\partial t}\int_{S_a}\vec{E}'\left(x',y';t-\frac{R}{c}\right)dS' + O\left(z^{-2}\right) \quad \text{as } z \to \infty$$

$$Z_0\,\vec{H}(\vec{r},t) = \vec{1}_z \times \vec{E}(\vec{r},t) + O\left(z^{-2}\right) \quad \text{as } z \to \infty \tag{2.2}$$

Here the dependence on R is retained in the retarded time in the temporal argument for the aperture field. Note that in this approximation the fields are TEM with respect to the z axis with errors in the fields of order z^{-2}. This is like the usual far field (taken in the z direction) except for the dependence on R in the aperture field.

Defining retarded time

$$t_r \equiv t - \frac{z}{c} \tag{2.3}$$

and keeping only up to quadratic terms in (2.1) for the expansion of R we have

$$\vec{E}(\vec{r},t) \simeq \vec{E}_i(\vec{r},t) = \frac{1}{2\pi cz}\frac{\partial}{\partial t}\int_{S_a} \vec{E}'\left(x',y';t_r - \frac{1}{2cz}\left[[x'-x]^2 + [y'-y]^2\right]\right)dS' \tag{2.4}$$

It is this that we call the intermediate field, a far field with quadratic time (or phase) correction on the aperture. The approximation here will give a small error as to when an event is seen at the observer (as compared to an error in amplitude).

Now let

$$\vec{E}'(x',y';t) = E_0\,\vec{e}_0(x',y')\,f(t) \tag{2.5}$$

so that there is a uniform waveform (or phase) for the entire aperture, giving focusing in the z direction (at ∞). (One can focus in a different direction, say $\vec{1}_0$, but then one should do the expansion in that direction[3] instead of $\vec{1}_z$.) Our canonical aperture excitation function is a step function.[3,5] Let us then write

$$\vec{E}_i(\vec{r},t) = E_0\left[\frac{df(t)}{dt}\right]\circ\vec{e}_i(\vec{r},t)\quad,\quad \circ \equiv \text{convolution with respect to time}$$

$$\begin{aligned}\vec{e}_i(\vec{r},t_r) &= \frac{1}{2\pi cz}\frac{\partial}{\partial t}\int_{S_a}\vec{e}_0(x',y')\,u\left(t_r - \frac{1}{2cz}\left[[x'-x]^2 + [y'-y]^2\right]\right)dS' \\ &= \frac{1}{2\pi cz}\int_{S_a}\vec{e}_0(x',y')\,\delta\left(t_r - \frac{1}{2cz}\left[[x'-x]^2 + [y'-y]^2\right]\right)dS'\end{aligned} \tag{2.6}$$

Define

$$\begin{aligned}&\xi \equiv 2czt_r \ \text{(units } m^2) \equiv \text{time parameter or normalized retarded time} \\ &C_\xi \equiv \left\{(x',y')\,\middle|\,[x'-x]^2 + [y'-y]^2 = \xi\ ,\ \xi>0\right\} \\ &C_c \equiv C_\xi \cap S_a \ \text{(a proper subset of } S_a \text{ when } C_\xi \text{ lies partly or entirely outside of } S_a)\end{aligned} \tag{2.7}$$

We then have the representations

$$\begin{aligned}\vec{e}_i(\vec{r},t_r) &= \frac{1}{2\pi}\frac{\partial}{\partial\xi}\int_{S_a}\vec{e}(x',y')\,u\left(\xi^{1/2} - \left[[x'-x]^2 - [y'-y]^2\right]^{1/2}\right)dS' \ (\text{for } \xi \geq 0) \\ &= \frac{1}{2\pi}\xi^{-\frac{1}{2}}\int_{S_a}\vec{e}_0(x',y')\,\delta\left(\xi^{\frac{1}{2}} - \left[[x'-x]^2 - [y'-y]^2\right]^{\frac{1}{2}}\right)dS' \\ &= \frac{1}{2\pi}\xi^{-\frac{1}{2}}\int_{C_c}\vec{e}_0(x',y')\,d\ell' = \vec{e}_i(x,y;\xi)\end{aligned} \tag{2.8}$$

Space z and time t_r are combined as a product. So for a given (x, y) this normalized field need be computed only as a function of ξ and this temporal waveform applies for all z (within the limits of our approximations). This scaling also applies to the frequency spectrum of the intermediate field as discussed in Appendix A.

As we go to larger and larger z the waveform becomes shorter and shorter in retarded time. If we take the complete time integral the circle of radius $\xi^{1/2}$ on the aperture plane expands to include all of S_a giving

$$\int_{-\infty}^{\infty} \vec{e}_i(x,y;\xi)\,dt_r = \frac{1}{2cz}\int_{-\infty}^{\infty} \vec{e}_i(x,y;\xi)\,d\xi = \frac{1}{2\pi cz}\int_{S_a} \vec{e}_0(x',y')\,dS' \tag{2.9}$$

This integral (the area of the intermediate-field waveform) is independent of x and y, and hence of whether or not $\vec{r}$ is contained in the aperture extension V_a (of course limited to $\Psi << z$). Note also that the vector orientation of this integral is the same as the average of $\vec{e}_0$ over S_a. For typical cases of one or more symmetry planes this direction is found by inspection (e.g., $\vec{1}_y$ for $\vec{e}_0$ *symmetric*[5, 8, 24] with respect to $x' = 0$.)

As a simple example one can consider the observer on a symmetry plane which we take as the (x, z) plane. With *symmetric* fields on the aperture the x component of the field is zero giving

$$\vec{e}_i(0,y;\xi) = e_{i_y}(0,y;\xi)\,\vec{1}_y \quad , \quad \int_{-\infty}^{\infty} e_{i_y}(0,y;\xi)\,dt = \frac{1}{2\pi cz}\int_{S_a} e_{0_y}(x',y')\,dS' \tag{2.10}$$

This scalarization allows us to find the approximate delta function[3]

$$\delta_a(t_r) = 2\pi cz\left[\int_{S_a} e_{0_y}(x',y')\,dS'\right]^{-1} e_{i_y}(0,y;\xi) \quad , \quad \int_{-\infty}^{\infty}\delta_a(t_r)\,dt = 1 \tag{2.11}$$

The present development extends this to include its vector properties and scaling with z and t_r.

3. INITIAL STEP IN INTERMEDIATE FIELD WITHIN APERTURE EXTENSION

If the observer is within the aperture extension V_a, the first contribution to the intermediate field comes from $(x',y') = (x,y)$. Our first approach follows a previous development.[6] In that case the TEM property on the aperture, which can be given by

$$\vec{e}_0(x',y') = \nabla' u(x',y') \quad , \quad \nabla'^2 u(x',y') = 0 \tag{3.1}$$

allows one to expand $\vec{e}_0$ around the origin ($\Psi' = 0$) in terms of non-negative integer powers of Ψ' (say $\Psi'^{m-1}, m \geq 1$) times $\cos(m\,\phi')$ and $\sin(m\,\phi')$. Then integrating over a circular aperture only the $m = 1$ term contributes for an observer on the z axis, and the field at the observer is initially the same as the field on S_a at $\Psi' = 0$.

Extending this result, consider some small $\xi > 0$. The contour C_ξ is a circle of radius $\xi^{1/2}$ lying completely within S_a for ξ sufficiently small, i.e., $\xi = 0_+$. The integration is only over the circular disk contained inside C_ξ. The aperture field is TEM there, and the observer initially sees ($t_r = 0_+$) that aperture field. This result applies not only on the z axis, but inside the aperture extension V_a (at least a small distance away from the cylindrical boundary), since the field is TEM at all (x',y') within S_a. (Equivalently the z axis can be shifted to any position within V_a.) So we therefore conclude

$$\vec{e}_i(x,y;0_+) = \vec{e}_0(x,y) \tag{3.2}$$

a remarkably simple result. Note then that the early-time $\vec{e}_i$ need not have the same orientation as its complete time integral as can be seen from the symmetric aperture in (2.10) and (2.11). Furthermore, note that this result does not depend on the intermediate-field approximation; it applies right up to the aperture.

Referring to fig. 3.1, we can see that as ξ increases the radius $\xi^{1/2}$ of C_ξ increases until a non-TEM domain is reached. This could be the aperture boundary C_a or some discontinuity at conductors forming boundaries for the assumed planar TEM mode on S_a. Defining

$$\xi^{(1)}(x,y) \equiv [\text{radius of largest circular disk wholly } \textit{within} \text{ the TEM region}]^2 = 2czt_r^{(1)} \tag{3.3}$$

we have the result

$$\vec{e}_i(x,y;\xi) = \vec{e}_0(x,y)\;u(\xi) \;\text{ for } \xi < \xi^{(1)}(x,y) \tag{3.4}$$

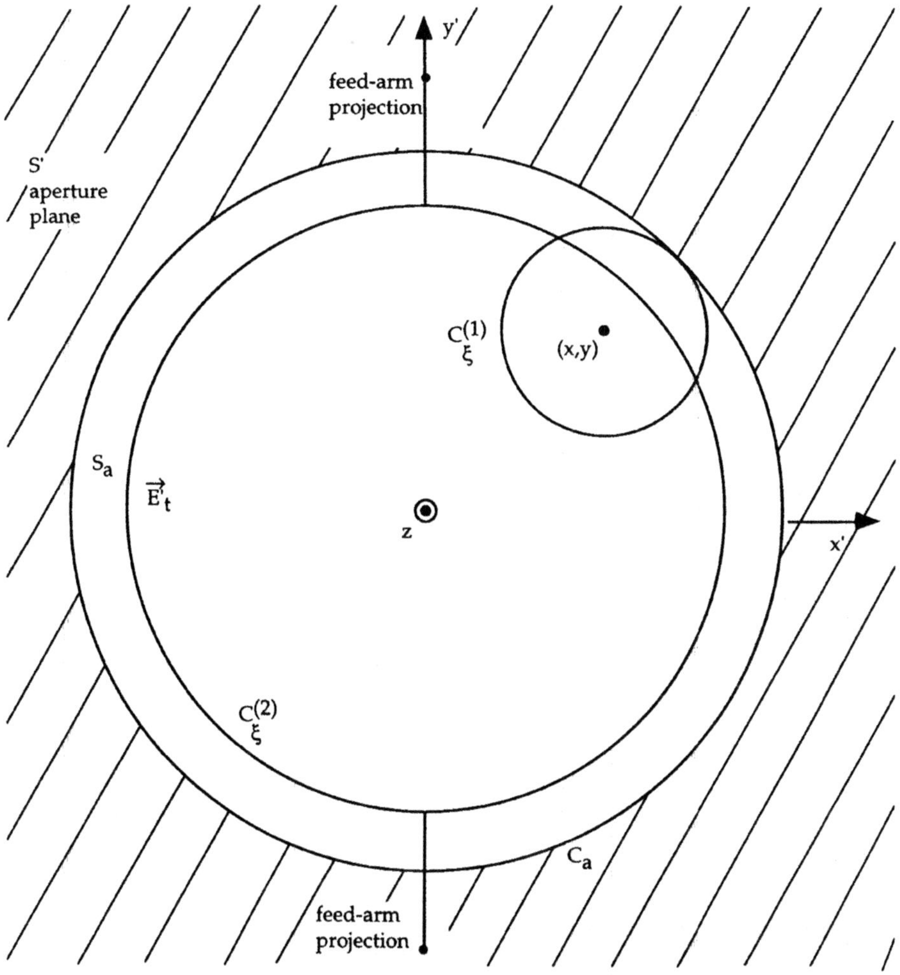

Fig. 3.1 Example Aperture: Circular with Projection of Two Coplanar Feed Arms

So we have a step function valid out to some readily calculable ξ (or equivalent retarded time). Note in fig. 3.1 we have shown two contours: $C_\xi^{(1)}$ centered on (x, y) and $C_\xi^{(2)}$ centered on the origin. In the first case $\xi^{(1)}$ is determined by the circular C_a; in the second case it is determined by the distance to the conductors introducing a singularity in the TEM mode. If one wishes $\xi^{(1)}$ can be more accurately calculated based on R as in (2.2), thereby avoiding the intermediate-field approximations for this early-time window.

Note that for observers outside the aperture extension (3.3) can be extended to

$$\xi^{(1)}(x,y) \equiv [\text{radius of largest circular disk wholly } \textit{outside} \text{ the TEM region (aperture)}]^2$$
$$= 2czt_r^{(1)} \tag{3.5}$$

In this case we have

$$\vec{e}_i(x,y;\xi) = \vec{0} \quad \text{for} \quad \xi < \xi^{(1)}(x,y) \tag{3.6}$$

so that $\xi^{(1)}$ now gives the first retarded time that the observer can detect any signal, the tangential electric field on the aperture outside S_a being assumed zero.

Going a step further we can define

$$\xi^{(2)}(x,y) \equiv [\text{radius of smallest circular disk wholly } \textit{containing} \text{ the TEM region (aperture)}]^2$$
$$= 2czt_r^{(2)} \tag{3.7}$$

In this case (2.8) gives

$$\vec{e}_i(x,y;\xi) = \vec{0} \quad \text{for } \xi > \xi^{(2)}(x,y) \tag{3.8}$$

Between $\xi^{(1)}$ and $\xi^{(2)}$ the waveform is more complicated and polarization may even rotate. One constraint that we can obtain concerns the time integral of this portion. Combining (2.9) with (3.2) gives

$$\int_{t_r^{(1)}}^{t_r^{(2)}} \vec{e}_i(x,y;\xi)\,dt_r = \frac{1}{2cz}\int_{\xi^{(1)}}^{\xi^{(2)}} \vec{e}_i(x,y;\xi)\,d\xi = \frac{1}{2\pi cz}\int_{S_a} \vec{e}_0(x',y')\,dS' - \vec{e}_0(x,y)\,t_r^{(1)} \tag{3.9}$$

A second approach to the early-time result in (3.4) involves a gedankenexperiment. Again, referring to fig. 3.1, consider first a boundary value problem in which the entire aperture plane S' ($z = 0$ plane) has specified on it the TEM tangential electric field given by (2.5) with $f(t)$ as the unit step at $t = 0$. Furthermore, let the fed-arm projections be extended as conductors (perfect) parallel to the z axis toward $z = +\infty$. Noting that the TEM mode $\vec{e}_0(x,y)$ is exactly the transverse field for such a mode on a TEM transmission line matching the requisite boundary condition on the above transmission-line conductors, then

$$\vec{E}(x,y;t) = E_0\,\vec{e}_0(x,y)\,u\left(t-\frac{z}{c}\right) \tag{3.10}$$

satisfies the Maxwell equations and boundary conditions (tangential $\vec{E}$ on S', zero tangential $\vec{E}$ on the conductors, and radiation condition (+z propagation)) and is thereby the unique solution for positive z. Next, change the field on S' to be zero outside S_a, but unchanged on S_a. Then by causality, this change cannot be observed inside the aperture extension V_a until some time after the wave in (3.10) first arrives at the observer. This retarded time is $t_r^{(1)}$ except that the non-TEM part of the aperture (feed-arm projection) is not considered to limit the expansion of C_c, now to the aperture boundary. Finally, remove the transmission-line conductors from the $z > 0$ half space. There being no currents on these to support the wave in (3.10), now an observer in the aperture extension can observe the fact of their absence after a retarded time $t_r^{(1)}$ with the feed-arm projection included as part of the aperture boundary (boundary of non-TEMness).

Since this gedankenexperiment does not rely on the asymptotic expansion of R for large z as in (2.1), then one can use exact values of R to determine a more accurate value of $t_r^{(1)}$. Thereby one can extend the early-time result in (3.4) right back to the aperture plane, i.e., z need not be considered large.

4. COMPLEX POTENTIALS AND FIELDS ON APERTURE

Having constrained the aperture fields to be TEM (as on a cylindrical transmission line) we can derive the aperture tangential electric field from a potential function as in (3.1). This allows us to use the complex-variable formulation.[5] Summarizing we have

$\zeta = x' + jy' \equiv$ complex coordinate (on aperture) , $\zeta_0 = x + jy \equiv$ complex observer coordinate

$w(\zeta) = u(\zeta) + jv(\zeta) \equiv$ complex potential , $\vec{e}_0(x',y') = \nabla' u(x',y')$, $\vec{h}_0(x',y') = \nabla' v(x',y')$

$$e_0(\zeta) = e_{0_x}(x',y') - je_{0_y}(x',y') \;,\; h_0(\zeta) = h_{0_x}(x',y') - jh_{0_y}(x',y')$$

$$e_{0_x}(x',y') = h_{0_y}(x',y') \;,\; e_{0_y}(x',y') = -h_{0_x}(x',y')$$

$$e_0(\zeta) = jh_0(\zeta) = \frac{dw(\zeta)}{d\zeta} \equiv \text{complex aperture field (normalized)} \tag{4.1}$$

$$E(\zeta) = jZ_0H(\zeta) = -\frac{V}{\Delta u}\frac{dw(\zeta)}{d\zeta} = -\frac{V}{\Delta u}e_0(\zeta) = E_x(x',y') - j\,E_y(x',y')$$

$$f_g = \frac{\Delta u \text{ (between appropriate conductors)}}{\Delta v \text{ (around appropriate conductor(s))}}$$

$\equiv$ geometric impedance factor for TEM transmission line feeding aperture

Here V is the voltage across the aperture (via the feed arms in the case of a reflector IRA); this can be a function of time to match (2.5) as desired.

In complex form (2.8) becomes

$$e_i(\zeta_0,\xi) = e_{i_x}(x,y;\xi) - j\,e_{i_y}(x,y;\xi) = \frac{1}{2\pi}\xi^{-\frac{1}{2}}\int_{C_c} e_0(\zeta)\,|d\zeta| \tag{4.2}$$

Consider first the early-time behavior. For $\xi<0$ the integrals for $\vec{e}_i$ give zero. For the first time interval C_ξ is a circle consistent with (3.3). One way to approach this problem is from Cauchy's theorem

$$e_0(\zeta) = \frac{1}{2\pi j}\oint_C \frac{e_0(\zeta')}{\zeta'-\zeta}d\zeta' \;,\quad \zeta \text{ inside } C \tag{4.3}$$

which requires that $e_0(\zeta)$ be analytic within and on C. Let us then choose C as a circle centered on ζ and change variables as

$$\zeta'-\zeta = \zeta'' = |\zeta''|\,e^{j\arg(\zeta'')} \;,\quad d\zeta' = je^{j\arg(\zeta'')}d\ell'' \text{ on } C$$
$$d\ell'' = \text{line element (real) on } C \;,\quad |\zeta''| = \text{radius of circle} \tag{4.4}$$

giving

$$e_0(\zeta) = \frac{1}{2\pi|\zeta''|}\oint_C e_0(\zeta')\,d\ell'' = \text{average value of } e_0(\zeta) \text{ on circular contour } C \tag{4.5}$$

Applying this general result to our circle of radius $\xi^{1/2}$ centered on ζ_0 gives

$$e_0(\zeta_0) = \frac{1}{2\pi}\xi^{-\frac{1}{2}}\oint_{C_\xi} e_0(\zeta)|d\zeta| \;,\quad d\ell' = |d\zeta| \tag{4.6}$$
$$\vec{e}_0(x,y) = \frac{1}{2\pi}\xi^{-\frac{1}{2}}\oint_{C_\xi}\vec{e}_0(x',y')d\ell' \quad \text{(real vector form)}$$

Applying this result to (4.2) gives

$$e_i(\zeta_0,\xi) = e_0(\zeta_0)u(\xi) \quad \text{for} \quad \xi<\xi^{(1)}(x,y) \tag{4.7}$$

This is the same result as (3.4) stated in complex form and derived by a different technique (Cauchy's theorem). For observers outside the aperture extension this gives the same result as (3.6) noting that $e_0(\zeta_0)$ is zero for ζ_0 outside S_a.

For the third time interval the circular contour C_ξ contains S_a completely inside (as per (3.5)). The contour C_c in (4.2) and (2.7) is then the null set and

$$e_i(\zeta_0,\xi) = 0 \quad \text{for} \quad \xi^{(2)}(x,y) < \xi \tag{4.8}$$

consistent with (3.6). For the second time interval we need to apply the contour integral to a circular arc C_c centered on ζ_0 in S_a. Consider then a circular arc of radius ξ which we can parameterize as

$$\zeta-\zeta_0 = \xi^{\frac{1}{2}}e^{j\psi} \;,\quad \psi_1 \le \psi \le \psi_2 \;,\quad \psi \text{ real}$$
$$d\zeta = j\xi^{\frac{1}{2}}e^{j\psi}\,d\psi \;,\quad d\ell' = |d\zeta| = \xi^{\frac{1}{2}}\,d\psi \tag{4.9}$$

Noting the positive sense of ψ, the range of ψ ($-\pi$ to π etc.) can be chosen at our convenience. Then (4.2) becomes

$$e_i(\zeta_0,\xi) = \frac{1}{2\pi}\int_{\psi_1}^{\psi_2} e_0\left(\xi\, e^{j\psi} + \zeta_0\right) d\psi \tag{4.10}$$

If C_c consists of more than one such arc (say due to interruption by any feed arm projection(s)), then other arc portions (e.g., ψ_3 to ψ_4) need to be included in (4.10).

A special case has ζ_0 on the aperture boundary. Then we have

$$\xi^{(1)}(x,y) = 0_+ \quad , \quad e_i(\zeta_0,0_+) = \frac{1}{2\pi}\left[\psi_2 - \psi_1\right] e_0\left(\zeta_{0\_}\right)$$
$$\zeta_{0\_} \in S_a \text{ (just inside from } \zeta_0) \tag{4.11}$$

The 0_+ value of $\xi^{(1)}$ indicates the limit as $\xi \to 0$ from positive values. Note that there is assumed a zero field just outside the boundary, so an infinitesimally thin feed-arm projection does not apply here, but one could include two such terms in that case. Assuming that the boundary is smooth (locally straight) at ζ_0, then we have the simple result

$$\psi_2 - \psi_1 = \pi \quad , \quad e_i(\zeta_0,0_+) = \frac{1}{2} e_0\left(\zeta_{0\_}\right) \tag{4.12}$$

5. SYMMETRICAL TWO-WIRE TRANSMISSION LINE AS FEED-ARM PROJECTION WITH CIRCULAR APERTURE

For a concrete example for which one can compute $e_i(\zeta_0,\xi)$ let us consider the canonical example[5] of the aperture fields as given by those of a symmetrical two-wire transmission line as illustrated in fig. 5.1. The positions of the equivalent line charges are made coincident with the edge of the circle of radius a (at $\zeta = \pm ja$), while the wires of radius b are centered at $\zeta = \pm ja_c$. Using previous results[1,5] we have

$$a_c^2 = a^2 + b^2$$

$$f_g = \frac{\Delta u}{\Delta v} = \frac{1}{\pi}\operatorname{arccosh}\left(\frac{a_c}{b}\right) = \frac{1}{\pi}\operatorname{arcsinh}\left(\frac{a}{b}\right) = \frac{1}{\pi}\ell n\left(\frac{2a}{b}\right) + O\left(\left(\frac{b}{a}\right)^2\right) \text{ as } \frac{b}{a} \to 0 \tag{5.1}$$

For the complex potential and field we use

$$w(\zeta) = w_0\, \ell n\left(\frac{\zeta + ja}{\zeta - ja}\right) \quad , \quad e_0(\zeta) = \frac{dw(\zeta)}{d\zeta} = w_0\left[[\zeta + ja]^{-1} - [\zeta - ja]^{-1}\right] = -\frac{j2w_0 a}{\zeta^2 + a^2} \tag{5.2}$$

For convenience we choose

$$\vec{e}_0(0,0) = \vec{1}_y \quad , \quad e_0(0) = -j \quad , \quad w_0 = \frac{a}{2} \tag{5.3}$$

so that E_0 in (2.5) represents the field amplitude at the aperture center. With this we have the potential difference between the conductors as

$$\Delta u = a \operatorname{arcsinh}\left(\frac{a}{b}\right) = \pi a f_g \quad , \quad \Delta v = a\pi \text{ (circling one conductor)} \tag{5.4}$$

We can then identify Δu with the voltage V between the conductors giving

$$E_0 = -\frac{\mathrm{V}}{\Delta u} = -\frac{V}{\pi a f_g} \tag{5.5}$$

where the upper conductor has convention positive and the bottom has negative. Note, however, the sign reversal in the case of a reflector IRA due to the negative reflection at the paraboloidal reflector.

In fig. 5.1, the observer location (projected on the aperture plane) is

$$\zeta_0 = x + jy = \Psi_0\, e^{j\phi_0} \tag{5.6}$$

As previously determined the intermediate field is initially (the first time interval) the aperture field, i.e.,

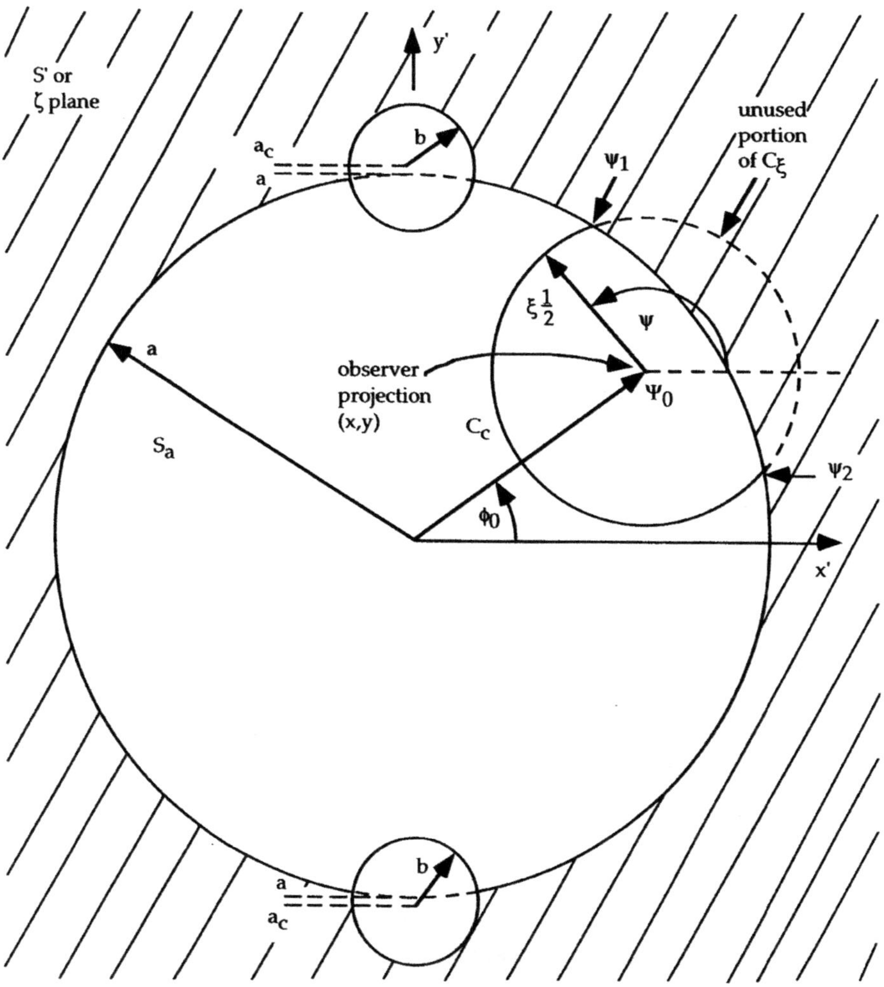

Fig. 5.1 Symmetrical Two-Wire Transmission Line Feeding Circular Aperture

$$e_i(\zeta_0,\xi) = e_0(\zeta_0) = -\frac{ja^2}{\zeta_0^2 + a^2} \quad \text{for } 0 < \xi < \xi^{(1)}(x,y) \ , \ \Psi_0 < a$$

$$\xi^{(1)}(x,y) = \left[a - \Psi_0\right]^2 \tag{5.7}$$

where b has been assumed negligible compared to a for certain ϕ_0 (e.g., $\pi/2$, $3\pi/2$) so that the feed projections do not significantly lessen $\xi^{(1)}$. On the two symmetry planes ($x = 0$ and $y = 0$) the field is polarized in the y direction. The field[8, 24] is *symmetric* with respect to $x = 0$ and *antisymmetric* with respect to $y = 0$. Note that for $\Psi_0 > a$ the observer is outside the aperture extension, and the field is initially zero consistent with the extended definition of $\xi^{(1)}$ in (3.5).

In the third time interval we have

$$e_i(\zeta_0,\xi) = 0 \quad \text{for } \xi^{(2)}(x,y) < \xi \ , \ \xi^{(2)}(x,y) = \left[a + \Psi_0\right]^2 \tag{5.8}$$

As in (2.9) the complete time integral is

$$\int_{-\infty}^{\infty} e_i(\zeta_0,\xi)dt_r = \frac{1}{2cz}\int_{-\infty}^{\infty} e_i(\zeta_0,\xi)d\xi = \frac{1}{2\pi cz}\int_{S_a} e_0(\xi)dS' \tag{5.9}$$

which varies as z^{-1}, but is independent of ζ_0 (or x and y). As in (3.9) the time integral over the second time interval is

$$\int_{t_r^{(1)}}^{t_r^{(2)}} e_i(\zeta_0,\xi)dt_r = \frac{1}{2cz}\int_{\xi^{(1)}}^{\xi^{(2)}} e_i(\zeta_0,\xi)d\xi = \frac{1}{2cz}\left[\frac{1}{\pi}\int_{S_a} e_0(\zeta)dS' - e_0(\zeta_0)\,\xi^{(1)}(x,y)\right] \tag{5.10}$$

For the special case of an observer on the z axis we have no second time interval. The waveform is explicitly

$$e_i(0,\xi) = e_0(0)\ u(\xi)\ u\left(a^2-\xi\right)\ ,\ \zeta_0 = 0\ ,\ \xi^{(1)} = \xi^{(2)} = a^2 \tag{5.11}$$

The complete time integral is

$$\int_0^{t_r^{(1)}} e_i(0,\xi)\,dt_r = \frac{1}{2cz}\int_0^{a^2} e_i(0,\xi)\,d\xi = \frac{a^2}{2cz}e_0(0) \tag{5.12}$$

This is consistent with the results[6] for the leading part (impulsive part) of the on-axis fields from a circular aperture. This applies to any TEM distribution on a circular aperture where there is no significant penetration of the aperture by feed-arm projections. Thus the results apply to four (or any number) of thin feed arms driving the aperture provided one calculates the tangential electric field at the aperture center.

Now we come to the second time interval for general ζ_0. As illustrated in fig. 5.1, we have the circular arc C_c of radius $\xi^{1/2}$ centered on ζ_0 on which to integrate as in (4.10). The end points of the contour are given by

$$\left|\Psi_0\, e^{j\phi_0} + \xi^{1/2}\, e^{j\psi}\right| = a\ ,\ \psi_{\substack{1\\2}} - \phi_0 = \pm\arccos\left(\frac{a^2-\Psi_0^2-\xi}{2\Psi_0\,\xi^{1/2}}\right) \tag{5.13}$$

For $\Psi_0 < a\left(\zeta_0 \notin S_a\right)$ the cosine goes from +1 to −1 as ξ increases. For $\Psi_0 > a\left(\zeta_0 \notin S_a\right)$, the cosine goes from −1 to some maximum value and then back to −1 as ξ increases.

For our aperture field as in (5.2) the contour integral for the second time interval as in (4.10) becomes

$$e_i(\zeta_0,\xi) = \frac{a}{4\pi}\left[\int_{\psi_1}^{\psi_2}\left[\xi^{1/2}\,e^{j\psi}+\zeta_0+ja\right]^{-1}d\psi - \int_{\psi_1}^{\psi_2}\left[\xi^{1/2}\,e^{j\psi}+\zeta_0-ja\right]^{-1}d\psi\right] \tag{5.14}$$

The general form of these integrals is given by the indefinite integral[22]

$$\int\frac{dv}{\alpha_1+\alpha_2 e^{\alpha_3 v}} = \frac{1}{\alpha_1\alpha_3}\left[\alpha_3\, v - \ell n\left(\alpha_1 + \alpha_2\, e^{\alpha_3 v}\right)\right] \tag{5.15}$$

which can be verified by differentiation. Noting that we have a logarithm of a complex argument we will need to observe the location of any branch cut(s) in evaluating the integrals. A good place for such a cut is where it does not cross C_c, but goes from ζ_0 outward in the $+\Psi$ direction, crossing C_ξ at an angle that is not part of C_c. Note also that as ψ_1 and ψ_2 sweep around C_a and go by the wires, the end points of C_c pass the wires slightly into the aperture (where $e_0(\zeta)$ is analytic).

Applying (5.17) to (5.16) we have

$$\begin{aligned} e_i(\zeta_0,\xi) = \frac{a}{4\pi}\Bigg[&-\frac{j}{\zeta_0+ja}\left[j[\psi_2-\psi_1]-\ell n\left(\frac{\zeta_0+ja+\xi^{1/2}e^{j\psi_2}}{\zeta_0-ja+\xi^{1/2}e^{j\psi_1}}\right)\right] \\ &+\frac{j}{\zeta_0-ja}\left[j[\psi_2-\psi_1]-\ell n\left(\frac{\zeta_0-ja+\xi^{1/2}e^{j\psi_2}}{\zeta_0+ja+\xi^{1/2}e^{j\psi_1}}\right)\right]\Bigg] \end{aligned} \tag{5.16}$$

with $\psi_1 - \psi_2$ progressing from 2π to 0 for increasing ξ with $|\zeta_0|<a$ (inside the aperture), but from 0 to some maximum value and then back to zero for $|\zeta_0|>a$ (outside the aperture).

To better understand this result, let us consider some special cases. At the beginning of the second time interval for ζ_0 within the aperture we have

$$\psi_2 - \psi_1 = 2\pi \quad \text{(full circular contour)}$$

$$e_i(\zeta_0,\xi) = -j\frac{a^2}{\zeta_0^2 + a^2} = e_0(\zeta_0) \tag{5.17}$$

Note that the logarithmic terms are zero. As we should expect the waveform begins the second time interval at the same value as at the end of the first time interval. For ζ_0 outside the aperture we have

$$\psi_2 - \psi_1 = 0 \quad \text{(contour just touching the aperture)} \;, \quad e_i(\zeta_0,\xi) = 0 \tag{5.18}$$

At the end of the second time interval for ζ_0 either inside or outside the aperture we have

$$\psi_2 - \psi_1 = 0 \;, \quad e_i(\zeta_0,\xi) = 0 \tag{5.19}$$

again the logarithmic terms being zero. We can also recall from (4.12) that on the circular boundary

$$\psi_2 - \psi_1 = \pi \;, \quad \xi = 0_+ \;, \quad e_i(\zeta_0,0_+) = -\frac{j}{2}\frac{a^2}{\zeta_0^2 + a^2} = \frac{1}{2}e_0(\zeta_{0\_}) \tag{5.20}$$

which is consistent with (5.16)

Choices of ϕ_0 as $\pi/4$ plus integer multiples of $\pi/2$ are of interest for the four-wire feed-arm configuration. If two thin feed-arm pairs (including the case of coplanar-plate pairs) are oriented on a circular aperture with their planes at right angles to each other, the two pairs do not interact by symmetry.[4] The fields of each may be added to give the total resulting field. With the fields given as in (5.16) we can construct those for the symmetric four-wire configuration from the two-wire results (superscript 2) via

$$e_i^{(4)}(\zeta_0,\xi) = \frac{1}{\sqrt{2}}\left[e^{-j\frac{\pi}{4}}e_i^{(2)}\left(e^{j\frac{\pi}{4}}\zeta_0,\xi\right) + e^{j\frac{\pi}{4}}e_i^{(2)}\left(e^{-j\frac{\pi}{4}}\zeta_0,\xi\right)\right] \tag{5.21}$$

The first term rotates coordinates 45° in a positive sense, accounting for the feed-arm pair at $\pm e^{j(\pi/4)}a$, and –45° for the other pair. The factor of $1/\sqrt{2}$ normalizes the field to be of unit amplitude in the y direction at the origin. One needs to allow for this in calculating the antenna response for a given voltage applied to the feed arms.[7]

6. JUMPS IN WAVEFORM AS CONTOUR SWEEPS BY FEED-WIRE PROJECTIONS

There are some peculiarities in the formulae for the intermediate field as the contour end points sweep past the wire projections where they touch S_a. For the observer in the first quadrant of the ζ_0 plane this happens first for ψ_1 when $\zeta \to ja$. From (5.13) and (5.16) this occurs when

$$\zeta_0 - ja + \xi^{1/2}e^{j\psi_1} \to 0 \tag{6.1}$$

Referring to fig. 5.1 we can see that the argument of the logarithm (denominator in first form in (5.16), noting the form in (6.1)) changes from 0 to π, noting that the contour passes *under* the line charge at $\zeta = ja$. This gives a jump in the intermediate field as

$$\Delta_1 e_i(\zeta_0,\xi) = \frac{1}{4}\frac{a}{\zeta_0 - ja} = \frac{1}{4}\left[\frac{\zeta_0}{a} - j\right]^{-1} \tag{6.2}$$

near the origin this is $\simeq j/4$ which is a decrease noting the minus sign with e_{i_y} in the complex field. Here Δ_1 is the change as value after minus value before. There is a logarithmic singularity from the log magnitude as well, but this is integrable without discontinuity in the integral.

Similarly when ψ_2 passes $\zeta = -ja$ we have

$$\zeta_0 + ja + \xi^{1/2}e^{j\psi_2} \to 0 \;, \quad \Delta_2 e_i(\zeta_0,\xi) = -\frac{1}{4}\frac{a}{\zeta_0 + ja} = \frac{1}{4}\left[-\frac{\zeta_0}{a} - j\right]^{-1} \tag{6.3}$$

Curiously enough adding the two changes gives

$$\Delta_1 e_i(\zeta_0,\xi) + \Delta_2 e_i(\zeta_0,\xi) = \frac{a}{4}\left[[\zeta_0 - ja]^{-1} - [\zeta_0 + ja]^{-1}\right]$$
$$= \frac{j}{2}\frac{a^2}{\zeta_0^2 + a^2} = -\frac{1}{2}e_0(\zeta_0) \text{ for } \zeta_0 \text{ in aperture projection} \tag{6.4}$$

which, in the aperture projection, accounts for one half of the initial fields at the observer (first time interval).

7. CONCLUDING REMARKS

So here we have some interesting ways to evaluate the intermediate field. The contour integrals for thin-wire TEM aperture distributions can be expressed in closed form. Other types of TEM fields are also of interest (e.g., for coplanar plates). So there is much that can be done to exploit the present results. Furthermore, the aperture need not be a circular disk (e.g., a semi-circular disk, as in a half IRA).

Here we have assumed that the TEM fields on the aperture are propagating parallel to the aperture normal. This need not be the case.[3] One can assume that the aperture fields take the form of a TEM plane wave propagating in some direction, say $\vec{1}_0$, into the positive half space $(z > 0)$. Then by taking the asymptotic expansion in the $\vec{1}_0$ direction (instead of z), one might expect to obtain similar results.

Note that the results here apply to only one part of the waveform, the "impulsive" part. For a reflector IRA one needs to include the prepulse associated with the TEM field. Then there is also what follows the impulse (the postpulse).

APPENDIX A. LAPLACE/FOURIER TRANSFORM OF THE INTERMEDIATE FIELD

The two-sided Laplace transform is expressed as

$s = \Omega + j\omega \equiv$ Laplace-transform variable or complex frequency

$\sim \equiv$ Laplace transform (two-sided)

$$\tilde{\vec{e}}_i(x,y;\Sigma) = \int_{-\infty}^{\infty} \vec{e}_i(x,y;\xi)\, e^{-st_r} dt_r = \frac{1}{2cz}\int_{-\infty}^{\infty} \vec{e}_i(x,y;\xi)\, e^{-\Sigma\xi} d\xi \tag{A.1}$$

$$\Sigma = \sigma + j\nu = \frac{s}{2cz} \equiv \text{normalized complex frequency (units m}^{-2}\text{)}$$

Here we have used the retarded time as the time variable to avoid a large phase shift (phase wrapping) for large z.

The important thing here is that the scaling in time domain represented by the parameter ξ goes over to a similar scaling in frequency domain via the parameter Σ. As we can see, if we double z the waveform has the same amplitude but is squeezed into half the time. In frequency domain this corresponds to a doubling of the frequency, but dividing the value of the spectrum by 2 (from the z^{-1} coefficient). So what one better calculates is

$$\tilde{\vec{e}}_i^{(0)}(x,y;\Sigma) = 2cz\,\tilde{\vec{e}}_i(x,y;\Sigma) = \int_{-\infty}^{\infty} \vec{e}(x,y;\Sigma)\, e^{-\Sigma\xi} d\xi \tag{A.2}$$

This then applies to any z for a given (x, y) and can be calculated once and scaled as above.

For observers in the aperture extension we have derived several different ways (Sections 3 and 4) the result

$$\vec{e}_i(x,y;0_+) = \vec{e}_0(x,y) \tag{A.3}$$

Noting that this is initially a step function we have

$$\tilde{\vec{e}}_i^{(0)}(x,y;\Sigma) = \frac{\vec{e}_0(x,y)}{\Sigma} = \frac{2cz\,\vec{e}_0(x,y)}{s} \quad , \quad \tilde{\vec{e}}_i(x,y;\Sigma) = \frac{\vec{e}_0(x,y)}{s} \quad \text{as } s \to \infty \text{ in RHP} \tag{A.4}$$

We can let s go to the $j\omega$ axis (on the way to $\pm j\infty$) provided there are no further discontinuities (steps) later in time in the waveform. Lower order discontinuities are allowed since they give contributions that go to zero faster than $1/s$.

REFERENCES

1. C. E. Baum, Impedances and Field Distributions for Symmetrical Two Wire and Four Wire Transmission Line Simulators, Sensor and Simulation Note 27, October 1966.
2. C. E. Baum, Focused Aperture Antennas, Sensor and Simulation Note 306, May 1987.
3. C. E. Baum, Radiation of Impulse-Like Transient Fields, Sensor and Simulation Note 321, November 1989.
4. C. E. Baum, Configurations of TEM Feed for an IRA, Sensor and Simulation Note 327, April 1991.
5. C. E. Baum, Aperture Efficiencies for IRAs, Sensor and Simulation Note 328, June 1991.
6. C. E. Baum, Circular Aperture Antennas in Time Domain, Sensor and Simulation Note 351, November 1992.
7. C. E. Baum, Some Topics Concerning Feed Arms of Reflector IRAs, Sensor and Simulation Note 414, October 1997.
8. C. E. Baum, Interaction of Electromagnetic Fields with an Object Which Has an Electromagnetic Symmetry Plane, Interaction Note 63, March 1971.
9. R. C. Rudduck and C.-L. Chen, New Plane Wave Spectrum Formulations for the Near-Fields of Circular and Strip Apertures, IEEE Trans. Antennas and Propagation, 1976, pp. 438–449.
10. A. D. Yaghjian, Efficient Computation of Antenna Coupling and Fields Within the Near-Field Region, IEEE Trans. Antennas and Propagation, 1982, pp. 113–128.
11. D. J. Blejer, On-Axis Transient Fields from a Uniform Circular Distribution of Electric or Magnetic Current, PIERS Proc., Boston, July 1989, pp. 362-363.
12. O. V. Mikheev et al, New Method for Calculating Pulse Radiation from an Antenna With a Reflector, IEEE Trans. EMC, 1997, pp. 48–53.
13. H.-T. Chou, P. H. Pathak, and P. R. Rousseau, Analytical Solution for Early-Time Transient Radiation from Pulse-Excited Parabolic Reflector Antennas, IEEE Trans. Antennas and Propagation, 1997, pp. 829–836.
14. C. E. Baum and E. G. Farr, Impulse Radiating Antennas, pp. 139–147, in H. Bertoni et al (eds.), *Ultra-Wideband, Short-Pulse Electromagnetics*, Plenum Press, 1993.
15. D. J. Blejer, R. C. Wittman, and A. D. Yaghjian (eds.), On-Axis Fields from a Circular Uniform Surface Current, pp. 285–292, in H. Bertoni et al (eds.) *Ultra-Wideband, Short-Pulse Electromagnetics*, Plenum Press, 1993.
16. S. P. Skulkin and V. I. Turchin, Radiation of Nonsinusoidal Waves by Aperture Antennas, pp. 1498–1504, in D. J. Serafin, J. Ch. Bolomey, and D. Dupony (eds.), Proc. EUROEM 94, Bordeaux, May 1994.
17. E. G. Farr, C. E. Baum, and C. J. Buchenaur, Impulse Radiating Antennas, Part II, pp. 159–170, in L. Carin and L. B. Felsen (eds.), *Ultra-Wideband, Short-Pulse Electromagnetics 2*, Plenum Press, 1995.
18. E. G. Farr and C. E. Baum, Impulse Radiating Antennas, Part III, pp. 43–56, in C. E. Baum et al (eds.), *Ultra-Wideband, Short-Pulse Electromagnetics 3*, Plenum Press, 1997.
19. S. P. Skulkin, Transient Fields of Rectangular Aperture Antennas, pp. 57–63, in C. E. Baum et al (eds.), *Ultra-Wideband, Short-Pulse Electromagnetics 3*, Plenum Press, 1997.
20. D. V. Giri and C. E. Baum, Temporal and Spectral Radiation on Boresight of a Reflector Type of Impulse-Radiating Antenna (IRA), pp. 65–72, in C. E. Baum et al (eds.), *Ultra-Wideband, Short-Pulse Electromagnetics 3*, Plenum Press, 1997.
21. S. P. Skulkin and V. I. Turchin, Transient Fields of Parabolic Reflector Antennas, pp. 81–87, in C. E. Baum et al (eds.), *Ultra-Wideband, Short-Pulse Electromagnetics 3*, Plenum Press 1997.
22. I. S. Gradshteyn and I. M. Ryzhik, *Table of Integrals, Series, and Products*, Academic Press, 1980.
23. Y. T. Lo and S. W. Lee (eds.), *Antenna Handbook*, van Nostrand Reinhold, 1988.
24. C. E. Baum and H. N. Kritikos, Symmetry in Electromagnetics, ch. 1, pp. 1–90, in C. E. Baum and H. N. Kritikos (eds.), *Electromagnetic Symmetry*, Taylor & Francis, 1995.

APERTURE EFFICIENCIES OF IMPULSE RADIATING ANTENNAS

C. Jerald Buchenauer,[1,2] J. Scott Tyo,[1] and Jon S. H. Schoenberg[1]

[1]Air Force Research Laboratory / DEHP
Kirtland AFB, NM 87117

[2]Los Alamos National Laboratory, NIS-9
Los Alamos, NM 87545

ABSTRACT

A concept of aperture efficiency is introduced for the purpose of comparing and optimizing the performance of impulse radiating antennas (IRAs). The aperture efficiencies of popular lens and reflector IRAs are computed as the ratios of peak radiated power densities on boresight compared with that produced by an ideal IRA with an aperture of equal area and equal total input power. Loss of aperture efficiency occurs through two distinct mechanisms: from power that falls outside the aperture and is lost; and from nonuniform power and polarization distributions within the aperture. Both loss mechanisms are addressed, and means for increasing efficiencies are identified. Aperture efficiencies approaching 100 % are feasible in TEM-horn arrays and similar structures.

APERTURE EFFICIENCY

Impulse radiating antennas are a class of focused aperture antenna that uses a lens or parabolic reflector to convert an outgoing inhomogeneous spherical TEM-wave, propagating on a conically symmetric feedline structure, into a plane wave. For such focused aperture systems, the radiated field on boresight at distance r in the far-field limit is given by [1]

$$\vec{E}_{rad}(r,t) = \frac{1}{2\pi rc}\frac{d}{dt}\iint_A \vec{E}(x,y,t-r/c)\,dx\,dy, \tag{1}$$

where $c = 1/\sqrt{\mu\varepsilon}$ is the wave velocity in the external medium, and the surface integral over the field distribution $\vec{E}(x,y,t)$ covers the planar aperture of area A lying just outside the antenna in the external medium. A well known property of focused aperture antennas is that they achieve the highest gain when the aperture is uniformly illuminated with uniform polarization [2]. Thus, the aperture of our ideal reference antenna is illuminated with a uniform field $\vec{E_0}$, and Eq. 1 for this antenna becomes

$$\vec{E}_{rad^*}(r,t) = \frac{A}{2\pi rc}\frac{d}{dt}\vec{E_0}(t-r/c)\,. \tag{2}$$

Ultra-Wideband, Short-Pulse Electromagnetics 4
Edited by Heyman *et al.*, Kluwer Academic / Plenum Publishers, New York, 1999.

To make a proper comparison between test and reference antennas, both must have certain parameters in common. They must share the same external medium or media with the same wave impedance Z_{med} and wave velocity c, the distance r to the point of measurement must be the same, they must have the same aperture area A, and they must have the same input drive waveform $f(t)$ and input drive power $P_{in}(t)$. The total antenna input power for the test antenna described by Eq. 1 is related to the input voltage $V_{in}(t)$ and feedline impedance Z_{line} by

$$P_{in}(t) = \frac{V_{in}^2(t)}{Z_{line}}. \tag{3}$$

The total input power for the ideal reference antenna is taken to be the power incident upon the aperture. The reference-antenna input power is then

$$P_{in^*}(t) = A\frac{E_0^2(t)}{Z_{med}}, \text{ with } Z_{med} = \sqrt{\frac{\mu}{\varepsilon}}. \tag{4}$$

For both antennas to be driven by the same waveforms, all pertinent quantities must be proportional to the same function of time $f(t)$, that is

$$V_{in}(t) = V_0 f(t), \ \vec{E_0}(t) = \vec{E_0} f(t), \text{ and} \tag{5}$$
$$\iint \vec{E}(x,y,t)\,dx\,dy = \iint \vec{E}(x,y)\,dx\,dy\,f(t) \ \text{ for } t << \sqrt{A}/c.$$

Equation 5 holds only for early times, before edge effects can propagate inward and modify the aperture fields. What lies outside the aperture is critical to what occurs later in time. To obtain consistent results, we are restricted to comparing only the prompt responses of different antennas.

We define the prompt aperture efficiency η_A to be the ratio of radiated power density on boresight due to the test antenna to that of an equivalent ideal reference antenna with the parameter constraints described above. Equating Eqs. 3 and 4 and combining Eqs. 1–5, the prompt aperture efficiency then becomes

$$\eta_A = \frac{E_{rad}^2(r,t)}{E_{rad^*}^2(r,t)} = \frac{1}{A}\frac{Z_{line}}{Z_{med}}\left[\frac{1}{V_{in}}\iint_A E_y(x,y)\,dx\,dy\right]^2, \tag{6}$$

where, without loss of generality, the y-axis is chosen parallel to the principal component of polarization, and $E_y(x,y)$ is the principal component of the prompt aperture field distribution for an applied antenna input voltage V_{in}. Another convenient way of expressing Eq. 6 is

$$\eta_A = \frac{A}{V_{in}^2}\frac{Z_{line}}{Z_{med}}\langle E_y\rangle^2 = \left[\frac{A\left\langle E_y^2\right\rangle}{Z_{med}}\Big/\frac{V_{in}^2}{Z_{line}}\right]\left[\frac{\langle E_y\rangle^2}{\left\langle E_y^2\right\rangle}\right], \text{ with } \left\langle E_y^k\right\rangle = \frac{1}{A}\iint_A E_y^k(x,y)\,dx\,dy, \tag{7}$$

where $\left\langle E_y^k\right\rangle$ is the average value of the k'th power of the y-component of the E-field over the aperture. The first term in brackets in Eq. 7 is exactly equal to the fraction of the total input power passing through the aperture in the y-component of the field. The second term in brackets in Eq. 7, which is always less than or equal to one by Schwarz's inequality, is due solely to the inhomogeneity in the distribution of the y-component of the E-field within the aperture. This second term is unity only for a uniform E-field distribution.

For apertures and feed electrodes with a well defined width W, as shown in Fig. 1, the prompt aperture efficiency simply reduces to

$$\eta_A = \frac{1}{A}\frac{Z_{line}}{Z_{med}}\left[\int_0^W dx\frac{1}{V_{in}}\int_{y_1(x)}^{y_2(x)} dy\,E_y(x,y)\right]^2 = \frac{1}{A}\frac{Z_{line}}{Z_{med}}\left[\int_0^W dx\right]^2 = \frac{W^2}{A}\frac{Z_{line}}{Z_{med}}. \tag{8}$$

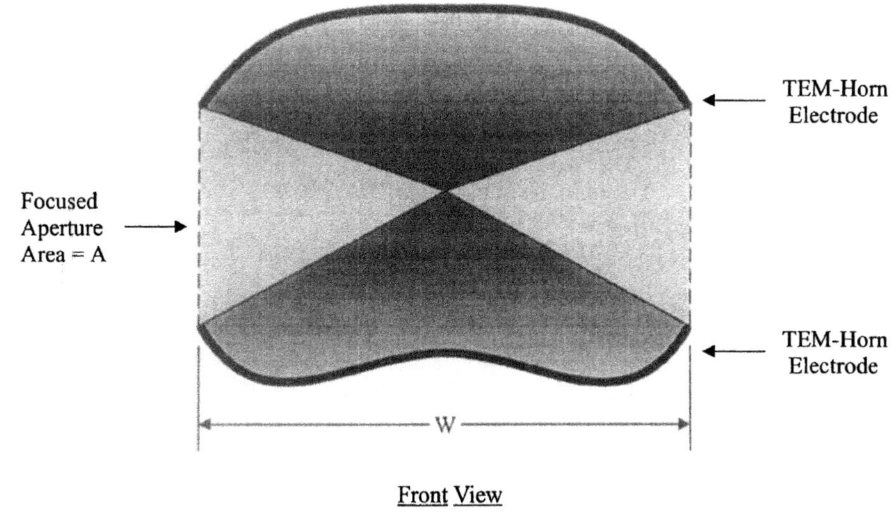

Transmission-Line Impedance = Z_{line}

Wave or Medium Impedance = Z_{med}

Aperture Efficiency: $\eta_A = [W^2/A]\,[Z_{line} / Z_{med}]$

Transmit Response: $E_{rad} = W\,[2\pi rc]^{-1}\,[d/dt]\,V_{in}$

Receive Response: $V_{out} = W\,[Z_{line} / Z_{med}]\,E_{rec}$

Restricted to Early Times: $A/W/c >> t << W/c$

Figure 1. Aperture geometry, prompt aperture efficiency, transmitting response, and receiving response of focused aperture antennas with a well defined feed-electrode and aperture width *W*.

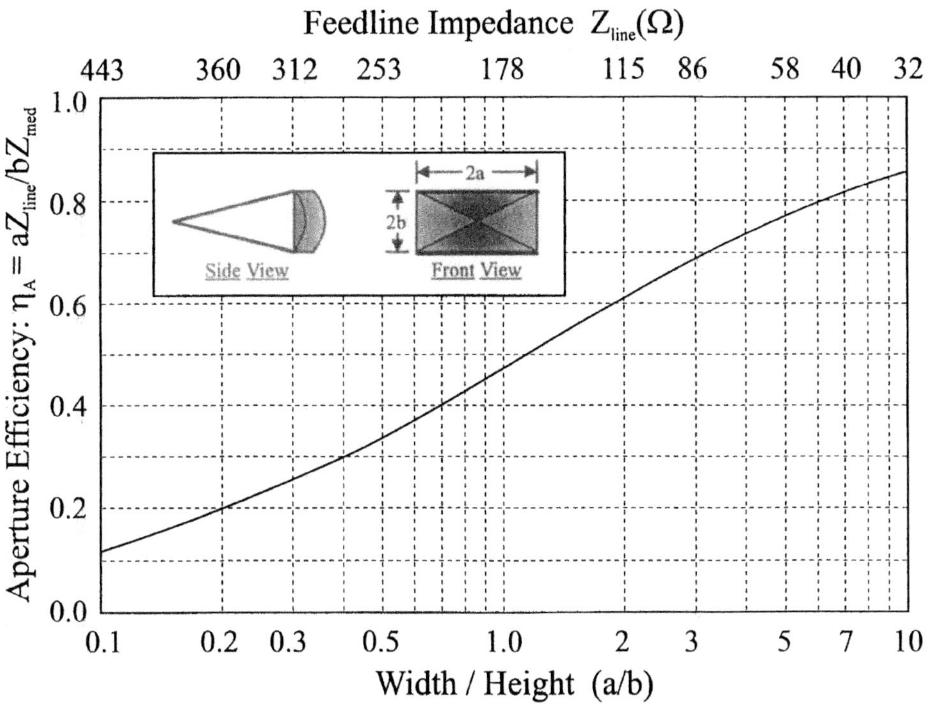

Figure 2. Prompt aperture efficiency versus impedance and aspect ratio of flat-plate horns with narrow rectangular focused apertures. Low-impedance horns yield the highest aperture efficiencies.

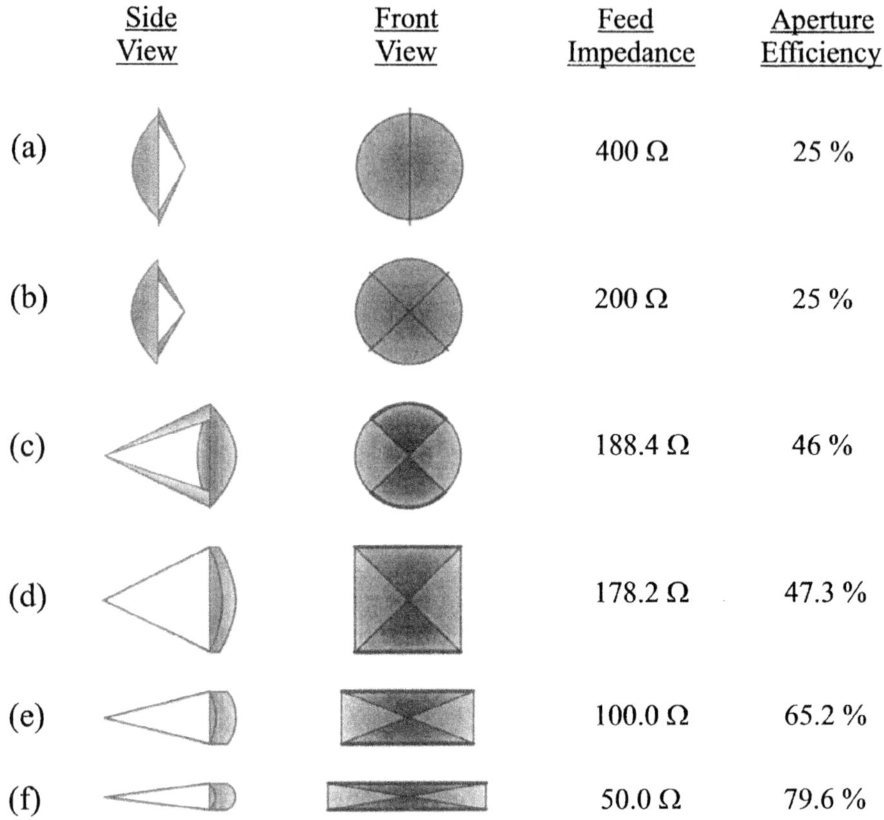

Figure 3. Prompt aperture efficiencies and feedline impedances of reflector and lens IRAs. Antennas are (a) single-coplanar- and (b) crossed-coplanar-feedline reflector IRAs, and (c) 90°-circular-conical- and (d), (e), and (f) parallel-flat-plate-feedline lens IRAs.

In this simple expression, all complexity is concealed in the determination of the transmission line impedance Z_{line}. Figure 2 shows a plot of the aperture efficiency versus feedline impedance Z_{line} and aspect ratio a/b for flat-plate TEM horns with rectangular focused apertures bounded by their electrode dimensions. We see that the lower-impedance planar TEM horns have the highest aperture efficiencies and will therefore produce the highest prompt radiated fields for a given input power and aperture area. This result is understandable because the low-impedance planar TEM horns have the most uniform fields within their apertures and the lowest field energies falling outside their apertures. Prompt aperture efficiency approaches 100 % as $a/b \to \infty$ and $Z_{line} \to 0$. However, as we will see in the next section, a close-fitting rectangular aperture is not the optimal focused aperture for a planar TEM horn.

Figure 3 shows computed aperture efficiencies and feed impedances for several reflector and lens IRAs. For circular-aperture antennas, η_A ranges from 25 % for reflector IRAs, with a single or crossed pair of 400 Ω feed arms, neglecting feed blockage, to 46 % for a lens IRA with an optimal 188-Ω circular-conical feed line. For rectangular-aperture lens antennas, η_A increases with increasing aspect ratio, reaching 80 % for a planar 50-Ω feed system with $a/b = 6$. Therefore, when it is important to achieve the highest prompt aperture efficiency, a low-impedance lens IRA is preferable to a high-impedance reflector IRA.

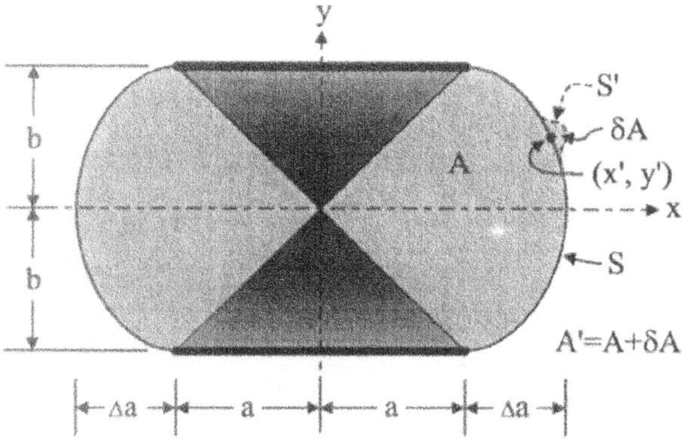

Figure 4. Frontal view of a focused-aperture TEM horn with optimal aperture of area A, showing a small perturbation in area δA at the boundary point (x', y').

OPTIMAL APERTURES

The question one might ask is "What is the most efficient focused aperture for a given feed structure?" We first imagine that such an optimal aperture exists with area A and boundary path S. We then add a small incremental area δA at some point $\left(x', y'\right)$ along S to form a new boundary path S' enclosing an area A', as shown in Fig. 4. The new aperture efficiency $\eta_{A'}$ for the aperture with boundary S' and area A' then becomes

$$\begin{aligned}
\eta_{A'} &= \eta_A + \delta\eta_A \\
&= \frac{1}{A'}\frac{Z_{line}}{Z_{med}}\left[\frac{1}{V_{in}}\iint_{A'} E_y(x,y)\,dx\,dy\right]^2 \\
&= \frac{1}{(A+\delta A)}\frac{1}{V_{in}^2}\frac{Z_{line}}{Z_{med}}\left[\iint_A E_y(x,y)\,dx\,dy + E_y(x',y')\,\delta A\right]^2 \\
&= \frac{A}{V_{in}^2}\frac{Z_{line}}{Z_{med}}\left[\langle E_y\rangle + E_y(x',y')\frac{\delta A}{A}\right]^2\left[1+\frac{\delta A}{A}\right]^{-1} \\
&= \frac{A}{V_{in}^2}\frac{Z_{line}}{Z_{med}}\langle E_y\rangle^2\left[1+2\frac{E_y(x',y')}{\langle E_y\rangle}\frac{\delta A}{A}+\frac{E_y^2(x',y')}{\langle E_y\rangle^2}\frac{\delta A^2}{A^2}\right]\left[1-\frac{\delta A}{A}+O\left(\frac{\delta A^2}{A^2}\right)\right] \\
&= \eta_A\left[1+2\frac{E_y(x',y')}{\langle E_y\rangle}\frac{\delta A}{A}-\frac{\delta A}{A}+O\left(\frac{\delta A^2}{A^2}\right)\right].
\end{aligned} \tag{9}$$

If the original aperture were optimal, the aperture efficiency must be stationary for any incremental alteration in area at the boundary. Therefore, as $\delta A \to 0$ we have

$$\delta\eta_A = \eta_{A'} - \eta_A \to \eta_A\left[2\frac{E_y(x',y')}{\langle E_y\rangle} - 1\right]\frac{\delta A}{A} = 0. \tag{10}$$

The above expression can hold only if the principal field components E_y at all points (x', y') on the boundary S obey the relationship

$$E_y(x',y') = \frac{1}{2}\langle E_y\rangle. \tag{11}$$

Thus, *for maximum aperture efficiency, the magnitude of the principal field component on the aperture boundary must be everywhere exactly half of the average of this field component within the aperture.* If the optimal aperture consists of several disconnected

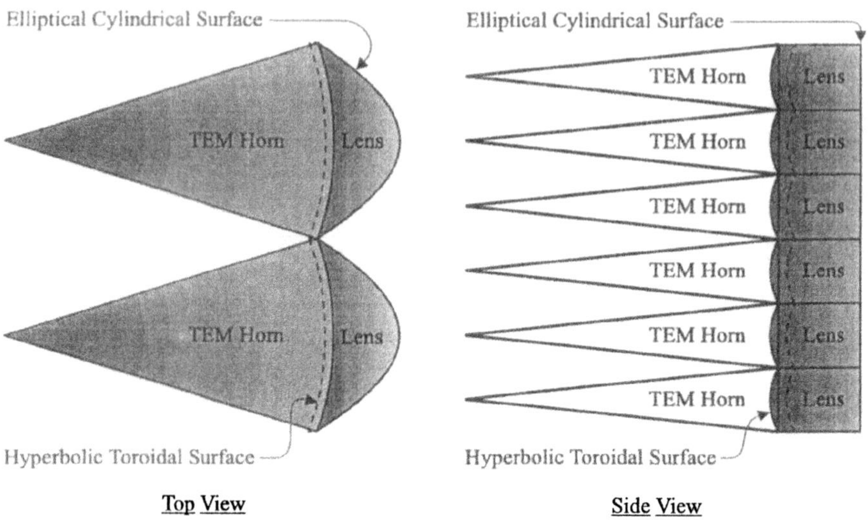

Figure 5. Conceptual design of a dense array of focused-aperture TEM horns that have relatively low impedances and high aperture efficiencies: $Z_{line} = 86\,\Omega$, and $\eta_A = 68\,\%$. All of the lens surfaces are analytic curves. For fixed total aperture area and antenna-element shape, the total lens mass varies inversely with the number of antenna elements.

zones, then Eq. 11 applies to each zone with the average field and area applying to all the zones collectively.

The simplicity of Eq. 11 for the optimal aperture does not reflect the computational difficulties that may be encountered in trying to determine it. In general, most solutions will be found only by delicate numerical methods. One important class of problems is treated in this work: the flat-plate TEM-horn focused aperture. Although low-impedance lens IRAs achieve the highest aperture efficiencies, the lens weight can be prohibitive. By arranging smaller lens IRAs in dense arrays, the weight of lens material may be greatly reduced or even eliminated. For fixed total aperture area and antenna-element shape, the total lens mass varies inversely with the number of antenna elements. Figure 5 shows one possible arrangement for a dense TEM-horn-lens array. The choice of horn impedances may be limited by the source impedances and other considerations, thus limiting the aperture efficiency. However, the best arrangement for horn placement and aperture construction within the array are yet to be determined. Figure 6 shows two different lattice arrangements for array construction: one with a rectangular and the other with a hexagonal unit cell and aperture shape. Note that the aperture areas of $6ab$ are 50 % greater than the areas $4ab$ within the electrode boundaries. In fact, this arrangement, with the approximate dimensions shown in Fig. 6, produces the highest aperture efficiencies for array elements with $a/b = 1$. The aperture efficiency increases from 47.3 % for dense packing, with $\Delta a = 0$, to about 54.6 % for the less dense packing shown in Fig. 6, with $\Delta a \cong b/2$ for the rectangular aperture and $\Delta a \cong b$ for the hexagonal aperture.

Numerical and analytical methods described in [3] were used to derive the optimal focused apertures for flat-plate TEM horns. For a wide range of electrode aspect ratios, the aperture efficiencies and dimensions were determined for the best curved, rectangular, and hexagonal apertures. Figure 7 shows the three optimal apertures for an electrode aspect ratio of one ($a/b = 1$). The aperture efficiencies for the optimal curved, rectangular, and hexagonal apertures were 55.3 %, 54.5 %, and 54.7 %, respectively. These aperture efficiencies are nearly the same but are noticeably larger than the 47.3 % value for the close-fitting rectangular aperture. Little is lost by choosing either the

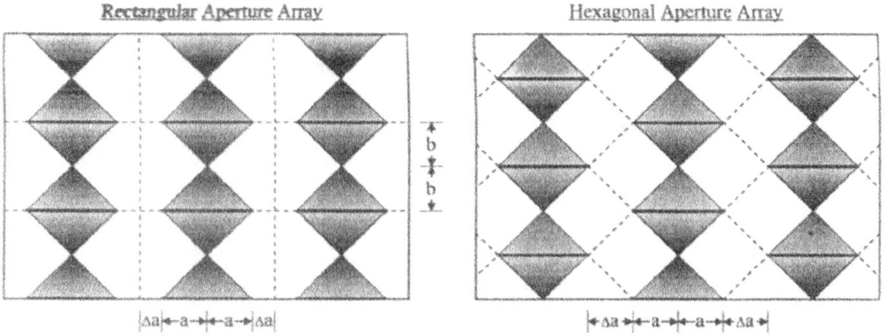

Figure 6. Front views of flat-plate TEM-horn arrays employing rectangular and hexagonal unit cells. For $a/b = 1$, the aperture efficiencies increase from 47.3 % for densely packed arrays with $\Delta a/b = 0$, to about 54.6 % for the less densely packed arrays shown above, with $\Delta a/b \cong 1/2$ for the rectangular apertures and $\Delta a/b \cong 1$ for the hexagonal apertures.

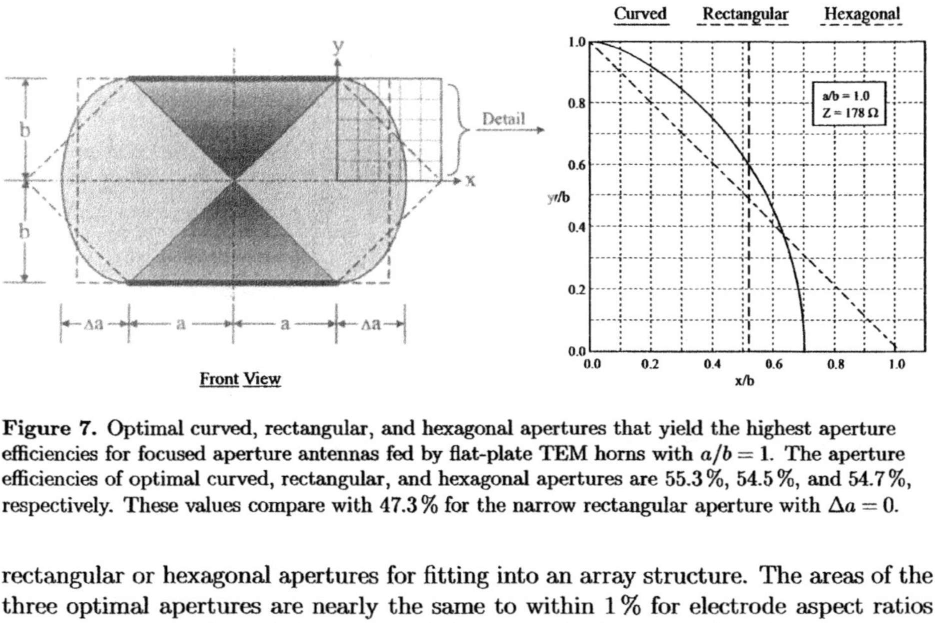

Figure 7. Optimal curved, rectangular, and hexagonal apertures that yield the highest aperture efficiencies for focused aperture antennas fed by flat-plate TEM horns with $a/b = 1$. The aperture efficiencies of optimal curved, rectangular, and hexagonal apertures are 55.3 %, 54.5 %, and 54.7 %, respectively. These values compare with 47.3 % for the narrow rectangular aperture with $\Delta a = 0$.

rectangular or hexagonal apertures for fitting into an array structure. The areas of the three optimal apertures are nearly the same to within 1 % for electrode aspect ratios of one or greater. This similarity results from the ideal aperture boundary lying on a curve of constant E_y.

Optimal curved apertures covering the entire range of electrode aspect ratios are shown in Fig. 8. All curves contact the electrodes horizontally. Curves for limiting cases were derived analytically and found to lie on field lines. These curves are given by

$$y = \frac{b}{\pi}\left[\arccos\left(\frac{\pi x}{b} - 1\right) + \sqrt{1 - \left(\frac{\pi x}{b} - 1\right)^2}\right] \text{ for } a/b \to \infty, \text{ and}$$
$$y = \sqrt{b^2 - x^2} \text{ for } a/b \to 0. \tag{12}$$

The intermediate cases were found by numerical methods described in [3]. Although the curves for these intermediate cases appear to lie very close to field lines, the correspondence has been shown to be inexact [4].

The aperture efficiency $\eta_A\,(a, b, \Delta a = 0)$ of flat-plate TEM horns with narrow rec-

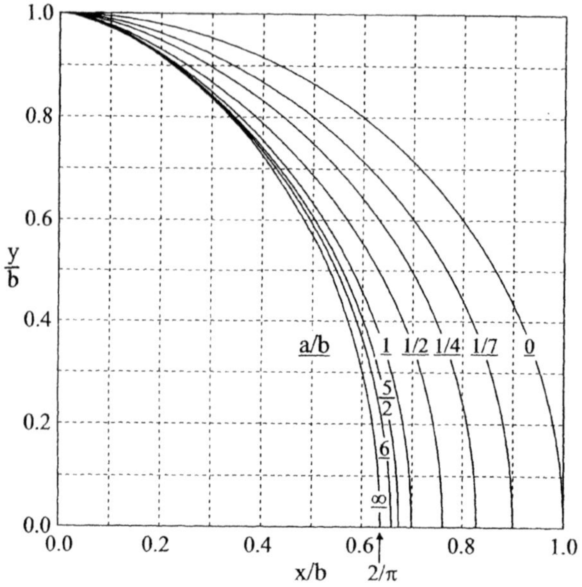

Figure 8. Optimal curved apertures for flat-plate TEM horns with different aspect ratios a/b, and transmission-line impedances Z_{line}. Parameter sets are $(a/b, Z_{line}) = (0, \infty\,\Omega)$; $(1/7, 400\,\Omega)$; $(1/4, 333\,\Omega)$; $(0.5, 253\,\Omega)$; $(1.0, 178\,\Omega)$; $(2.5, 100\,\Omega)$; $(6.0, 50\,\Omega)$; and $(\infty, 0\,\Omega)$.

tangular focused apertures bounded by their electrode dimensions is chosen as a reference function for comparison purposes. This function of a/b, which was shown in Fig. 2, may be computed from Eq. 8 and the transmission-line impedance Z_{line}. The transmission-line impedance is a rather tedious function involving incomplete elliptic integrals and their inverses, and it is not easily calculated [5]. We approximate this aperture efficiency with the sigmoidal function

$$\eta_A(a,b,0) = \frac{a}{b}\frac{Z_{line}}{Z_{med}} = \frac{1}{1+\mu_\eta\,(a/b)^{-k_\eta}} + \Delta\eta, \text{ with} \tag{13}$$

$$\mu_\eta = 1.1158,\ k_\eta = 0.8300, \text{ and } \left|\frac{\Delta\eta}{\eta}\right| \leq 0.005, \text{ for } 0.01 \leq a/b \leq 10.$$

Figure 9 presents numerically computed values of aperture efficiency $\eta_A(a,b,\Delta a)$ versus aspect ratio a/b for the three optimal aperture types (using the optimal values of $\Delta a/b$) as well as $\eta_A(a,b,0)$ for the narrow rectangular aperture. In this plot, numerically computed values of $\eta_A(a,b,0)$ are used instead of the values derived from Eq. 8 or Eq. 13. This is done, for the purpose of comparison, to equalize any bias caused by the numerical computational process, such as the finite plate spacing to thickness ratio, which was 60 or greater. The curves are least-squares fits to the numerically

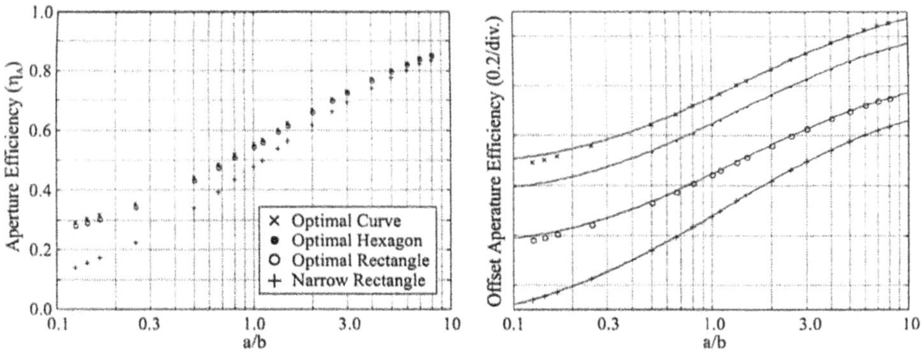

Figure 9. Plots of numerically computed values of aperture efficiency $\eta_A(a,b,\Delta a)$ versus aspect ratio a/b using the optimal values of $\Delta a(a/b)$ from Fig. 10, for the three aperture types, rectangular, hexagonal, and curved, and for the narrow rectangular aperture with $\Delta a = 0$. The plot on the right shows the same data sets vertically offset along with the solid curves given by Eqs. 13 and 14 that best fit the data.

derived data points. The fitted curve for $\eta_A(a,b,0)$ has the form given by Eq. 13 with slightly different coefficients ($\mu_\eta = 1.0938$ and $k_\eta = 0.8289$). The functional forms of the $\eta_A(a,b,\Delta a)$ curves are given by

$$\eta_A(a,b,\Delta a) = \left[1 + \frac{\nu_\eta}{(a/b)}\right] \eta_A(a,b,0), \tag{14}$$

where the single parameter ν_η was determined by a least-squares fit to the data using the numerically computed values of $\eta_A(a,b,0)$. The values of ν_η range from 0.14 to 0.16 for the three optimal apertures. The curves fit the data well with a maximum deviation of ± 0.0035 except for the three end points at very small a/b where the accuracy of the numerical method is suspect.

Figure 10 presents plots of computed optimal aperture edge widths Δa versus aspect ratio a/b for the three optimal aperture types. The curves are best fits to the data using sigmoidal functions of the form

$$\log[\Delta a(a,b)] = \frac{\log[\Delta a(0,b)]}{1 + u_\Delta (a/b)^{k_\Delta}} + \frac{\log[\Delta a(\infty,b)]}{1 + u_\Delta (a/b)^{-k_\Delta}}. \tag{15}$$

The parameters u_Δ and k_Δ are chosen to minimize the mean-square deviation of the data, whereas the asymptotic quantities $\Delta a(0,b)$ and $\Delta a(\infty,b)$ are derived. The absolute accuracy of the fit to the data is about $\pm 1.5\,\%$. Equation 12 for the ideal curved aperture yields $\Delta a(0,b)/b = 1$ and $\Delta a(\infty,b)/b = 2/\pi$. For large a/b, the areas of the edge regions shown in Figs. 6–8 are the same to within $1\,\%$ of that of the ideal curved aperture, which as $a/b \to \infty$ becomes $A(x \geq 0) = 3b^2/\pi$. Assuming an exact equality in this limit provides an estimate of $\Delta a(\infty,b)/b = 3/(2\pi)$ for the optimal rectangle and $\Delta a(\infty,b)/b = 3/\pi$ for the optimal hexagon. Values of $\Delta a(0,b)$ for the optimal rectangular and hexagonal apertures are obtained analytically for the limiting case as η_A and $a/b \to 0$. For the rectangular aperture as $a/b \to 0$, the leading term in the aperture efficiency is

$$\eta(a,b,z = \Delta a/b) \to \left[8\pi z \log\left(\frac{2b}{a}\right)\right]^{-1} \left[z \log\left(1 + \frac{4}{z^2}\right) + 4\arctan\left(\frac{z}{2}\right)\right]^2, \tag{16}$$

which has a maximum at $\Delta a/b = z = 0.8422$, where z is the solution to the equation

$$4\arctan(z/2) = z\log\left(1 + 4/z^2\right). \tag{17}$$

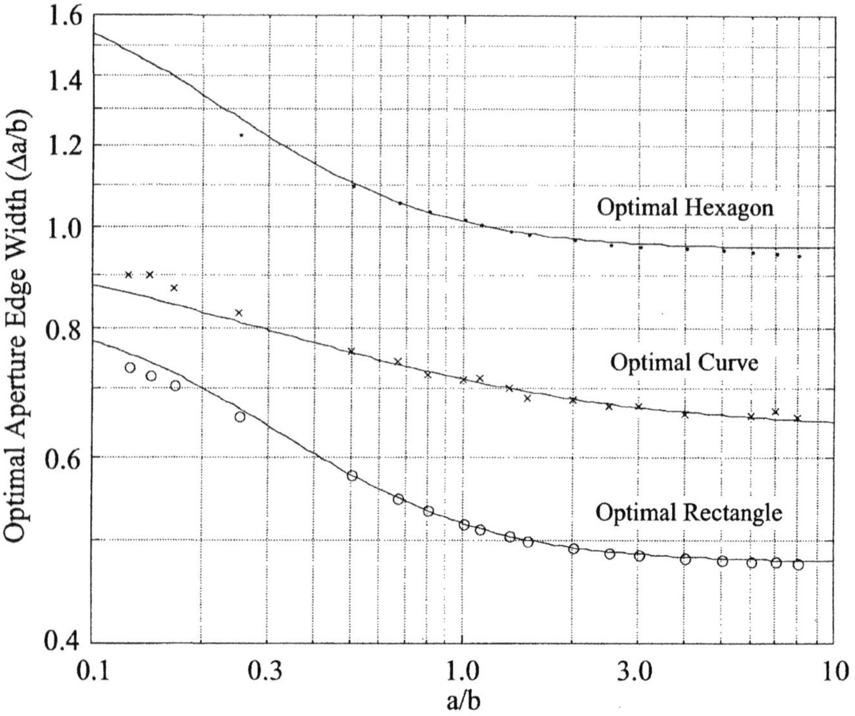

Figure 10. Plots of numerically computed values of the edge-width parameter $\Delta(a/b)/b$ versus aspect ratio a/b that yield the maximum aperture efficiencies $\eta_A\,(a, b, \Delta a)$ for the three aperture types, rectangular, hexagonal, and curved. The solid curves are the functions given by Eq. 15 that best fit the data.

For the hexagonal aperture as $a/b \to 0$, the leading term in the aperture efficiency is

$$\eta\,(a, b, z = \Delta a/b) \to \left[2\pi z \log\left(\frac{2b}{a}\right)\right]^{-1} \left[\left(\frac{z}{1+z^2}\right)\left(\log\left(\frac{4}{1+z^2}\right) + 2z\arctan(z)\right)\right]^2, \tag{18}$$

which has a maximum at $\Delta a/b = z = \sqrt{3}$, where z is the solution to the equation

$$2z\left(z^2 - 3\right)\arctan(z) = \left(3z^2 - 1\right)\log\left[\left(1 + z^2\right)/4\right]. \tag{19}$$

The parameters used to characterize the computed data are summarized in Table 1.

	$\Delta a(0,b)/b$	$\Delta a(\infty,b)/b$	u_η	k_η	ν_η	u_Δ	k_Δ
Narrow Rectangle*	0*	0*	1.1158*	0.8300*	-	-	-
Narrow Rectangle	0*	0*	1.0938	0.8289	-	-	-
Optimal Rectangle	0.8422*	$\sqrt{3}/(2\pi)$*	1.0938	0.8289	0.1411	0.1799	1.4039
Optimal Hexagon	$\sqrt{3}$*	$\sqrt{3}/\pi$*	1.0938	0.8289	0.1453	0.1136	1.3907
Optimal Curve	1*	$2/\pi$*	1.0938	0.8289	0.1568	0.3532	1.0395

Table 1. Parameter values used in Eqs. 13–15 to fit the numerically computed values of η_A and Δa except where (*) denotes quantities derived from or fit to an analytic function or a derived value.

Figure 11 shows curves of $\eta_A\,(a, b, \Delta a)$ normalized to unity peak value versus $\Delta a/b$ for rectangular and hexagonal apertures. Curves generated by numerical computation

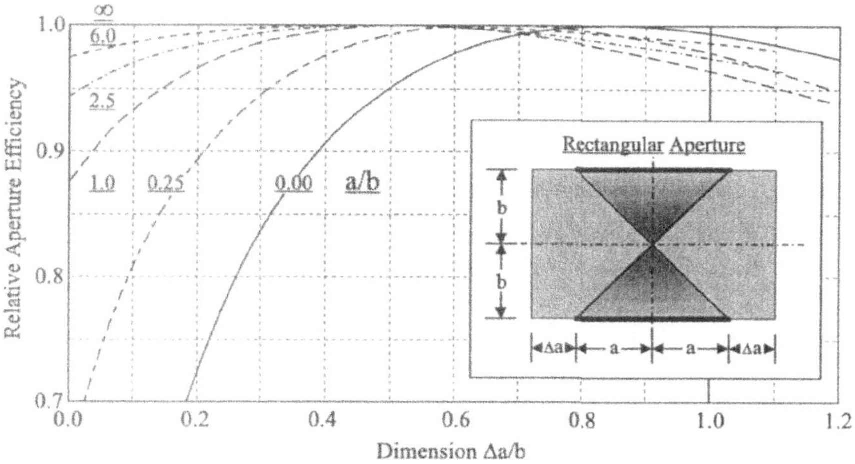

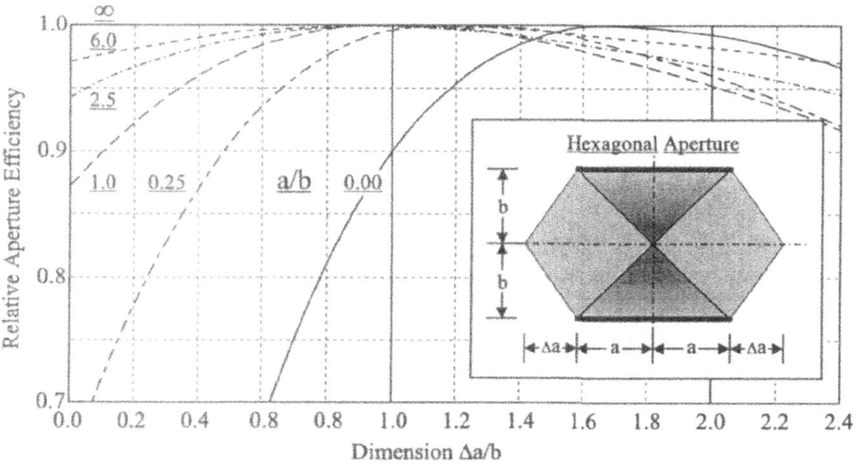

Figure 11. Relative aperture efficiencies of rectangular apertures (top) and hexagonal apertures (bottom) versus edge-width parameter $\Delta a/b$ for flat-plate TEM horns with different aspect ratios a/b and transmission-line impedances Z_{line}. Parameter sets are $(a/b, Z_{line}) = (0, \infty\,\Omega)$; $(1/4, 333\,\Omega)$; $(1.0, 178\,\Omega)$; $(2.5, 100\,\Omega)$; $(6.0, 50\,\Omega)$; and $(\infty, 0\,\Omega)$.

cover the important impedance range from 50 to 300 Ω as well as the limiting cases as $a/b \to 0$ from Eqs. 16–19 and as $a/b \to \infty$. The aperture efficiency curves have rather broad peaks. The improvement in aperture efficiency achieved by optimizing Δa is greatest at the higher impedances, as might be expected. However, for the important case where $a/b = 1$, optimizing Δa increases the relative aperture efficiency by 15.4 % and the absolute aperture efficiency from 47.3 % to 54.6 %. Further increases in are possible using methods described in later sections.

ISOREFRACTIVE MEDIA

So far we have seen that energy that falls outside the aperture is lost and can be reduced only by using lower impedance feed structures. There is another method to reduce this energy loss that utilizes isorefractive media. Isorefractive media will have

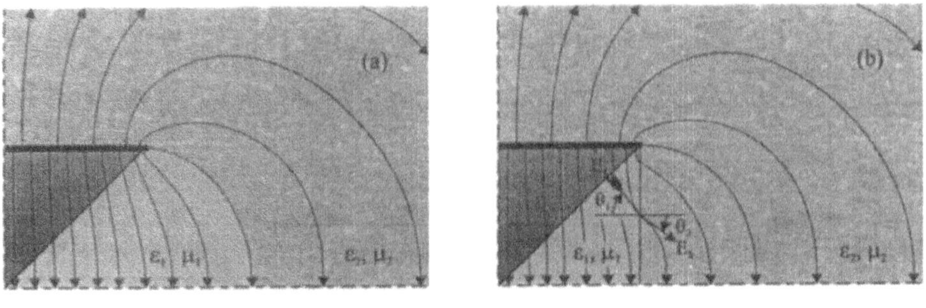

Figure 12. Single-quadrant field-line maps of the image plane of focused-aperture TEM horns immersed in isorefractive media, when the boundary between the two media (a) does or (b) does not lie on a field line of the single-medium E-field distribution. In the latter case, the boundary fields satisfy the relations: $E_1 \sin\theta_1 = E_2 \sin\theta_2$ and $H_1 \cos\theta_1 = H_2 \cos\theta_2$, where $Z_1 \tan\theta_1 = Z_2 \tan\theta_2$. TEM modes will propagate parallel to the dielectric boundary if the isorefractive condition is met, i.e. $c_1 = 1/\sqrt{\mu_1\varepsilon_1} = 1/\sqrt{\mu_2\varepsilon_2} = c_2$. Aperture efficiency is enhanced if the space impedance of medium 1 is less than that of medium 2, i.e. $Z_1 = \sqrt{\mu_1/\varepsilon_1} < \sqrt{\mu_2/\varepsilon_2} = Z_2$.

the same wave velocity but different wave impedances. For example

$$c_1 = \frac{1}{\sqrt{\mu_1\varepsilon_1}} = \frac{1}{\sqrt{\mu_2\varepsilon_2}} = c_2, \text{ and } Z_{med1} = \sqrt{\frac{\mu_1}{\varepsilon_1}} < \sqrt{\frac{\mu_2}{\varepsilon_2}} = Z_{med2}. \tag{20}$$

By using medium 1 in the region that illuminates the aperture and medium 2 in the region that is blocked by the aperture, the proportion of the energy blocked by the aperture is reduced. If the aperture boundary lies on E-field lines of the single-medium feedline structure shown in Fig. 12, the introduction of medium 2 will leave the E-fields unchanged but the H-fields reduced in medium 2. Thus, without altering the fields in medium 1, the power flow $\vec{P} = \vec{E} \times \vec{H}$ in medium 2 is everywhere reduced. The surface integral in Eq. 6 for η_A will be unchanged, but the transmission line impedance $Z_{line}(Z_1)$ will increase to $Z_{line}(Z_1, Z_2)$, and the dual-medium aperture efficiency $\eta_A(Z_1, Z_2)$ will be related to the single-medium aperture efficiency η_A by

$$\eta_A(Z_1, Z_2) = \frac{Z_{line}(Z_1, Z_2)}{Z_{line}(Z_1)}\eta_A. \tag{21}$$

That such a structure is nondispersive and can support TEM-modes is shown in [3], and the general requirements for TEM-mode propagation in inhomogeneous media are given in [6].

The circular-conical lens IRAs shown in Fig. 13 is often used in ultra-wideband applications. The natural aperture for this system is the circle at the edge of the lens and circular conical electrodes, which is also the boundary of self-reciprocal symmetry, which always lies on a field line [7]. Also, for such self-reciprocal antennas, exactly half of the field energy lies outside of the circular aperture. Furthermore, the charge densities on the inner and outer surfaces of the thin electrodes are identical, and the inside and outside surfaces contribute equally to the transmission-line impedance. The aperture efficiency of this system may be increased using isorefractive media, with the higher impedance medium lying outside the circular aperture. The new aperture efficiency may be derived by inspection using Eq. 21, once the single-medium aperture efficiency is known.

When θ_0 is small, the prompt fields in the aperture can be approximated by the fields on a parallel transmission line with the same cross section. For an infinitely long transmission line, the three-dimensional wave equation reduces to the two-dimensional Laplace's equation in the transverse plane, and the resulting fields can be obtained by

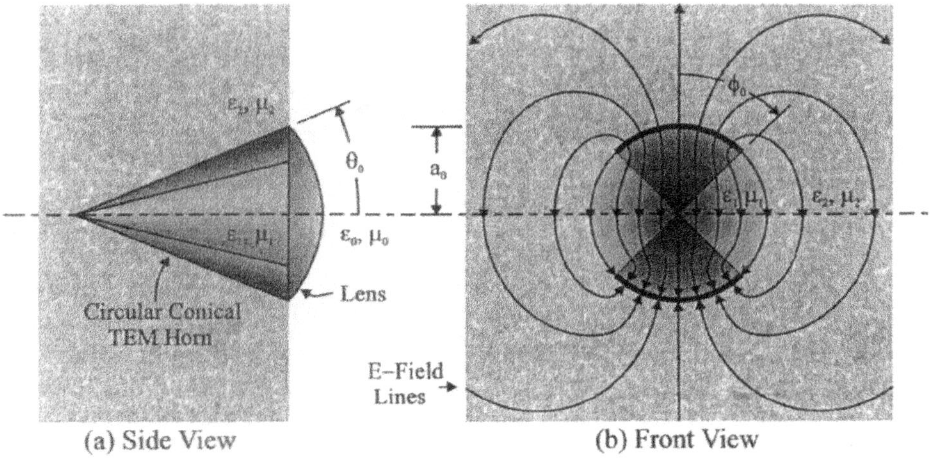

Figure 13. Circular conical lens IRA employing isorefractive media. The boundary between medium 1 and 2 is the circular cone containing the electrodes and the edge of the circular aperture at the edge of the lens. E-field lines are shown in the aperture plane.

conformal transformation. The characteristic impedance of this transmission line in a single medium is [8]

$$Z_{line}\left(Z_{med}, m\right) = \frac{K\left(m\right)}{K\left(1-m\right)} Z_{med}, \tag{22}$$

and the aperture efficiency for the circular aperture is [9]

$$\eta_A\left(m\right) = \pi \frac{Z_{line}}{Z_{med}} \left[\left(1+\sqrt{m}\right) K\left(m\right)\right]^{-2} = \pi \left[\left(1+\sqrt{m}\right)^2 K\left(m\right) K\left(1-m\right)\right]^{-1} \tag{23}$$

$$\text{with} \tan\phi_0 = \frac{1-\sqrt{m}}{2\,m^{1/4}}, \text{ or } m = \left[\sec\phi_0 - \tan\phi_0\right]^4, \tag{24}$$

where $K\left(m\right)$ is the complete elliptic integral of the first kind. Maximizing η_A with respect to m, the optimum electrode half-angle becomes $\phi_0\left(m\right) = 45^{\circ}$, and the optimum aperture efficiency is 46%. It was previously shown [8] that the aperture fields are most uniform for $\phi_0 = 45^{\circ}$, so it is understandable that this configuration should have the highest aperture efficiency. However, even with this optimum configuration, the aperture efficiency is less than half. Half of the aperture efficiency is lost because half of the power lies outside the aperture, and the remaining loss is due to the nonuniform field distribution within the aperture.

To improve the aperture efficiency, isorefractive media may be used to concentrate the transmission-line power within the circular aperture. The modified impedance of the TEM horn due to the presence of the isorefractive media is given by taking the impedances of the two halves of the transmission line described by Eq. 22 in parallel:

$$Z_{line}\left(Z_1, Z_2, m\right) = \left[\frac{1}{2\,Z_{line}\left(Z_1, m\right)} + \frac{1}{2Z_{line}\left(Z_2, m\right)}\right]^{-1} = \frac{K\left(m\right)}{K\left(1-m\right)} \left[\frac{2\,Z_1 Z_2}{Z_1+Z_2}\right]. \tag{25}$$

Generally the inner medium with impedance Z_1 will not be the same as free space with impedance Z_0. Graded-dielectric antireflection coatings may be applied to the lens boundary that will operate efficiently for times appropriate to the prompt aperture-efficiency concept. Under these lossless conditions, Eq. 21 applies, and using Eq. 25 in Eq. 21, the two-medium aperture efficiency becomes

$$\eta_A\left(Z_1, Z_2, m\right) = \frac{2\,Z_2}{Z_1+Z_2} \eta_A\left(m\right), \tag{26}$$

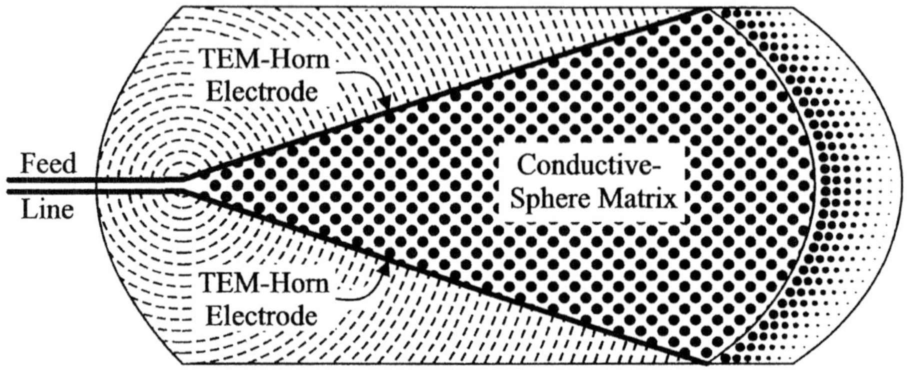

Figure 14. Cross-sectional view of a TEM-horn array element employing isorefractive media. The inner medium is a cubic matrix of conductive hollow spheres. The outer medium is a matrix of conductive disks or strips aligned parallel to both the E- and H-fields. The broadband antireflection layer is a graded-index matrix of conductive spheres with spatially varying radii.

where $\eta_A(m)$ is given by Eq. 23 for a single medium. We see that by increasing Z_2, much of the 50 % loss in aperture efficiency due to power lying outside the aperture can be recovered. For $\phi_0 = 45^o$, the limiting value of Eq. 26 is 92 %. The remaining 8 % loss is due to nonuniform aperture filling.

Finding low-loss, nondispersive materials with arbitrary impedance and wave velocity is nearly impossible. It is particularly difficult to increase μ without significant loss at microwave frequencies. However, synthetic dielectrics allow a limited degree of independent manipulation of ε and μ. For example, using synthetic dielectrics containing suspended conducting spheres, it is possible to decrease μ by a fractional factor of about half of that by which ε is increased [10]. Also, ε may be increased without changing μ using thin metal disks aligned parallel to both the E- and H-fields. For example, if medium 1 is constructed of metal spheres in a cubic matrix having a packing fraction of 0.32 (diameters equal to 85 % of the lattice spacing) with $\mu_1 = 0.584\mu_0$ and $\varepsilon_1 = 2.42\varepsilon_0$, and medium 2 is a matrix of aligned metal disks with $\mu_2 \cong \mu_0$ and $\varepsilon_2 = 1.42\varepsilon_0$, the media will be isorefractive with $Z_1 = 0.49Z_0$, $Z_2 = 0.84Z_0$, and $c_1 = c_2 = 0.84c$. Using these values in Eq. 26, the aperture efficiency increases from 46 % for a single-medium to 58 % for these synthetic media. Although the media properties are not optimized, this increase in aperture efficiency is significant, and the saving in lens weight would have a profound impact on the practicality of larger lens IRAs. Figure 14 shows a cross-sectional view of a TEM-horn array element using these metal-particle synthetic isorefractive media. A wideband graded-index antireflection layer is used to transition the impedance of medium one to that of free space. This is accomplished using a matrix of conductive spheres with spatially varying radii. The antireflecting behavior of this layer can be maintained for a time appropriate to the prompt-response time of the antenna elements.

One major problem remains, however. Synthetic media for time-domain applications must exhibit minimal dispersion. This can be accomplished by making the metal particles small. Smaller particles will necessarily have greater loss due to skin resistance. Whether an acceptable compromise between dispersion and loss can be found is unknown at this time. Unusual particle construction that employs special shapes and lumped-element electrical components may be required to achieve the desired results. Optimizing particle size, shape, density, composition, and lattice structure remains a topic for future work.

SEPTUM PLATES

The final factor affecting loss of aperture efficiency is field nonuniformity. We have seen that low-impedance TEM horns have superior field uniformity. Improved field uniformity may be achieved with higher impedance horns by using septum plates, as shown in Fig. 15(a). The septum plates divide the aperture into zones that act as low impedance horns, which are all electrically in series. However, these interstitial plates are passive, in that they are not actively driven. They originate near the throat of the horn or within the feedline where the transmission-line approximation is valid (i.e. where transmission-line dimensions are smaller than the shortest radian wavelength of interest) and extend to the aperture plane.

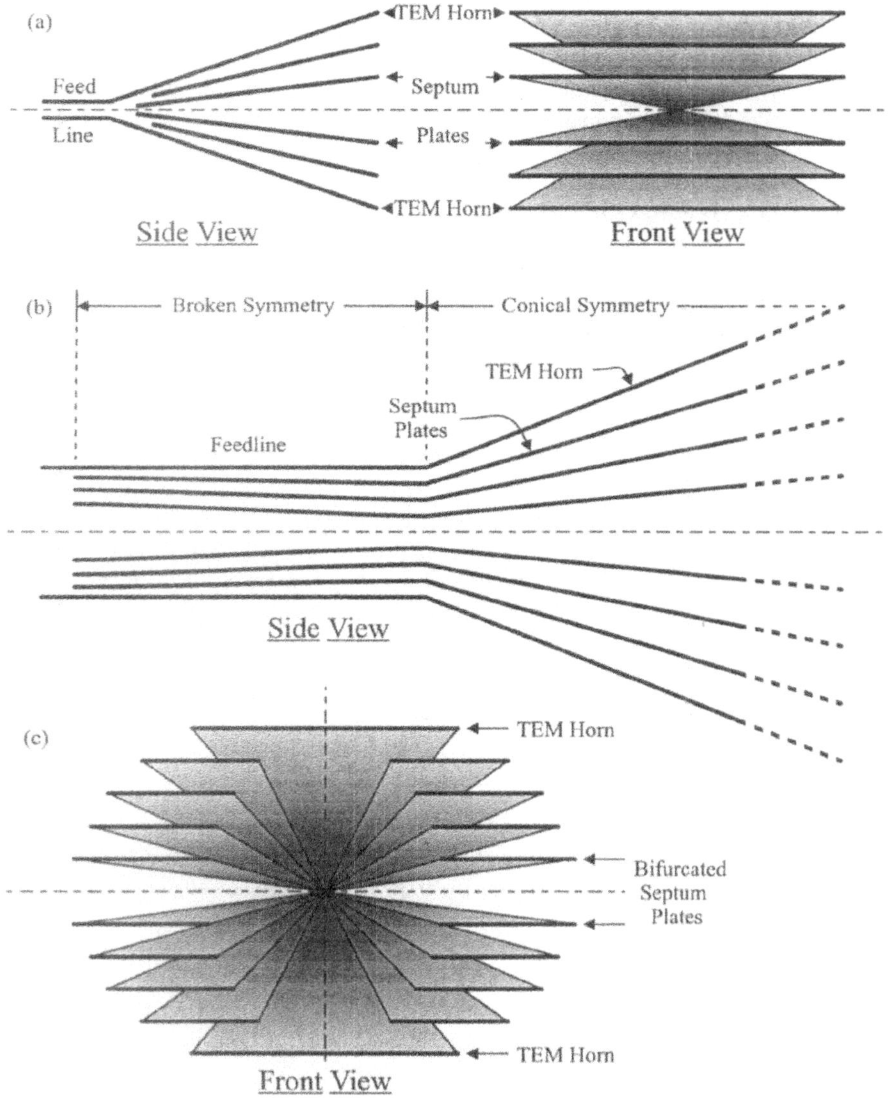

Figure 15. Conceptual designs of TEM horns using septum plates to improve field uniformity within the apertures. Examples include systems with (a) conically symmetric septum plates, (b) altered symmetry near feedlines to modify potential distributions, and (c) bifurcated septum plates.

If the antenna structure has strictly conical symmetry, the septum plates will carry no net charge near the aperture plane, but merely become transversely polarized in the plane in a way that makes the fields more uniform. Within each zone, the fields are more uniform, just as they are in similarly-shaped low-impedance horns. However, the field magnitudes in adjacent zones will differ in the higher impedance horns because the fields are more concentrated near the actively driven electrodes. This effect may be counteracted by allowing the plates to deviate from conical symmetry in a way that polarizes the plates longitudinally as well. This geometry alteration must occur in a region where the transmission-line approximation is still applicable, as shown in Fig. 15(b). On the plates near the aperture plane, a net charge will then appear that compensates for the natural field gradients near the driven electrodes. Because no charge density is needed to maintain uniform fields in the interior of the aperture, the septum plates may be omitted there, as shown in Fig. 15(c).

Similar methods may be used to accomplish the opposite result, i.e. to passively roll-off the fields near the aperture edge for controlling off-boresight radiation, an action that necessarily degrades aperture efficiency.

OTHER CONSIDERATIONS

There are other performance criteria used in the design of antennas and sensors for time-domain applications that can interfere with the achievement of high aperture efficiency. As mentioned earlier, control of off-boresight radiation is one. Other considerations are array-element interactions, reflected powers, resonances, and late-time responses. The aperture efficiency concept mentioned thus far pertains only to the prompt response. Controlling what follows this prompt response is a much greater task that will inevitably affect many aspects of the antenna designs. This subject is left for future work.

There is one important example of late-time response effects that has been studied. It is possible to construct E-field sensors that produce signals that precisely replicate the incident field for a finite clear time [11],[12]. They consist of TEM horns, possibly with lenses, to the front of which are added linear transmission-line extensions. The late-time aperture efficiency of these antennas can never exceed 25 %. To maintain a flat replicating response, it is not desirable that the prompt aperture efficiency be optimized in these devices. Precision replicating sensors of this type can never have high aperture efficiencies.

Antennas with high aperture efficiencies that are sustained over prolonged time scales may yet be possible to construct. Instead of maintaining the late-time response of a lens IRA with a long transmission-line extension, the response may be prolonged by using a large-aperture array, where the prompt response is maintained until effects propagate inward from the edge of the entire array rather than the edge of each element. The rudiments of a design that may approach this objective are shown in Fig. 16. We start with low-impedance horn elements with optimal rectangular edge regions having $\Delta a/b \cong 1/2$. At $a/b = 6$ with $Z_{line} = 50\,\Omega$, the prompt aperture efficiency is a respectable 82.6 %, without using isorefractive media. To maintain the response at late times, the forward aperture fields must be sustained. Radiation in the backward direction would eventually occur, reducing the aperture efficiency by half. The backward radiation is prevented by the conducting back plane, and the detrimental effects resulting from the subsequent short circuiting of the array are minimized with feromagnetic cores between the back plane and the array. High-frequency low-μ ferrites are used near the location where the horns join together. This is where the prompt

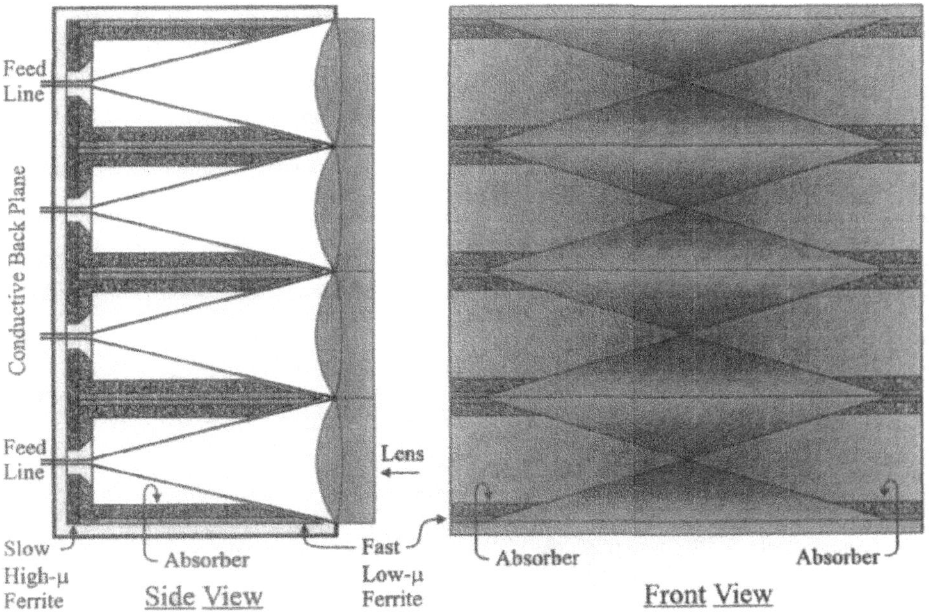

Figure 16. Conceptual design of antenna elements for a large TEM-horn array that can maintain a high aperture efficiency over a prolonged time scale. The array response is sustained for times comparable to the time for light to traverse the entire array rather than the individual elements. The post-prompt response is controlled by careful placement of absorber and feromagnetic materials.

interactions between the horns occur and where rapid magnetization is required. At later times, a slower high-μ magnetic material near the back plane absorbs the build-up of magnetic flux with a minimal energy expenditure. The radiated field energy blocked by the individual antenna apertures must be absorbed to prevent resonances. This is accomplished with carefully placed absorber material as well as the ferrites. Neither material should be located near the individual feed points. While prompt and late-time step responses are relatively easily to predict, the intermediate-time response, during which the E- and H-fields inside the antenna structure are evolving toward an electromagnetic equilibrium, will be very difficult to predict and will be very sensitive to the exact details of the antenna-system design. The fine tuning of this type of antenna design will most likely be accomplished by empirical and numerical methods.

CONCLUSIONS

The concept of aperture efficiency introduced in this report provides a quantitative measure of performance for comparing and optimizing the prompt responses of impulse radiating antennas. Aperture efficiencies can be substantially increased by choosing feed structures and aperture shapes that maximize field uniformity within the aperture and minimize energy that is blocked by the aperture. The most direct way to do this is through the use of low-impedance planar feed structures. Other methods that might be exploited utilize isorefractive media and septum plates. Antenna designs may be optimized under constraints such as restrictions on source impedance, aperture shape, off-boresight radiation, array-element interactions, and late-time responses.

REFERENCES

[1] C. E. Baum, "Radiation of Impulse-Like Transient Fields," Sensor and Simulation Note 321, USAF Phillips Lab, Albuquerque, NM, Nov. 1989.

[2] W. L. Stutzman and G. A. Thiele, Antenna Theory and Design, pp. 395, John Wiley & Sons, Inc., New York, 1981.

[3] C. J. Buchenauer, J. S. Tyo, and J. S. H. Schoenberg, "Aperture Efficiencies of Impulse Radiating Antennas," Sensor and Simulation Note 421, USAF Phillips Lab, Albuquerque, NM, Nov. 1997.

[4] J. S. Tyo, "Estimating the Optimum Aperture for Maximizing Prompt Aperture Efficiency in an IRA," Sensor and Simulation Note 422, USAF Phillips Lab, Albuquerque, NM, Jan. 1998.

[5] T. L. Brown and K. D. Granzow, "A Parameter Study of Two-Parallel-Plate Transmission-Line Simulators of EMP Sensor and Simulation Note 21," Sensor and Simulation Note 52, USAF Phillips Lab, Albuquerque, NM, Apr. 1968.

[6] C. E. Baum, "Use of Generalized Inhomogeneous TEM Plane Waves in Differential Geometric Lens Synthesis," Sensor and Simulation Note 405, USAF Phillips Lab, Albuquerque, NM, Dec. 1996.

[7] E. G. Farr and C. E. Baum, "Radiation from Self-Reciprocal Apertures," in Electromagnetic Symmetry, C. E. Baum and H. N. Kritikos, Editors., pp. 281-308, Taylor and Francis, Bristol, PA, 1995.

[8] T. K. Liu, "Impedances and Field Distributions of Curved Parallel-Plate Transmission-Line Simulators," Sensor and Simulation Note 170, USAF Phillips Lab, Albuquerque, NM, Feb. 1973.

[9] J. S. Tyo, C. J. Buchenauer, and J. S. H. Schoenberg, "Use of Isorefractive Media to Improve Prompt Aperture Efficiency in a Lens IRA," IEEE Transactions on Antennas and Propagation, vol. 46, pp. 1114-1115, July 1998.

[10] S. B. Cohn, "Lens-Type Radiators," in Antenna Engineering Handbook, H. Jansik, Editor, pp. 14-22 to 14-30, McGraw-Hill, New York, 1961.

[11] C. J. Buchenauer and J. R. Marek, "Antennas and Electric Field Sensors for Time Domain Measurements: an Experimental Investigation," in Ultra-Wiceband Short-Pulse Electromagnetics 2, L. Carin, and L. B. Felsen, Editors., pp. 197-208, Plenum Press, New York, 1995.

[12] C. J. Buchenauer, J. S. Tyo, and J. S. H. Schoenberg, "Antennas and Electric Field Sensors for Ultra-Wideband Transient Time-Domain Measurements: Applications and Methods," in Ultra-Wideband, Short-Pulse Electromagnetics 3, C. E. Baum, L. Carin, and A. P. Stone, Editors, pp. 405-420, Plenum Press, New York, 1997.

TRANSIENT GAIN OF ANTENNAS RELATED TO THE TRADITIONAL CONTINUOUS-WAVE (CW) DEFINITION OF GAIN

Carl E. Baum
Air Force Research Laboratory
AFRL/DEHP, Bldg. 909
3550 Aberdeen Avenue S. E.
Kirtland AFB, NM 87117

Everett G. Farr
Farr Research, Inc.
614 Paseo Del Mar N. E.
Albuquerque, NM 87123

Charles A. Frost
Pulse Power Physics, Inc.
1039 Red Oaks Loop N. E.
Albuquerque, NM 87122

1. INTRODUCTION

In previous papers[3, 10] the problem of appropriate definition of antenna gain and radiation pattern in time domain is considered in some detail. Using time-domain reciprocity the definitions are made to apply in both transmission and reception. Since the time-domain parameters are equivalent to the related frequency-domain parameters considered over all frequencies, the time-domain parameters are not simple numbers, but in general vector convolution operators. Specifically the gain is related to an appropriately normalized effective height convolution operator. This can in turn be reduced to a scalar number by application of appropriate mathematical norms. (A summary of these results, with an example, is included as Appendix A.)

Figure 1.1 indicates various parameters appropriate to an antenna in transmission and reception. Summarizing, in transmission we have the radiated far electric field[2]

$$\begin{gathered}\tilde{\vec{E}}_f(\vec{r},s) = \frac{e^{-\gamma r}}{r}\tilde{\vec{F}}_t(\vec{1}_0,s)\tilde{V}_t(s) \quad , \quad \tilde{V}_t(s) = \frac{\tilde{Z}_{in}(s)}{\tilde{Z}_L(s)+\tilde{Z}_{in}(s)}\tilde{V}_s(s) \equiv \text{transmitted voltage} \\ \tilde{V}_s(s) \equiv \text{source voltage} \quad , \quad \tilde{Z}_L(s) \equiv \text{load impedance} \\ \tilde{Z}_{in}(s) = \frac{\tilde{V}_t(s)}{\tilde{I}_t(s)} = \tilde{Y}_{in}^{-1}(s) \equiv \text{antenna input impedance} \quad , \quad \sim \; \equiv \; \text{Laplace transform (two-sided)} \\ s \equiv \Omega + j\omega \equiv \text{complex frequency or Laplace - transform variable} \\ \vec{1}_0 \equiv \text{direction to far - field observer} \quad , \quad r \equiv \text{distance to far - field observer} \\ \gamma = \frac{s}{c} \equiv \text{propagation constant} \quad , \quad c = [\mu_0\varepsilon_0]^{-\frac{1}{2}} \equiv \text{speed of light}\end{gathered} \tag{1.1}$$

Note that the antenna may have a transmission line (e.g., a coaxial cable or strip line) of characteristic impedance Z_c, connecting the antenna to the source/receiver. This input impedance is in general not the same as this, except in cases where this is part of the specific design. In time domain, however, if one puts a pulse into the port then $V(t)/I(t)$ will be Z_c for a time given by the round-trip transit time.

In reception we have

Ultra-Wideband, Short-Pulse Electromagnetics 4
Edited by Heyman *et al.*, Kluwer Academic / Plenum Publishers, New York, 1999.

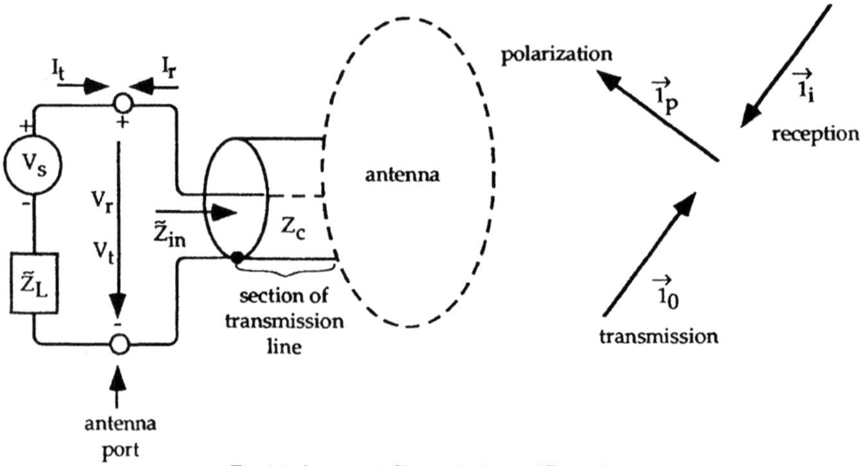

Fig. 1.1 Antenna in Transmission and Reception

$$\tilde{\vec{E}}^{(inc)}(\vec{r},s) = \tilde{\vec{E}}_0(s)e^{-\gamma \vec{1}_i \cdot \vec{r}} \equiv \text{incident plane wave} \quad , \quad \vec{1}_i \equiv \text{direction of incidence}$$

$$\tilde{\vec{E}}_0(s)\cdot\vec{1}_i = 0 \quad , \quad \tilde{V}_r(s) = \tilde{\vec{h}}_t(\vec{1}_i,s)\cdot\tilde{\vec{E}}^{(inc)}(\vec{0},s) = \tilde{\vec{h}}_t(\vec{1}_i,s)\cdot\tilde{\vec{E}}_0(s) \tag{1.2}$$

$$\tilde{\vec{h}}_t(\vec{1}_i,s) \equiv \text{effective height}$$

$$\tilde{\vec{h}}_t(\vec{1}_i,s)\cdot\vec{1}_i = 0 \text{ (no } \vec{1}_i \text{ component of incident electric field)}$$

$$\tilde{V}_r(s) \equiv \text{voltage across load } \tilde{Z}_L(s) \quad , \quad \tilde{Z}_L(s) = \frac{\tilde{V}_r(s)}{\tilde{I}_r(s)}$$

Note the opposite direction conventions of I_t and I_r.

Now applying reciprocity[2] we have the fundamental result

$$\tilde{\vec{F}}_t(\vec{1}_0,s) = \frac{s\mu_0}{4\pi}\left[\tilde{Y}_{in}(s) + \tilde{Y}_L(s)\right]\tilde{\vec{h}}_t(-\vec{1}_0,s) \tag{1.3}$$

where we have here defined $\tilde{\vec{h}}_t$ such that it has only a transverse part with respect to $\vec{1}_i = -\vec{1}_0$. Note that for the special, but useful, case that

$$\tilde{Z}_{in}(s) = \tilde{Z}_L(s) = R \tag{1.4}$$

i.e., a frequency-independent input impedance with a matched load, we have

$$\tilde{\vec{F}}_t(\vec{1}_0,s) = \frac{s\mu_0}{2\pi R}\tilde{\vec{h}}_t(-\vec{1}_0,s) \tag{1.5}$$

2. EFFECTIVE AREA (CW) OF ANTENNA IN RECEPTION

The effective area or effective aperture of an antenna [8,9] is traditionally defined via

$$\tilde{P}_r(j\omega) \equiv \tilde{A}_{eff}(j\omega)\tilde{S}^{(inc)}(j\omega) \tag{2.1}$$

where the received power $P_r(j\omega)$ is taken as the real power (or real part of the power) into the antenna load $\tilde{Z}_L(j\omega)$. Note that the above form has been used on the $j\omega$ axis of the s plane. Furthermore there is the polarization and angle-of-incidence dependence of these parameters to consider.

In analytic (s-plane) form the received power can be defined in alternate ways as

$$\tilde{P}_r^{(1)}(s) \equiv \tilde{V}_r(s)\tilde{I}_r(-s) = \tilde{V}_r(s)\tilde{V}_r(-s)\tilde{Y}_L(-s) \quad , \; \tilde{P}_r^{(2)}(s) \equiv \tilde{V}_r(-s)\tilde{I}_r(s) = \tilde{V}_r(-s)\tilde{V}_r(s)\tilde{Y}_L(s) \tag{2.2}$$

Another interesting form then averages these as

$$\tilde{P}_r(s) \equiv \tilde{V}_r(s)\tilde{V}_r(-s)\frac{1}{2}\left[\tilde{Y}_L(s) + \tilde{Y}_L(-s)\right] \tag{2.3}$$

which gives an even function of s. Of course if the load is a frequency-independent resistance R, then the above all give the same result as

$$\tilde{P}_r(s) = \frac{1}{R}\tilde{V}_r(s)\tilde{V}_r(-s) \tag{2.4}$$

This is an important case for application with transient waveforms (pulses). Note that this power can also be defined with a factor of 1/2 by averaging over a cycle, but this is a common factor with the Poynting vector, not influencing the effective area.

The incident plane wave in (1.2) has a magnetic field

$$\tilde{\vec{H}}^{(inc)}(\vec{r},s) = \frac{1}{Z_0}\vec{1}_i \times \tilde{\vec{E}}^{(inc)}(\vec{r},s) = \frac{1}{Z_0}\vec{1}_i \times \tilde{\vec{E}}_0(s)e^{-\gamma\vec{1}_i\cdot\vec{r}}$$

$$Z_0 = \frac{1}{Y_0} = \left[\frac{\mu_0}{\varepsilon_0}\right]^{\frac{1}{2}} \equiv \text{wave impedance of free space} \tag{2.5}$$

Now define an analytic form of the Poynting vector as

$$\tilde{\vec{S}}^{(inc)}(\vec{r},s) = \tilde{\vec{E}}^{(inc)}(\vec{r},s) \times \tilde{\vec{H}}^{(inc)}(\vec{r},-s) = \frac{1}{Z_0}\vec{1}_i\,\tilde{\vec{E}}^{(inc)}(\vec{r},s)\cdot\tilde{\vec{E}}^{(inc)}(\vec{r},-s) = \frac{1}{Z_0}\vec{1}_i\,\tilde{\vec{E}}_0(s)\cdot\tilde{\vec{E}}_0(-s) \tag{2.6}$$

which is conveniently independent of r. The above form then allows us to define a scalar form of the Poynting vector as

$$\tilde{\vec{S}}^{(inc)}(s) = \vec{1}_i\,\tilde{S}^{(inc)}(s) \quad , \quad \tilde{S}^{(inc)}(s) = \frac{1}{Z_0}\tilde{\vec{E}}_0(s)\cdot\tilde{\vec{E}}(-s) \tag{2.7}$$

which is also an even function of s.

This allows us to write

$$\tilde{A}_{eff}(\vec{1}_i,s) = \frac{Z_0}{2}\left[\tilde{Y}_L(s) + \tilde{Y}_L(-s)\right]\frac{\tilde{V}_r(s)\tilde{V}_r(-s)}{\tilde{\vec{E}}_0(s)\cdot\tilde{\vec{E}}_0(-s)} \tag{2.8}$$

which is an even function of s. From (2.1) we then have

$$\tilde{A}_{eff}(\vec{1}_i,s) = \frac{Z_0}{2}\left[\tilde{Y}_L(s) + \tilde{Y}_L(-s)\right]\frac{\tilde{\vec{E}}_0(s)\cdot\tilde{\vec{h}}_t(\vec{1}_i,s)\,\tilde{\vec{h}}_t(\vec{1}_i,-s)\cdot\tilde{\vec{E}}_0(-s)}{\tilde{\vec{E}}_0(s)\cdot\tilde{\vec{E}}_0(-s)} \tag{2.9}$$

Note that

$$\overset{\leftrightarrow}{\tilde{H}}_t(\vec{1}_i,j\omega) \equiv \overset{\rightarrow}{\tilde{h}}_t(\vec{1}_i,j\omega)\,\overset{\rightarrow}{\tilde{h}}_t(\vec{1}_i,-j\omega) = \overset{\rightarrow}{\tilde{h}}_t(\vec{1}_i,j\omega)\,\overset{\rightarrow}{\tilde{h}}{}^*_t(\vec{1}_i,j\omega) \tag{2.10}$$

is obviously Hermitian. With $s = j\omega$ then (2.9) is maximized by choosing

$$\begin{gathered}\overset{\rightarrow}{\tilde{E}}_0(j\omega) \equiv C_0^*(j\omega)\,\overset{\rightarrow}{\tilde{h}}{}^*_t(\vec{1}_i,j\omega) = C_0(-j\omega)\,\overset{\rightarrow}{\tilde{h}}_t(\vec{1}_i,-j\omega)\\ C_0 = \text{scalar function} \neq 0\end{gathered} \tag{2.11}$$

This C_0 can be a simple real constant if desired. Then we have a maximum

$$\begin{gathered}\tilde{A}^{(0)}_{eff}(\vec{1}_i,j\omega) = Z_0\,\mathrm{Re}\left[\tilde{Y}_L(j\omega)\right]\overset{\rightarrow}{\tilde{h}}_t(j\omega)\cdot\overset{\rightarrow}{\tilde{h}}_t(-j\omega) = Z_0\,\mathrm{Re}\left[\tilde{Y}_L(j\omega)\right]\left|\overset{\rightarrow}{\tilde{h}}_t(j\omega)\right|^2\\ 0 \le \tilde{A}_{eff}(\vec{1}_i,j\omega) \le \tilde{A}^{(0)}_{eff}(\vec{1}_i,j\omega)\end{gathered} \tag{2.12}$$

This suggests that we define an analytic form in the s plane as

$$\tilde{A}^{(0)}_{eff}(\vec{1}_i,s) \equiv \frac{Z_0}{2}\left[\tilde{Y}_L(s)+\tilde{Y}_L(-s)\right]\overset{\rightarrow}{\tilde{h}}_t(\vec{1}_i,s)\cdot\overset{\rightarrow}{\tilde{h}}_t(\vec{1}_i,-s) \tag{2.13}$$

Note that there also exists a polarization of the incident electric field such that

$$\overset{\rightarrow}{\tilde{A}}_{eff}(\vec{1}_i,s) = 0 \quad , \quad \overset{\rightarrow}{\tilde{E}}_0(j\omega)\cdot\overset{\rightarrow}{\tilde{h}}_t(\vec{1}_i,j\omega) = 0 \tag{2.14}$$

So the effective area can be considered as a function of the polarization $\vec{1}_p$ with

$$\tilde{A}^{(0)}_{eff}(\vec{1}_i,j\omega) = \sup_{\vec{1}_p}\tilde{A}_{eff}(\vec{1}_i,\vec{1}_p,j\omega) \tag{2.15}$$

where the polarization can even be complex as

$$\vec{1}_p = \overset{\rightarrow}{\tilde{E}}_0(j\omega)\quad|\overset{\rightarrow}{\tilde{E}}_0(j\omega)|^{-1} \tag{2.16}$$

Here we have maximized over $\vec{1}_p$. We can also maximize over $\vec{1}_i$ if we desire as

$$\tilde{A}^{(1)}_{eff}(j\omega) = \sup_{\vec{1}_i\ \text{real}}\ \tilde{A}^{(0)}_{eff}(\vec{1}_i,j\omega) \tag{2.17}$$

The corresponding $\vec{1}_i$ is opposite to the $\vec{1}_0$ for maximum gain in transmission.

3. CW ANTENNA GAIN

Antenna gain is usually defined as

$$\tilde{G}(\vec{1}_0,j\omega) = \frac{4\pi r^2}{Z_0\,\mathrm{Re}\left[P_{in}(j\omega)\right]}\overset{\rightarrow}{\tilde{E}}_f(\vec{1}_0,j\omega)\cdot\overset{\rightarrow}{\tilde{E}}{}^*_f(\vec{1}_0,j\omega) \tag{3.1}$$

where the usual far-field behavior is assumed. The power into the antenna in analytic form can be defined as

$$\tilde{P}_{in}^{(1)}(s) \equiv \tilde{V}_t(s)\tilde{I}_t(-s) = \tilde{V}_t(s)\tilde{V}_t(-s)\tilde{Y}_{in}(-s) \tag{3.2}$$

analogous to the received power in (2.2). Similarly we have alternate forms as

$$\tilde{P}_{in}^{(2)}(s) \equiv \tilde{V}_t(-s)\tilde{I}_t(s) = \tilde{V}_t(s)\tilde{V}_t(-s)\tilde{Y}_{in}(s) \quad , \quad \tilde{P}_{in}(s) \equiv \tilde{V}_t(s)\tilde{V}_t(-s)\frac{1}{2}\left[\tilde{Y}_{in}(s)+\tilde{Y}_{in}(-s)\right] \tag{3.3}$$

If the antenna has a frequency-independent input resistance R (which is the case for special antennas of interest), then the above all reduce to

$$\tilde{P}_{in}(s) = \frac{1}{R}\tilde{V}_t(s)\tilde{V}_t(-s) \tag{3.4}$$

The gain now takes the analytic form

$$\tilde{G}(\overrightarrow{1}_0,s) = \frac{8\pi r^2}{Z_0\left[\tilde{Y}_{in}(s)+\tilde{Y}_{in}(-s)\right]}\frac{\tilde{\vec{E}}_f(\overrightarrow{1}_0,s)\cdot\tilde{\vec{E}}_f(\overrightarrow{1}_0,-s)}{\tilde{V}_t(s)\tilde{V}_t(-s)} \tag{3.5}$$

which is an even function of s, as well as real and positive for $s = j\omega$. From (1.1) and (1.3) we have

$$\begin{aligned}\tilde{G}(\overrightarrow{1}_0,s) &= \frac{8\pi}{Z_0\left[\tilde{Y}_{in}(s)+\tilde{Y}_{in}(s)\right]}\tilde{\vec{F}}_t(\overrightarrow{1}_0,s)\cdot\tilde{\vec{F}}_t(\overrightarrow{1}_0,-s)\\ &= -\frac{[s\mu_0]^2}{2\pi Z_0}\frac{\left[\tilde{Y}_{in}(s)+\tilde{Y}_L(s)\right]\left[\tilde{Y}_{in}(-s)+\tilde{Y}_L(-s)\right]}{\tilde{Y}_{in}(s)+\tilde{Y}_{in}(-s)}\tilde{\vec{h}}_t(-\overrightarrow{1}_0,s)\cdot\tilde{\vec{h}}_t(-\overrightarrow{1}_0,-s)\end{aligned} \tag{3.6}$$

One can now define

$$\tilde{\vec{g}}(\overrightarrow{1}_0,s) \equiv \frac{s\mu_0}{\sqrt{2\pi Z_0}}\frac{\tilde{Y}_{in}(s)+\tilde{Y}_L(s)}{\left[\tilde{Y}_{in}(s)+\tilde{Y}_{in}(s)\right]^{\frac{1}{2}}}\tilde{\vec{h}}_t(-\overrightarrow{1}_0,s) \quad , \quad \tilde{G}(\overrightarrow{1}_0,s) = \tilde{\vec{g}}(\overrightarrow{1}_0,s)\cdot\tilde{\vec{g}}(\overrightarrow{1}_0,-s) \tag{3.7}$$

In this form $\tilde{\vec{g}}$ represents a relationship between voltage or current and field (linear) while $\tilde{G}$ is in terms of power.

In terms of the effective area we have

$$\begin{aligned}\tilde{G}(\overrightarrow{1}_0,s) &= -\frac{[s\mu_0]^2}{\pi Z_0^2}\frac{\left[\tilde{Y}_{in}(s)+\tilde{Y}_L(s)\right]\left[\tilde{Y}_{in}(-s)+\tilde{Y}_L(-s)\right]}{\left[\tilde{Y}_{in}(s)+\tilde{Y}_{in}(-s)\right]\left[\tilde{Y}_L(s)+\tilde{Y}_L(-s)\right]}\tilde{A}_{eff}^{(0)}(-\overrightarrow{1}_0,s)\\ &= -\frac{\gamma^2}{\pi}\frac{\left[\tilde{Y}_{in}(s)+\tilde{Y}_L(s)\right]\left[\tilde{Y}_{in}(-s)+\tilde{Y}_L(-s)\right]}{\left[\tilde{Y}_{in}(s)+\tilde{Y}_{in}(-s)\right]\left[\tilde{Y}_L(s)+\tilde{Y}_L(-s)\right]}\tilde{A}_{eff}^{(0)}(-\overrightarrow{1}_0,s)\end{aligned} \tag{3.8}$$

For $s = j\omega$ this reduces to

$$\tilde{G}(\vec{1}_0,j\omega)=\frac{\pi}{\lambda^2}\frac{\left|\tilde{Y}_{in}(j\omega)+\tilde{Y}_L(j\omega)\right|^2}{\mathrm{Re}\left[\tilde{Y}_{in}(j\omega)\right]\mathrm{Re}\left[\tilde{Y}_L(j\omega)\right]}\tilde{A}_{eff}^{(0)}(-\vec{1}_0,j\omega)\quad,\quad \lambda=\frac{f}{c}=\frac{2\pi\omega}{c}\equiv \text{wavelength} \tag{3.9}$$

4. SPECIAL CASE OF RESISTIVE MATCHED LOAD

The foregoing results simplify considerably for a resistive matched load. Returning to fig. 1.1 we note that we can now restrict

$$\tilde{Z}_{in}(s)=\tilde{Z}_L(s)=Z_c\equiv f_g Z_0 \tag{4.1}$$

so that the resistive impedance is the same as the characteristic impedance of a transmission-line antenna feed. This of course also requires more of the antenna design so that the antenna presents a resistive "termination" to the feed line of resistance Z_c. Fortunately some kinds of antennas have this property, at least to a reasonable approximation. (See, e.g., the TDR measurements.[4])

From (2.9) the effective area becomes

$$\tilde{A}_{eff}(\vec{1}_i,s)=\frac{1}{f_g}\frac{\tilde{\vec{E}}_0(s)\cdot\tilde{\vec{h}}_t(\vec{1}_i,s)\,\tilde{\vec{h}}_t(\vec{1}_i,-s)\cdot\tilde{\vec{E}}_0(-s)}{\tilde{\vec{E}}_0(s)\cdot\tilde{\vec{E}}_0(-s)} \tag{4.2}$$

with the frequency-dependent admittances now removed. The maximum over polarization as in (2.13) now becomes

$$\tilde{A}_{eff}^{(0)}(\vec{1}_i,s)=\frac{1}{f_g}\tilde{\vec{h}}_t(\vec{1}_i,s)\cdot\tilde{\vec{h}}_t(\vec{1}_i,-s) \tag{4.3}$$

The gain in (3.7) and (3.8) is now

$$\begin{aligned}\tilde{G}(\vec{1}_0,s)&=-\frac{\gamma^2}{\pi}\tilde{A}_{eff}^{(0)}(-\vec{1}_0,s)=-\frac{\gamma^2}{\pi f_g}\tilde{\vec{h}}_t(-\vec{1}_0,s)\cdot\tilde{\vec{h}}_t(-\vec{1}_0,-s)\\ \tilde{G}(\vec{1}_0,j\omega)&=\frac{4\pi}{\lambda^2}\tilde{A}_{eff}^{(0)}(-\vec{1}_0,j\omega)=\frac{4\pi}{\lambda^2 f_g}\tilde{\vec{h}}_t(-\vec{1}_0,j\omega)\cdot\tilde{\vec{h}}_t(-\vec{1}_0,-j\omega)\end{aligned} \tag{4.4}$$

So this case of resistive matched load considerably simplifies the formulae by the removal of some of the frequency-dependent factors.

5. RELATION OF EFFECTIVE HEIGHT TO EFFECTIVE AREA AND GAIN UNDER RESTRICTIVE ASSUMPTIONS

Note first that effective area and gain are scalars, whereas effective height is a vector (related to the polarization). On the $j\omega$ axis we have

$$\begin{aligned}\tilde{A}_{eff}^{(0)}(\vec{1}_i,j\omega)&=\frac{1}{f_g}\tilde{\vec{h}}_t(\vec{1}_i,j\omega)\cdot\tilde{\vec{h}}_t^{*}(\vec{1}_i,j\omega)=\frac{1}{f_g}\left|\tilde{\vec{h}}_t(\vec{1}_i,j\omega)\right|^2\\ \tilde{G}(\vec{1}_0,j\omega)&=\frac{4\pi}{\lambda^2 f_g}\tilde{\vec{h}}_t(-\vec{1}_0,j\omega)\cdot\tilde{\vec{h}}_t^{*}(-\vec{1}_0,j\omega)=\frac{4\pi}{\lambda^2 f_g}\left|\tilde{\vec{h}}_t(-\vec{1}_0,j\omega)\right|^2\end{aligned} \tag{5.1}$$

Thus the best that can be inferred from $\tilde{A}_{eff}^{(0)}$ or $\tilde{G}$ is the magnitude of the effective height. If one has polarization dependent information as in (4.2) (including complex polarizations perpendicular to $\vec{1}_i$) then the orientation of $\vec{\tilde{h}}_t$ can also be inferred.

At this point one can impose additional assumptions. Suppose that the orientation of $\vec{\tilde{h}}_t$ is known for $\vec{1}_i$ of interest (say from symmetry in the antenna) and this orientation is real so that we can write

$$\vec{\tilde{h}}_t(j\omega) = \vec{1}_t \tilde{h}_t(j\omega) \quad , \quad \tilde{A}_{eff}(\vec{1}_i, j\omega) = \frac{1}{f_g}\left|\tilde{h}_t(j\omega)\right|^2 \quad , \quad \tilde{G}(\vec{1}_0, j\omega) = \frac{4\pi}{\lambda^2 f_g}\left|\tilde{h}_t(j\omega)\right|^2 \tag{5.2}$$

Then $\tilde{G}$ and $\tilde{A}_{eff}^{(0)}$ can infer $\left|\tilde{h}_t(j\omega)\right|$ which does not include phase. If $\vec{1}_t$ is frequency-independent (which it is in some cases) then the problem is reduced to determining $\tilde{h}(j\omega)$ (and hence $\tilde{h}(s)$) as our transfer function of interest.

Even with the foregoing assumptions there still is the problem of phase. From (5.2) we have

$$\left|\tilde{h}_t(\vec{1}_i, j\omega)\right| = \left[f_g \tilde{A}_{eff}^{(0)}(\vec{1}_i, j\omega)\right]^{\frac{1}{2}} = \frac{\lambda}{2}\left[\frac{f_g}{\pi}\tilde{G}(\vec{1}_0, j\omega)\right]^{\frac{1}{2}} \tag{5.3}$$

However, in the way gain is defined (Section 3), it has no phase information (on the $j\omega$ axis). So various possible effective heights can correspond to the same gain function. The simplest case has

$$\tilde{h}_t(\vec{1}_i, s) = \tilde{h}_t^{(0)}(\vec{1}_i, s)e^{-st_0} \tag{5.4}$$

which is just an added delay of a time t_0 to some other effective height. Such a delay (such as by a length of transmission line) is not important here. However, for transfer functions (effective height being such a function) one can have more generally

$$\tilde{h}_t(\vec{1}_i, s) = \tilde{h}_t^{(0)}(\vec{1}_i, s)\tilde{a}(s) \quad , \quad \left|\tilde{a}(j\omega)\right| = 1 \tag{5.5}$$

Such a function $\tilde{a}(s)$ is called an all-pass function[6,7] which can be realized various ways in an idealized circuit synthesis sense (but perhaps with practical difficulties involving real circuit elements over a very large band of frequencies).

The effect of this phase ambiguity is to give various transient responses to the various antennas with the same gain function of frequency. Some antennas are less dispersive than others. This can be interpreted loosely as keeping a pulse more together (not spreading out in time). This also tends to increase the amplitude of the radiated transient field. An interesting question concerns the construction of some optimal $\tilde{h}_t$ from a given $\tilde{G}$. This is something like a minimum-phase problem. From (3.8) for a fixed polarization $\dot{1}_t$ as in (5.2) we have

$$\tilde{G}(\vec{1}_0, j\omega) = \left|\tilde{g}(\vec{1}_0, j\omega)\right|^2 \quad , \quad \left|\tilde{g}(\vec{1}_0, j\omega)\right| = \tilde{G}^{\frac{1}{2}}(\vec{1}_0, j\omega) = \frac{2}{\lambda}\left[\frac{\pi}{f_g}\right]^{\frac{1}{2}}\left|\tilde{h}_t(j\omega)\right| = \frac{\omega}{c}\left[\pi f_g\right]^{-\frac{1}{2}}\left|\tilde{h}_t(j\omega)\right| \tag{5.6}$$

Such functions can then be Hilbert transformed to obtain a phase function which is a minimum phase function, thereby giving at least one form of the complex function(s). This may give some kind of "best" effective-height function for a given gain function. Realizing this in a practical antenna is another matter.

One type of ideal effective-height function is

$$\vec{h}_t(\vec{1}_i,s)=h_t e^{-st_0}\,\vec{1}_t \tag{5.7}$$

for a particular choice of $\vec{1}_i=-\vec{1}_0$ (boresight). This approximately characterizes an impulse radiating antenna (IRA)[1] in reception[2]. (This holds over a large band of frequencies, limited by antenna size at low frequencies and construction details and conductor losses at high frequencies.) Note the inclusion of a simple delay t_0, such as occurs on ideal transmission lines (cables). By appropriate choice of reference for time or phase, this can be made zero. In this case we have

$$\begin{gathered}\tilde{A}^{(0)}_{eff}(\vec{1}_i,s)=\frac{1}{f_g}h_t^2 \quad , \quad \tilde{\vec{g}}(\vec{1}_0,s)=\frac{s}{c}\left[\pi f_g\right]^{-\frac{1}{2}}h_t e^{-st_0}\,\vec{1}_t \\ \tilde{G}(\vec{1}_0,s)=-\frac{s^2}{c^2}\frac{1}{\pi f_g}h_t^2 \quad , \quad \tilde{G}(\vec{1}_0,j\omega)=-\frac{\omega^2}{c^2}\frac{1}{\pi f_g}h_t^2\end{gathered} \tag{5.8}$$

As we can see the effective height gives the effective area and gain, but these latter eliminate the phase information and make the inference of the effective height have an unknown delay which could be a function of frequency.

6. CONCLUDING REMARKS

As we have seen, it is possible to relate the usual CW antenna gain (IEEE) to the effective height operator which more adequately expresses the time-domain response in both transmission and reception. This relationship results in some loss of information. There is a requirement first of a constant-resistance load at the antenna port, matched to the same constant-resistance input impedance of the antenna, unless one is willing to carry these as extra factors in the relationship. Second, one needs a frequency-independent polarization for the given angle of incidence (e.g., boresight), and a corresponding constant (frequency independent) orientation of the effective height. Third, the loss of phase information results in an unknown dispersion of a pulse. If one is willing to place additional constraints (assumptions, as in the case of an IRA of various types) then this phase ambiguity can be resolved.

APPENDIX A: SUMMARY OF THE GAIN EQUATIONS WITH AN EXAMPLE

We begin with the standard expressions in the frequency domain. Thus, the received power is

$$\tilde{P}_r=\tilde{A}_{eff}\tilde{S}^{(inc)} \quad , \tilde{A}_{eff}=\frac{\lambda^2}{4\pi}\tilde{G} \quad , \quad \tilde{P}_r=\frac{\lambda^2\tilde{G}}{4\pi}\tilde{S}^{(inc)} \tag{A.1}$$

Take the square root, and recast into voltages

$$\frac{\tilde{V}_r}{\sqrt{\tilde{Z}_c}}=\frac{\lambda\sqrt{\tilde{G}}}{2\sqrt{\pi}}\frac{\tilde{E}^{(inc)}}{\sqrt{Z_0}} \tag{A.2}$$

where Z_c, the feed or input impedance, is assumed a positive constant. Thus, the final result in the frequency domain is

$$\tilde{V}_r = \frac{\lambda\sqrt{\tilde{G}f_g}}{2\sqrt{\pi}}\tilde{E}^{(inc)} \quad , \quad f_g = \tilde{Z}_c/Z_0 \tag{A.3}$$

Let us now compare the above equation to one that we have been using in the time domain, i.e.,

$$V_r(t) = h_t(t) \circ E^{(inc)}(t) \ , \quad \circ \equiv \text{convolution} \tag{A.4}$$

We routinely have already measured $h(t)$, so we just have to rescale to get gain. Converting this to the frequency domain, we have

$$\tilde{V}_r(j\omega) = \tilde{h}(j\omega)\tilde{E}^{(inc)}(j\omega) \tag{A.5}$$

Now compare (A.3) and (A.4), to get

$$\boxed{\tilde{G}(j\omega) = \frac{4\pi}{\lambda^2}\frac{\left|\tilde{h}_t(j\omega)\right|^2}{f_g} = \frac{4\pi f^2}{c^2}\frac{\left|h_t(j\omega)\right|^2}{f_g}} \tag{A.6}$$

Using this theory, we can calculate the measured gain[5] of the 9-inch diameter reflector and lens IRAs. The gain of the 9-inch diameter reflector IRA is shown in fig. A.1 and the gain of the 9-inch diameter lens IRA is shown in fig. A.2. When comparing the two gains, the most striking difference is apparent at the low frequencies, between 1 and 3 GHz. The dip in $h(j\omega)$ in this region becomes quite apparent when it is multiplied by frequency squared. We presume this dip occurs because the feed arms are not terminated in the reflector design. Future designs will include terminations of the feed arms. Even without the dip there is a 3 dB advantage for the lens design.

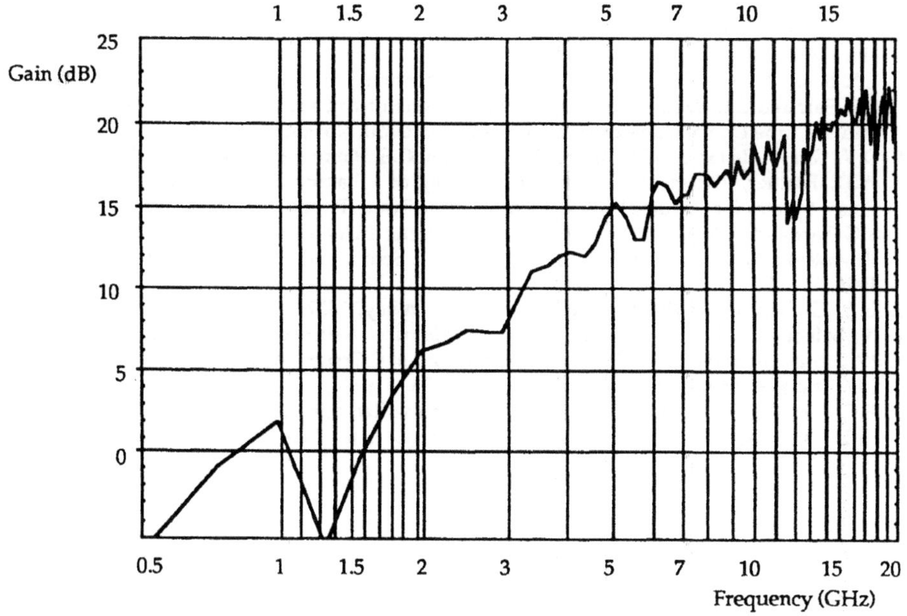

Figure A.1. Gain of the 9-inch reflector IRA.

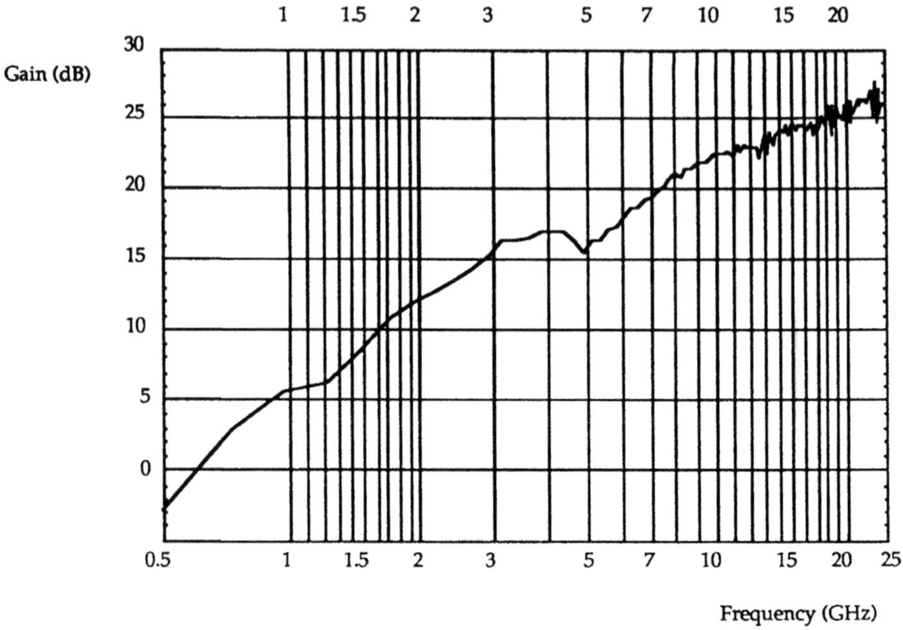

Figure A.2. Gain of the 9-inch lens IRA.

REFERENCES

1. C.E. Baum, Radiation of Impulse-Like Transient Fields, Sensor and Simulation Note 321, November 1989.
2. C.E. Baum, General Properties of Antennas, Sensor and Simulation Note 330, July 1991.
3. E.G. Farr and C.E. Baum, Extending the Definitions of Antenna Gain and Radiation Pattern Into the Time Domain, Sensor and Simulation Note 350, November 1992.
4. D.V. Giri, H. Lackner, I.D. Smith, D.W. Morton, C.E. Baum, J.R. Marek, D. Scholfield, and W.D. Prather, Design, Fabrication and Testing of a Paraboloidal Reflector Antenna and Pulser System for Impulse Like Waveforms, IEEE Trans. Plasma Science, 1997, pp. 318–326.
5. E.G. Farr and C.A. Frost, Development of a Reflector IRA and Solid Dielectric Lens IRA part II: Antenna Measurements and Signal Processing, Sensor and Simulation Note 401, October, 1996.
6. E.A. Guillemin, *Synthesis of Passive Networks*, Wiley, 1957.
7. N. Balabanian, T.A. Bickart, and S. Seshu, *Electrical Network Theory*, Wiley, 1969.
8. W.L. Stutzman and G.A. Thiele, *Antenna Theory and Design*, Wiley, 1981.
9. J.D. Kraus, *Antennas, 2nd Ed.*, McGraw Hill, 1988.
10. E.G. Farr, C.E. Baum, and C.J. Buchenauer, Impulse Radiating Antennas, Part II, pp. 159-170, in L. Carin and L.B. Felsen (eds.), *Ultra-Wideband, Short-Pulse Electromagnetics* 2, Plenum Press, 1995.

Ultra-Wideband Sources and Antennas

William D. Prather, Forrest J. Agee, Carl E. Baum, Jane M. Lehr, James P. O'Loughlin, Jeffrey W. Burger, Jon S.H. Schoenberg, David W. Scholfield, Robert J. Torres, Jonathan P. Hull, John A. Gaudet

Air Force Research Laboratory
Kirtland AFB, New Mexico

INTRODUCTION

Ultra-Wideband (UWB) sources and antennas are of interest for a variety of potential applications that range from transient radar systems to communications systems. Pre-dating the recent technology developments of the last several years were a number of important discoveries in the 1980's that led to the first UWB pulsers capable of transmitting high peak field signals through a variety of antennas. The Bournlea pulse generator is a good example of an instrument that is capable of fast rise times and a high repetition rate [1]. Using a CX1599 hydrogen thyratron as its basis, the original Bornlea pulser, made in 1986, could produce a 5 kV pulse with a rise time of 3 ns and a pulse repetition frequency (PRF) of 1 kHz. Later, when Bournlea researchers employed a ferrite line pulse sharpener for pulse compression, the rise time of the signal was reduced to 500 ps [2]. As a result, the promise of very fast, high voltage, UWB pulser was realized.

Ultra-wideband technology is of current interest to the Air Force Research Laboratory where research efforts have been underway for a number of years [3]-[5]. Sandia National Laboratories (SNL) has also produced a number of UWB sources and solid-state devices that have demonstrated substantial increases in power and repetition rates [6]-[10].

The development of new sources and antennas for the production of high power UWB fields has proceeded along several distinct lines in the past 6 to 8 years and is now beginning to branch out into other lines of technology. This work has brought new knowledge about gas switching and insulation [11], materials breakdown [12], shock line pulser technology [13], propagation [14], and optics [15]. This research into UWB transient antennas in our laboratory and in others around the world has also contributed significantly to the development and improvement of extremely wideband continuous wave (cw) and transient (non-dispersive) antenna designs [16]-[18].

The research has followed several distinct thrusts: high pressure gas switching, oil switching, and solid state switches. Recent improvements in high voltage insulating oil [19], [20] and short-pulse gas switching have made significant contributions to the UWB technology. As the technology progresses, of course, it is also making possible experiments that are teaching us a great deal about material properties under high stress, short pulse conditions. The new parameter space poses significant research challenges to the attainment of more powerful UWB sources. The solid-state switched UWB sources are also showing

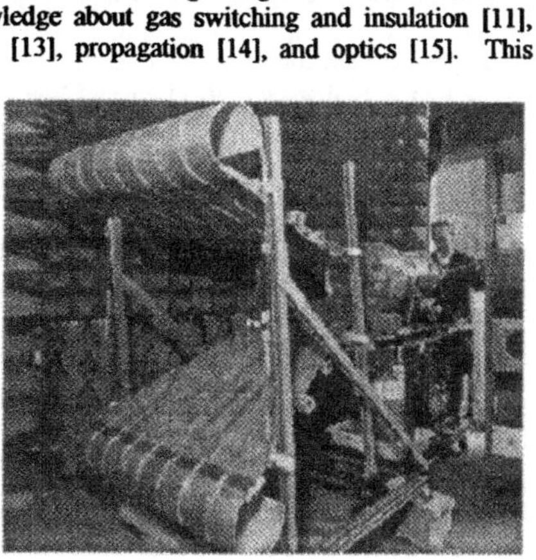

Figure 1. H-2 with Large TEM Horn

significant promise. Solid state arrays presently hold the record for peak radiated field strength in the far field of the antenna.

In the following sections, we discuss the research issues and progress being made in gas, oil, and solid state switching, UWB sources, and antennas.

GAS-SWITCHED UWB SOURCES

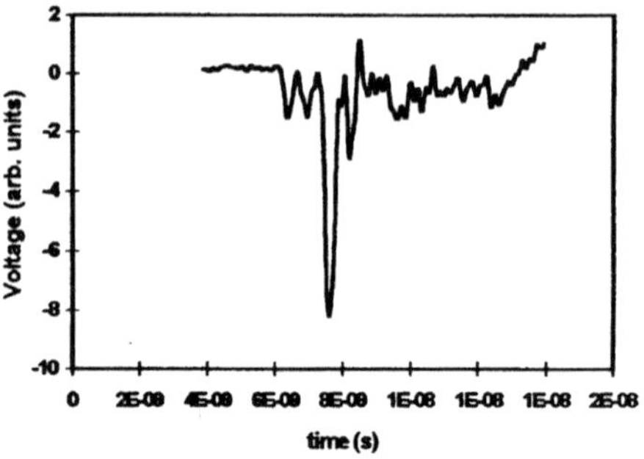

Figure 2. Radiated Field from H-3 with 18" Horn

Gas Switches: The switching element is a major component of any power conditioning system and ultra-fast closing capability, along with fast voltage recovery are desired for UWB HPM generation. A fast-rising pulse is critical because the high frequency components of the transmitted spectrum are determined by the rise time. To sharpen the rise time on a pulse, a peaking switch may be used. Such switches have been part of UWB systems since the early 1970s, when they were included in EMP simulators. The crux of the peaking gap is the establishment of very high electric fields between very closely spaced electrodes. The inter-electrode spacing is chosen to be as small as possible to minimize the intrinsic inductance of the spark channel, since the channel inductance will obviously limit the rise time. Spark channel inductance of less than 1 nH has been achieved with gap lengths of 1 mm and less. Because of the short gap lengths, even charge voltages of less than 100 kV can produce inter-electrode electric fields in the MV/cm range with voltage doubling at the spark gap. Short gaps and high electric fields also minimize the resistive phase, which can also limit rise time. According to J.C. Martin's widely used result [21], it decreases as $E^{-4/3}$, where E is the electric field. The accuracy of this result for the extreme of UWB switching conditions is, however, uncertain. These small inter-electrode distances do yield high capacitance, even for relatively small diameter electrodes. Moreover, this high spark gap capacitance, and the fast charging times lead to a strong displacement current which manifests itself as an undesirable prepulse. To produce ultra-fast switching, the spark gap is dramatically overvolted; that is, the spark gap is charged far in excess of its self-breakdown voltage. Modern peaking gaps typically operate at gas pressures in the range of 100 atm (1400 psi) and electric fields in the MV/cm range. The self-breakdown curve for gases is known to saturate in the vicinity of 100 MV/m for pressures to 50 atm [22]. To achieve overvolting without switching at the self-breakdown voltage, the spark gap is pulse charged very quickly. This allows a large overvoltage to be achieved, and overvoltages of over 300% are achievable.

H-series of UWB Sources: The H-Series devices use high-pressure hydrogen switches to produce powerful and very compact UWB sources. The first H-series device, the H-1, was completed in 1991, and the development has continued to the recent development of the H-5. The H-1 produced a peak voltage of 150 kV into a 2" diameter, 50 Ω coaxial line. Although the power output was a modest 450 MW, it successfully demonstrated the concept of a high pressure (1400 psi), hydrogen-switched source. The H-2, completed in late 1992, represented a substantial improvement over the earlier version. It generated a 300 kV pulse in a 4" diameter, 40Ω line with a rise time of about 250 ps and a total pulse length of 1.5 to 2 ns. This provided an ultra-wide spectrum with its center frequency around 150 MHz. A picture of the H-2 mated to a large TEM horn is shown in Figure 1. The next two pulsers in the series, H-3 and H-4, were designed to produce a much higher frequency spectrum with the center frequency between 1 and 2 GHz.

Figure 3. The H-5 Pulser

The H-3 device generates nearly 1 MV into a 40 Ω coaxial line. The radiated field from H-3 has a rise time of 130 ps and a total pulse duration of only 300 ps, as shown in Figure 2. In order to achieve the very fast rise time, the H-3 makes use of a patented charging technique [25], [26] and a multi-channel output switch. The unique transmission line-to-transmission line charging technique compresses the pulse and increases the voltage. The dual transmission line allows the first transmission line to be charged relatively slowly. When fully charged, it is switched into the second line using a high-pressure hydrogen spark gap. The fast charging of the second transmission line creates a large overvoltage on the high-pressure hydrogen, multi-channel, ring gap switch. The full potential of H-3 has not been realized to date as it has only been used with small antennas. The radiated field data shown in Figure 4 was generated using an 18-inch TEM horn with a peak voltage of 130kV. The field at 6 m was 60 kV/m.

The H-4, which is essentially a larger version of the H-3, is still in design. The goal for the H-4 is to produce 2 MV into a 40 Ω coaxial line. The waveform will have a rise time of 130 ps and a pulse width (FWHM) greater than 250 ps. The H-4 design is near completion, but we postponed work on the H-4 to devote our energies to the completion of a new source, the H-5 which is basically a higher power version of the H-2. The H-5, shown in Figure 3, was first demonstrated in March 1997 and found to be a robust, high power, ultra-wide band source that is mobile and self-contained. The design goal is to generate 500 kV pulses with a rise time of < 250 ps, and a pulse width of 2.5 ns. The system is still in development, and thus far has reached about 250 kV. It is our goal to achieve full design voltage by the end of this year. The radiated field waveform is shown in Figure 4. In this case, the antenna was the 20" unbalanced TEM horn shown in Figure 10, with the addition of a 20" diameter polyethylene focusing lens at the aperture. At 30 kV charge, the peak field strength radiated was 60 kV/m at a range of 4.5 m. The H-5 was also used with the large TEM horn, and the radiated field in this configuration is shown in Figure 5. The peak field strength measured was 43 kV/m at a range of 10 m. The rise time was 238 ps.

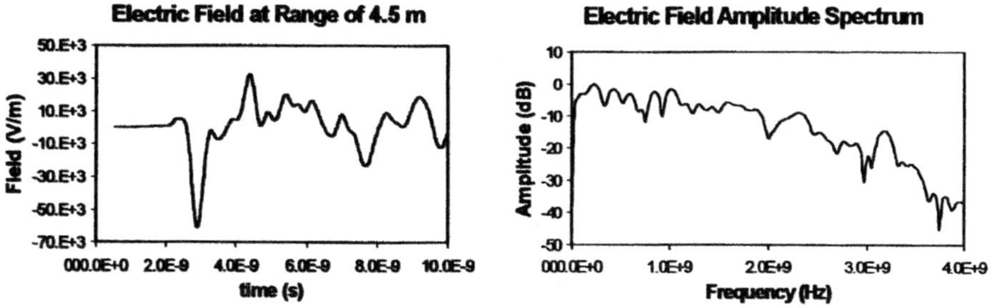

Figure 4. Radiated Field of H-5 with a 20" TEM Horn with a Polyethylene Lens

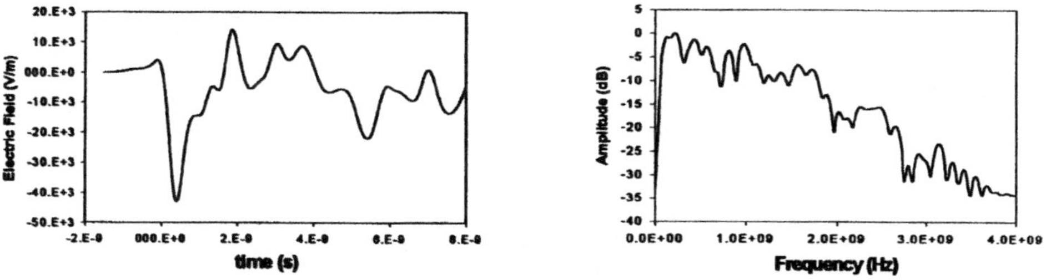

Figure 5. Radiated Field of H-5 at 10m with Large TEM Horn.

The IRA: The Reflector Impulse Radiating Antenna (IRA) shown in Figure 6 also uses high-pressure hydrogen switching to produce an extremely powerful UWB pulse [16]-[18]. With a charge of only ±60 kV, the system generates a transient signal that was measured to be 4.25 kV/m at 305 m. In the next version, we plan to increase the charge voltage to ± 180 kV which will increase the radiated field proportionately. Also, if a smaller reflector IRA is desired, this can be done as long as it has the same focal length-to-diameter (F/D) ratio. The voltage pulse from the 4m IRA has a rise time of less than 100 ps and 1/e fall time of about 20 ns. As a result, the radiated field has a rise time of 85 ps and a pulse width (FWHM) of 130 ps plus a long undershoot which contains most of the low frequency energy. The radiated spectrum of the IRA is flat within 3 dB from 35 MHz to 4 GHz. The high frequency roll-off is governed by the voltage and rise time of the gas switch and the optical correction afforded by a very precise oil and polyethylene lens at the apex. The lower frequency point is a primarily a function of the diameter of the antenna aperture and, in this case, the **p x m** compensation network at the base of each arm that allows wavelengths larger than the diameter of the antenna to be radiated [16], [27].

At late times or low frequencies, the antenna is dominated by electric and/or magnetic dipole moments, which we represent as **p** and **m**. The electric dipole results from the late-time charge separation (voltage) appearing at the ends of the feed horns. The magnetic dipole is due to the current flowing around the loop formed by the feed horns and the dish itself. In order to minimize oscillations, the feed arms are terminated against the parabolic reflector in their characteristic impedance. The resulting balance of **p** and **m** serves to direct the low frequency radiation vector (**p x m**) in the direction of the reflector's main beam. Thus, at low frequencies where the wavelength is of the order of the reflector diameter, the antenna will not radiate omni-directionally, but rather in a cardioid pattern with the maximum on the boresight of the reflector. The net effect of this is to increase the amplitude of the low frequency spectrum by about 6 dB and increases the low frequency bandwidth by a factor of two. In the case of the IRA, the low frequency limit (-3 dB) is lowered from 70 to about 35 MHz. These design features gave the IRA a 3dB beamwidth of ±1.2° as determined from the measured peak electric field. The performance of this antenna and pulser system compares favorably with that of our most powerful UWB radiating system, the GEM II array, despite the comparatively modest pulser used to drive it. Even with this kind of performance, there is substantial growth potential in the combination of the gas-switched H-Series pulsers and variants of the IRA.

Figure 6. The 4-meter IRA

Challenges For Gas-Switched Source Research

Oil vs. Gas Switches: High-pressure hydrogen is the gas of choice for high repetition rate, sealed, peaking switches. Hydrogen has the advantage in fast recovery time over other atomic gas. Pressures on the order of 100 atm, however, are necessary to get the required breakdown voltage. The very high pressures impose severe constraints on the geometry of the source, which in turn may introduce impedance mismatches into the transmission of electrical energy within the source.

The use of transformer oil in the peaking switch instead of high pressure gas alleviates many of the mechanical design problems. Switches using transformer oil are able to achieve higher inter-electrode electric fields than those using gaseous dielectrics. The time dependency of threshold electric fields is much greater in liquids than in gases. Thus, liquid switches can be pulse charged to much higher fields than can generally be achieved in gases. The equations which estimate the rise time are strongly dependent on the electric field. Thus, the higher electric field strengths achievable in liquids can result in the shorter switch rise times.

The use of various types of hydrocarbon oil as a switching media for the generation of sub-nanosecond rise time pulses at high repetition rates was first seriously proposed, then demonstrated, by the late Phil Champney and his team at Titan-Pulse Sciences [23]. When oil fails electrically, it leaves long-lived, electrically charged streamer byproducts which decrease the effective breakdown strength of the oil. This inhibits the recovery rate of the oil, and discourages its use for repetitively pulsed systems. The Titan-Pulse Sciences team undertook a systematic investigation of flowing oil switching systems for the generation of sub-nanosecond rise time, high repetition rate pulses. This work of Titan-Pulse Sciences is widely referenced and is the basis for all for all wideband switches utilizing oil switches developed to date. The Titan-Pulse Sciences effort resulted in a modulator that contained three oil switches: a transfer switch, a sharpening switch and a peaking/sharpening switch. The modulator produced pulses with a 10-90% rise time of 100 ps and performed at 200 kV, 1500 Hz and 450 kV, 500 Hz [24]. Voltages of 800 kV were achieved with single shot conditions.

This work investigated the recovery of switches with the direction and rate of oil flow as parameters. If the flow displaces a sufficient amount of oil in the gap region, the switch breakdown strength, even in repetitive mode, is as for a single pulse. Thus, the oil flow requirements are a strong function of the switch gap, *d*, and a useful relationship between the switch gap spacing and the flow requirements has been determined [23]. The stressed oil region must be displaced between pulses. The volume to be displaced, then, is proportional to d^3. The displacement distance, and hence the velocity, is proportional to *d*. Thus, the volumetric rate goes as d^4 and the power to drive the flow goes as d^5. Thus, the flow rate requirements for peaking gaps are generally realizable, while flow requirements for switches located earlier in the pulse compression sequence may be prohibitively large. Two examples of flowing oil UWB transmitters are the Sandia Labs' SNIPER and the AFRL Phoenix.

Multi-Channel Switching: The most powerful UWB pulsers involving gas switching have been the H-series sources that have been developed in a sequence that now includes the H-5. The designs for the H-series pulsers, H-3 and H-4 used a low inductance, annular multi-channel switch for the final stage of the pulser. In theory, this should work because the uniform multi-channel switching should provide very fast rise time. However, we found it to be difficult in practice. The dimensions of the gap and the tolerances required were found to be critical and hard to maintain in the large switch geometry at the high pressures required. As a result, our experience to date with multi-channel switches has been mixed. The switch tends to break down at only one point along the circumference resulting in non-uniform launching of the electromagnetic wave (launching of a non-TEM mode). Also, a higher inductance than expected was noted. These factors tended to make the performance erratic and to increase the rise time at the output section. At this time, the H-3 is being redesigned to attempt to improve the switch performance. There is interest in investigating other means to insure multi-channel switching for these pulsers. Some techniques involving the use of UV sources have been widely used to enhance the performance of gas spark gaps; however, they do not work at pressures above ten atmospheres [28] and therefore are not appropriate here. Other methods of reliably inducing multi-channeling are still being pursued.

Repetition Rates: Repetition rates of 2 kHz have been achieved with the H-2 and H-3 devices. This is apparently limited by the recombination rate of the gas. However, we note with interest that personnel at the Naval Surface Warfare Center, Dahlgren VA have suggested that pulse rates of up to 10 kHz are possible by flowing the high-pressure hydrogen gas [29]. Higher repetition rates for UWB gas switched pulsers obviously can be done, but engineering it into a compact source presents a challenge. The range of repetition rates from 2 kHz to 10 kHz is a research interest for the near term for reliable switching in configurations that preserve the low inductance required to achieve the rise times of 100 ps to 1 ns. We plan also to study a number of gases to develop data for repetitively pulsed switch use. We anticipate that further progress in UWB switching at high repetition rates is possible.

Insulators: For our radiating systems, insulators have been used in two ways: as dielectric standoffs and as lenses. The desirable electrical properties for both uses are low dispersion and low loss. Typically, the material of choice for dielectric standoffs in the H-series of pulsers has been Lexan™, a brand of polycarbonate resin manufactured by General Electric. Lexan has good dielectric strength and is easy to machine. While Lexan™ is strong enough that it will not flow when placed under high pressure, it does tend to creep. Moreover, Lexan™ tends to craze when it comes into contact with transformer oil, which results in a reduction of mechanical strength. Electrically, Lexan™ has the relatively high loss tangent of 0.006 at 10 GHz. The loss tangent in a transmission line structure is modeled as a conductance, which is dispersive. For a device such as H-3, with a fast rising output pulse, the propagation through the Lexan™ standoff degrades the rise time. For these reasons, alternate materials are being sought. Among the interesting candidates are Rexolite and Torlon. Rexolite 1422 is a plastic with a dielectric constant of 2.55 at 100 MHz and 2.54 at 10 GHz and a loss tangent of 0.00047 [30] . Rexolite also exhibits considerable mechanical strength, with a tensile strength of 7,000 psi [31]. Torlon, with a dielectric constant of 3.9 at 1 MHz, is another strong material which was successfully used for the high-pressure hydrogen gas enclosure on the Prototype IRA [16]. Some other materials, namely G-10 and Kel-f, are also being studied for use in switch and insulator design. Their mechanical strength and dielectric properties make them very attractive for use in high pressure switches. In addition, Kel-f is a very dense material that is impervious to gases like hydrogen.

For lenses, there is usually little requirement for mechanical strength. For transient optical applications, the materials of choice have been polypropylene and polyethylene. Both of these materials are relatively dispersionless with a low loss tangent. For high power applications, the lens is generally paired with a feed-through which is insulated from high voltage with transformer oil. Both polypropylene and polyethylene have a relative dielectric constant which is close to that of transformer oil, making the oil/lens interface electrically invisible. Typically, high power lenses are large, and therefore constructed from polyethylene that can be obtained in large castings. Currently, the Air Force Research Laboratory is also investigating artificial dielectrics for applications as lenses [32]. Artificial dielectrics are synthetic media which may be composed of conductive elements embedded in a natural dielectric binder. The effective dielectric constant of artificial dielectrics can be designed to be much greater than natural dielectrics. It is expected that the use of artificial dielectrics for lenses would result in considerable weight savings.

SOLID-STATE UWB SOURCES

Stacked, Semiconductor-Based Pulsers: A proven technology for UWB applications is based on ultra-fast, semiconductor-based stacked avalanche switches. An example of a pulser employing stacked avalanche transistors is the Kentech HMP. This device demonstrates a rise time less than 100 ps, has selectable pulse widths of up to 5 ns, an rms jitter of 10 ps, and a pulse amplitude of 6 kV per module into a

50 Ω load. The modules may be stacked in series and parallel to provide higher voltages without loss of timing performance. Kentech has demonstrated a stack that produces 24 kV into a 50 Ω load and they believe that 50 kV is achievable [33]. In the early 1990's, a module was used to conduct experiments to determine the effectiveness of UWB radar for target identification [34]. A significant advantage of stacked avalanche devices is that a laser is not required to trigger the avalanche process. These devices employ well-characterized, low-cost silicon which has four times more thermal conductivity than the gallium arsenide used in the avalanche devices.

An alternative, two-terminal avalanche technology uses a Drift Step Recovery Diode (DSRD) driving a Silicon Avalanche Shaper (SAS) in a trapped plasma avalanche triggered transit (TRAPATT) mode. Ivor Grekhov and his colleagues at the Ioffe Physical-Technical Institute in St. Petersburg, Russia, have pioneered this technology [35], [36] since the late 1970's. Recently, the Air Force Research Laboratory, in conjunction with the University of New Mexico and Old Dominion University, has studied the physical processes of the silicon avalanche sharpeners and its operating characteristics [37]-[39]. This study confirmed the robustness of the technology and validated all the operating parameters. The DSRD opening switch achieved a rise time of 740 ps and 1.9 kV pulse amplitude, providing the 10^{12} V/s transient required to rapidly drive the SAS into reverse bias breakdown. The SAS sharpens the transient to a 100 ps rise time at the DSRD pulse amplitude level into a 50 Ω load. Life times greater than 10^{11} shots have been achieved for this system. The DSRDs have been stacked in such a way as to produce high voltage pulsers with fast transients using a SAS shaping head. One such system which was tested was an Ioffe Institute PPG-10-5 pulse generator which achieved an output amplitude of 10 kV, a rise time of less than 100 ps (with shaping head), a pulse width of 2 ns, and a pulse repetition rate of 5 kHz into a 50 Ω load. It is generally agreed that the stacking limits of this technology have not yet been reached, and in response, Ioffe Institute has proposed to develop a design which will produce 100 kV pulses with rise times of less than 100 ps.

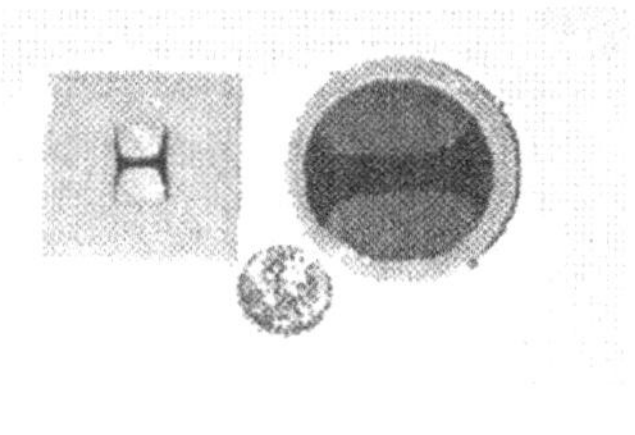

Figure 8. Lateral PCSS Switches

Photoconductive Solid-State Switched Arrays: Photoconductive switching technology offers the potential for very low turn-on jitter, fast rise times, extremely compact packaging, and ease of phasing numerous modules. The Air Force Research Laboratory involvement in development of photoconductive solid state (PCSS) switching technology began in the late 1980's shortly after investigators at Sandia National Laboratory made a new discovery while studying fast recovery switches. They demonstrated that a "lock-on" phenomenon existed in optically activated Gallium Arsenide (GaAs) switches at fields above 8 kV/cm [40]. This non-linear effect caused the switch to remain conducting even after the optical pulse had turned off, and the switch rise time was not limited by the laser rise time. In addition, the optical energy required to trigger the "lock-on" mode was less than 0.1% of the energy required to trigger linear switching. This new effect opened the door for many exciting new applications that require high current switching in the fast, sub-nanosecond time regime including the generation of high-power UWB microwave sources and other applications.

The first effort with PCSS switch development was in collaboration with SNL to assess the feasibility of using Photoconductive switching to improve the performance of electromagnetic pulse (EMP) simulators. Two candidate semiconductor materials were investigated for this purpose. The first set of experiments clearly eliminated silicon as a candidate material because of the energy and rise time requirements of the laser imposed by linear switching. GaAs became the material of choice because it requires less laser energy, and the switch rise times were independent of the laser pulse rise times because of the lock-on phenomena making these devices an ideal candidate for the EMP simulator application.

Figure 7. The GEM II Source.

A second effort involved a joint government and industry program to develop solid-state-switched High Power Microwave (HPM) sources using GaAs. This effort's scope has two parallel developments that can be characterized by the switch conduction path dictated by the switch contact layout on the GaAs substrate. For switch contacts formed on opposite sides of the substrate such that the gap is *across* the substrate thickness, the conduction mechanism is through the bulk of the GaAs. This is termed bulk switching. Notable examples are the Bistable Optically-controlled Semiconductor Switch (BOSS) developed by the Naval Surface Warfare Center at Dahlgren VA [41] and the Bulk Avalanche Semiconductor Switch (BASS™) produced by Power Spectra, Inc [42].

The BASS™ device is extremely compact, and several of them can be switched with sub-100 ps rise time performance with a GaAs laser diode coupled to an optical fiber splitter. Additionally, these devices achieve 10 ps rms jitter and greater than 3 x 10^9 shot lifetime, making them particularly suitable for parallel arraying to achieve greater current handling capability. The maximum switch gap is constrained by optical absorption and thermal dissipation in the GaAs, thereby capping the voltage hold-off to about 17 kV. Under contract to the U.S. Air Force, Power Spectra, Inc. designed a series of UWB laboratory sources using BASS™ devices called the GEM series, which have demonstrated the unique capabilities of low-jitter, solid-state switching technology. With very low jitter, on the order of one tenth of the risetime, one can then develop arrays of independently triggered, low voltage sources. It also allow for electronic beam steering and waveform variation which can be achieved by varying the triggering sequence of the individual elements. By "ripple" firing the array elements in time sequence one can generate short bursts of pulses at a very high PRF. The culmination of this series of sources is the 144-element GEM-II array, shown in Figure 7, that achieved 22.3 kV/m at 74 m range and a PRF of 3 kHz [4]. The field-range product of 1,650 kV generated by the GEM-II array, the highest ever demonstrated by an UWB source, was produced with a charge voltage of only 17 kV.

Another important switch design currently in use is the lateral photo-conducting switch. In this case, the switch contacts are formed on the substrate such that the gap and the main conduction path are *parallel* to the substrate surface. The development of lateral switches has been conducted through a cooperative internal DoD effort involving several service laboratories and universities. The lateral PCSS technology is being pursued in order to increase considerably the operating voltage and hence the radiated energy of solid-state switched arrays. Refractory-metal contacts shaped in a Rogowski profile, used to produce a uniform electric field across the gap, are formed on an undoped, GaAs substrate surface as shown in Figure 8. The Air Force Research Laboratory has tested two switch gap widths, 1.0 cm and 0.25 cm, to demonstrate the gap width effect on rise time, voltage hold off capability and lock-on voltage drop [43]. The switch topology is compatible with parallel plate transmission lines and Blumleins, which serve as energy storage for the PCSS prior to switching and a low-dispersion transmission line to an UWB antenna or resistive load upon switching. The 1.0 cm gap demonstrated the ability to hold off up to 135 kV, but a working potential of 50 kV enables greater than 2 x 10^6 shot life time in a 75 Ω resistively terminated, parallel-plate transmission line. However, the large gap increases the rise time (430 ps demonstrated), establishes a 5 kV voltage drop across the switch during lock-on, and only one switch can be fabricated on a 2" GaAs substrate. The 0.25 cm gap has demonstrated voltage hold-off as high as 32 kV, but a working potential of 20 kV is generally used in order to increase switch lifetime in a 75Ω resistively terminated, parallel-plate transmission line. The smaller gap reduces the rise time to 350 ps, reduces the voltage drop across the switch during lock-on to 1 kV, and enables four switches to be fabricated from a 2" GaAs disc, thus reducing manufacturing cost. Recently, the four-element array shown in Figure 9 was designed, fabricated, and tested at the Air Force Research Laboratory. It uses four TEM horns stacked together to form a 30 x 30 cm output aperture. Each 70 Ω element is matched to a parallel plate Blumlein commuted with a lateral PCSS switch having a 0.25 cm gap. With a charge voltage of 17 kV, the array achieved a field-range product (Er) of 20 kV [44]. A sample radiated waveform is shown in Figure 10. Expected further improvements in lateral PCSS switch speed and voltage hold-off will improve the performance of the array.

Figure 9. Four Element UWB Array with Lateral PCSS Switches

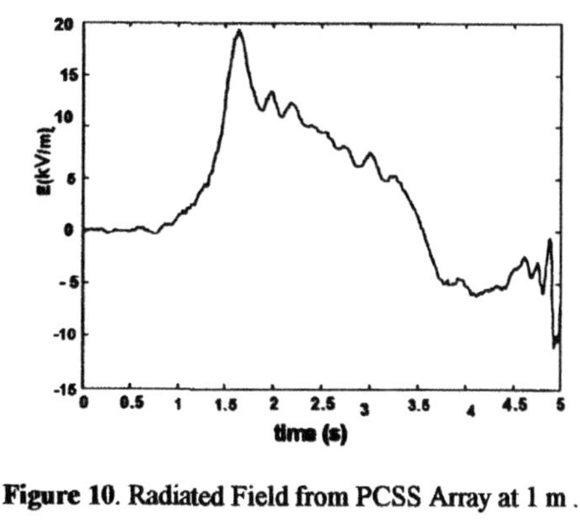

Figure 10. Radiated Field from PCSS Array at 1 m.

Presently, solid-state switching technology is being assessed for suitability in a number of applications such as Impulse Radar. Depending on the device and the application, switch lifetimes up to 10^9 shots can be achieved. Also pulse repetition frequencies (PRF's) of 10kHz and power handling capability in the hundreds of megawatts have been achieved. System jitter in the prototype systems allows timing control within 10 ps. The ability to array large numbers of independently controlled elements is possible and electronic beam steering is possible has been demonstrated. Further development of this technology is still required to address lifetime and power-handling issues; however, PCSS technology has reached a level of maturity where applications are feasible [3].

Research Issues and Technical Challenges for Solid State Switched Sources: The building and demonstration of advance array designs will provide an increased understanding of the issues involved with phasing of high power UWB signals. It is clear that improvements can be made in antennas to allow the radiation over a wider spectrum of RF energy in arrays over the relatively limited range made possible by the dual finline antenna used to date in the GEM series waveguide arrays. New designs that incorporate better radiating elements and better configurations that allow the late time energy to be radiated rather than being cut off will allow for a more dynamic interaction of RF radiated pulses with scatterers with the potential for obtaining more target response. This can make applications from radar to electronic warfare more robust. Compact packaging will be accomplished for remote prospecting, airborne, and space applications.

Lifetime and power handling issues continue to be the major concerns associated with PCSS technology. The "lock-on" mechanism leaves a substantial percentage of the switching field present during the conduction phase. This can lead to large power dissipation in the devices and filamentation that shortens switch lifetime. Lifetime and power handling tend to be inversely proportionate; however a careful systems design approach can greatly enhance the lifetime and power-handling capability of a PCSS. In general, as systems and application requirements increase (i.e. PRF, power, lifetime), the difficulty in maintaining the optimum operating parameters for the PCSS increases substantially. A number of experiments are planned or underway to improve the high-voltage lateral switch design, these changes are expected to increase lifetimes to an acceptable level. Power handling issues can be addressed through series-parallel arraying.

System efficiency has become the critical issue in future application of the PCSS technology. The requirement in HPM sources is toward compact, lightweight, and high-power sources. This requires that we reduce the size of all the subsystems that make up the source while increasing the power output. At first glance, this would seem to be the impossible task; however, improvements in coupling and radiation efficiency would enhance the output power while reducing prime power requirements. Compact UWB antenna designs that preserve radiation efficiency and frequency spectrum are by far the greatest problem facing compact source design. The PCSS device will not survive a poorly designed pulse charging system or an inefficient radiating element. Great care must be taken to eliminate any excess energy, either stored or reflected, when the PCSS device is entering a high resistance, open switch state. Neglecting to consider stray energy in the system after a shot will severely shorten the life of the PCSS device.

UWB ANTENNAS

UWB Antenna Design Considerations: For radiating or receiving UWB transients, we need antennas that are specifically designed and optimized for that task. Ordinary "wideband" antennas will generally not transmit fast transients because they have not been corrected for dispersion. As a result, there has been a considerable amount of research into fast transient antennas in the past five years. There are many articles and technical reports available now and more than a few very impressive antenna designs available. The IRA is one of the best [16-18] but there are others that are also exciting in terms of their potential for operating with CW or UWB transient signals.

An important consideration for transient antennas is minimizing both frequency and spatial dispersion. As demonstrated by Farr, Baum, and Buchenauer [18], frequency dispersion is most easily overcome by a conical TEM horn or a biconic antenna, both of which are frequency independent structures which produce spherical wave fronts. However, because of the finite size of high voltage switches, it is often necessary to add some corrective lenses to the structure to optimize the quality of the spherical wavefront. The spatial dispersion caused by the non-planarity of the wavefront can then be corrected with the use of a lens or a reflector. Both of these concepts have recently brought new standards of performance to the antenna world. A TEM horn with a polyethylene lens from SNL has been measured to emit a pulse with a FWHM of 23 ps [45]. Also, the team of Farr and Frost have recently built and demonstrated a 9" diameter reflector IRA with a rise time of 25 ps, an improvement over the 18" IRA which they had previously shown to radiate a pulse with a FWHM of 35 ps [46]. This corresponds to upper frequency limits of around 20 GHz.

Radio Frequency Lenses: As pulse rise times get faster, specifically $t_r < 250$ ps, we have found it necessary to employ optics as well as antenna and transmission theory in the designs of both pulsers and antennas. The materials most commonly used for lenses include polyethylene and polypropylene which have a relative dielectric constant near that of transformer oil ($\varepsilon_r = 2.3$). Where there are interfaces between two different dielectric materials such as polyethylene and air, it is effective to use a Brewster angle to minimize reflections [47]. Notice in Figure 11 that the Brewster angle appears as a curved surface since it was designed to match the spherical wavefront emanating from the throat of the horn. At the Air Force Research Laboratory, lenses have been used for several specific purposes. In the IRA, a lens is used at the apex of the antenna to correct the astigmatism introduced by the finite size of the gas switch. With the lens in place, the wave is corrected to a near-perfect spherical wavefront so that upon reflecting from the parabolic surface, it emerges as a plane wave [48].

Figure 11. TEM Horn with Polyethylene Brewster Angle Window.

Focusing lenses are also used at the output of TEM horns to correct the non-planarity of the wavefront (create a plane wave) and, in doing so, reduce the spatial dispersion which tends to limit the rise time Figure 12 is a picture a very fast horn filled with a solid dielectric which is shaped into a focusing lens at the output [46].

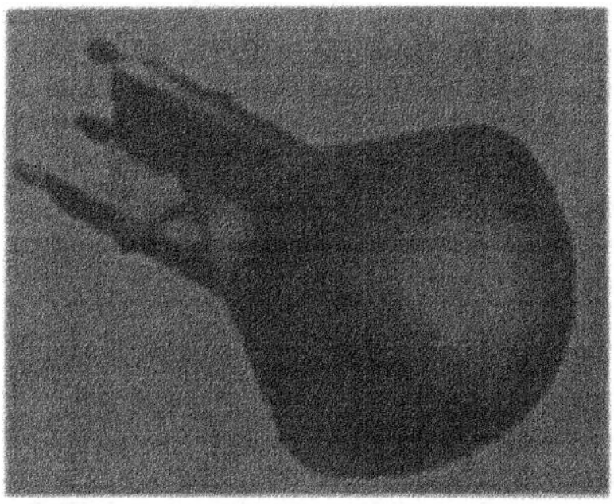

Figure 12. UWB Horn with Solid Dielectric.

New Antenna Concepts: Sometimes, it is desirable to be able to change the beamwidth of an antenna. As a result, a new variation of the IRA has been created which has demonstrated that this is possible with a reflector IRA. Called the Multi-purpose IRA or MIRA, the design achieves the variation of the beamwidth by mechanically varying the position of the horn apex with respect to the focal point of the parabolic reflector [49]. Figure 14 is a picture of the prototype MIRA that was demonstrated in the summer of 1997. The next step will be to demonstrate that the beam can be steered by moving the apex from side to side with respect to the focal point. Bigelow, Farr and Baum are developing graded dielectrics using mixtures of epoxy and TiO_2 which can be used to guide fast waves around bends in waveguides without causing dispersion [50]. The Air Force Research Laboratory is also working on the development of artificial dielectrics in an attempt to reduce the weight of the lensing material [32].

Research in UWB antennas is beginning to also have an effect in other areas of the electromagnetics world. The research into non-dispersive, UWB antennas has made possible multi-band antenna designs that offer up tantalizing possibilities for replacing several individual antennas with one multi-function antenna, thus reducing the space and weight. Likewise, the **p x m** concept, which increases the forward- directed, low frequency radiation from a small antenna, offers many exciting possibilities for reducing antenna size in air vehicles.

Technology Challenges in UWB Antennas: Several areas of antenna design offer challenges which must be tackled in coming years. An important one is that of making and demonstrating a compact array of TEM horns. While many strides in arraying of UWB antennas have been made to date, several characteristics

Figure 13. Multi-Function IRA.

remain to be examined such as the amount of cross coupling between antennas and switches and the effect it will have on switch and trigger circuit operation. Arrays driven by low-jitter, pulsed power sources such as photoconductive solid-state switches and small antennas with wideband characteristics are of interest.

Another exciting area of research is arrays of IRA structures. Theoretical considerations have indicated that such designs are feasible [51]. Such arrays would consist of many elements so that the voltage on any individual antenna is relatively small. Yet, the characteristic directionality of the IRA and the uniform response over an extremely large frequency range would be maintained. With a large number of antennas in the array, the signal received could be small, or, the transmitted signal large for the case of an active array. The "IRA Array" takes advantage of self-complementary principles (symmetry in design), aperture plane synthesis, and **p x m** resistive terminations to optimize the array's low frequency response. An IRA array will make steerable UWB transient emitters a feasible technology for radar and other purposes that can be installed on ships, aircraft and spacecraft of the future.

SUMMARY

In the last 6 years, there has been steady advancement in oil, gas and solid-state switched UWB sources and antennas. Numerous sources have been built and fielded by the Air Force, Army, Navy, and the National Laboratories. We have developed a range of pulsers and antennas that enable us to provide a variety of energies, rise times, pulse widths, and radiated waveforms that provide a flexible range of emission parameters available for experimentation. For a fairly complete list of the sources which are currently available, see the 1998 Special Issue on HPM [55].

The high voltage properties of insulators and the breakdown properties of gases and oils have not been systematically investigated and documented for very fast rise times. As a result, the available data is that gathered for a single engineering application. Generalized models and a thorough understanding of the physics do not exist. There is a limited database on liquid breakdown [11], [20], [54] that we plan to expand. Through the sponsorship of AFOSR, the Air Force Research Laboratory and collaborating universities have begun a program to fill in this parameter space.

A portion of the research has reached for higher voltages and shorter rise times until we are now pushing the limits in material properties. In particular, the dielectric breakdown properties of insulating materials at these extremely high voltages and extremely fast rise times are being challenged. Likewise, with the fast rise times, we are now seeing the high frequency loss tangents of some materials show up where they were not a problem before.

The future promises many more advances including the eventual realization of the full potential offered by our work in gas and oil switched pulsers and novel antennas. The technology we have in hand can substantially increase the radiated field strength at range from these sources, as we combine more powerful and highest gain antennas. Research into more capable photoconductive solid-state switches and UWB sources driven by them continues. The accomplishments to date by this class of UWB source encourages further work on switch lifetime and voltage performance.

ACKNOWLEDGMENTS

The authors are indebted to Malcolm Buttram and Larry Rinehart for helpful insight into UWB research at Sandia National Laboratories and to Ian Smith for very useful information on the state of the art in oil switching. This work was sponsored in part by the Air Force Office of Scientific Research.

REFERENCES*†

1. H. D. Kitchin and D. M. Parkes, "A Fast Rise Time, High PRF Pulse Generator Using a Hydrogen Thyraton," *Proceedings of the 17th IEEE Power Modulator Symposium*, Seattle WA, June 1986.
2. R. Pouladian-Kari, T. M. Benson, A. J. Shapland and D. M. Parkes, "The Electrical Simulation of Pulse Sharpening by Dynamic Lines," *Proc. 7th IEEE Pulsed Power Conf.*, Monterey CA, June 1989.
3. F.J. Agee, D.W. Scholfield, W.D. Prather, and J.W. Burger, "Powerful Ultra-Wide Band RF Emitters: Status and Challenges," *Proceedings of the SPIE*, **2557**, 1995.
4. W.D. Prather, C.E. Baum, F.J. Agee, J.P. O'Loughlin, D.W. Scholfield, J. Hull, J.S.H. Schoenberg, and R. Copeland, "Ultra-Wideband Sources and Antennas: Present Technology, Future Challenges," *Ultra-*

*All references to the *Digest of Technical Papers, International Workshop on High Power Microwave Generation and Pulse Shortening* can be found at http://library.plk.af.mil/edinburgh/WEB.html, or http://library.plk.af.mil/edinburgh/digest.html..

† Copies of Sensor and Simulation Notes are available from the authors or the editor, Dr. C.E. Baum, Air Force Research Laboratory, AFRL/DEHP, 3550 Aberdeen Ave, SE, Kirtland AFB, NM, 87117, USA.

Wideband, Short Pulse Electromagnetics 3, C.E. Baum, L. Carin, and A.P. Stone (eds.), Plenum Press, New York, 1997.
5. W.D. Prather and F.J. Agee, "Research in Ultra-Wideband/Short Pulse Sources and Antennas," *Digest of Technical Papers, International Workshop on High Power Microwave Generation and Pulse Shortening*, F.J. Agee, D.M. Parkes, J.A. Gaudet, E. Schamiloglu, Y.Y. Lau, Y. Saveliev (eds.), Edinburgh, UK, June 1997.
6. A.R.S. Clark, L.F. Rinehart, M.T. Buttram, J.F. Aurand, J.M. Lundstrom, T.E. Patterson, and L.D. Roose, "Repetitive Ultra-Wideband Impulse Source Development at Sandia National Laboratories," *Proceedings of the 20th IEEE Power Modulator Symposium*, 1992.
7. L.F. Rinehart, M.T. Buttam, W.R. Crowe, R.F. Clark, J.M. Lundstrom, and P.E. Patterson, "An Enantiomorphic Blumlein Impulse Generator," *Proc. 20th IEEE Power Modulator Symposium*, 1992.
8. C.A. Frost, T.H. Martin, P.E. Patterson, L.F. Rinehart, G.J. Rohwein, L.D. Roose, J.F. Aurand, and M.T. Buttram, "Ultra-fast Gas Switching Experiments," *Proc. of the 9th IEEE International Pulsed Power Conference*, Albuquerque, NM, 1993.
9. L.F. Rinehart, M.T. Buttram, G.J. Denison, J.M. Lundstrom, W.R. Crowe, J.F. Aurand, and P.E. Patterson, "Sandia National Laboratories' High Power Electromagnetic Impulse Sources," *Proceedings of the EUROEM '94 International Symposium*, D.J. Serafin, J.Ch. Bolomey, and D..DuPouy (eds), Bordeaux, France, May 1994.
10. F.J. Zutavern, M.W. O'Malley, and G.M. Loubriel, "Recent Developments in Opening Photoconductive Semiconductor Switches," *Proc. 6th IEEE Pulsed Power Conference*, Arlington VA, 1987.
11. F.J. Agee, D.W. Scholfield, R.P. Copeland, T.H. Martin, J.J. Carroll, J.J. Mankowski, M. Kristiansen, and L.L. Hatfield, "A Review of Catastrophic Electromagnetic Breakdown for Short Pulsewidths," *Proceedings of the SPIE*, **2843**, 1996.
12. V.G. Shpak, M.I. Yalandin, S.A. Shunailov, M.R. Oulmascoulov, S.N. Ivanov, A.D.R. Phelps, K. Ronald, W. He, and A.W. Cross, "Preliminary Investigation of the Dynamics of Initiating Processes in Vacuum Breakdown in High Sub-nanosecond Electric Fields," *Digest of Technical Papers, International Workshop on High Power Microwave Generation and Pulse Shortening*, F.J. Agee, D.M. Parkes, J.A. Gaudet, E. Schamiloglu, Y.Y. Lau, Y. Saveliev (eds.), Edinburgh, UK, June 1997.
13. D.M. Parkes, "Ultra-Wideband Pulser Technology," *Ultra-Wideband, Short Pulse Electromagnetics 3*, C. E. Baum, L. Carin, and A.P. Stone (eds.), Plenum Press, New York, 1997.
14. "Pulse Propagation and Guidance," Chapter in *Ultra-Wideband, Short Pulse Electromagnetics 3*, C.E. Baum, L. Carin, and A.P. Stone (eds.), Plenum Press, New York, 1997.
15. C.E. Baum and A.P. Stone, *Transient Lens Synthesis: Differential Geometry in Electromagnetic Theory*, Hemisphere Publishing Corp, 1991.
16. D.V. Giri, H. Lackner, I.D. Smith, D.W. Morton, C.E. Baum, J.R. Marek, W.D. Prather, and D.W. Scholfield, "Design, Fabrication, and Testing of a Paraboloidal Reflector Antenna and Pulser System for Impulse-Like Waveforms," *IEEE Trans. on Plasma Science*, **25**, 1997.
17. C.E. Baum and E.G. Farr, "Impulse Radiating Antennas," *Ultra-Wideband, Short Pulse Electromagnetics*, H.L. Bertoni, L. Carin, and L.B. Felsen (eds.), Plenum Press, New York, 1993.
18. E.G. Farr, C.E. Baum, and C.J. Buchenauer, "Impulse Radiating Antennas, Part II," *Ultra-Wideband, Short Pulse Electromagnetics*, H.L. Bertoni, L.Carin, and L.B. Felsen, (eds.), Plenum Press, N .Y., 1993.
19. G.J. Rohwein, "Corona Processing of Insulating Oil," *Proc. 22nd International Power Modulator Symposium*, L.H. Luessen and M. Gunderson, (eds.), 1996.
20. J.M. Lehr, F.J. Agee, R.P. Copeland, and W.D. Prather, "Measurement of the Electric Breakdown Strength of Transformer Oil in the Sub-nanosecond Regime," IEEE Trans. Dielectrics & Insulation, 1998, to be published.
21. J.C. Martin, "Solid, Liquid and Gaseous Switches," *Pulsed Power Lecture Series*, Lecture No. 30, Dept. of Electrical Engineering, Texas Technical University, Lubbock, TX, 1978.
22. E.A. Avilov and N.V. Belkin, "Electrical Strength of Nitrogen and Hydrogen at High Pressures," *Soviet Physics, Technical Physics*, **19** (12), June 1975.
23. P. D'S Champney, et al., "The development and testing of subnanosecond-rise kilohertz oil switches," *Proceedings of the 8th IEEE Pulse Power Conf*erence, June 1991.
24. Ian D. Smith, private communication, February 1998.
25. J.P. O'Loughlin, "A Nanosecond Transmission Line Charging Technique," U.S. Patent No. 5,444,308.
26. J.P. O'Loughlin and R.P. Copeland, "Subnanosecond Power Conditioning Technique Using Transmission Line-to-Transmission Line Charging," *Proc. of the 20th Power Modulator Symposium*, 1992.
27. C.E. Baum, "Radiation of Impulse-Like Transient Fields," Sensor and Simulation Note 321, Air Force Research Laboratory, Kirtland AFB, NM, November 25, 1989.
28. M. Kristiansen and J. Mankowski, Private Communication, 1996.
29. M.G. Grothaus, S.L. Morgan, and L.W. Hardesty, "Recovery Characteristics of Hydrogen Spark Gap Switches," *Proceedings of the 9th IEEE Pulsed Power Conference*, K. Prestwich and W. Baker (eds.), Albuquerque, NM, 1993.
30. A.R. Von Hippel, (ed.), *Dielectric Materials and Applications*, M.I.T. Press, Cambridge MA, 1954.

31. Manufacturer's Data Sheet, C-Lec Plastics, Beverly, NJ.
32. J.S.H. Schoenberg and C.J. Buchenauer, "Artificial Dielectrics for Ultra-Wideband Application," *Ultra Wide-band/Short-Pulse Electromagnetics 3*, C.E. Baum, L. Carin, and A.P. Stone (eds.), Plenum Press, New York, 1997.
33. Kentech Web Site, www.kentech.co.uk, 1998.
344. D.M. Parkes, M.F. Lewis, R.L.S. Devine, and K. Trafford, "Practical Measurements using Ultrawideband Radar," *Proceedings of the SPIE*, **1631**, 1992.
35. I.V. Grekhov, and A.F. Kardo-Sysoev, "Subnanosecond Current Drops in Delayed Breakdown of Silicon p-n Junctions," *Sov. Tech. Phys. Lett.* **5** (8), 1979.
36 I.V. Grekhov, "New Principles of High Power Switching with Semiconductor Devices," *Solid State Electronics*, **32** (11), 1989.
37. R.J. Focia, E. Schamiloglu, C.B. Fleddermann, F.J. Agee, and J. Gaudet, "Silicon Diodes in Avalanche Pulse Sharpening Applications," *IEEE Trans. Plasma Sci.*, **25** (2), April 1997.
38. R.J. Focia, E. Shamiloglu, and C.B. Fleddermann, "Simple Techniques for the Generation of High Peak Power Pulses with Nanosecond and Sub-nanosecond Risetimes," *Rev. Sci. Instrum.*, **67** (7), July 1996.
39. H. Jalali, R. P. Joshi, and J. Gaudet, "Effects of Transverse Doping Variations on the Transient Response of Silicon Avalanche Shaper (SAS) Devices," *IEEE Trans. Electron Devices*, to be published.
40. F.J. Zutavern and G.M. Loubriel, "High-Voltage Lateral Switches From Silicon or Gallium Arsenide," *High-Power Optically Activated Solid-State Switches*, A. Rosen and F. Zutavern (eds.), Artech House, Boston, 1994.
41. D.C. Stoudt, "Demonstration of a Frequency-Agile RF Source Configuration Using Bistable Optically Controlled Semiconductor Switches (BOSS)," *IEEE International Pulse Power Conference*, 1995.
42. J. Oicles, M. Staskus, and P. Brunemeier, "High-Power Impulse Generators for UWB Applica-tions," *Ultra-Wideband, Short-Pulse Electromagnetics 2*, L. Carin and L.B. Felson (eds.), Plenum Press, New York, 1995.
43. Jon S.H. Schoenberg, Jeffrey W. Burger, J. Scott Tyo, Michael D. Abdalla, Michael C. Skipper, and Walter R. Buchwald, "Ultra-Wideband Source Using Gallium Arsenide Photoconductive Semiconductor Switches," *IEEE Trans. Plasma Science*, Vol 25, No 2, April 1997.
44. Jon S.H. Schoenberg, J. Scott Tyo, Jeffrey W. Burger, Michael D. Abdalla, and Michael C. Skipper, "A Linear Array of Ultra-Wideband Radiators Driven by Photoconductive Semiconductor Switched Blumleins," *Ultra-Wideband/Short Pulse Electromagnetics 4*, this volume.
45. J.F. Aurand, "A TEM-Horn Antenna with Dielectric Lens for Fast Impulse Response," *Ultra-Wideband/Short Pulse Electromagnetics 3*, C.E. Baum, L. Carin, and A.P. Stone, (eds.), Plenum Press, New York, 1997.
46. E.G. Farr and C.A. Frost, "Compact Ultra-Short Pulse Fuzing Antenna Design and Measurements," Sensor and Simulation Note 380, Air Force Phillips Laboratory, Kirtland AFB, NM, June 1995.
47. J. Wells, C.E. Baum, N. Keator, and W.D. Prather, "A Radiating Structure Incorporating and Extended Ground Plane and a Brewster Angle Window," *Ultra Wideband/Short-Pulse Electromagnetics 3*, C.E. Baum, L. Carin, and A.P. Stone (eds.), Plenum Press, New York, 1997.
48. C.E. Baum, et al, "A Uniform Lens for Launching a Spherical Wave into a Paraboloidal Reflector," Sensor and Simulation Note 360, Air Force Phillips Laboratory, Kirtland AFB NM, July 1993.
49. E.G. Farr, C.E. Baum and W.D. Prather, "Multi-Function Impulse Radiating Antenna," *Ultra Wideband/Short-Pulse Electromagnetics 4*, this volume.
50. W.S. Bigelow and E.G. Farr, "Minimizing Dispersion in a TEM Waveguide Bend by a Layered Approximation of a Graded Dielectric Material," Sensor and Simulation Note 416, Air Force Research Laboratory, Kirtland AFB, NM, Jan 1998.
51. C.E. Baum, "Transient Arrays," *Ultra-Wideband, Short-Pulse Electromagnetics 3*, C. E. Baum, L. Carin, and A.P. Stone (eds.), Plenum Press, New York, 1997.
52. Larry Rinehart, private communication, September 1997.
53. Mark Litz, private communication, October 1997.
54. J. Mankowski, M. Kristiansen, and L. Hatfield, "Nanosecond Breakdown of Liquid Dielectrics," *Proc. 11th Inter-national Conf. on High Power Particle Beams*, Prague, Czech Republic, June 1996.
55. F.J. Agee, et al., "Ultra-Wideband Transmitter Research," *IEEE Trans. Plasma Science, 7th Special Issue on High Power Microwave Generation*, June 1998.

MULTIFUNCTION IMPULSE RADIATING ANTENNAS: THEORY AND EXPERIMENT

Everett G. Farr[1], Carl E. Baum[2], William D. Prather[2] and Leland H. Bowen[1]
[1]Farr Research, Inc.
Albuquerque, NM 87123
[2]Air Force Research Laboratory
Kirtland AFB, NM 87117

I. ABSTRACT

A Multifunction IRA is an extension of a standard Impulse Radiating Antenna that has the additional flexibility of an adjustable beamwidth. This adjustability is implemented by defocusing the feed, in order to select between a narrow or broad beam. We provide here the theory of operation of the antenna, for both in-focus and out-of-focus situations. Furthermore, we built and tested a design with a 46 cm diameter. We found good agreement of the experiment with theory.

II. INTRODUCTION

A reflector Impulse Radiating Antenna (IRA) consists of a parabolic reflector with a TEM feed. This class of antenna has a considerable body of literature associated with both its analysis and measurements [1-3]. One issue that has been raised concerning this type of antenna, however, is that the beamwidth is too narrow for many applications. In order to broaden the beam, we introduce the Multifunction IRA, or MIRA.

The principle behind the MIRA is quite simple. The feed point of an IRA is normally at the geometric focus of a parabolic reflector. In a MIRA, we defocus the feed arms slightly, by placing the feed point somewhat closer to the dish than its normal position at the focus of the reflector.

If one can add a mechanical control to the feed point location, then one can have a single antenna with a narrow or broad beam, as required. This results in a single antenna with very broad bandwidth and beamwidth control. Such an antenna may be useful in applications where a single antenna must serve multiple functions due to limited aperture space. In this note we develop the theory of such a device, and we describe the fabrication and testing of a prototype design.

The theory developed here is based on the quasistatic electric field in an aperture. In previous papers [2-8], we have developed a number of techniques for calculating the radiated field when an aperture field is turned on uniformly, all at the same time. With a defocused aperture, however, the aperture is turned on gradually, beginning at the center. The theory is modified here that takes this gradual turnon into account. We provide here theoretical

Ultra-Wideband, Short-Pulse Electromagnetics 4
Edited by Heyman *et al.*, Kluwer Academic / Plenum Publishers, New York, 1999.

calculations of the radiated impulse in the H- and E-planes, for both focused and defocused aperture positions.

In addition, an experimental prototype was developed with a 46 cm (18 in) diameter reflector using four feed arms with an adjustable position. The position is controlled by a servo mechanism that is controlled by a personal computer.

By including computer control in the design, we allow a great deal of flexibility in system design. For example, one might use the MIRA as part of a radar system that can operate in either search mode, which requires a broad beamwidth, or tracking mode, which requires a narrow beamwidth. A block diagram of such an arrangement is shown in Figure 2.1. The controller would select which of two radar systems would be fed into the antenna. The controller would also set the antenna feed position, in order to control the beamwidth.

The field is measured using TEM sensors. These were developed based on an idea by C. J. Buchenauer[9] to enhance signal-to-noise ratio with very fast, low-voltage pulsers. These sensors are replicating sensors, not the derivative sensors that are perhaps more commonly used. We calibrate these sensors using two identical sensors.

Note that this work was first reported in [14]. It is refined here by using improved sensors and by building an improved feed point.

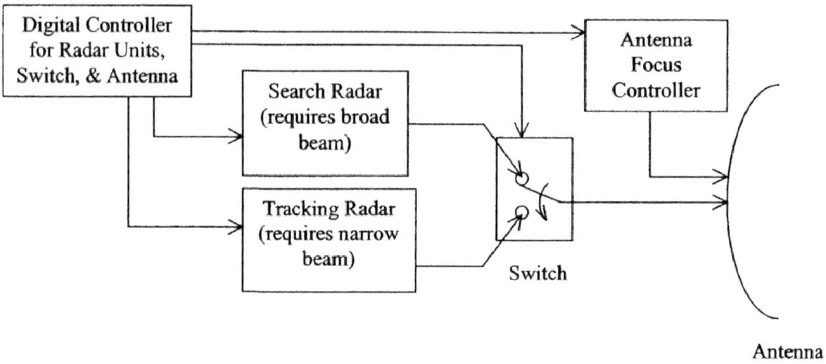

Figure 2.1. Control of the RF switch and antenna focus using a digital controller.

III. Modification of Theory to Include Defocused Aperture

To calculate the radiated field for the defocused aperture, it is first necessary to review the theory for the focused aperture radiation [2,3]. The theory for the defocused aperture is a straightforward perturbation on the theory for the focused aperture. To do so, we need to return to more general expressions for the field radiated from an aperture. Thus, we have the field radiated from an aperture, from [10, eqn 2.7]

$$\begin{aligned}\vec{E}(\vec{r},s) &= \frac{1}{2\pi r}\iint_{S_a} \gamma\left[\vec{1}_z \times \vec{E}_t(x',y',s)\times \vec{1}_R\right] e^{-\gamma R}\, dA \\ &= \frac{s}{2\pi r c}\iint_{S_a}\left[\vec{1}_z \times \vec{E}_t(x',y',s)\times \vec{1}_R\right] e^{-sR/c}\, dA\end{aligned} \qquad (3.1)$$

where primed coordinates refer to the aperture plane, and unprimed coordinates refer to the observation point. Furthermore, $\vec{E}_t(x',y',s)$ is the tangential aperture field in the aperture plane, and S_a is the aperture surface. The aperture plane is the x'-y' plane (at $z'=0$). Note that the above equation differs from [10 eqn 2.7] by retaining only the far-field ($1/r$) term, and by substituting $\gamma = s/c$. The expressions for r and R are

$$r = \sqrt{x^2 + y^2 + z^2}$$
$$R = \sqrt{(x-x')^2 + (y-y')^2 + (z-z')^2} \tag{3.2}$$

We will shortly specialize the above expressions for the radiated field to the H- and E-planes, but before we do so, we first wish to describe the form of the aperture field.

The form of the tangential aperture electric field is

$$E_y(t) = -\frac{V_o}{\Delta u}\frac{\partial u(x', y')}{\partial y'} W(t) \qquad u(\zeta) = \mathrm{Re}(w_4)$$
$$w_4(\zeta) = w_2(\sqrt{2}\zeta / a + 1) + w_2(\sqrt{2}\zeta / a - 1) \qquad w_2 = \ln\left(\frac{\zeta - ja}{\zeta + ja}\right) \tag{3.3}$$

where $u(x', y')$ is the potential function described previously in [2, eqns. (2.2)-(2.3)]. (Note that there was an error in the original expression for $w_4(\zeta)$ appearing in [14 eqn 4.2], which is corrected here.) Furthermore, $W(t)$ is a window function that takes into account the gradual illumination of the aperture. This is illustrated in Figure 3.1. Thus, $W(t)$ is a window that is circular in shape, with a radius described by $\Psi_a(t')$. We assume that the curved phase front first touches the center of the aperture at $t' = 0$, so at $\Psi_a(t'=0) = 0$. Furthermore, the instant where the aperture is first completely filled is denoted by t'_{max}, so $\Psi_a(t'_{max}) = \Psi_{a,max}$.

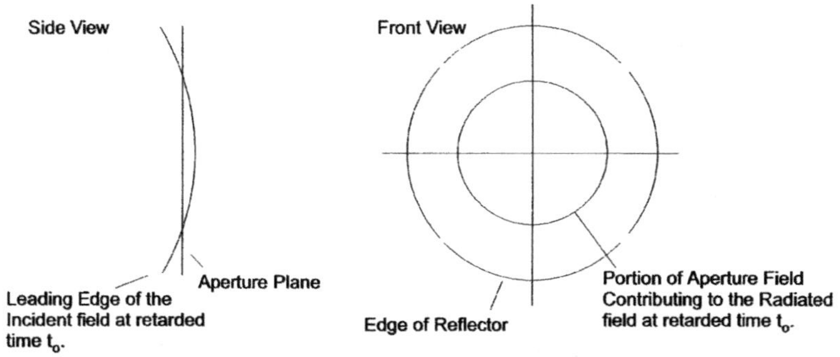

Figure 3.1. Illustration of how a defocused aperture is filled gradually by the incident field (left). The portion of the aperture plane contributing to the aperture field is shown on the right.

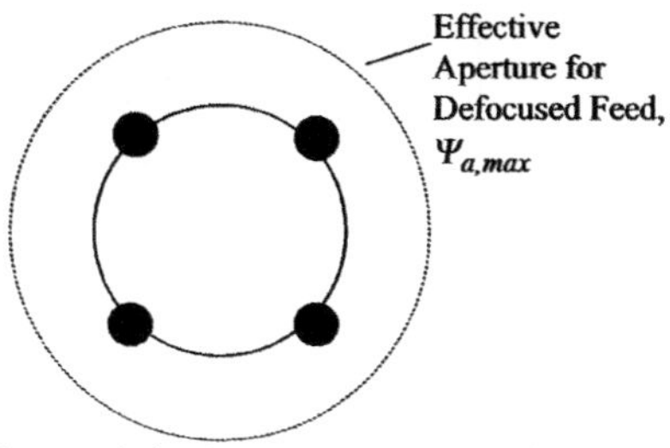

Figure 3.2. The effective aperture when a rigid feed arm assembly is pushed toward the reflector.

Note that some care must be taken when determining the position where the above potential function is evaluated. The position in the aperture plane, (x', y'), must be mapped onto a position in a flat virtual plane (x_v, y_v), using stereographic projections. If we describe a position on the aperture in the polar (cylindrical) coordinates, (Ψ',ϕ'), it is necessary to map the position on the aperture onto a virtual plane, described by (Ψ_v,ϕ_v), shown in Figure 3.2. To do so, we use geometric optics to trace the ray that arrives at (ϕ',Ψ'). This ray has an origin at the feed point, and it points in the direction $(\theta_{inc}, \phi_{inc})$. Due to symmetry, we have

$$\phi' = \phi_{inc} = \phi_v \tag{3.4}$$

so we have only left to find Ψ_v.

To find Ψ_v, we first find θ_{inc} by geometric optics and ray tracing. In other words, we find the angle of departure of the ray that begins at the feed point and arrives at (Ψ', ϕ'). Thus, having found θ_{inc}, we need only find a relationship between θ_{inc} and Ψ_v. The geometry is shown in Figure 3.3.

To complete the process of calculating Ψ_v, we compare the stereographic projection of (Ψ', ϕ') to that of the charge center of the feed arms. We know the charge center of the feed arms projects onto the unit circle in the virtual plane, so $\Psi_{v,o} = 1$. Now by taking a ratio of stereographic projections,

$$\Psi = 2F \tan(\theta / 2) \tag{3.5}$$

we obtain

$$\begin{aligned} \Psi_v &= \Psi_{v,o} \frac{\tan(\theta_{inc} / 2)}{\tan(\theta_{inc,o} / 2)} \\ \phi_v &= \phi' \end{aligned} \tag{3.6}$$

The only remaining unknown in this equation is the incident angle of the charge center. This is found from the F/D ratio of the reflector, f_d, so

$$\theta_{inc,o} = \arctan\left[\frac{1}{2 f_d - 1/(8 f_d)}\right] \tag{3.7}$$

and we now have both Ψ_v,and ϕ_v. From these two, we can calculate (x_v,y_v), so we have completed our procedure for mapping (x',y') into (x_v,y_v).

Having described the aperture fields, we now restrict the expression for the radiated field, (3.1), to the H-plane and E-plane. We begin with the H-plane.

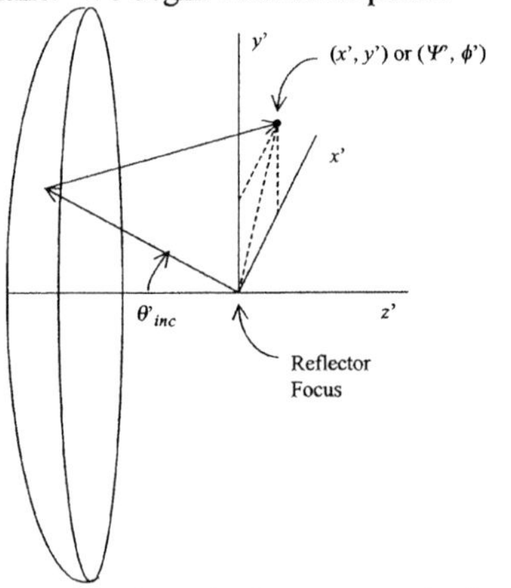

Figure 3.3. The ray trace from the focus to a point in the aperture plane.

A. H-Plane Calculation

In the H-plane the dominant component of the radiated field is the y-component. Furthermore, only the y-component of the aperture field contributes to the y-component of the radiated field due to symmetry. Thus, after restricting the tangential field to the y-component, we have

$$\begin{aligned} \vec{1}_z \times \tilde{\vec{E}}_t(x', y', s) \times \vec{1}_R &= \vec{1}_y \; \tilde{E}_y(x', y', s) \; \cos(\theta) \\ R &\cong r - x' \sin(\theta) \end{aligned} \tag{3.8}$$

Substituting the above relationships into (3.1), we have, in both frequency and time domains,

$$\begin{aligned} \tilde{E}_y^{(h)}(\theta, s) &= \frac{s \cos(\theta)}{2 \pi r c} \iint_{S_a} \tilde{E}_y(x', y', s) \; e^{-(s/c)(r - x' \sin(\theta))} \, dA \\ E_y^{(h)}(\theta, t) &= \frac{\cos(\theta)}{2 \pi r c} \frac{d}{dt} \iint_{S_a} \tilde{E}_y(x', y', t - (r - x' \sin(\theta)) / c) \, dA \end{aligned} \tag{3.9}$$

where the superscript (h) indicates that the expression is specific to the H-plane. For simplicity, we convert to retarded time, $t' = t - r / c$, so

$$E_y^{(h)}(\theta, t') = \frac{\cos(\theta)}{2 \pi r c} \frac{d}{dt} \int dx' \int dy' \; E_y(x', y', t' + (x' / c) \sin(\theta)) \tag{3.10}$$

We now need to account for the time-varying nature of S_a. As stated previously, the aperture expands as a function of time, with a radius $\Psi_a(t')$. Thus, the limits of integration over the aperture change with time. Let us define

$$\begin{aligned} \Phi^{(h)}(x',t') &= -\frac{1}{V_o} \int_{-y_{max}(x't')}^{y_{max}(x',t')} dy' \; E_y(x', y') \\ y_{max}(x',t') &= \begin{cases} 0 & t' < 0 \\ \sqrt{\Psi_a^2(t') - x'^2} & 0 \le t' \le t_{max} \\ \Psi_{a,max} & t' > 0 \end{cases} \end{aligned} \tag{3.11}$$

This is closely analogous to [3, eqn. (4.5)], with the exception that there is an additional time dependence. The integration is carried out along a straight line of constant x', with limits that are determined by $W(t)$, the window function that describes the effective aperture radius as a function of t'. If we now combine the above expression with (3.3), we have

$$\Phi^{(h)}(x',t') = \frac{2}{\Delta u} \; u(x', y_{max}(x',t')) \tag{3.12}$$

By combing equations (3.10-3.12) we have

$$E_{y,step}^{(h)}(\theta, t') = -\frac{V_o}{2 \pi r c} \cos(\theta) \frac{d}{dt} \int dx' \; \Phi^{(h)}(x', t' - (x' / c) \sin(\theta)) \tag{3.13}$$

This is now in a form that can be implemented numerically. Note that one can tabulate $\Phi^{(h)}(x',t')$ in order to avoid repetitive calculations. Then, $E_{y,step}^{(h)}(\theta,t')$ is calculated by a one-dimensional integral. Thus, we have successfully converted a double integral over the aperture to a single integral, which should be more numerically efficient. Note also that when calculating $u(x',y')$ it is necessary to calculate it into the virtual plane, as shown in equation (3.6). To find the response to a standard Gaussian pulse, we convolve the above step response with the derivative of the Gaussian, as shown previously in [2, eqns. 4.5-4.7].

B. E-Plane Calculation

Next, we calculate the radiated field in the E-plane. The derivation is very similar to that in the H-plane. The dominant component is the θ- component , and once again, we need only calculate the y-component of the aperture field due to symmetry. This leads to

$$\vec{1}_z \times \tilde{\vec{E}}_t(x', y', s) \times \vec{1}_R = \vec{1}_y \; \tilde{E}_y(x', y', s)$$
$$R \cong r - y' \sin(\theta) \tag{3.14}$$

Substituting the above relationships into (3.1), we have, in both frequency and time domains,

$$\tilde{E}_\theta^{(e)}(\theta, s) = \frac{s}{2\,\pi\, r\, c} \iint_{S_a} \tilde{E}_y(x', y', s)\; e^{-(s/c)\,(r - x' \sin(\theta))}\; dA$$
$$E_\theta^{(e)}(\theta, t) = \frac{1}{2\,\pi\, r\, c} \frac{d}{dt} \iint_{S_a} \tilde{E}_y(x', y', t - (r - y' \sin(\theta)) \,/\, c)\; dA \tag{3.15}$$

where the superscript (e) indicates that the expression is specific to the E-plane. As before, we convert to retarded time, $t' = t - r / c$, so

$$E_\theta^{(e)}(\theta, t') = \frac{1}{2\,\pi\, r\, c} \frac{d}{dt} \int dy' \int dx'\; E_y(x', y', t' + (y' / c) \sin(\theta)) \tag{3.16}$$

We now need to account for the time-varying nature of S_a. As stated previously, the aperture expands as a function of time, with a radius $\Psi_a(t')$. Thus, the limits of integration over the aperture change with time. Let us define

$$\Phi^{(e)}(y',t') = -\frac{1}{V_o} \int_{-x_{max}(y',t')}^{x_{max}(y',t')} dy'\; E_y(x', y')$$
$$x_{max}(y',t') = \begin{cases} 0 & t' < 0 \\ \sqrt{\Psi_a^{\,2}(t') - y'^2} & 0 \le t' \le t_{max} \\ \Psi_{a,max} & t' > 0 \end{cases} \tag{3.17}$$

This is closely analogous to [3, eqn. (4.7)], with the exception that there is an additional time dependence. The integration is carried out along a straight line of constant y', with limits that are determined by $W(t)$, the window function that describes the effective aperture radius as a function of t'. Note that one must take care in calculating $\Phi^{(e)}(x',t')$, due to the branch cut in the contour map. If we now combine the above with (3.3), we have

$$\Phi^{(e)}(y',t') = \frac{1}{\Delta u} \int_{-x_{max}(t')}^{x_{max}(t')} \frac{\partial u}{\partial y} dx = -\frac{1}{\Delta u} \int_{-x_{max}(t')}^{x_{max}(t')} \frac{\partial v}{\partial x} dx$$
$$= -\frac{2}{\Delta u} \left[v(x'_{max}(y',t'), y') - v(0, y') \right] \tag{3.18}$$

where we have used the relationship $\partial u / \partial x = -\partial v / \partial y$. Care must be used in the application of the above formula for the difference of two v's if a branch cut is crossed. In that case, a term of 2π must be subtracted from the difference in square brackets above. If we combine the above three equations, we have

$$E_{\theta,step}^{(e)}(\theta,t') = -\frac{V_o}{2\pi r c} \frac{d}{dt} \int dy'\; \Phi^{(e)}(y', t' - (y' / c)\sin(\theta)) \tag{3.19}$$

This is now in a form that can be implemented numerically. Once again, we have converted a double integral over the aperture into a single integral, which is faster to calculate. Note that the position where the potential function $v(x',y')$ is calculated must be mapped onto the virtual plane as described in equation (3.6).

Having calculated the step response in both the H-and E-planes, we now convolve with a Gaussian, as shown previously in [2, eqns. 4.5-4.7], to obtain the far field As an example, we calculated the radiated field in the H- and E-planes for a 46 cm (18 in) diameter reflector MIRA with $F/D = 0.5$, a risetime $t_d = 50$ ps and a feed impedance of 200 Ω.

Let us define now a ratio that describes the focal position. Let F be the focus of the reflector, and F_2 be the distance from the feed point to the reflector. Then we define a ratio

$$f_f = \frac{F_2}{F} \tag{3.20}$$

When the feed is located at the focus, $f_f = 1$. In our MIRA, we can adjust the focal point from $f_f = 0.7$ to $f_f = 1.0$. In the data we present here, $f_f = 0.7$, 0.85, and 1.0. The definition of f_f is shown in Figure 3.4.

The radiated field for $f_f = 1.0$ (feed at the focus) is shown in Figure 3.5. Both the H-plane and E-plane cuts are shown, for values off-boresight of 0, 7.5, and 15 degrees. We can clearly see that the field goes down as we proceed off-boresight in either plane. Furthermore, the radiated field for $f_f = 0.85$ and 0.7 are shown in Figures 3.6 and 3.7, respectively. From these plots, we can see that the peak field on boresight is reduced as we progressively defocus the antenna.

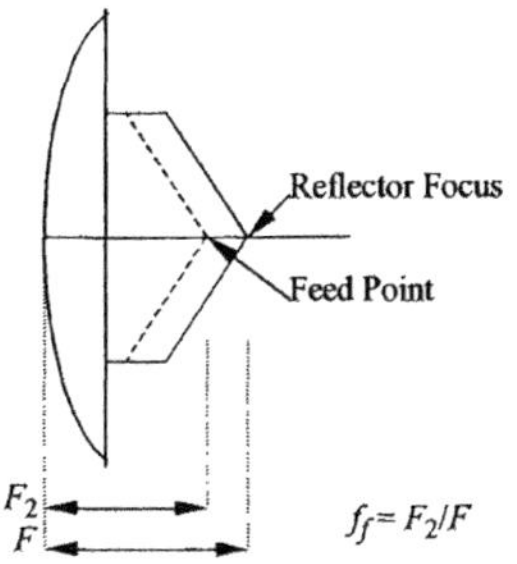

Figure 3.4. Definition of f_f.

Next, we need to define a beamwidth for this antenna, in order to compactly describe the data. We define the half field beamwidth (HFBW) as the angle between the two locations in a pattern cut where the field is down by half from the peak. Since we ran data at discreet angles off-boresight, the best we can do is estimate beamwidths. Furthermore, note that these beamwidths are specific to the particular choice of risetime, i.e., $t_d = 50$ ps. Finally, note that we estimate full beamwidths, not half beamwidths from boresight.

The HFBWs are provided in Table 3.1. From the data it is clear that the MIRA's beam broadens from about 8° when in focus, to about 36° when at the maximum defocus of $f_f = 0.7$. This is true in both the H- and E-planes. This is precisely the behavior we were hoping to find in this antenna.

Having demonstrated the relevant properties in theory, we now turn to experiment to verify the result.

Table 3.1. MIRA Half Field Beam Widths

	Half Field Beam Width (HFBW)	
$f_f = F_2/F$	H-Plane	E-Plane
1.00	8°	10°
0.85	15°	15°
0.70	36°	36°

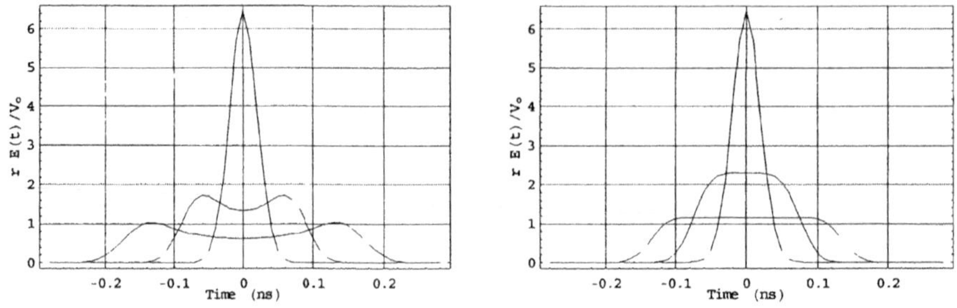

Figure 3.5. MIRA radiated field, $f_f = 1.0$ (in focus), H-plane (left) and E-plane (right), at angles of 0, 7.5, and 15 degrees off-boresight.

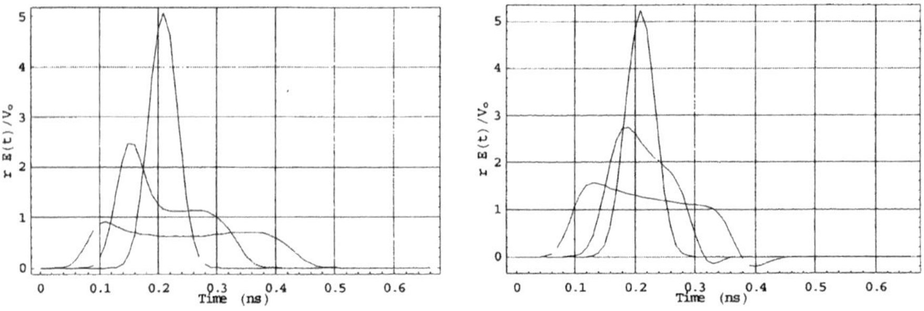

Figure 3.6. MIRA radiated field, $f_f = 0.85$, H-plane (left) and E-plane (right), at angles of 0, 7.5, and 15 degrees off-boresight.

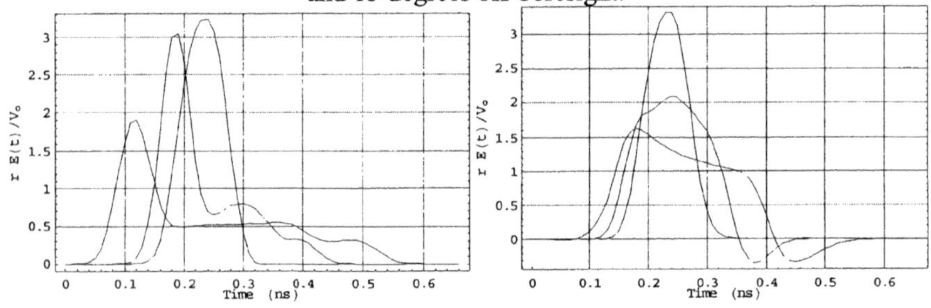

Figure 3.7. MIRA radiated field, $f_f = 0.7$, H-plane (left) and E-plane (right), at angles of 0, 7.5, and 15 degrees off-boresight.

IV. Experimental Validation

In order to demonstrate the MIRA experimentally, we built a model with a diameter of 46 cm (18 in), with adjustable feed arms. A diagram of the configuration as built is shown in Figures 4.1 and 4.2. It includes a solid 18-inch diameter parabolic reflector that slides along a set of four fixed feed arms. A servo motor controls the position of the reflector with respect to the feed, and a laptop computer communicates with the servo controller using a serial port.

The reflector was made of solid aluminum, and the feed arms were tin-coated copper. The F/D ratio was 0.5, and the feed impedance was 190 Ω. The antenna was fed with the so-called "splitter balun," of the type first described by C. Baum in [11]. This type of balun normally uses two 100 Ω cables connected in parallel at the feed end and in series at the antenna apex. However, it was difficult to find 0.141-inch diameter semi-rigid cable with 100 Ω impedance, so we had to use 95 Ω cables. We believe the error is small enough for our purposes. Note also that the feed arms were designed to form a 190 Ω transmission line, which is consistent with the two 95 Ω lengths of feed cable.

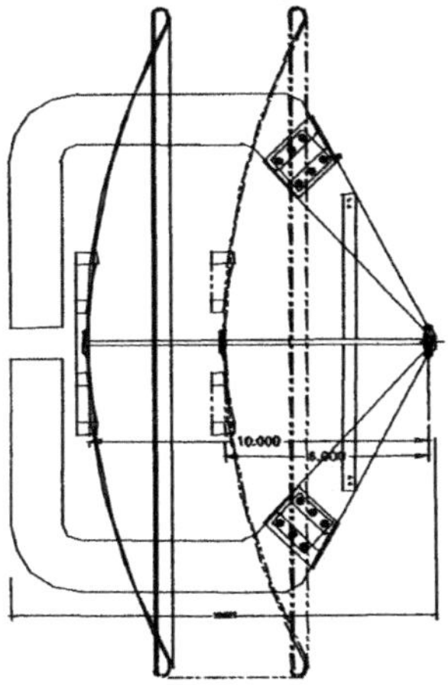

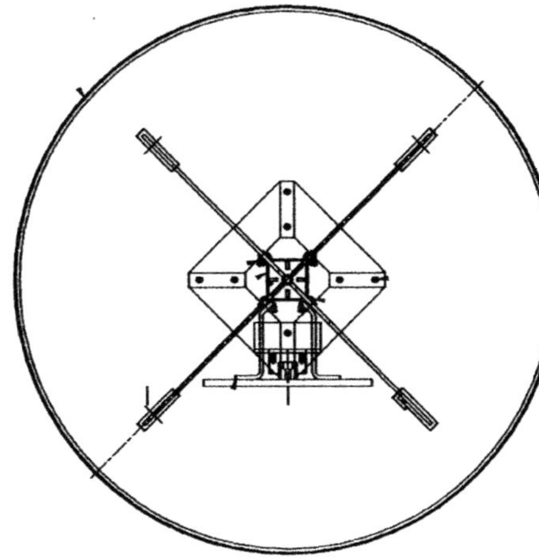

Figure 4.1. The MIRA, side view (left). Figure 4.2. The MIRA, front view (right)

It was necessary to develop sensors that could be used to measure the radiated field. Standard derivative-type sensors have a very low sensitivity, so any measurements we made with our four-volt source would have been very noisy. Thus, we decided to develop a replicating sensor, which would replicate the incident electric field from the boresight direction. The design was essentially a half TEM horn mounted against a truncated ground plane (Figure 4.3). The ground plane was solid aluminum, and the conical plate was tin-plated copper. The impedance of the horn was 50 Ω, in order to avoid a mismatch to the 50 Ω electronics. The length of the horn was 45 cm. The top plate was supported with polystyrene foam for support.

The experimental test configuration for the antenna measurements is shown in Figure 4.4. In includes a Picosecond Pulse Labs 4015C step generator, which drives a TEM sensor. On the receive end, the Multifunction IRA receives the signal, which is then sampled by the SD24 sampling head and the Tektronix 11801B Digital Sampling Oscilloscope (DSO). Data is then downloaded to a computer for processing by way of a GPIB connection.

Before measuring the response of the MIRA, it is necessary to calibrate the TEM sensor. To do so, we measured the response of two identical TEM sensors, and then backed out the impulse response, using

$$V_{rec}(t) = \frac{1}{2\pi r c f_{g,tem}} h_{tem}(t) \circ h_{tem}(t) \circ \frac{d V_{src}(t)}{dt} \tag{4.1}$$

where $h_{tem}(t)$ is the impulse response of the TEM sensor, $V_{src}(t)$ is the source voltage, and $V_{rec}(t)$ is the raw received voltage. Furthermore, $f_{g,tem}$ is the normalized impedance of the TEM sensor, c is the speed of light in free space, and the " $\circ$ " symbol indicates convolution. The only unknown in the above equation is $h_{tem}(t)$, and we solve for that using the techniques described in [12, Section III]. Note that we are using r to be the distance between the front edges of the two TEM sensors. There is some ambiguity here, since the sensors are large enough that there is some field variation over the length of the sensors.

First, we measured the source voltage, which is a clean step function with risetime of 28 ps. Next we took a measurement with the two identical TEM sensors, measured 3.35 meters apart. The resulting measured waveform consists of a relatively clean impulse, with t_{FWHM} = 50 ps. Next, the data is processed to extract $h_{tem}(t)$, and the result is shown in Figure 4.5 in the frequency domain, and Figure 4.6 in the time domain. The result in the time domain is an almost-clean impulse with $t_{FWHM} = 33$ ps.

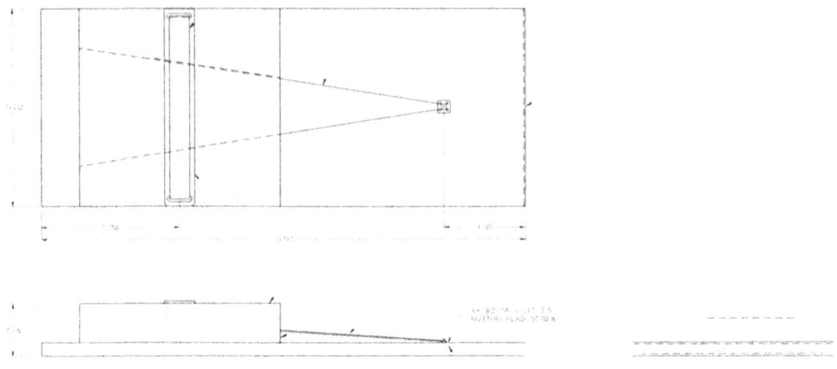

Figure 4.3. TEM Sensor. (Dimensions are inches.)

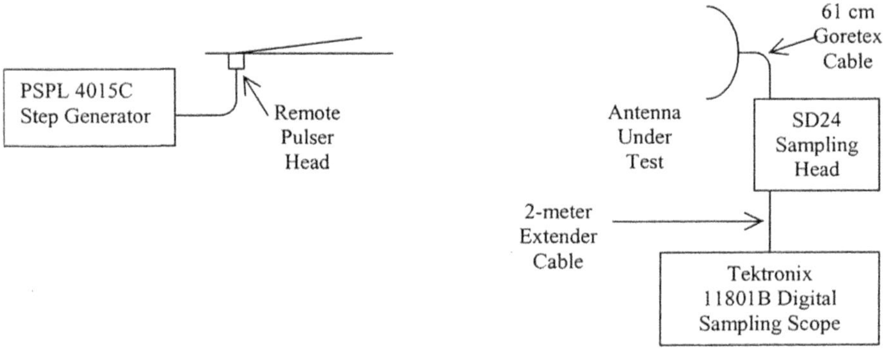

Figure 4.4.Experimental test setup

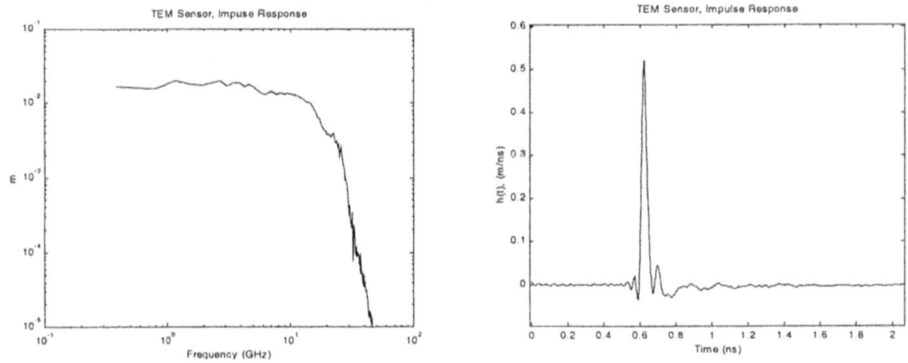

Figures 4.5 and 4.6. TEM sensor impulse response, $h_{tem}(\omega)$, in the frequency domain (left) and in the time domain (right).

Next, we provide the integral of $h_{tem}(t)$, shown in Figure 4.7. The degree to which this integral is a clean step function is a measure of whether we can use the TEM sensor data raw, without correction. The integral is a step with some sag for the first 1.5 ns. (Note that some sag is expected, due to the characteristics of all TEM horns.) The jump in this integral gives us the effective height of the sensor. That is, assuming the voltage out of the sensor is (approximately) proportional to the incident field, the jump in this integral gives us the proportionality factor, or effective height, h_{eff}. We find that the magnitude of this jump is about 17 mm. Note that the plate is 31 mm above the ground plane in the aperture, so the measured h_{eff} is a little more than half of the physical aperture height. For an ideal geometry, we expect h_{eff} to be exactly half the physical aperture height [13]. Thus, our measurement of h_{eff} is consistent with theory.

Finally, we provide the TDR of the TEM sensor. The data is shown in Figure 4.10. A flat TDR indicates a good match to the 50 Ω input. We find the TDR is flat to within a few percent.

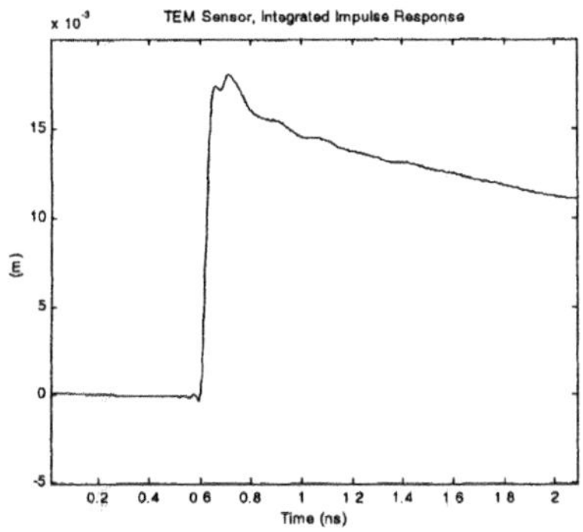

Figure 4.7. Integral of the TEM sensor impulse response.

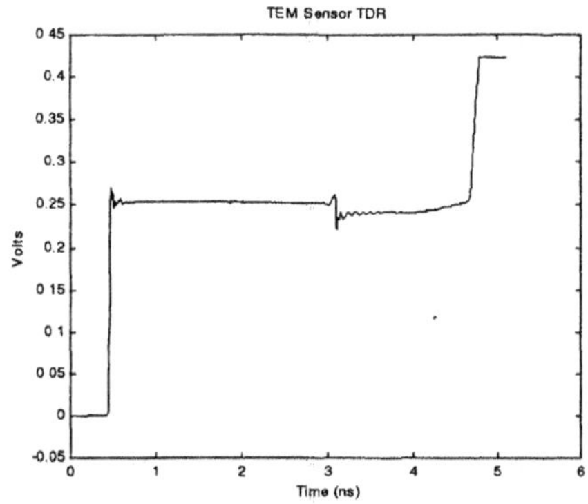

Figure 4.8. TDR of the TEM sensor (right).

Having characterized the sensor, we next measured the response of the MIRA. Note that we provide measurements of the MIRA that are slightly refined from those in [14], with various improvements to both the MIRA and the sensor. The sensor we used in the MIRA measurement presented here is slightly different that the sensor characterized above. The improvements to the sensor include a change in impedance to 100 Ω, and an improved support structure using teflon support posts instead of polystyrene foam. The new sensor, with its new impedance, has an improved sensitivity. Thus, for this sensor, h_a = 33 mm, and $h_{eff} = \tau \times h_a$, where τ is the transmission coefficient from a 100 Ω TEM sensor to a 50 Ω cable (=0.667). So for the new sensor, h_{eff} = 0.667 × 33 mm = 22 mm, which is a 30% improvement over the previous value of 17 mm. Note that h_{eff} is roughly the proportionality factor between the incident field and the measured voltage in a 50 Ω cable. In addition, the FWHM of the one-way impulse response of 35 ps, which compares closely to 33 ps above. The length of the sensor is the same as the one characterized above.

The MIRA was also improved for this paper by reducing the size of the feed point, and by shifting the structural support away from the feed point. These improvements led to an improvement in the shape of the radiated field, as we will see.

The MIRA data was taken at a distance of 3.25 meters, using the 100 Ω TEM sensor to detect the signal. The boresight response for the MIRA is shown in Figure 4.9, in both the time and frequency domains. It displays the classic time-domain waveform of a low-level prepulse, followed by a sharp impulse, with Full Width Half Max of 80 ps. Note that this is raw received voltage. The impulse is clean, without any additional humps, so this is an improvement over the results of [14].

Next, we consider the antenna pattern as a function of angle off-boresight, and as a function of focus position. Using the definition of focal position provided earlier in equation

(3.20), we made our measurements at focal positions of f_f = 1.0, 0.85, and 0.7. Furthermore, data were taken at positions of 0°, 7.5°, and 15° off boresight in the H-plane. The raw data for all nine waveforms is shown in Figure 4.10. Once again, we see the peak field is reduced as we go off-boresight, and as we become more defocused.

With the above data, we can estimate the beamwidths of this antenna for various focus settings in the H-plane. We use here the definition of Half Field Beam Width (HFBW) as described in Section III. These beam widths are tabulated in Table 4.1, and they are compared to the theoretical values calculated earlier in Section III. We find the trends agree, but the measured values are generally larger than the predictions.

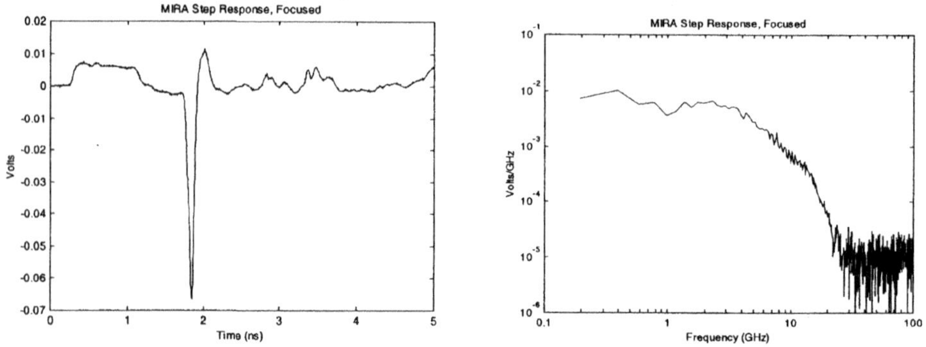

Figure 4.9. Boresight step response of the Multifunction IRA, in the time domain (left) and frequency domain (right).

Table 4.1. MIRA Half Field Beam Widths

	H-Plane Half Field Beam Width (HFBW)	
$f_f = F_2/F$	Theory (Section III)	Measurement
1.00	8°	12°
0.85	15°	28°
0.70	36°	60°

Next, we provide the TDR of the MIRA in Figure 4.12, using the 0.25-Volt source built into the SD-24 sampling head. We want to see a smooth match to 50 Ohms, but there are some reflections at the feed point. This represents some improvement over the TDR of the first version of the MIRA in [14], and we believe this is about the best we will be able to achieve.

Finally, we deconvolve the radiated field in focus from the sensor response and measurement system. The deconvolved $h_{IRA}(t)$ is shown in Figure 4.12, in both the frequency and time domains. The FWHM of the resulting waveform is 53 ps. When we integrate the time domain function, we should obtain an area under the impulse of $a/2^{1/2}$, where a is the radius. After integrating the pulse, we find an area of 103 mm, compared to a theoretical value of 161 mm. Thus, we find a radiated impulse area of 64% of the expected value. We consider this to be reasonable agreement, although we would like to see better agreement in future designs.

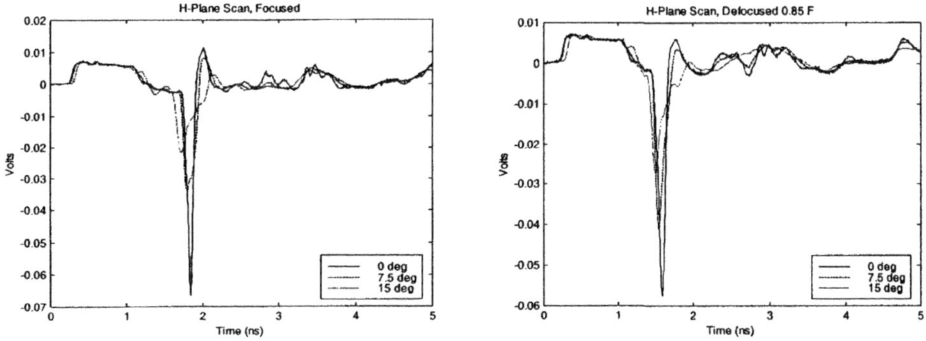

Figure 4.10. Antenna response for MIRA with f_f = 1.0 (left), and f_f = 0.85 (right). Antenna responses are for 0, 7.5, and 15 degrees off boresight in the H-plane.

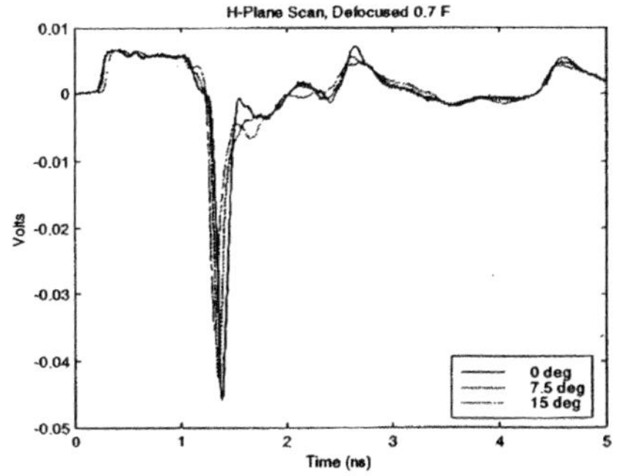

Figure 4.10 (cont'd) Same as above with f_f = 0.75.

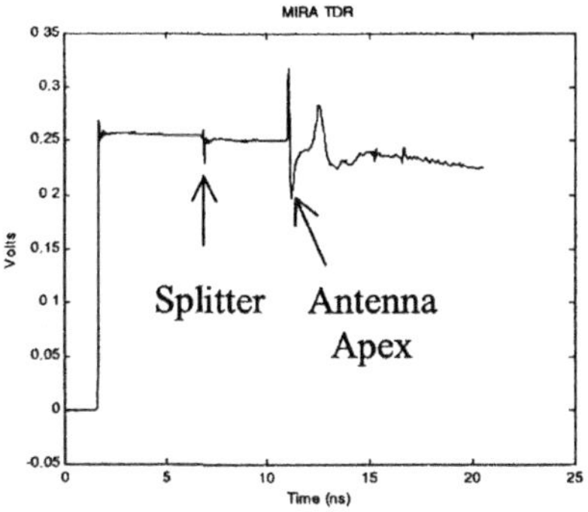

Figure 4.11. TDR of the MIRA.

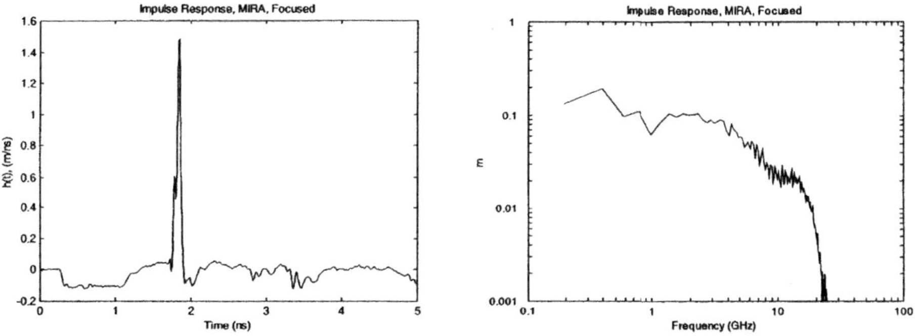

Figure 4.12. MIRA $h(t)$ after correction for the sensor, measurement system, and cabling, in the time domain (left) and frequency domain (right).

V. Conclusions

We have provided here the theory that describes the MIRA, and we have built and tested a model with 46 cm diameter. Satisfactory agreement was obtained between theory and measurements. The general technology of defocusing an IRA has been demonstrated, thereby increasing the flexibility of a single antenna to handle multiple missions.

References

1. C. E. Baum and E. G. Farr , Impulse Radiating Antennas, pp. 139-148 in H. L. Bertoni *et al* (eds.), *Ultra-Wideband, Short-Pulse Electromagnetics*, New York, Plenum Press, 1993.
2. E. G. Farr, C. E. Baum, and C. J. Buchenauer, Impulse Radiating Antennas, Part II, pp. 159-170 in L. Carin and L. B. Felsen (eds.), *Ultra-Wideband, Short-Pulse Electromagnetics 2*, New York, Plenum Press, 1995.
3. E. G. Farr, C. E. Baum, and C. J. Buchenauer, Impulse Radiating Antennas, Part III, pp. 43-56 in C. E. Baum *et al* (eds.), *Ultra-Wideband, Short-Pulse Electromagnetics 3*, New York, Plenum Press, 1997.
4. C. E. Baum, Radiation of Impulse-Like Transient Fields, Sensor and Simulation Note 321, November 25, 1989.
5. E. G. Farr and C. E. Baum, The Radiation Pattern of Reflector Impulse Radiating Antennas: Early-Time Response, Sensor and Simulation Note 358, June, 1993.
6. E. G. Farr, Off-Boresight Field of a Lens IRA, Sensor and Simulation Note 370, October 1994.
7. E. G. Farr and G. D. Sower, Design Principles of Half Impulse Radiating Antennas, Sensor and Simulation Note 390, December 1995.
8. E. G. Farr and C. A. Frost, Development of a Reflector IRA and a Solid Dielectric Lens IRA, Part I: Design, Predictions and Construction, Sensor and Simulation Note 396, April 1996.
9. C. J. Buchenauer and R. Marek, Antennas and Field Sensors for Time Domain Measurements: An Experimental Investigation, pp. 197-208 in L. Carin and L. B. Felsen (eds.) *Ultra-Wideband, Short-Pulse Electromagnetics 2*, New York, Plenum Press, 1995.
10. C. E. Baum, Focused Aperture Antennas, Sensor and Simulation Note 306, May 1987.
11. C. E. Baum, Configurations of TEM Feed for an IRA, Sensor and Simulation Note 327, April 1991.
12. E. G. Farr and C. A. Frost, Compact Ultra Short Pulse Fuzing Antenna Design and Measurements, Sensor and Simulation Note 380, June 1995.
13. C. E. Baum, Aperture Efficiencies for IRAs, Sensor and Simulation Note 328, June 1991.
14. E. G. Farr, C. E. Baum, and W. D. Prather, Multifunction Impulse Radiating Antennas: Theory and Experiment, Sensor and Simulation Note 413, November 1997.

ANTENNA-SOURCE INTEGRATED ULTRA-WIDEBAND ELECTROMAGNETIC PULSES RADIATING DEVICE

Xiaoxin Song, Guozhi Liu, Yajun Fan, Xiaolong Liu, Feng Liu

Northwest Institute of Nuclear Technology P.O.BOX 69-13, 710024,Xi'an, P. R. China

INTRODUCTION

In 1901 Marconi tested the propagation property of electromagnetic fields using the folded monopole antenna, which was 150ft in height, 200ft in width, and excited at the tip of the monopole by the shorting of a spark gap. In the study of Ultra-wideband electromagnetic pulse technology, we noted that with some improvements, the prototype method Marconi used takes great advantages over the popular source-antenna-separated microwave launching system, especially for producing and launching Ultra-wideband(UWB), high power level, and impulsive electromagnetic fields. As the power level increasing and risetime decreasing, the transmission line approximation turns out to be unsuitable for analyzing the field distribution of UWB systems with baluns and transmission lines, owing to the transit-time dispersion, skin and dielectric losses, and electrical breakdown effects. Therefore, the introduction of baluns and transmission lines into high power level UWB systems will degrade the system performance, and make the theoretical analysis of the loss mechanism extremely difficult. By integrating the source and antenna directly, one can reduce the loss significantly and design a compact, simple, and robust device that can be used to produce and radiate UWB pulses simultaneously. This paper presents the design and experiment of the so-called "source-antenna-integrated" device, including the theoretical analysis and numerical simulation of the conical antenna.

THEORETICAL ANALYSIS OF CONICAL ANTENNA

The conical antenna is shown in figure 1 and consists of two conducting cones with half angle θ_0. It is excited at the center. The conical antenna structure will support a transverse spherical(TEM) electromagnetic wave, similar to that on a conventional transmission line. For an infinite length conical antenna, the radiated field at distance r, elevation angle θ is expressed as

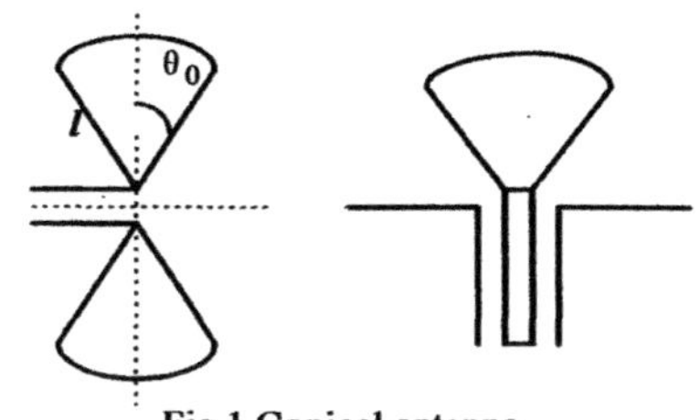

Fig.1 Conical antenna

$$E_\theta = \frac{1}{r\sin\theta}\left(E_+ e^{-jkr} + E_- e^{jkr}\right)$$

$$H_\phi = \frac{1}{r\sin\theta}\left(\frac{E_+}{\eta} e^{-jkr} - \frac{E_-}{\eta} e^{jkr}\right)$$

Where E_+ and E_- decided by the excitation, and η is the wave impedance of the space. It is obvious

Ultra-Wideband, Short-Pulse Electromagnetics 4
Edited by Heyman *et al.*, Kluwer Academic / Plenum Publishers, New York, 1999.

that the fundamental solution consists of radially outward and inward-propagating spherical waves with amplitude E+ and E-, also clear is that the solution is frequency independent. So if the excitation at the feed point is $f(t)$, then in a homogeneous dielectric media, the tangentical electric field E_θ at time t, elevation angle θ and distance r from the feed point is

$$E_\theta(t,r,\theta)=\frac{f(t-r\sqrt{\varepsilon_r}/c)}{(r\sin\theta)(\ln ctg\,\theta_0/2)} \quad \text{for } \theta_0\le\theta\le\pi-\theta_0 \qquad (1)$$

Equation (2) will hold as long as no signal can propagate to the observer by way of a point of broken symmetry. In the case of infinite length cone with a medium of dielectric constant ε_r filling the region $r<r_0$, the outgoing waves will remain spherical, with the modification of transmission coefficient for the dielectric boundary, thus the resulting field will be

$$E_\theta(t,r,\theta)=\frac{2\sqrt{\varepsilon_r}}{\sqrt{\varepsilon_r}+1}\frac{f\left(t-\frac{r+(\sqrt{\varepsilon_r}-1)r_0}{c}\right)}{r\sin\theta\ln ctg\frac{\theta_0}{2}} \qquad (2)$$

For conical antenna of finite length l, the solution becomes much more complicated. Let the angle of the cones being θ_1 and θ_2, respectively, we start the analysis with dividing the space into two regions: $\theta_1<\theta<\theta_2$, $0<r<l$ for the inner region, and $r>l$ for the outer region. By evaluating the Marxwell's equations for a sinusoidal excitation we have

$$\left.\begin{aligned}E_{1r}&=\sum_\nu \nu(\nu+1)T_\nu(\theta)\tilde{J}_\nu(k_0r)/j\omega\mu_0\varepsilon_0r^2\\ E_{1\theta}&=\sum_\nu T'_\nu(\theta)\tilde{J}'_\nu(k_0r)/jr\sqrt{\mu_0\varepsilon_0}\\ H_{1\varphi}&=-\sum_\nu T'_\nu(\theta)\tilde{J}_\nu(k_0r)/\mu_0r\end{aligned}\right|\quad \left.\begin{aligned}E_{2r}&=\sum_n B_nn(n+1)P_n(\cos\theta)\tilde{H}_n^{(2)}(k_0r)\,j\omega\mu_0\varepsilon_0r^2\\ E_{2\theta}&=\sum_n B_nP'_n(\cos\theta)\sin\theta\tilde{H}_n^{(2)\prime}(k_0r)/jr\sqrt{\mu_0\varepsilon_0}\\ H_{2\varphi}&=-\sum_n B_nP'_n(\cos\theta)\sin\theta\tilde{H}_n^{(2)}(k_0r)/\mu_0r\end{aligned}\right\}\qquad (3)$$

Where $T(\theta)=C_\upsilon[p_\upsilon(\cos\theta)-p_\upsilon(-\cos\theta)]$ $\tilde{J}_\nu(k_0r)=k_0rj_\nu(k_0r)$, $\tilde{H}_n^{(2)}(k_0r)=k_0rh_n^{(2)}(k_0r)$, $j_\nu(k_0r)$ and $h_n^{(2)}(k_0r)$ are the spherical Bessel and Hankel functions, respectively. ν is chosen to be the roots of the characteristic equation, $P_\nu(\cos\theta)-P_\nu(-\cos\theta)=0$.

The investigation of the solution for conical antenna of finite length shows that because of the termination of the cones at $r=l$, a perturbing field is excited in the inner region corresponding to other non-TEM modes as well as a radiated field in the outer region. This causes an effective terminating impedance Z_t at $r=l$ that is different from an ideal open circuit for infinite length cones. It is obvious that the additional modes excited at $r=l$ decay in amplitude as r decreases. The antenna input impedance is quite similar to the usual transmission line formula:

$$Z_{in}=\frac{1}{Y_{in}}=Z_c\frac{Z(l)\cos k_0l+jZ_c\sin k_0l}{Z_c\cos k_0l+jZ(l)\sin k_0l}$$

where Z_c is the characteristic impedance of the antenna given by $Z_c=120\ln ctg\,{}^{\theta_0}\!/_2\ (\Omega)$.

THE SIMULATION OF THE CONICAL ANTENNA

To better understand the conical antenna of finite length, The FDTD(finite difference of time domain) method was employed, allowing us to simulate the performance of the antenna of various dimension excited by different time-domain pulses. Figure 2 shows the simulation structure, the 30cm monocone antenna is mounted perpendicular to a grounded conducting plane of 400cm radius functioning the mirror for the antenna, an additional 50Ω, 100cm coaxial transmission line feeding the excitation to the antenna. In time domain, the excitation is a 500ps rise-and-fall time Gaussian pulse, as shown in figure 3. An absorbing boundary is exerted at r=400cm and z=800cm to analogue the infinite free space. The E-field distributions at 7, 11,.13ns in the space, as well as the time slices of *Er* was drawn and shown in figure 4(a)~(d).

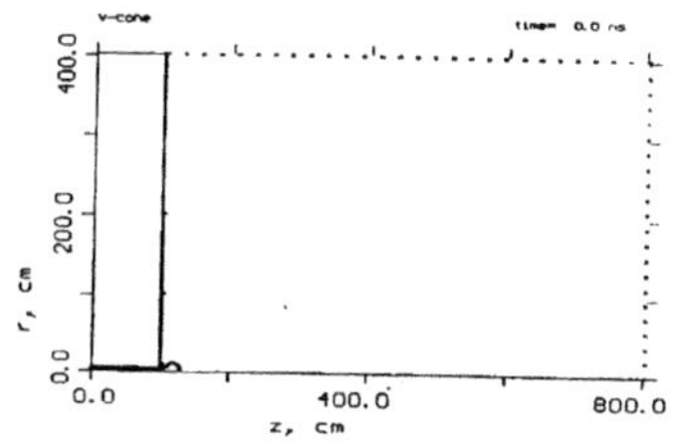

Fig.2. Simulation structure

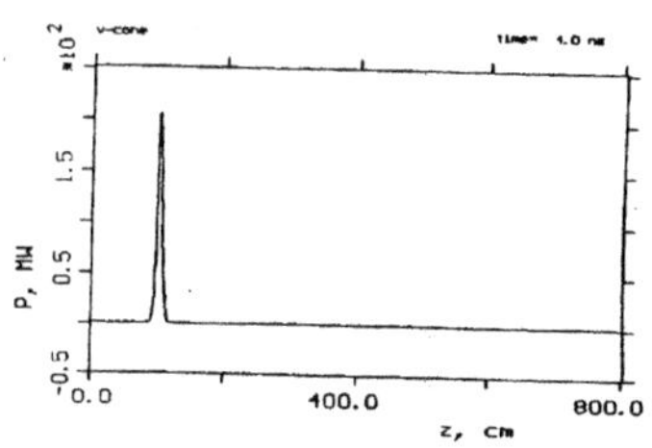

Fig.3. The excitation pulse to the antenna

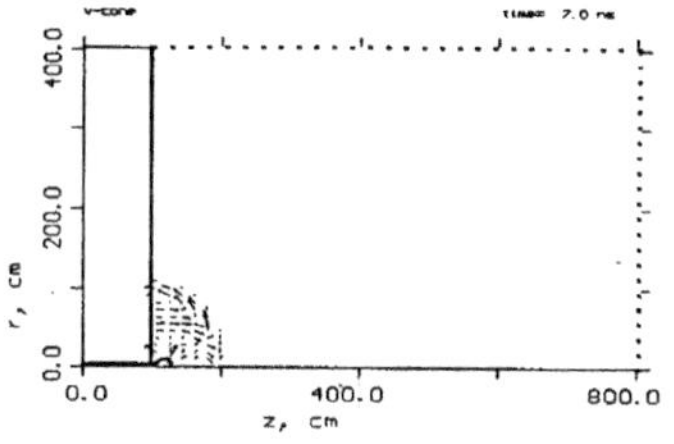

Fig.4(a) E-field distribution at 7ns

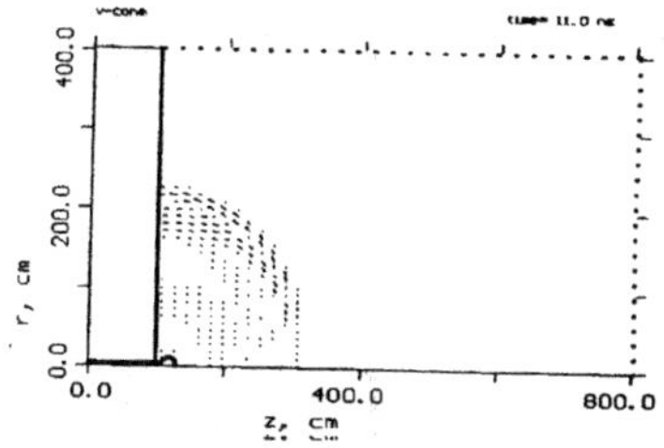

Fig.4(b) E-field distribution at 11ns

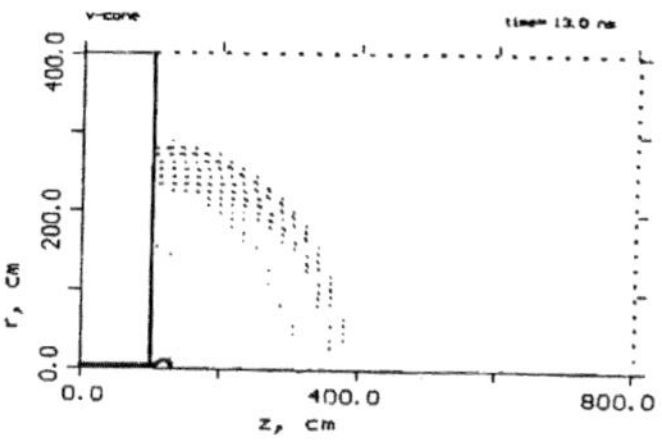

Fig.4(c) E-field distribution at 13ns

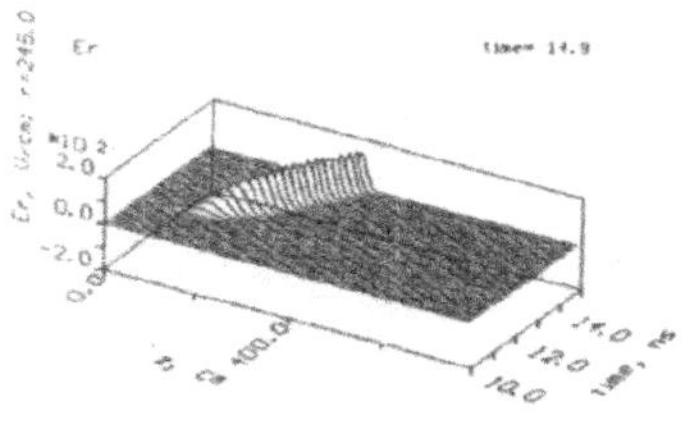

Fig.4(d) Time slices of the E-fields

The reflected field from the antenna was monitored inside the transmission line at the distance of 50cm from the feed point and shown in figure 6. Radiated fields at different elevation angle with respect to r=0 for a fixed distance 300cm from the feed point were concluded and presented in figure 6.

The simulation results suggests that the conical antenna is suitable for radiating UWB signals, however, two unmatched points could be notified and greatly affect the antenna performance. The first one is the feed point of the antenna, which arises the front of reflection monitored in the transmission(see fig.5). The second unmatched point is the end of the cone, giving rise to the complex field distribution of the inner region, as well as the "tailing" of the reflected pulse.

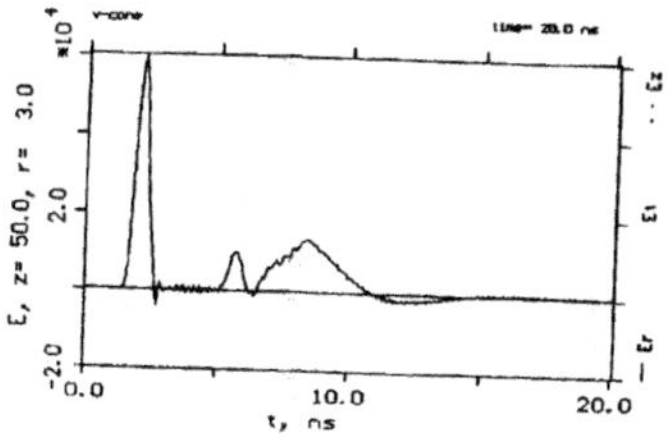

Fig.5 Reflected pulse from the antenna

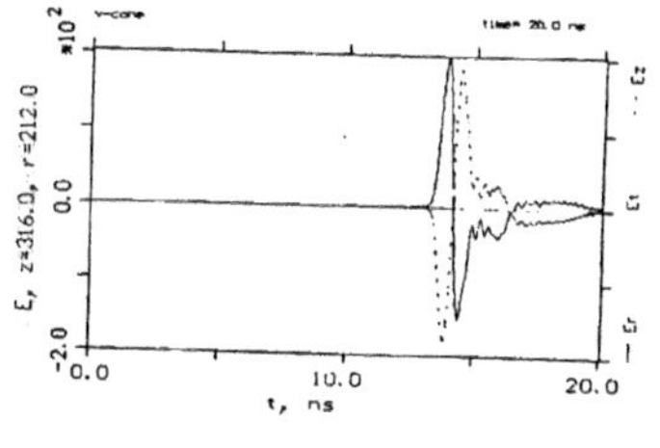

Fig.6(a) E-field at 300cm, 20 degree elevation angle

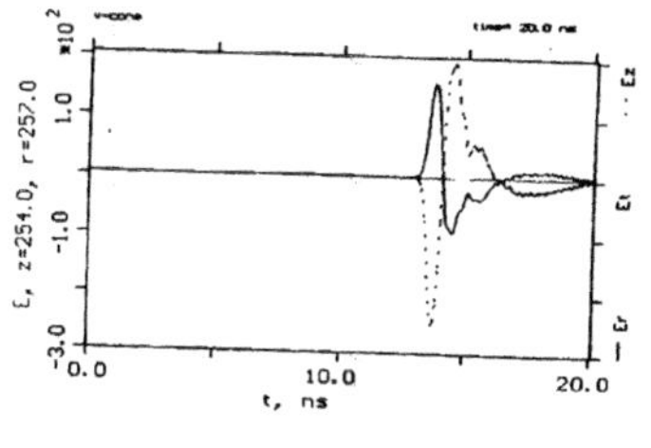

Fig.6(b) E-field at 300cm, 60 degree elevation angle

EXPERIMENTAL SETUP AND RESULTS

The source-antenna-integrated device for generating and radiating UWB pulses is consists of a pulse transformer and a 30-degree monocone antenna. The pulse transformer outputs a 140kV, 1 μs pulse to charge the antenna through an inductor. An oil spark switch located at the feed point of the antenna. The experimental setup of the monocone antenna ware shown in figure 7. After the oil switch is shorted, a high power short pulse begins to form and radiate. The pulse width is determined by the length of the antenna to be approximately 2ns correspondent to the 30cm height of the cone. The rise-and-fall time of the radiated pulse is determined by the combination of the oil spark gap switch performance and the radiation characteristics of the antenna.

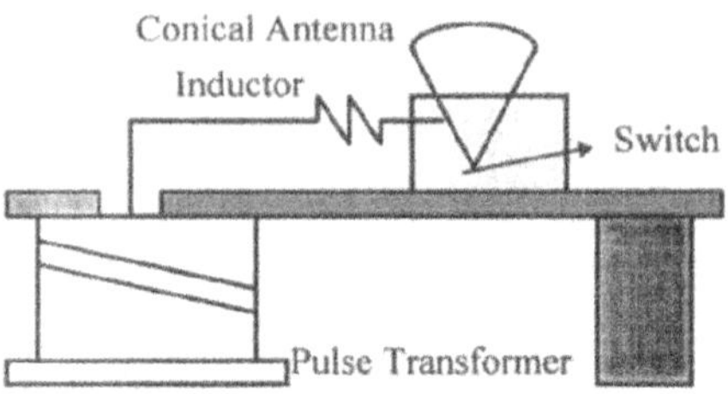

Fig.7 The experimental setup

The measured output pulse of the PT by a capacitor divider locates at its output end was shown in Figure 8, the peak output voltage of the PT is 120kV. The radiated fields were detected by a TEM horn in dark room. As shown in Figure 9, the peak radiated E-field intensity away from the monocone antenna by 300cm, with the elevation angle being 20 degrees is 19kV/m, the pulse risetime and width is 600ps and 2ns respectively.

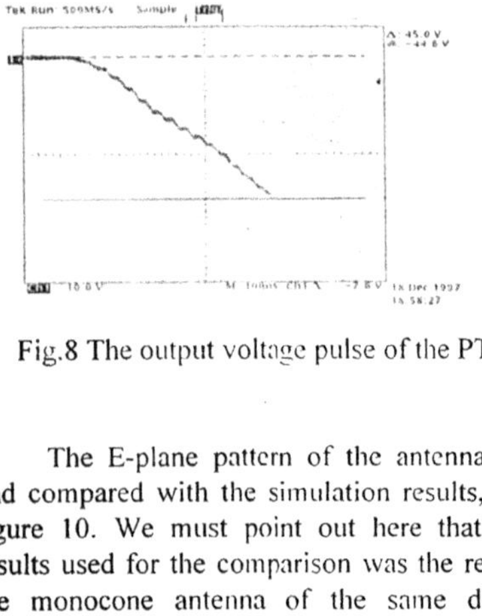

Fig.8 The output voltage pulse of the PT

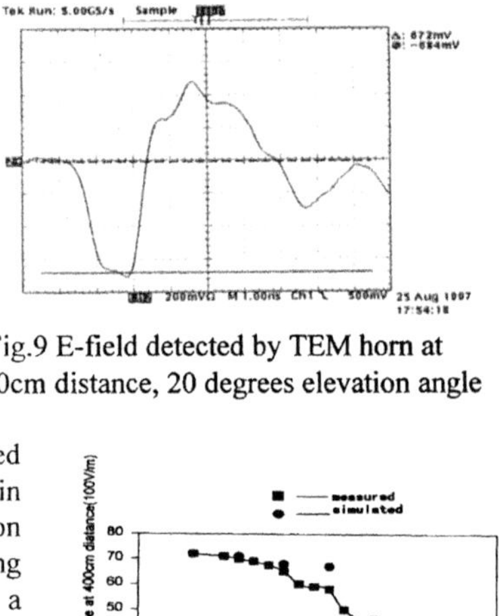

Fig.9 E-field detected by TEM horn at 300cm distance, 20 degrees elevation angle

The E-plane pattern of the antenna was measured and compared with the simulation results, both shown in figure 10. We must point out here that the simulation results used for the comparison was the results of feeding the monocone antenna of the same dimension by a Gaussian pulse of 2ns width. Since the shortening of the oil spark switch was replaced by the Gaussian pulse excitation in the simulation, owing to the difficulty in constructing a proper model for the switch, the comparison between the experiment and simulation only has limited meaning.

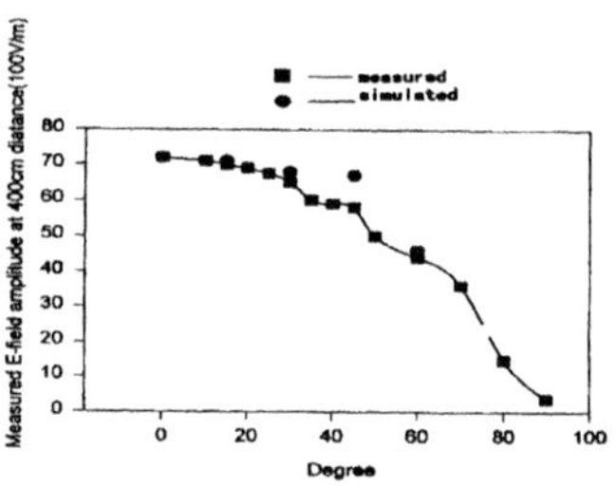

Fig.10 Comparison between the measured and simulated radiation pattern

CONCLUSIONS

The source-antenna-integrated device can produce and radiate high power level fast risetime pulses, with carefully maintaining the symmetry structure of the monocone antenna and designing of the oil spark switch. It is compact, simple, and robust. The device could be used to feed impulsive radiating antenna and for other applications requiring high peak power level electromagnetic pulse(EMP) radiation.

Acknowledgments

The authors are grateful to Mr. Zhimin Song , Mr. Qiushi, and Mr. Hao Shao for their generous help.

LARGE CURRENT RADIATOR FOR THE SHORT ELECTROMAGNETIC PULSES RADIATION [1]

Gennadiy P.Pochanin

Institute for Radiophysics and Electronics
National Academy of Sciences of Ukraine,
12 Acad. Proskura St., Kharkov 310085,
Ukraine.
E-mail: gpp@ire.kharkov.ua

INTRODUCTION

One of the most important problems in modern ground penetrating radar designing is creating the ultra wide band (UWB) antennas for short sounding electromagnetic pulses (SP) radiation. The most important requirement to antenna of the kind is elimination of "ringing". It deals with a necessity of the reflected (scattered) signals reception directly after sounding signals radiation. So, any afterpulse "ringing" is very strong clutter which decrease the quality of radar. Thus the "clear" pulse of radiated field is needs for near zone radiolocation.

Usually the wide band dipole is used to radiate SP signals. The field radiated by pulses of the current reflected from the ends of the dipole neutralizes the field radiated when driving current flew from pulse generator to the ends (fig.1.a). Therefore, in order to increase the antenna efficiency the spatial duration of electromagnetic field radiated by traditional dipole antenna must be less than half length of dipole. But in this case there is a number of resonant frequencies for antenna in the spectrum of signal. It leads to onset of oscillations in radiated field.

One of the promising UWB/SP radiators is proposed by Henning F. Harmuth[1,2,3]. This radiator named Large Current Radiator (LCR) because it is possible to create a large amplitude current in radiating element with relatively small driving voltage. The LCR is inherently non-resonating radiator. Principally the LCR is a segment of a line conductor through which a current pulse I(t) of short duration but large amplitude is driven (fig.1.b).

---

[1] This work was supported by Geophysical Survey System Inc. of North Salem, NH and by Prof. Henning. F.Harmuth of Catholic University of America, Washington DC and partly by Science and Technology Center in Ukraine (project no.366).

Driving voltage applies to ends of the LCR. There is no reflected current if the radiating element is matched in driving area with pulse generators G1 and G2. So, the cause of afterpulse oscillation excite is absent. Therefore the LCR allows to radiate electromagnetic waves with an arbitrary time dependence.

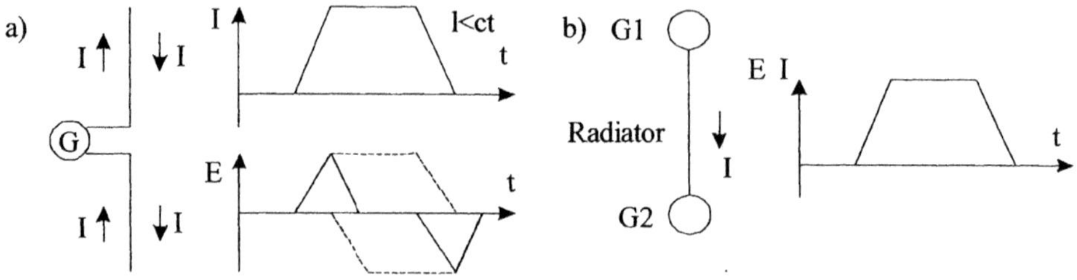

Figure 1. a) Radiation by dipole; b) Radiation by LCR.

The actual implementation of a LCR was proposed by Harmuth[1]. It is a conducting closed loop shown in Fig.2. A current pulse flowing through the radiator produces the field strength. In order to suppress radiation from the other conductors it was proposed to shield it by the ferrite or metal ferrite plate.

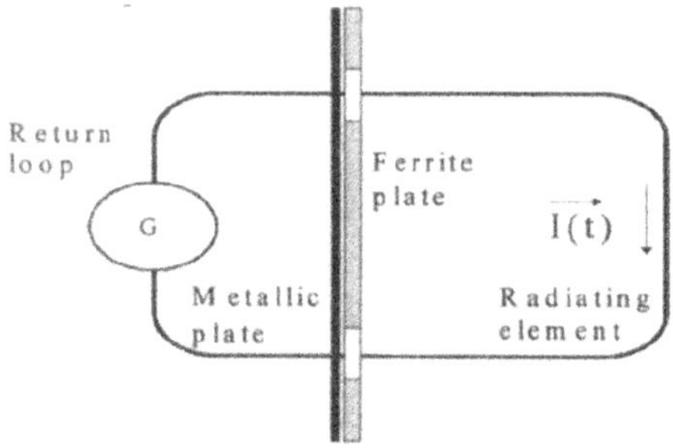

Figure 2. Large current radiator. Basic circuit.

The report discusses the performance of improved design of LCR.

THE LCR DESIGN

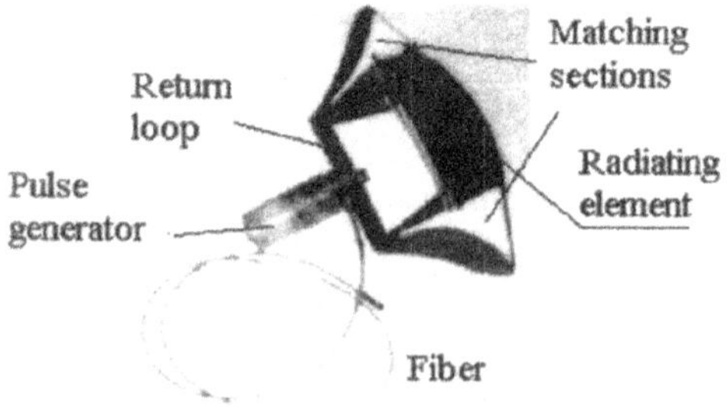

Figure 3. LCR design.

The LCR developed is shown at the figure 3. We distinguished a number of functional elements of the LCR design: radiating plate, pulse generator, return loop, matching sections, and analyzed each of them in order to design the effective LCR.

Radiating Element. The basic equations for the electric and magnetic fields strengths E and H radiated by a small dipole are as follows[1]:

$$\vec{\mathbf{E}} = \frac{\mathbf{Z_0}\ell}{4\pi\mathbf{c}}\left[\frac{\mathbf{1}}{\mathbf{r}}\frac{\mathbf{dI}}{\mathbf{dt}}\frac{\vec{\mathbf{r}}\times(\vec{\mathbf{r}}\times\vec{\ell})}{\ell\cdot\mathbf{r}^2}+\left(\frac{\mathbf{c}}{\mathbf{r}^2}\mathbf{I}+\frac{\mathbf{c}^2}{\mathbf{r}^3}\int\mathbf{I}\,\mathbf{dt}\right)\cdot\left(\frac{3(\vec{\ell}\cdot\vec{\mathbf{r}})\cdot\vec{\mathbf{r}}}{\ell\cdot\mathbf{r}^2}-\frac{\vec{\ell}}{\ell}\right)\right] \tag{1}$$

$$\vec{\mathbf{H}} = \frac{\ell}{4\pi\mathbf{c}}\left(\frac{\mathbf{1}}{\mathbf{r}}\frac{\mathbf{dI}}{\mathbf{dt}}+\frac{\mathbf{c}}{\mathbf{r}^2}\mathbf{I}\right)\frac{\vec{\ell}\times\vec{\mathbf{r}}}{\ell\cdot\mathbf{r}} \tag{2}$$

Here ℓ is the length of the dipole, Z_0 is the wave impedance of free space, and r is the location vector. For large values of r the field strengths E and H are determined by the terms in (1) and (2) varying like $1/r$. Poynting's vector $P = E \times H$ at large distances equals

$$P\left(\frac{1}{r^2}\right) = E\left(\frac{1}{r}\right)\times H\left(\frac{1}{r}\right) = Z_0\left(\frac{1}{4\pi rc}\right)^2\left(\frac{dI}{dt}\right)^2 \sin^2\alpha\frac{r}{r} \tag{3}$$

where $\mathbf{sin}\,\alpha = \frac{|\ell\times\mathbf{r}|}{\ell\mathbf{r}}$. Integrating (3) over a sphere yields the radiated power

$$\mathbf{P} = \frac{\mathbf{Z_0}\ell^2}{6\pi\mathbf{c}^2}\left(\frac{\mathbf{dI}}{\mathbf{dt}}\right)^2. \tag{4}$$

Expressions (1)-(4) point that efficiency of radiation depends on radiator length ℓ, amplitude of the current pulse $I(t)$ flowing through the radiator, and the rate of change $I(t)$ of the current.

For the current pulse with a certain duration driven through the radiator there is an optimal radiator length ℓ that permits the most efficient radiation without distortion of the radiated pulse. If the length ℓ of the radiator is fixed one may increase the radiation efficiency by choosing a radiator structure with small inductance L.

The radiating element of the LCR consists of 5 plates. Each of them is 23cm long. They are 7cm wide in the middle and 5cm at the ends. The shape of the radiating elements permitted the reduction of the inductance to 30nH.

Generator for Short Pulses Using the S-Diode. Pulse generator with S-diode AA742B[4] and a relaxation circuit is used to produce current pulses with a peak amplitude of 12A, rise time of 0.5ns, decay time of 30ns, and pulse repetition rate up to 1kHz.

Electrical circuit of the LCR is shown at the figure 4. In this LCR design the driving voltage is provided by a storage battery V. It is fed to a voltage stabilizer SV that maintains a constant voltage output during the discharge of the storage battery. A DC/DC high-voltage converter is used to charge storage capacitors C1 and C2 up to the avalanche breakdown threshold of the S-diode. Avalanche breakdown of the S-diode occurs when this threshold is exceeded. The discharge current flows through the LCR circuit and produces a short radiated electromagnetic pulse. Resistors R1 and R2 with capacitors C1 and C2 are needed as repetition rate controlling circuit.

The characteristics of S-diode make it possible to drive the radiator symmetrically, to keep the energy consumption of the relaxation circuit low, and to use a storage battery embedded into the LCR as a power supply. The triggering of S-diode with an infrared semiconductor laser is possible. This design allows to eliminate the interfering objects - such as wires and cables - from the surroundings of the radiator and the resulting reduction of distortions of radiated pulses.

All these allow to achieve the large field strengths and at the same time to improve the space-time characteristics of the radiated signals.

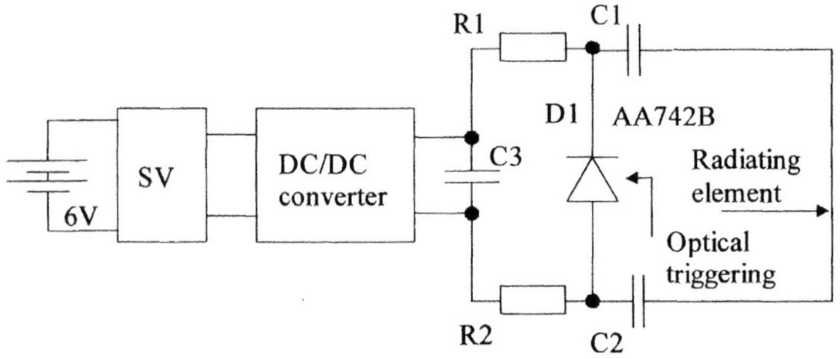

Figure 4. Electrical circuit of the LCR.

The pulse generator that drives the radiator is designed to be part of the radiator. This eliminates problems of transmission of the driving pulse and energy losses due to any mismatch between pulse generator, transmission link, and radiator; ringing at the end of the pulses is thus greatly reduced.

Return Loop. A cylindrical combined double-layer shield[5] is used to separate the space with radiator from the space with return loop and thus to eliminate the interfering radiation from the return loop. If the return loop is within the cylinder along its axis, the energy of the pulse driver is localized in the internal space and its effect on the radiated electromagnetic field strengths is eliminated.

The inner shield is made of metal and the outer shield is made of ferrite. The pulse generator **G** drives the current pulse $I(t)$ through the radiating plate. In addition the current pulse $J(t)$ is produced in the outer surface of the metallic shield. But the amplitude of $J(t)$ is small because the metal shield is covered by a ferrite layer. Most of the current $J(t)$ energy is absorbed by the ferrite shield and the radiation from the return loop is suppressed effectively.

The electromagnetic wave radiated by the current $J(t)$ is further reduced by the large inductance of the outer surface of the metallic shield that reduces the derivative dJ/dt. According to Eqs.(1) and (2) a small time derivative of the current implies small field strengths.

Hence, the combination of a metallic and a ferrite shield is very effective for the suppression of the return loop radiation.

Matching Sections. Metallic plates connected to the ends of the metallic cylinder shield are used as additional matching plates.

The use of matching sections to match the impedance of the radiating plate and the return loop provides the following benefits:

- A decrease of the total inductance of the LCR.
- Smoothing of the discontinuities of the impedance in the radiator driving circuit and the resulting reduction of the energy reflected back to the current generator.

Therefore, the additional matching sections permit to increase the energy radiated by the LCR, to decrease amplitude and duration of ringing following a pulse. It is possible due to 1-concentration of the energy in the space between the matching sections and the additional matching plates; 2-reduction of the energy losses due to radiation from the matching sections and a resulting increase of the amplitude of the current flowing through the radiating plate; 3-reduction of the amplitude of the driving current pulse reflected from opposite ends of the radiating element.

The impedance variation of the additional matching sections was chosen experimentally to obtain optimal matching between the return loop and the radiator for fixed dimensions of the matching sections.

EXPERIMENTAL RESULTS

Angular Variation of Field Strength

Figures 5 a) and b) show the time variation of the electric field strength of a radiated signal. In the direction of the azimuth angle $\beta = 0°$ and the elevation angle $\alpha = 0°$ the variation of the field strengths is similar to one cycle of a sinusoidal oscillation. The peak amplitude of the electric field strength in the first half cycle is 56V/m, which is 1.2 times the peak in the second half cycle. The duration of the whole signal is 3.5ns.

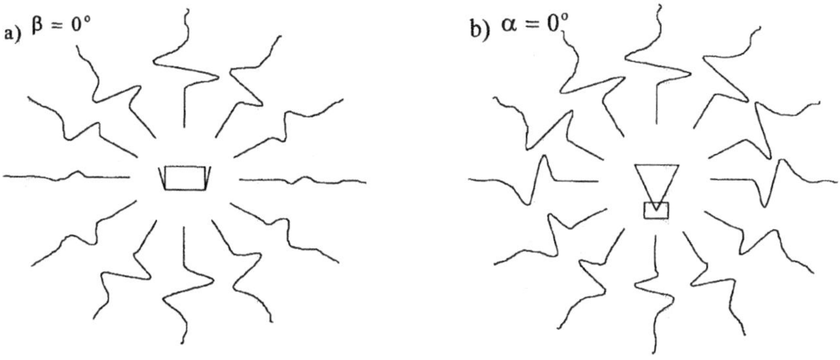

Figure 5. Form of radiated signal a) in the E-plane; b) in the H-plane.

E- plane. The maximum amplitude of the electric field strength in Fig.5a is obtained for an elevation angle $\alpha = 0°$. A minimum of the amplitude is observed for the two angles $\alpha = 90°$ and $\alpha = 270°$. When rotating through the elevation angles α the electric field strength changes polarity twice at $\alpha = 90°$ and $\alpha = 270°$. This is due to the change of the direction of the current flow in the radiator relative to the polarity of the receiving antenna.

There should be no radiation in Fig.5a for $\alpha = 90°$ and $\alpha = 270°$. However, the length of the radiating plate creates a delay of about 0.75ns between the radiation coming from opposite ends of the plate. This prevents a complete cancellation of the electric field strength for $\alpha = 90^\circ$ and $\alpha = 270^\circ$.

H- plane. The direction of radiator current flowing does not depend on the azimuth angle β, which is quite different from the rotation through the elevation angles for the field strength in Fig.5a.

This makes the time variation and the peak amplitude of the field strength in Fig.5b more independent of the azimuth angle. The observable dependence of amplitude of field strength takes place due to the matching sections. In the direction $\beta = 0^\circ$ the peak amplitude is 1.5 times the peak amplitude in the direction $\beta = 180^\circ$.

Figures 5 a) and b) demonstrate that the angular variation of the electric and magnetic field strengths produced by an LCR is similar very much to that of a Hertzian electric dipole.

Radiation Patterns of the LCR

Figure 6 shows the radiation patterns of the LCR normalized to be 1 in the direction of maximum radiation:

$$F_A^H(\alpha) = \frac{U_a(\alpha, t = t_{max})}{U_a(\alpha = \alpha_{max}, t = t_{max})}$$

$$F_A^E(\beta) = \frac{U_a(\beta, t = t_{max})}{U_a(\beta = \beta_{max}, t = t_{max})}$$

Here t_{max} is the time when the field strength has its maximum, while α_{max} and β_{max} are the angles of maximum field strength.

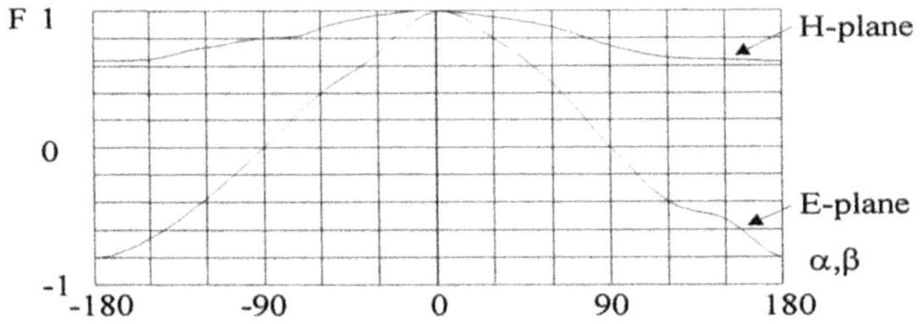

Figure 6. Radiation patterns.

The beam width of the LCR where the field strength drops to one half its peak value is 70° for the E-plane pattern. The radiation pattern for the H-plane varies between 1 and 0.63 for the whole range of β from 0° to 360°.

CONCLUSIONS

The results obtained show that the LCR is a very promising radiator for nonsinusoidal carrier-free signals.

The use of a relaxation pulse generator with an S-diode provides good energy efficiency. This makes it possible to use a battery as the power supply. The use of a battery and the ability to trigger the S-diode via infrared radiation removes many sources of distortion from the vicinity of the radiator. A three layer metal-ferrite-ferrite shield and additional matching sections suppress radiation from the return loop significantly. Two matching sections reduce distortions and increase the amplitude of the current pulses driving the radiating plates.

The designed LCR has the following features:

- peak current driven through the LCR is 12A;
- time variation of the radiated electric field strength resembles one period of a sinusoidal oscillation with the following characteristics:
  - rise time of the electric field strength from 0 to $|E_{max}|$ is 0.5ns;
  - duration of the first half period measured at $0.5E_{max}$ is 1ns;
  - total duration of first and second half period is 3.5ns;
  - pulse repetition rate can be varied from 150Hz to 1000Hz;
  - peak amplitude of the electric field strength is about 56V/m at a distance of 3m from the radiator;
- beam width of the peak amplitude pattern of the LCR for the electric field strength at the level of 0.5 E_{max} measured are:
  - in the E-plane 70°;
  - in the H-plane 360°
- a power supply for 6V, and a current of 0.15A is required;
- size of the LCR is $0.4\times0.3\times0.15$ m^3;
- weight is 1 kg.

Acknowledgment

The author is very grateful to H.F.Harmuth of The Catholic University of America, Washington DC for his helpful suggestions.

REFERENCES

1 H.F.Harmuth "Sequency Theory. Foundations and Applications," Academic Press, New York, (1977).

2 H.F.Harmuth "Nonsinusoidal Waves for Radar and Radio Communication," Academic Press. New York, (1981).

3 H.F.Harmuth "Antennas and Waveguides for Nonsinusoidal Waves," Academic Press, New York, (1984).

4B.A.Nalivaiko et al., "Semiconductor Devices. Microwave Diodes. Handbook," MGP "RASKO", Tomsk: (1992).

5 K.A.Lukin, G.P.Pochanin and S.A.Masalov, Large Current Radiator with Avalanche Transistor Switch, *IEEE Trans. Electromagn. Compat.* 39:156, (1997).

A NEW BROAD BAND RESISTIVE WIRE ANTENNA FOR ULTRA-WIDE-BAND APPLICATIONS

Y. Chevalier, Y. Imbs, B. Beillard, J. Andrieu, M. Jouvet, B. Jecko (1)
M. Le Goff, E. Legros (2)

(1) Institut de Recherche en Communications Optiques et Microondes
IRCOM - 7 rue Jules Vallès 19100 Brive la gaillarde (France)
E-mail : imbs@brive.unilim.fr
(2) CELAR (DGA) GEOS/SDM - BP 7419 - 35174 Bruz Cedex (France)

INTRODUCTION

This article presents the theoretical analysis and the experimental results of a wire antenna. The wire antenna has been particularly designed to illuminate complex targets with an electromagnetic impulse. This study was supervised by the CELAR (French Technical Centre for Armament Electronics) and the IRCOM (Research Institute of Microwave and Optical Communications). The aim of the study is to realise a low frequency (200 MHz - 1.2 GHz) Radar Cross Section (RCS) measurement facility in the time domain. Then, not only the antenna has been designed to radiate and to receive short pulses without distortion but also to illuminate the test zone with an homogeneous and uniform electric field. The antenna must be fairly directive with maximum field in the axial direction.

WIRE ANTENNA : PRINCIPLE AND STUDY

The set-up design is based on the radial line modelling with perfectly conducting wire layers. Its geometrical characteristics are presented on the figure 1.

The antenna electromagnetic characteristics (antenna pattern, gain, frequency range, characteristic impedance) depend essentially on the geometrical dimensions such length, width and height. So, the antenna design consists to study the effect of these several parameters. The theoretical study uses the space-time integral equations method[1]. This rigorous method allows analysis of currents induced by the excitation pulse on the wire structure.

Ultra-Wideband, Short-Pulse Electromagnetics 4
Edited by Heyman *et al.*, Kluwer Academic / Plenum Publishers, New York, 1999.

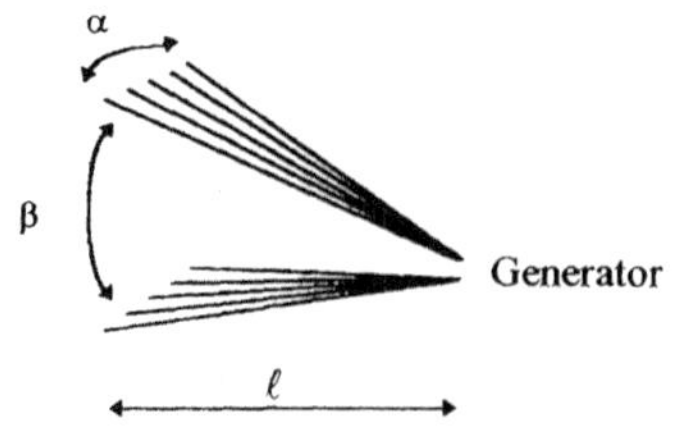

- ℓ : length of the antenna,
- α, β : characteristic angles,
- r : wire radius,
- N : wire number.

Figure 1. Wire antenna.

NUMERICAL TOOLS - INTEGRAL FORMULATION

The theoretical study of perfectly conducting wire structure behaviour can be resolved like a problem of transient electromagnetic wave diffraction by an obstacle[1].

This leads a space-time integral equations making out[2] verified by the induce currents. For the wire structure, currents are obtained by solving the equation (1) where the last three terms express the presence of the perfectly conducting ground[3].

$$\vec{s}.\vec{E}^{a}(s,t)=\frac{\mu_o}{4\pi}\int_{c_0}\left\{\begin{array}{l}\frac{\vec{s}.\vec{s}_o}{R}\frac{\partial I}{\partial to}(s_o,t_o)+c\frac{\vec{s}.\vec{R}}{R^2}\frac{\partial I}{\partial s_o}(s_o,t_o)+c^2\frac{\vec{s}.\vec{R}}{R^3}\int^{t_o}\frac{\partial I}{\partial s_o}(s_o,\tau).d\tau\\ -\frac{\vec{s}.\vec{s}_o^{*}}{R^{*}}\frac{\partial I}{\partial t_o^{*}}(s_o,t_o^{*})-c\frac{\vec{s}.\vec{R}^{*}}{R^{*2}}\frac{\partial I}{\partial s_o}(s_o,t_o^{*})-c^2\frac{\vec{s}.\vec{R}^{*}}{R^{*3}}\int^{t_o^{*}}\frac{\partial I}{\partial s_o}(s_o,\tau).d\tau\end{array}\right\}ds_o \qquad (1)$$

with $R=\sqrt{(s-s_o)^2+a^2}$ and $t_o=t-\frac{R}{c}$ and $R^{*}=\sqrt{(s-s_o^{*})^2+a^2}$ and $t_o^{*}=t-\frac{R^{*}}{c}$

s and s_0 are respectively the observation points and the source points. s_0* represents the image source points, a and c_0 are the radius and the wire outline. $\vec{s}, \vec{s}_0, \vec{s}_0^{\,*}$ are the unit vectors. $E_a(S, t)$ is the applied field corresponding to the source voltage. The electric field is deduced from currents with the equation (2).

NUMERICAL PROCESSING

The integral equation (1) is discretised with the method-of-moments. It uses a second order Lagrange interpolation polynome and is numerically resolved in the time domain[2]. It can be written in a matrix system as :

$$|I_{iv}|=|Z_{iu}|^{-1}\left\{|E^{a}_{uv}|-|E^{d}_{ux}|\right\}$$

with : $i = 1, N_p$; $u = 1, N_p$; $v = 1, N_T$; $x = 1, v\text{-}1$
N_S : the wire structure is divided into N_S segments of length Δ_i,
N_T : the time is divided into N_T equal intervals Δ_t,
The unknown is the colon vector $|I_{iv}|$ at the time $t_v = v \cdot \Delta_t$.

$$\vec{E}(s,t) = -\frac{\mu_o}{4\pi}\int_{c_o}\left\{\begin{aligned}&\frac{\vec{s}_o}{R}\frac{\partial I}{\partial o}(s_o,t_o)+c\frac{\vec{R}}{R^2}\frac{\partial I}{\partial s_o}(s_o,t_o)+c^2\frac{\vec{R}}{R^3}\int^{t_o}\frac{\partial I}{\partial s_o}(s_o,\tau).d\tau\\&-\frac{\vec{s}_o^*}{R^*}\frac{\partial I}{\partial t_o^*}(s_o,t_o^*)-c\frac{\vec{R}^*}{R^{*2}}\frac{\partial I}{\partial s_o}(s_o,t_o^*)-c^2\frac{\vec{R}^*}{R^{*3}}\int^{t_o^*}\frac{\partial I}{\partial s_o}(s_o,\tau).d\tau\end{aligned}\right\}ds_o \tag{2}$$

The current at the ith structure segment at the time t_v depends on diffracted field $E^u_{u,v-1}$ at the time t_{v-1}. The latter is calculated from the induce currents on the structure at the anterior times t_x and from the incident electric field E_{uv} which is known at this time. Then, the matricial system is solved numerically by successive time steps.

APPLICATION TO THE WIRE ANTENNA

The study of the antenna consists to determine the geometrical dimensions (α, β, length, wire number) to obtain the required electromagnetic characteristics. The basic model used in simulation is shown in figure 2. This one is composed of four one metre long arms. The upper half-lengths of each arms are resistively loaded. This loads allow to eliminate progressively the reflected currents from the end of the antenna. This technique uses the principle of the Wu-King non-reflecting dipole [4] and it allows to improve the quality of the radiated pulse. The required distributed resistance Z(ρ) at a distance ρ from the wire terminals is given by :

$$Z(\rho) = \frac{Z_0}{1-\frac{\rho}{\ell}} \quad \text{with } 0 \le \rho < \ell \tag{3}$$

ρ : position of the resistive element on the wire,
Z_0 : first resistance in $\rho = 0$ m.

Figure 3 shows the theoretical pulse radiated by an antenna with and without the resistive loads. The reflection from the end of the antenna is correctly eliminated to the detriment of the amplitude of the pulse which is fairly reduced.

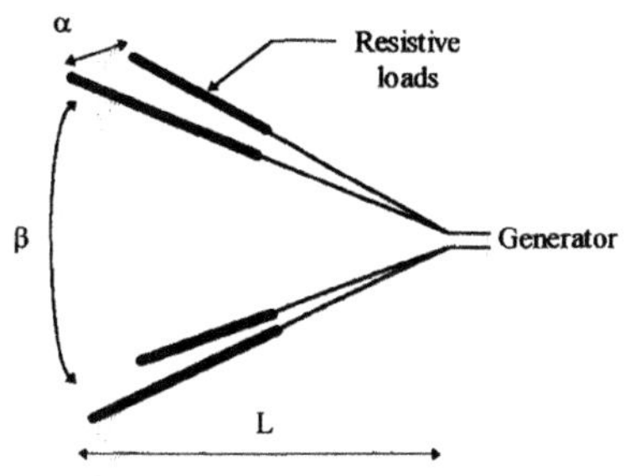

Figure 2. Dimensions of the 4 wires antenna.

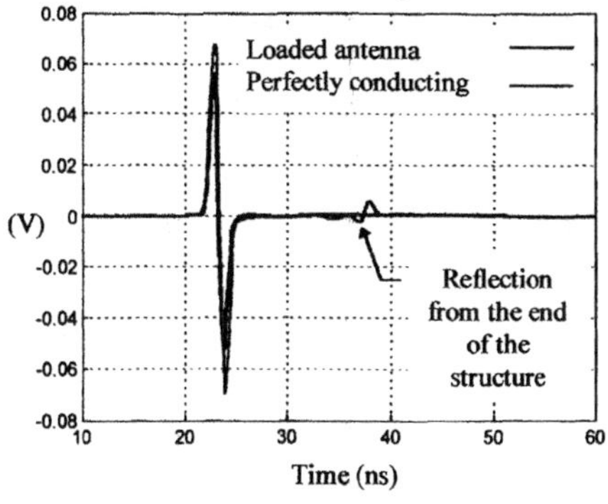

Figure 3. Theoretical pulse radiated by a loaded antenna and a perfectly conducting antenna.

The electromagnetic antenna behaviour is first studied according to the principal angles α and β, then according to the length and the number of wires.

Figures 4 and 5 show the maximum gain computed in the axial direction and the correspondent frequency for different values of α and β. This curves show that the width of the antenna (α angle) must be preferably smaller than the height (β angle) to obtain a high gain and a frequency range centred around 1 GHz.

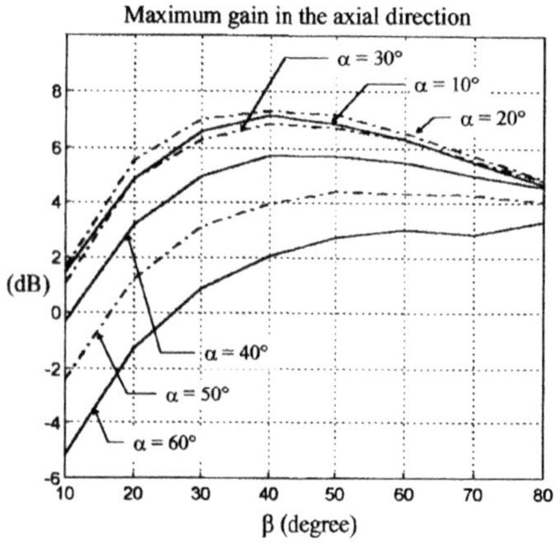

Figure 4. Maximum gain in the axial direction.

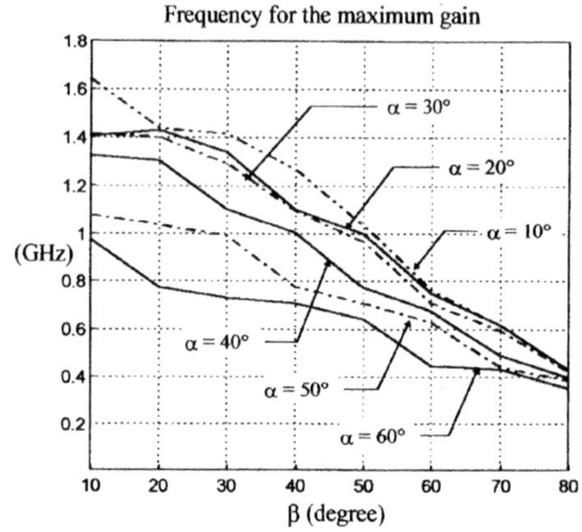

Figure 5. Frequency for the maximum gain.

Figures 6-a and 6-b show that the antenna gain increases with the wire length and with the number of wires. A more important length favours the lower frequencies radiation but the upper spectral components are not correctly guided by the line because in this case the width of the antenna becomes more important. To improve this phenomenon, one solution consists to increase the wire number to create a wire grid. So the radiation of high frequencies are privileged.

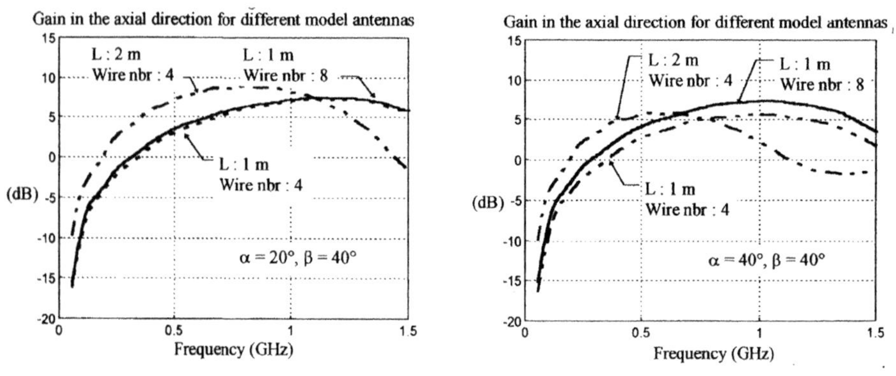

(a) α = 20°, β = 40° (b) α = 40°, β = 40°

Figures 6-a and 6-b. Gain in the axial direction for antennas with different length and wire number and various angles (α = 20°, β = 40°) (α = 40°, β = 40°).

THE EXPERIMENTAL ANTENNA

Two wire antennas were constructed to satisfy the specifications (maximum gain in the frequency range 200 MHz - 1.2 GHz). Only four conducting wires compose the antenna in order to simplify manufacturing (figure 7). The wires radius is 2 centimetres. The antenna angles are $\alpha = 20°$ and $\beta = 40°$. The length is 2 metres. Ten carbon resistances are uniformly distributed on the upper half-length of each arms. These dimensions allow to obtain a high gain and a frequency range centred around 800 MHz.

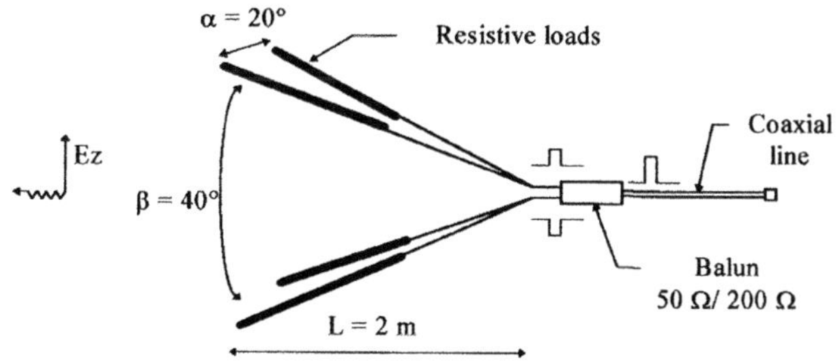

Figure 7. Four-wires antenna dimensions

A differential balun[5] was especially designed to feed the antenna with a symmetrical pulse. This balun transforms the 50 Ω line impedance to the 200 Ω antenna feed point impedance.

The generator used for the experiments provides a pulse output voltage of 4 KV. Two high voltage attenuators of 20 and 6 dB are used to decrease the peak voltage at 2 KV or 200 V. The delivered signal has a 10% - 90 % risetime of 220 ps and a 50 % pulse duration (full-width at half-maximum) of 700 ps (1.4 ns at 10% of the maximum). This gaussian pulse has a frequency range from DC to about 1.5 GHz (-20 dB/maximum) (figures 8 and 9).

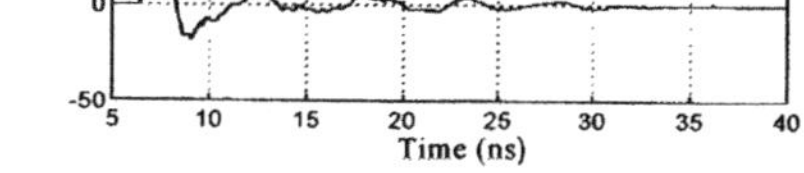

Figure 8. Generator pulse output

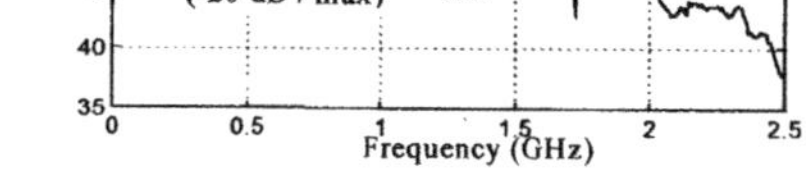

Figure 9. Spectral waveform

REFLECTIONS MEASUREMENT

Figure 10 compares the reflected voltage in the feeding transmission line for the antenna with the balun and for a short circuit. The first reflection appears at the balun feeding connector. The second is due to the balun entrance and the third corresponds to the entrance line of the wire antenna. This last reflection corresponds to the low frequency components of the excitation pulse which are not radiated by the antenna.

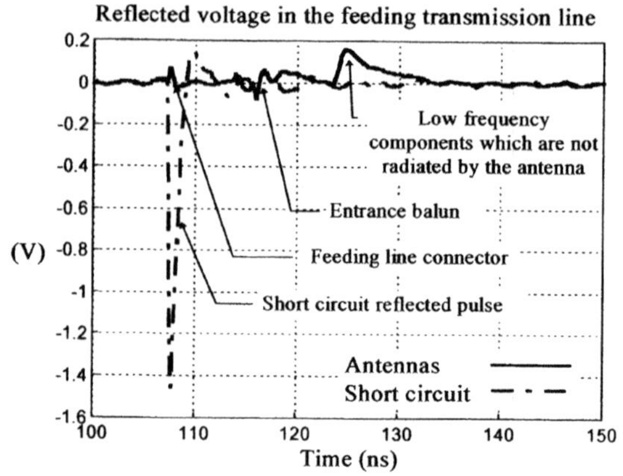

Figure 10 - Reflected voltage in the feeding transmission line.

RADIATED PULSE MEASUREMENT

The electromagnetic performance evaluations were realised in one of the anechoic chamber of the CELAR. This shielded chamber named CHEOPS (Chambre Hyperfréquence pour l'Evaluation Optimale de Signatures) has inside dimensions of 25 m long by 12 m square. The antennas are identical and are positioned facing each other (figure 11).

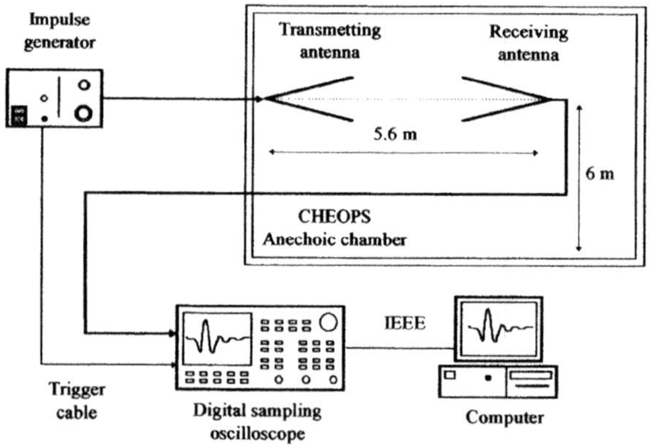

Figure 11. Impulse antenna laboratory

This experimental set-up allows to measure the pulses radiated by the antenna for different azimuthal angles. From these pulses measured in time domain, it is possible to determine the antenna gain and the antenna pattern via a discrete Fourier transform.

Figure 12 shows the pulses radiated by the wire antenna. The top of this figure shows the measured pulses in the horizontal plane or H plane (angle φ variation for $\theta = 90°$), while the bottom of this figure shows the measured pulses in the vertical plane or E plane (angle θ variation for $\varphi = 0°$). In the boresight direction, the antenna is seen to radiate ($\varphi = 0°$) a fairly large pulse that resembles the derivative of the incident voltage. The ground reflection is visible at the time $t = 40$ ns. In the backward direction, ($\varphi = 180°$) two distinct small contributions are observable. The first one corresponds to a radiation directly from the feed point of the antenna while the second one is a small reflection from the first resistance used to eliminate the reflection from the open end of the antenna. These measurements show that the antenna is fairly directive.

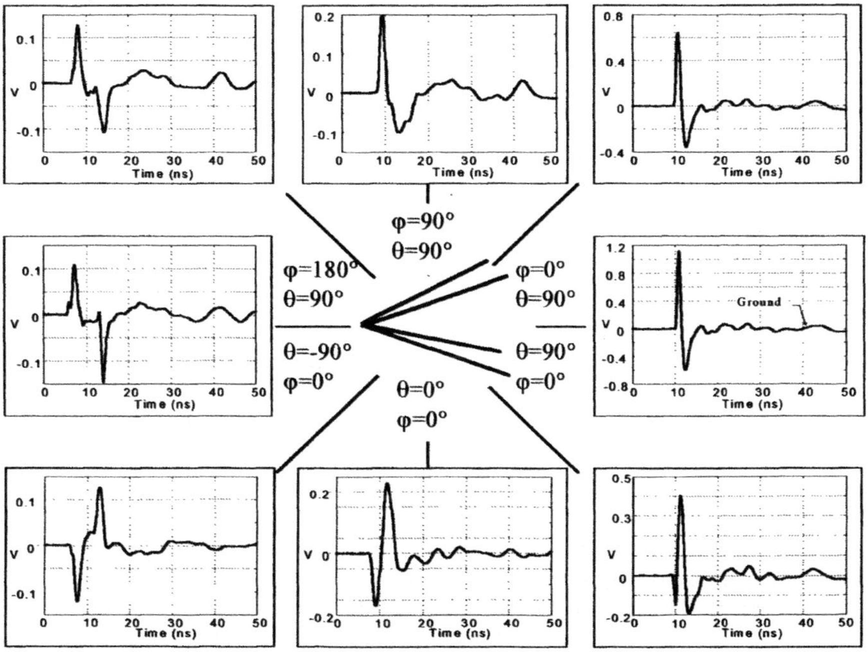

Figure 12. Radiated pulses in the E (bottom half of figure) and H (top half of figure) plane

FREQUENTIAL GAIN DETERMINATION

As the two antennas are perfectly identical, the axial gains will be identical too. Thus, it is possible to determine the gain for different azimuthal and elevation angles by the following expression :

$$G(f,\varphi) = \left(\frac{4.\pi.R}{\lambda}\right).\left(\frac{V_r(f,\varphi)^2}{V_g(f).V_r(f,axis\ measurement)}\right) \tag{4}$$

with V_g : excitation signal spectrum at the drive point of the emitting antenna,
V_r : measured pulse spectrum at the drive point of the receiving antenna,
R : distance between antennas.

So, the determined gain depends on the measured pulses at the drive point of the emitting and receiving antennas, on the electrical length and on the distance between antennas. Figure 13 shows the gain determined from the time domain measurements and from the frequency domain measurements. The accuracy is about 1 dB for both curves.

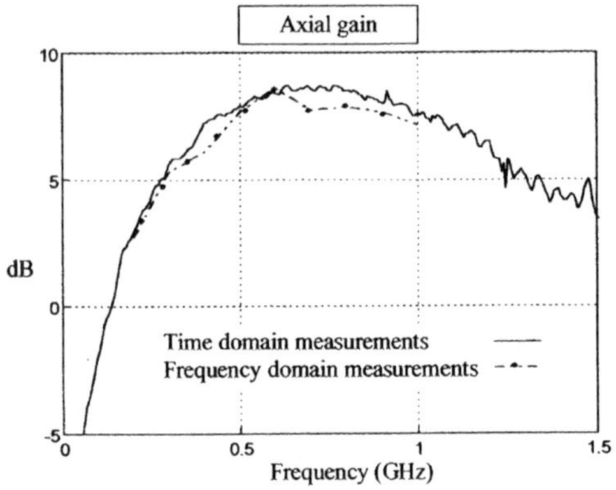

Figure 13. Gain in the axial direction

CONCLUSION

A particular study was led to design antennas capable to radiate and to receive short pulses with a minimum distortion in a frequency range from 200 MHz to 1.2 GHz. The computed results are in very good agreement with the experimental measurements. So, the wire antenna is a relatively easy to use testing facility insofar as only four arms are used to build the line. This antenna has been integrated in an experimental set-up whose aim is to measure the relatively low frequency RCS in the time domain.

REFERENCES

1. Miller E.K., Poggio A.J. and Burke G.J.; « An integrodifferential equation technique for time domain analysis of thin wire structures ». Journ. of Comp. Physics, pp.12-24, n° 48 (1973).
2. « Transient EM fields » Topics in Applied Physics . Ed. : L.B. Felsen - Vol. 10 - Springer Verlag, New-York (1976).
3. Jecko B., « Diffraction d'ondes EM impulsionnelles par des obstacles métalliques ». Annales. des Télécommunications - Vol. 38, n°5-6, pp. 215-225 (1983).
4. Maloney J.G., Smith G.S., « A study of transient radiation from the Wu-King resistive monopole - FDTD analysis and experimental measurements ». IEEE Transactions on Antennas and Propagation - Vol. 41 N°5 - May 1993.
5. The baluns and the antennas were made by the EUROPULSE company (Cressensac Lot, France)

TRANSIENT BEHAVIOR OF RADIATED AND RECEIVED FIELDS ASSOCIATED WITH A RESISTIVELY LOADED DIPOLE

S. N. Samaddar and E. L. Mokole

Naval Research Laboratory
Radar Division
4555 Overlook Avenue, S.W.
Washington, DC 20375-5336, USA

INTRODUCTION

In their frequency-domain analysis, Wu and King [1] have shown that a resistively loaded, thin dipole under the zero-order approximation of its current is a broadband antenna. Based on their results, various authors [2–6] have investigated this problem in the time domain from analytical, numerical, and experimental perspectives. The main reason why a resistively loaded antenna becomes broadband is that proper loading induces a traveling wave for the antenna's current distribution; that is, the reflections from the antenna's end points vanish or are negligible. In the present discussion, investigations of the time-dependent behaviors of the transmitted (radiated) field and the open-circuit received voltage by resistively loaded dipoles are addressed. Both the transmitted field and the received voltage are expressed in terms of an arbitrary exciting input voltage $V_g(t)$ that is applied to the transmitting dipole. The frequency-domain expressions for the current distribution and the radiated field that Wu and King derived are used to obtain the corresponding time-dependent representations by evaluating the associated inverse Fourier transforms. The derived time-domain expressions are considered because they reveal physical insights about the radiation mechanisms of this problem.

For Wu and King's model of the resistive loading, properties of the Fourier transform are used to represent the time-dependent current along the radiating dipole as the sum of two terms: (1) a time-delayed replica of $V_g(t)$, which is a pure traveling wave along the dipole, and (2) a convolution involving $V_g(t)$ and an exponential decay that arises from the loading. This form of the current is used to derive the temporal radiated field, which is expressed as the sum of four signals: two pulses from the dipole's feed and one pulse from each of its ends. With respect to the signals associated with the feed point,

Ultra-Wideband, Short-Pulse Electromagnetics 4
Edited by Heyman *et al.*, Kluwer Academic / Plenum Publishers, New York, 1999.

the two pulses emanate simultaneously. One of these pulses is a retarded replica of $V_g(t)$, and the other pulse is a time integral of the product of $V_g(t)$ and an exponential attenuation factor that depends on the transit time $\tau_t = h_t/c$, where $2h_t$ is the dipole's length and c is the speed of light in free space. As mentioned earlier, the attenuation factor is a consequence of the resistive loading. The pulses from the dipole's ends also are time integrals of $V_g(t)$ with the same attenuation factor; however, both radiate at different later times. After obtaining this new decomposition of the general expression for the transmitted field, examples are presented for three special mathematical representations of $V_g(t)$: a Gaussian pulse, a double exponential pulse, and a single cycle of the sine function.

When a pulsed field is incident on a resistively loaded dipole, the open-circuit received voltage is induced in four different ways in a similar fashion to that of a transmitted field. Therefore, if the the incident field is transmitted by a similarly loaded dipole, each of the four pulses that comprise the incident field induces an open-circuit voltage at the receiving dipole's input in four different ways; that is, the total open-circuit induced voltage is the sum of sixteen pulses, some of which exist on different intervals of time.

Some pulses which contribute to the induced voltage at the receiving dipole appear to be unrealistic, since the induced voltage contains d-c terms and terms that linearly increase with time. Evidently, the reason for the existence of these terms is the choice of the mathematical representation of the loading, which implies an infinite value for the internal impedance of the dipoles at their end points. However, in practice this situation never arises. Another reason for these unphysical terms may be that a real antenna has a finite functional band ($\omega_{min} \leq \omega \leq \omega_{max}$). Hence, allowing infinite support for the integrand in the inverse Fourier transform of the received voltage is probably inappropriate. To avoid unrealistic results, the inverse transform should be evaluated over the antenna's band. In spite of these shortcomings in the mathematical model of the loading, one can infer useful insights on the effective locations of radiation and reception, as well as, that the internal resistance of the dipole should increase progressively to very high values near the end points. In this way, one may design and build a broadband antenna.

OUTLINE OF WU AND KING'S GENERAL THEORY OF A LOADED DIPOLE

Since the behaviors of the transmitted field from a resistively loaded dipole and the induced received voltage of a similarly loaded dipole are treated, it is expedient to associate the subscripts t and r with the transmission and reception situations, respectively. For example, the coordinates (r_t, θ_t, ϕ_t) of a point (x_t, y_t, z_t) in space represent the spherical coordinates of that point relative to the origin at the center of the transmitting dipole of length $2h_t$ and radius a_t [Fig. 1]. Similarly, the coordinates (r_r, θ_r, ϕ_r) refer to a point (x_r, y_r, z_r) that is measured from the center of the receiving dipole of length $2h_r$ and radius a_r. The transmitting and receiving dipoles are oriented along the z_t and z_r axes, respectively.

In the radiation case, the antenna is excited by a slice generator at the origin. The assumed resistive loading is represented by a variable internal impedance,

$$Z^i(z_t) = \frac{\zeta_0 \Psi(0)}{2\pi} \left(\frac{1}{h_t - |z_t|} \right), \tag{1}$$

per unit length z_t along the dipole. The free-space impedance is $\zeta_0 = \sqrt{\mu_0/\epsilon_0} = 120\pi$ ohms, and $\Psi(0)$ is assumed to be constant (independent of frequency). In passing, it may be noted that the loading model given by (1) is too idealistic and may cause certain peculiar behaviors of the radiated field. These peculiarities are discussed subsequently. Under this loading condition, Wu and King have shown that the zero-order current distribution along the dipole ($-h_t \leq z_t \leq h_t$) in the frequency domain takes the form,

$$\hat{I}_t(z_t, \omega) = \frac{2\pi}{\zeta_0 \Psi(0)} \hat{V}(\omega) \left(\frac{ikh_t}{1 + ikh_t} \right) \left(1 - \frac{|z_t|}{h_t} \right) e^{-ik|z_t|}, \tag{2}$$

where $k = \omega/c$ and $\hat{V}(\omega)$ is the voltage applied across the center (feed) point of the dipole.

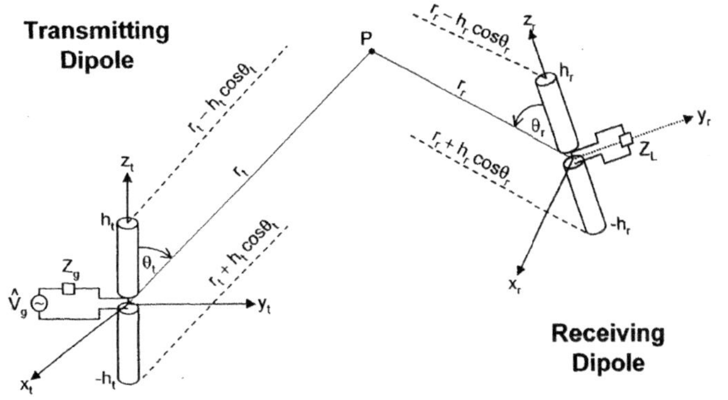

Figure 1. Transmit-receive pair of resistively loaded dipoles. The observation point P has coordinates (r_t, θ_t, ϕ_t) and (r_r, θ_r, ϕ_r) for the transmitter frame and the receiver frame, respectively.

The input current of the transmitting dipole can be written as

$$\hat{I}_t(0, \omega) = \frac{2\pi(ikh_t)\hat{V}(\omega)}{\zeta_0 \Psi(0)(1 + ikh_t)} = \frac{\hat{V}_g(\omega)}{Z_{in}(\omega) + Z_g(\omega)}, \tag{3}$$

where $Z_g(\omega)$ is the generator impedance and $Z_{in}(\omega)$ is the input impedance of the resistively loaded dipole. Assuming $Z_g(\omega) = 0$ implies that $\hat{V}_g(\omega) = \hat{V}(\omega)$. The appropriateness of this assumption was suggested by King [7]. Also, since this antenna sustains a progressive wave of current for $\Psi(0)/(2\pi) \geq 5.3$ [2,3], this inequality is assumed in the present analysis. Consequently, in the frequency domain, the radiated field can be expressed as

$$\hat{E}_{\theta t}(r_t, \theta_t, \omega) = -\frac{kh_t \hat{V}_g(\omega)}{\Psi(0)(1 + ikh_t)} \frac{e^{-ikr_t}}{r_t} F_t(kh_t, \theta_t). \tag{4}$$

The pattern function F_t is related to the antenna's effective height $h_{\theta t}$ via the relation

$$h_{\theta t}(kh_t, \theta_t) = -2\frac{\sin\theta_t}{k} F_t(kh_t, \theta_t), \tag{5}$$

where

$$\begin{aligned} F_t(kh_t, \theta_t) &= \frac{1}{kh_t \sin^3\theta_t}\Big[-ikh_t \sin^2\theta_t + (1 + \cos^2\theta_t) - \frac{1}{2}(1 + \cos\theta_t)^2 e^{-ikh_t(1-\cos\theta_t)} \\ &\quad - \frac{1}{2}(1 - \cos\theta_t)^2 e^{-ikh_t(1+\cos\theta_t)}\Big]. \end{aligned} \tag{6}$$

For the receiving antenna, the corresponding pattern function $F_r(kh_r, \theta_r)$ and the effective height $h_{\theta r}(kh_r, \theta_r)$ are obtained from (5) and (6) by replacing h_t by h_r and θ_t by θ_r.

TIME-DEPENDENT CURRENT AND RADIATED ELECTRIC FIELD

The time-dependent current is the inverse Fourier transform of (2) and is given by

$$\begin{aligned} I_t(z_t, t) &= \mathcal{F}^{-1}\{\hat{I}_t(z_t, \omega)\} = \frac{1}{2\pi}\int_{-\infty}^{\infty} \hat{I}_t(z_t, \omega) e^{i\omega t}\, d\omega \\ &= \frac{2\pi}{\zeta_0 \Psi(0)}\left(1 - \frac{|z_t|}{h_t}\right)\frac{1}{2\pi}\int_{-\infty}^{\infty} \frac{\omega \hat{V}_g(\omega)}{\omega - (i/\tau_t)} e^{i\omega t_z}\, d\omega, \end{aligned} \tag{7}$$

where $\tau_t = h_t/c$, $kh_t = \omega\tau_t$, $t_z = t - (|z_t|/c)$, the symbol $\mathcal{F}^{-1}$ represents the inverse Fourier transform operator, and $\hat{V}_g(\omega)$ is the Fourier transform of the generator voltage $V_g(t)$. The delay τ_t is the time it takes for the voltage and the current to travel from the dipole's feed point to either of its end points. On looking at the integral in (7), one may attempt to evaluate it via complex contour integration. However, such an approach requires knowledge of the behavior of $\hat{V}_g(\omega)$ over the entire complex ω-plane and the guarantee of convergence properties along the contour, which can be very difficult for some representations of $\hat{V}_g$. For example, for the Gaussian pulse, one cannot close the path along the real axis with a semicircle in the upper half of the ω-plane, because the integral along the semicircle does not converge to zero as $|\omega| \to +\infty$. Since results for arbitrary $\hat{V}_g$ are desired, this difficulty is avoided by applying the convolution theorem to obtain

$$\mathcal{F}^{-1}\left\{\frac{\hat{V}_g(\omega)}{\omega-(i/\tau_t)}\right\}=i\int_0^\infty V_g(t_z-t')e^{-t'/\tau_t}\,dt'=i\Big[V_g(t_z)*\left(e^{-t_z/\tau_t}U(t_z)\right)\Big], \qquad (8)$$

where the symbol $*$ in (8) indicates convolution and $U(t)$ is the unit step function. Applying the derivative property of Fourier transforms to (8) yields

$$\begin{aligned}\mathcal{F}^{-1}\left\{\frac{\omega\hat{V}_g(\omega)}{\omega-(i/\tau_t)}\right\} &= \frac{1}{2\pi}\int_{-\infty}^{\infty}\frac{\omega\hat{V}_g(\omega)}{\omega-(i/\tau_t)}e^{i\omega t_z}\,d\omega=\frac{\partial}{\partial t}\left[\int_0^\infty V_g(t_z-t')e^{-t'/\tau_t}\,dt'\right]\\ &= V_g(t_z)-\frac{1}{\tau_t}\int_0^\infty V_g(t_z-t')e^{-t'/\tau_t}\,dt', \qquad (9)\end{aligned}$$

which implies that

$$I_t(z_t,t)=\frac{2\pi}{\zeta_0\Psi(0)}\left[1-\frac{|z_t|}{h_t}\right]\left[V_g(t_z)-\frac{1}{\tau_t}\int_0^\infty V_g(t_z-t')e^{-t'/\tau_t}\,dt'\right]. \qquad (10)$$

The time-dependent current distribution consists of two terms. The factor $1-(|z_t|/h_t)$, common to both terms, decreases from 1 at the feed point ($z_t=0$) to 0 at the end points ($z_t=\pm h_t$) and is responsible for making the current vanish at the dipole's end points. The first term $V_g(t_z)$ represents a pure traveling wave and is, in fact, a delayed replica of the input signal as it propagates along the dipole. The second term is proportional to the time integral of the product of a delayed exciting voltage, an exponential decay, and the unit step. This integral is the convolution of the input voltage and the product of the decay and the unit step at delay time t_z. The exponential decay is due to the resistive loading through the presence of τ_t in the exponent, which arises from the distribution of the loading in (1). In addition, the second term becomes very small in the high-frequency region. This observation is supported by the fact that the factor $ikh_t/(1+ikh_t)$ in (2) approaches unity for high frequencies.

The time-dependent radiated electric field is the inverse Fourier transform of (4) and is given by

$$E_{\theta t}(r_t,\theta_t,t)=\frac{1}{2\pi}\int_{-\infty}^{\infty}\hat{E}_{\theta t}(r_t,\theta_t,\omega)e^{i\omega t}\,d\omega. \qquad (11)$$

Substituting (4) and (6) into (11) and applying (8) and (9) lead to

$$\begin{aligned}E_{\theta t}(r_t,\theta_t,t) &= A(r_t,\theta_t)\Bigg[\sin^2\theta_t V_g(t^*)-\frac{2}{\tau_t}\int_0^\infty V_g(t^*-t')e^{-t'/\tau_t}\,dt'\\ &\quad+\frac{(1+\cos\theta_t)^2}{2\tau_t}\int_0^\infty V_g(t_1-t')e^{-t'/\tau_t}\,dt'\\ &\quad+\frac{(1-\cos\theta_t)^2}{2\tau_t}\int_0^\infty V_g(t_2-t')e^{-t'/\tau_t}\,dt'\Bigg], \qquad (12)\end{aligned}$$

where $A(r_t,\theta_t)=1/[r_t\Psi(0)\sin^3\theta_t]$ and the three time delays are

$$t^*=t-\frac{r_t}{c},\qquad t_1=t^*-\tau_t(1-\cos\theta_t),\qquad t_2=t^*-\tau_t(1+\cos\theta_t). \qquad (13)$$

The first term of the radiated field in (12) is a delayed undistorted replica of the exciting input voltage V_g. Because the appearance of the retarded time t^* in V_g suggests that this term is not observed until the signal travels the distance r_t to the observation point, this component of the field is interpreted as radiating from the feed point of the resistive dipole. In contrast, the second term of (12) is proportional to the time integral of the product of the exciting voltage and an exponentially decaying factor. In fact, this integral is the convolution of the voltage and the decay at time t^*, which implies that it also radiates from the feed point of the dipole. As mentioned in the discussion of (10), the exponential decay is due to the resistive loading and is manifested by the presence of τ_t in the exponent. Similarly, the third and fourth terms are interpreted as pulses that are radiated from the top and bottom ends, respectively, of the dipole. This observation follows from the dependence of these terms on the delay times t_1 and t_2. The third and fourth pulses are also time integrals (convolutions) of the exciting voltage and the exponential attenuation factor that appear in the second pulse, but the convolutions are evaluated at the later times t_1 and t_2 instead of t^*. With respect to the three convolutions, note that they differ only in the times at which they are evaluated. Consequently, once one of the integrals is calculated, the rest are determined by inspection. Although it is not obvious from (12), $E_{\theta t}$ approaches zero as the observation point approaches either endfire direction. Moreover, the field is symmetric about $\theta_t = 90°$.

Previous authors [3,4,5,6] have considered specific mathematical expressions for $V_g(t)$; however, none has presented a result for an arbitrary exciting voltage as displayed in (12). Moreover, this decomposition is new and offers physical insights that are missing in the existing literature, for example, representing the radiated field as the sum of components that correspond to radiation from various points of the loaded dipole.

EXAMPLES OF TIME-DEPENDENT RADIATED ELECTRIC FIELDS FOR THREE EXCITING VOLTAGES

To illustrate the behavior of the time-dependent radiated field, the general result of (12) is now applied for three exciting voltages: a Gaussian pulse, a double exponential pulse, and a single-cycle-sine pulse.

Gaussian Pulse

Since the Gaussian excitation has been treated previously for the resistively loaded dipole [3], the authors merely wish to note in this case that the total radiated field is easily obtained after evaluating one convolution integral. For $V_g(t) = V_{g1}e^{-t^2/(2\sigma^2)}$, the basic convolution integral in (12) is

$$\int_0^{\infty} V_g(t - t')e^{-t'/\tau_t}\,dt' = V_{g1}\sigma\sqrt{\frac{\pi}{2}}e^{\sigma^2/(2\tau_t^2)}e^{-t/\tau_t}erfc\left[\frac{1}{\sqrt{2}}\left(\frac{\sigma}{\tau_t} - \frac{t}{\sigma}\right)\right], \qquad (14)$$

where $erfc$ is the complementary error function [8]. When the preceding equation is substituted into (12), one obtains the result of Sengupta and Liu [3].

Double Exponential Pulse

In this example, the exciting voltage is the double exponential function,

$$V_g(t) = V_{g2}\left(e^{-\alpha t} - e^{-\beta t}\right)U(t), \tag{15}$$

with $\alpha < \beta$. The double exponential function has been used to model the voltage associated with a lightning discharge [9], the nuclear-electromagnetic-pulse (NEMP) field from a high-altitude nuclear explosion that is incident on a two-wire transmission line [10], and the pulsed voltage that is delivered by a fast risetime pulser [11].

This exciting voltage is a continuous function with a single absolute maximum value at $t_m = [\ln(\beta/\alpha)]/(\beta - \alpha)$; however, the derivative of V_g has a jump discontinuity at $t = 0$. The jump discontinuity introduces higher frequency content in the double exponential function's spectrum than would appear in a real exciting voltage. As [12] observes, a physically realizable input voltage to a radiator should satisfy the differentiability conditions: $V_g^{(n)}(0) = 0$ for $n = 1, 2$.

Evaluation of the integral in (8) yields

$$\int_0^\infty V_g(t - t')e^{-t'/\tau_t}\,dt' = V_{g2}[Be^{-t/\tau_t} - D(t)]U(t), \tag{16}$$

where the quantities B and $D(t)$ are

$$B = \frac{\beta - \alpha}{(\alpha - \tau_t^{-1})(\beta - \tau_t^{-1})}, \qquad D(t) = \frac{e^{-\alpha t}}{\alpha - \tau_t^{-1}} - \frac{e^{-\beta t}}{\beta - \tau_t^{-1}}. \tag{17}$$

Therefore, the radiated electric field for the double exponential source is

$$\begin{aligned} E_{\theta t}(r_t, \theta_t, t) &= V_{g2}A(r_t, \theta_t)\Big\{\sin^2\theta_t\left[e^{-\alpha t^*} - e^{-\beta t^*}\right]U(t^*) - \frac{2}{\tau_t}\left[Be^{-t^*/\tau_t} - D(t^*)\right]U(t^*) \\ &+ \frac{(1+\cos\theta_t)^2}{2\tau_t}\left[Be^{-t_1/\tau_t} - D(t_1)\right]U(t_1) \\ &+ \frac{(1-\cos\theta_t)^2}{2\tau_t}\left[Be^{-t_2/\tau_t} - D(t_2)\right]U(t_2)\Big\}. \end{aligned} \tag{18}$$

For $\tau_t = 25$ ps, $\alpha = 600$ GHz, $\beta = 800$ GHz, and six observation angles ($\theta_t = 90°, 75°, 60°, 45°, 30°, 15°$), the electric field is plotted versus normalized time t^*/τ_t in Figure 2. Since the radiated field is symmetric about $\theta_t = 90°$, only plots for $0 < \theta_t \leq 90°$ are displayed. To get a sense of how the shape of the exciting voltage is affected, V_g/V_{g2} is also shown (solid curve). The radiated field at the observation point (r_t, θ_t, ϕ_t) for all θ_t is zero until t^* exceeds 0 (t exceeds r_t/c), that is, until the signal has traveled from the feed point to the observation point. For each θ_t, the field initially increases to a global maximum value, then it decreases to a negative local minimum value, then it increases toward 0, and finally it asymptotically approaches 0 as t^* increases without bound. At broadside ($\theta_t = 90°$), the duration of the field is 300% longer than the duration of the exciting voltage. As θ_t decreases from broadside to the upper endfire direction ($\theta_t = 0°$), the field's duration steadily decreases to zero, and the global max-

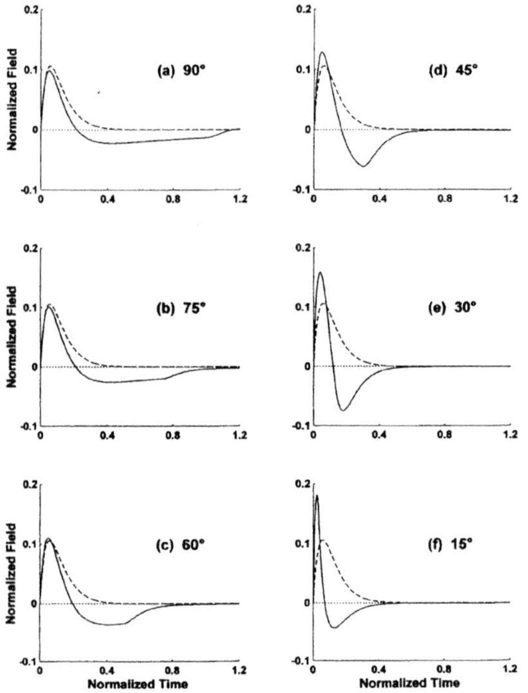

Figure 2. Plots of the normalized radiated electric field $r_t\Psi(0)E_{\theta t}/V_{g2}$ (dashed curve) and the normalized exciting voltage V_g/V_{g2} (solid curve) versus normalized time t^*/τ_t when the resistively loaded dipole is excited by a double-exponential pulse for $\alpha = 600$ GHz, $\beta = 800$ GHz, $\tau_t = 25$ ps, and six values of the observation angle θ_t: (a) 90°, (b) 75°, (c) 60°, (d) 45°, (e) 30°, (f) 15°.

imum value increases until θ_t is roughly 15° and then decreases to zero (which is not apparent from Figure 2 because plots for $\theta_t < 15°$ are not shown).

Single-Cycle-Sine Pulse

In this example, the exciting voltage is a single cycle of the $\sin(\omega_0 t)$ with fundamental period $T = 2\pi/\omega_0$; that is,

$$V_g(t) = V_{g3}\sin(\omega_0 t)[U(t) - U(t-T)]. \tag{19}$$

The carrier frequency ω_0 is chosen so that $h_t/c = \pi/(2\omega_0)$; that is, the antenna is a half-wave dipole relative to the carrier. After evaluating the integrals in (12) for (19), the radiated field is written as

$$\begin{aligned}
E_{\theta t}(r_t,\theta_t,t) &= V_{g3}A(r_t,\theta_t)\Bigg\{\sin^2\theta_t\sin(\omega_0 t^*)[U(t^*) - U(t^*-T)] \\
&\quad - \frac{2}{\tau_t(\omega_0^2+\tau_t^{-2})}\Bigg[\left[\frac{1}{\tau_t}\sin(\omega_0 t^*) - \omega_0\cos(\omega_0 t^*)\right][U(t^*) - U(t^*-T)] \\
&\quad + \omega_0 e^{-t^*/\tau_t}U(t^*) - \omega_0 e^{-(t^*-T)/\tau_t}U(t^*-T) \\
&\quad - \frac{(1+\cos\theta_t)^2}{4}\left(\left[\frac{1}{\tau_t}\sin(\omega_0 t_1) - \omega_0\cos(\omega_0 t_1)\right][U(t_1) - U(t_1-T)]\right. \\
&\quad \left. + \omega_0 e^{-t_1/\tau_t}U(t_1) - \omega_0 e^{-(t_1-T)/\tau_t}U(t_1-T)\right) \\
&\quad - \frac{(1-\cos\theta_t)^2}{4}\left(\left[\frac{1}{\tau_t}\sin(\omega_0 t_2) - \omega_0\cos(\omega_0 t_2)\right][U(t_2) - U(t_2-T)]\right. \\
&\quad \left. + \omega_0 e^{-t_2/\tau_t}U(t_2) - \omega_0 e^{-(t_2-T)/\tau_t}U(t_2-T)\right)\Bigg]\Bigg\}.
\end{aligned} \tag{20}$$

Although the single-cycle-sine input voltage has finite duration T, the radiated pulse in (20) has a much longer time duration as a consequence of the exponentially damped terms. This temporal stretching of V_g is caused by a combination of the dipole's loading and delayed radiations from its discontinuities. The stretching that is caused by the delayed radiations from the discontinuities has been discussed previously [13]. Specifically, for an unloaded half-wave dipole under the zero-order approximation of the current, it has been shown that the duration of the radiated pulse is 50% longer than that of the input. However, in the present situation, the mathematical representation of the loading in (1) is responsible for the longer extent of the radiated field, which is manifested in (20) by the presence of the six exponential terms. This choice of the loading indicates that the internal impedance of the dipole at its end points is infinite, which is an ideal academic situation rather than a real one. Furthermore, a single-cycle-sine pulse may not be realizable. In connection with other analysis [12], it has been observed that a real exciting voltage $V_g(t)$ has finite duration and must be at least twice differentiable with respect to time. Moreover, the voltage pulse through its second derivative must vanish at the beginning and the end of the pulse. It should also be pointed out that assuming $Z_g = 0$ may influence the behavior of the radiated field.

For $\tau_t = 25$ ps and a carrier frequency of 10 GHz, the electric field is plotted versus t^*/τ_t in Figure 3 for six observation angles, along with V_g/V_{g3} (solid curve). For each θ_t, the field is zero until t^* exceeds 0, initially increases to a local maximum value, decreases through 0 to a negative absolute minimum value, increases through 0 to a positive absolute maximum value, decreases through 0 to a negative local minimum value, and finally asymptotically approaches 0 from below as t^* increases without bound. At broadside, the field has its longest duration, which is approximately 50% longer than that of the input voltage. Regardless of the observation angle, the field has four zero crossings, one more than V_g. As θ_t varies from 90° to 0°, the field's duration decreases to zero, and its extreme values approach zero. For $|90° - \theta_t| < 20°$, the field behaves like $\cos(\omega_0 t^*)$ with a steep rise in the leading edge.

OPEN-CIRCUIT RECEIVED VOLTAGE FOR A RESISTIVELY LOADED DIPOLE

The open-circuit voltage induced at the input terminals of the receiving dipole by an incident electric field, the source of which may be arbitrary, is given in the frequency domain by the dot product

$$\hat{V}_{oc}(\omega) = \hat{\mathbf{E}}^{\mathbf{i}}(\mathbf{r}_0, \omega) \bullet \mathbf{h}_{\mathbf{r}}(\theta_r, \omega), \tag{21}$$

where $\hat{\mathbf{E}}^{\mathbf{i}}$ is the incident field that is transmitted by an arbitrary source and $\mathbf{h}_{\mathbf{r}}$ is the vector effective height of the receiving resistive dipole. The coordinates, $\mathbf{r}_0 = (r_0, \theta_0, \phi_0)$, of the input terminal of the receiving dipole are measured with respect to the coordinates (x_t, y_t, z_t) of the transmitting antenna (Fig. 1).

The effective height of the receiving dipole is given by

$$\mathbf{h}_{\mathbf{r}}(\theta_r, \omega) = \hat{\theta}_{\mathbf{r}} h_{\theta r}(\theta_r, \omega), \tag{22}$$

where $\hat{\theta}_{\mathbf{r}}$ is the unit vector in the θ_r-direction. After replacing the subscript t by r in (5) and (6) to obtain $h_{\theta r}$, substituting this result into (21), and inverting the subsequent expression, the time-dependent open-circuit voltage $V_{oc}(t)$ can be expressed as

$$\begin{aligned} V_{oc}(t) \quad = \quad & -\frac{2c}{\sin^2\theta_r}\Bigg\{\sin^2\theta_r \int_{-\infty}^{t} \hat{\theta}_{\mathbf{r}} \bullet \mathbf{E}^{\mathbf{i}}(\mathbf{r}_0, t')\, dt' \\ & - \frac{(1+\cos^2\theta_r)}{\tau_r} \int_{-\infty}^{t} \left[\int_{-\infty}^{t'} \hat{\theta}_{\mathbf{r}} \bullet \mathbf{E}^{\mathbf{i}}(\mathbf{r}_0, t'')\, dt''\right] dt' \\ & + \frac{(1+\cos\theta_r)^2}{2\tau_r} \int_{-\infty}^{t_3} \left[\int_{-\infty}^{t'} \hat{\theta}_{\mathbf{r}} \bullet \mathbf{E}^{\mathbf{i}}(\mathbf{r}_0, t'')\, dt''\right] dt' \\ & + \frac{(1-\cos\theta_r)^2}{2\tau_r} \int_{-\infty}^{t_4} \left[\int_{-\infty}^{t'} \hat{\theta}_{\mathbf{r}} \bullet \mathbf{E}^{\mathbf{i}}(\mathbf{r}_0, t'')\, dt''\right] dt' \Bigg\}, \end{aligned} \tag{23}$$

where $\mathbf{E}^{\mathbf{i}}(\mathbf{r}_0, t)$ is the inverse transform of $\hat{\mathbf{E}}^{\mathbf{i}}(\mathbf{r}_0, \omega)$, $\tau_r = h_r/c$, and t is the time of arrival of the incident field at the input terminal of the receiving dipole. The delay τ_r is the time it takes for the voltage and current to travel from either of the receiving dipole's end points to its center point (input terminal), and the retarded times t_3 and t_4 are given by

$$t_3 = t - \tau_r(1 - \cos\theta_r), \qquad t_4 = t - \tau_r(1 + \cos\theta_r). \tag{24}$$

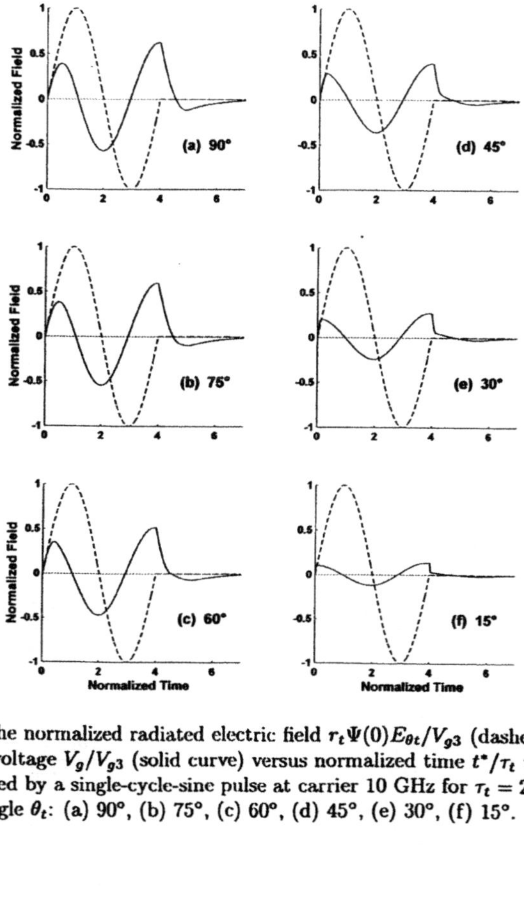

Figure 3. Plots of the normalized radiated electric field $r_t\Psi(0)E_{\theta t}/V_{g3}$ (dashed curve) and the normalized exciting voltage V_g/V_{g3} (solid curve) versus normalized time t^*/τ_t when the resistively loaded dipole is excited by a single-cycle-sine pulse at carrier 10 GHz for $\tau_t = 25$ ps and for six values of the observation angle θ_t: (a) 90°, (b) 75°, (c) 60°, (d) 45°, (e) 30°, (f) 15°.

The first term of the right side of (23) is proportional to the time integral of the incident field and depends on t, implying that the induced voltage is entering at the center of the receiving, open-circuited, resistive dipole. Similarly, the second term is also a voltage induced at the dipole's center since it depends on t; however, this term is a double time integral of the incident field. In contrast, the presence of retarded times t_3 and t_4 as the upper limits of the outer of the double iterated integrals of the third and fourth terms implies that these terms represent voltages induced at the upper end and the lower end, respectively, of the receiving dipole.

In particular, if the incident field is radiated from another resistively loaded dipole, then $\hat{\mathbf{E}}^{\mathbf{i}}(\mathbf{r}_0, \omega)$ is given by (4) with $r_0 = r_t$ and $\theta_0 = \theta_t$. Since the radiated field is axially symmetric, it is independent of $\phi_0 = \phi_t$. Consequently,

$$\hat{\mathbf{E}}^{\mathbf{i}}(\mathbf{r}_0, \omega) = \hat{\theta}_{\mathbf{t}} \hat{E}_{\theta t}(r_t, \theta_t, \omega). \tag{25}$$

Substituting (25) into (21) and using (4) through (6) lead to an expression for $\hat{V}_{oc}(\omega)$ when both the transmitting and receiving dipoles are similarly loaded. By taking the inverse Fourier transform of $\hat{V}_{oc}(\omega)$, the corresponding time-dependent, open-circuit, induced voltage can be represented in terms of the exciting voltage V_g of the transmitter as

$$\begin{aligned}
V_{oc}(t) \;=\; & B(r_t, \theta_t, \theta_r) \Bigg[\sin^2\theta_t \sin^2\theta_r \int_0^\infty V_g(t^* - t') e^{-t'/\tau_t}\, dt' \\
& - \frac{\tau_t \sin^2\theta_t (1+\cos^2\theta_r)}{\tau_r} V_1(t^*) + \frac{\tau_t \sin^2\theta_t (1+\cos\theta_r)^2}{2\tau_r} V_1(t_5) \\
& + \frac{\tau_t \sin^2\theta_t (1-\cos\theta_r)^2}{2\tau_r} V_1(t_6) - (1+\cos^2\theta_t)\sin^2\theta_r V_1(t_5) \\
& + \frac{(1+\cos^2\theta_t)(1+\cos^2\theta_r)}{\tau_t\tau_r} V_2(t^*) - \frac{(1+\cos^2\theta_t)(1+\cos\theta_r)^2}{2\tau_t\tau_r} V_2(t_5) \\
& - \frac{(1+\cos^2\theta_t)(1-\cos\theta_r)^2}{2\tau_t\tau_r} V_2(t_6) + \frac{1}{2}(1+\cos\theta_t)^2 \sin^2\theta_r V_1(t_7) \\
& - \frac{(1+\cos\theta_t)^2(1+\cos^2\theta_r)}{2\tau_t\tau_r} V_2(t_7) + \frac{(1+\cos\theta_t)^2(1+\cos\theta_r)^2}{2\tau_t\tau_r} V_2(t_9) \\
& + \frac{(1+\cos\theta_t)^2(1-\cos\theta_r)^2}{2\tau_t\tau_r} V_2(t_{10}) + \frac{1}{2}(1-\cos\theta_t)^2 \sin^2\theta_r V_1(t_8) \\
& - \frac{(1-\cos\theta_t)^2(1+\cos^2\theta_r)}{2\tau_t\tau_r} V_2(t_8) + \frac{(1-\cos\theta_t)^2(1+\cos\theta_r)^2}{2\tau_t\tau_r} V_2(t_{11}) \\
& + \frac{(1-\cos\theta_t)^2(1-\cos\theta_r)^2}{2\tau_t\tau_r} V_2(t_{12}) \Bigg],
\end{aligned} \tag{26}$$

where

$$B(r_t, \theta_t, \theta_r) = -\frac{2c(\hat{\theta}_{\mathbf{t}} \bullet \hat{\theta}_{\mathbf{r}})}{r_t \Psi(0) \sin^3\theta_t \sin^3\theta_r}, \tag{27}$$

$$V_1(t) = \int_{-\infty}^{t} V_g(t')\, dt' - \int_0^\infty V_g(t - t') e^{-t'/\tau_t}\, dt', \tag{28}$$

$$V_2(t) = -\tau_t^2 V_1(t) + \tau_t \int_{-\infty}^{t} \left[\int_{-\infty}^{t'} V_g(t'')\, dt'' \right] dt'. \tag{29}$$

The various retarded times are given by

$$\begin{aligned}
t_5 &= t^* - \tau_r(1-\cos\theta_r),\\
t_6 &= t^* - \tau_r(1+\cos\theta_r),\\
t_7 &= t^* - \tau_t(1-\cos\theta_t),\\
t_8 &= t^* - \tau_t(1+\cos\theta_t),\\
t_9 &= t^* - \tau_t(1-\cos\theta_t) - \tau_r(1-\cos\theta_r),\\
t_{10} &= t^* - \tau_t(1-\cos\theta_t) - \tau_r(1+\cos\theta_r),\\
t_{11} &= t^* - \tau_t(1+\cos\theta_t) - \tau_r(1-\cos\theta_r),\\
t_{12} &= t^* - \tau_t(1+\cos\theta_t) - \tau_r(1+\cos\theta_r).
\end{aligned} \tag{30}$$

The expression in (23) shows that an incident pulse induces voltages at the open-circuit input of a resistively loaded receiving dipole in four different ways. Further, (12) indicates that a resistively loaded transmitting dipole radiates four distinct pulses associated with the discontinuities of the dipole. Therefore, it is expected that when the transmitting and receiving dipoles are the same type, the induced open-circuit voltage of the receiving dipole will consist of sixteen pulses which appear at different retarded times. The retarded times t_n for $n = 5, \ldots, 12$ contain information which specifies the transmit-receive path that each of the pulses in (26) follows between the radiating dipole's feed and the receiving dipole's input terminal. The meanings attached to three of the terms in (26) are addressed to avoid a lengthy discussion of how to interpret each term of (26) and to provide insight on how to interpret the remaining terms. Specifically, a pulse associated with t^* travels directly from the input of the transmitting dipole to the input of the receiving dipole. In contrast, a pulse depending on t_{10} travels from the radiator's feed point to the upper end of the radiator to the lower end of the receiving dipole to the input terminal of the receiving dipole. As a final example, a pulse corresponding to t_{12} travels from the radiator's feed point to its lower end to the lower end of the receiving dipole to the input terminal of the receiving dipole. In this manner, one may infer the paths associated with each pulse of (26). Further discussion is found in [14], where a more detailed exposition of interpreting the individual pulses of the received voltage is provided for a transmit-receive pair of unloaded dipoles.

Instead of discussing the receiving situation completely for the three specific choices of V_g of the earlier examples, $V_{oc}(t)$ is considered in terms of the three generic integrals that appear in (26) as they relate to these examples:

$$\int_{-\infty}^{t} V_g(t')\,dt', \qquad \int_{0}^{\infty} V_g(t-t')e^{-t'/\tau_t}\,dt', \qquad \int_{-\infty}^{t}\left[\int_{-\infty}^{t'} V_g(t'')\,dt''\right]dt'. \tag{31}$$

Recall that the second integral of (31) also appears in (12). Regardless of whether V_g has finite duration, this term resulted in damped exponential terms that cause the radiated field to be significantly stretched when compared to the duration of V_g. On the other hand, pulses containing the first integral of (31) do not show any unusual behavior when V_g has finite duration. However, if V_g has infinite duration like the double exponential function of (15), then

$$\int_{-\infty}^{t} V_g(t')\,dt' = V_{g2}\left[\left(\frac{1}{\alpha} - \frac{1}{\beta}\right) - \left(\frac{e^{-\alpha t}}{\alpha} - \frac{e^{-\beta t}}{\beta}\right)\right]U(t), \tag{32}$$

which contains the d-c term $(\alpha^{-1} - \beta^{-1})$. Since a d-c term is not a component of the radiated field, the appearance of $(\alpha^{-1} - \beta^{-1})$ apparently results from the receiving dipole artificially rectifying a portion of the incident field, thereby, injecting an unphysical d-c term. When the third integral of (31) is evaluated for the double-exponential pulse, V_{oc} has terms of the form

$$\int_{-\infty}^{t}\left[\int_{-\infty}^{t'} V_g(t'')\,dt''\right]dt' = V_{g2}\left[\left(\frac{1}{\alpha} - \frac{1}{\beta}\right)t - \left(\frac{1}{\alpha^2} - \frac{1}{\beta^2}\right) + \left(\frac{e^{-\alpha t}}{\alpha^2} - \frac{e^{-\beta t}}{\beta^2}\right)\right]U(t), \tag{33}$$

which includes a term that increases linearly with time, as well as a d-c term. Even when V_g is the single-cycle-sine pulse, the iterated integral,

$$\int_{-\infty}^{t}\left[\int_{-\infty}^{t'} V_g(t'')\,dt''\right]dt' = \frac{V_{g3}}{\omega_0}\left[\left(t - \frac{\sin(\omega_0 t)}{\omega_0}\right)\Big(U(t) - U(t-T)\Big) + TU(t-T)\right], \tag{34}$$

also contains a linear term in t and the d-c term $TU(t-T)$, although the linear term has finite duration.

As the preceding discussion indicates, two types of unrealistic behavior occur when (31) is applied to double-exponential and single-cycle-sine pulses: appearances of d-c components and terms that linearly increase with t. These unrealistic behaviors of V_{oc} apparently stem from the theoretical model of the resistive loading, which leads to infinite values for the internal impedance of the dipole at its end points. In addition, these ideal exciting voltages are not realizable in a strict sense and may be partly responsible for such behaviors. Nonetheless, although analyzing ideal signals can lead to unphysical results, they still provide valuable insights.

SUMMARY

The mathematical model of the resistively loaded dipole, introduced by Wu and King for obtaining the current distribution along the dipole and the radiated electric field in the frequency domain, is used here to study the corresponding time-dependent current and field. In addition, the characteristics of the open-circuit voltage of a similarly loaded receiving dipole are investigated. The main results are presented in terms of an arbitrary applied exciting voltage V_g at the input of the transmitting dipole. For example, the temporal radiated field is expressed as the sum of four signals which are interpreted as follows: two pulses from the transmitting dipole's feed point and one pulse from each of its end points. One of the pulses from the feed point is an undistorted, retarded replica of V_g, and the other pulse is proportional to the time integral of the product of V_g and an exponentially decaying time function. The pulses radiating from the end points are also proportional to the time integral of V_g and the exponential factor, but they occur at later times. The exponential decay arises from the resistive loading, and the three exponentially decaying pulses increase the duration of the radiated field when compared to the duration of V_g. The exponential decay terms become negligible if the frequency content of V_g is very high, which implies that the losses are reduced at high frequencies, thereby increasing the antenna's efficiency.

The theoretical model of the resistive loading leads to infinite values for the internal impedance at the dipole's end points and, consequently, appears to be an academic model, since it leads to some unrealistic behavior of the open-circuit voltage at the input terminal of the receiving dipole. For example, if both the transmitting and receiving dipoles are resistively loaded in the same manner, although the radiated field contains no d-c term, a d-c voltage appears at the input of the receiving dipole. In an actual situation, no d-c voltage is present at the receiving dipole's input, since the internal resistance of the loaded antenna will never be infinite at the dipole's end points. Despite this shortcoming in the mathematical model of the loading, one can still gain useful information by arguing in the following way. Since, in an actual loading of the dipole, the dipole's internal impedance can never achieve an infinite value, one needs to design the loading by progressively increasing the impedance from the feed point to a very high finite value near the end points, in order to make the dipole broadband.

Acknowledgments. The authors thank Edmond Tomas for providing the figures. This work was supported by the Naval Research Laboratory and the Office of Naval Research in Washington DC.

REFERENCES

1. T. T. Wu and R. W. P. King, "The cylindrical antenna with non-reflecting resistive loading," *IEEE Trans. Antennas Propagat.* **13**:369 (1965).

2. L-C. Shen and T. T. Wu, "Cylindrical antenna with tapered resistive loading," *Radio Sci.* **2**:191 (1967).

3. D. L. Sengupta and Y-P. Liu, "Analytical investigation of waveforms radiated by a resistively loaded linear antenna excited by a Gaussian pulse," *Radio Sci.* **9**:621 (1974).

4. M. Kanda, "Time domain sensors for radiated impulsive measurements," *IEEE Trans. Antennas Propagat.* **31**:438 (1983).

5. K. P. Esselle and S. S. Stuchly, "Pulse-receiving characteristics of resistively loaded dipole antennas," *IEEE Trans. Antennas Propagat.* **38**:1677 (1990).

6. J. G. Maloney and G. S. Smith, "A study of transient radiation from the Wu-King resistive monopole – FDTD analysis and experimental measurements," *IEEE Trans. Antennas Propagat.* **31**:668 (1993).

7. R. W. P. King, private communication (1997).

8. W. Gautschi, "Error function and Fresnel integrals," in *Handbook of Mathematical Functions with Formulas, Graphs, and Mathematical Tables*, M. Abramowitz and I. A. Stegun, eds., Dover (ninth printing), New York (1964), p. 295.

9. J. E. Nanevicz, E. F. Vance, and J. M. Hamm, "Observation of lightning in the frequency and time domains," in *Lightning Electromagnetics*, R. L. Gardner, ed., Hemisphere, New York (1990), pp. 191–210.

10. N. Ari and W. Blumer, "Analytic formulation of the response of a two-wire transmission line excited by a plane wave," *IEEE Trans. Electromag. Compat.* **30** (4):437 (1988).

11. J. T. Naff and J. W. Douglas, "A 3-kV, 3-Ω, fast risetime pulser," Digest of 5th IEEE Pulsed Power Conference, Arlington VA (10–12 June 1985), pp. 133-134.

12. S. N. Samaddar and E. L. Mokole, "Some basic properties of antennas associated with ultrawideband radiation," in: *Ultra-Wideband, Short-Pulse Electromagnetics 3*, eds., Plenum, New York (1997), p. 147.

13. S. N. Samaddar, "Transient radiation of single-cycle sinusoidal pulse from a thin dipole," *J. Franklin Inst.* **329** (2):259 (1992).

14. S. N. Samaddar, "Behavior of a received pulse radiated by a half-wave dipole excited by a single-cycle sinusoidal voltage," *J. Franklin Inst.* **330** (1):17 (1993).

MULTICHANNEL ANTENNA SYSTEMS FOR RADIATION OF HIGH-POWER ULTRAWIDEBAND PULSES

Yuri A. Andreev, Yuri I. Buyanov, Vladimir I. Koshelev, Vyacheslav V. Plisko, and Konstantin N. Sukhushin

High Current Electronics Institute RAS,
4, Akademichesky Ave., 634055 Tomsk, Russia

INTRODUCTION

To realize attractive possibilities of ultrawideband (UWB) radars, it is necessary to develop multichannel antenna systems for radiation and control by characteristics of high-power (~ 10^9 W) electromagnetic pulses localized in space and time. As an element of such arrays we suggested to use radiators[1] presenting a combination of an electrical monopole and magnetic dipole. Small-dimensional combined radiators have a pattern of linearly polarized radiation close to a cardioid one and a stable phase center in a wide frequency band. The latter is especially important for electromagnetic pulse synthesis with an arbitrary wave form in free space at the excitation of the array element by electrical pulses of different wave form and length. We have chosen a bipolar pulse as a basic wave form that allows to realize the efficiency of a single radiator by power and energy up to 90%.

In the given paper, new results of a single combined antenna investigations are presented. However, the main consideration is given to the investigation of interaction of combined radiators and plane electrical monopoles in linear arrays including the investigation under conditions of steering by the wave beam. Possibility to synthesize the electromagnetic pulse in free space in order to expand radiation spectrum has been investigated theoretically.

ELECTROMAGNETIC PULSE SYNTHESIS

To investigate spatial-temporal characteristics of short electromagnetic pulses formed in a uniform medium by linear array of radiators the following mathematical model was used. In the far zone, the strength of the electromagnetic field radiated by a single array element with a cardioid pattern without taking into account interaction was determined by the expression:

Ultra-Wideband, Short-Pulse Electromagnetics 4
Edited by Heyman *et al.*, Kluwer Academic / Plenum Publishers, New York, 1999.

$$E_n(t,r_n) = \frac{A}{r_n}\frac{dI_n}{dt}(1+\cos\varphi),$$

where A is the dimensional constant, r_n is the distance from the array element to the field point, c is the velocity of electromagnetic wave propagation, $I_n=I(t-r_n/c)$ is the current value of the n-th radiator taken at the time moment t with a delay for r_n/c, φ determines the direction to the field point. Full field is a superposition of the fields radiated by the array elements. Spatial-temporal characteristics of electromagnetic pulse were investigated when exciting the radiators by a bipolar current pulse described by the expression

$$I(t) = e^{-\left(\frac{5t}{\tau}-2\right)^2} - e^{-\left(\frac{5t}{\tau}-4\right)^2},$$

where τ is the pulse length at the level of 0.1 of its amplitude.

As a result of numerical simulation[2], the following conclusions can be made for the array of radiators excited by the same bipolar pulses:

- radiated pulse form depends on the direction to the field point;
- side-lobe radiation level decreases with increase of number of radiators;
- side-lobe radiation minimum level is achieved at the same amplitudes of exciting pulses.

To expand radiated pulse spectrum it is suggested to use an array in which radiators are excited by bipolar pulses of different duration. Let's consider a linear array consisting of 8 radiators. Radiators n=1, 8, n=2, 7, n=3 ÷ 6 were excited by the pulses with the lengths 5ns, 3ns, 1ns, respectively. The distances between the radiators were equal to the half of the spatial length of a corresponding pulse. The pulses had equal amplitude and were synchronized by the time of changing the signal polarity. The form of a synthesized electromagnetic pulse in the pattern maximum is presented in Fig.1. Pattern width by field at a half-maximum was less than 30°. Fig. 2 presents the spectrum of a resulting pulse (solid line) and the pulses composing it (shaded line). Ratio of the frequencies at the spectrum half-maximum for a synthesized pulse increased by a factor of 3-4 and made up 10.5. Calculations have shown that the suggested synchronization variant has enough good steadiness to the pulse scattering. At a jitter composing ±5% of the length of a corresponding pulse, radiated pulse wave form, spectrum width and pattern are changed unessentially.

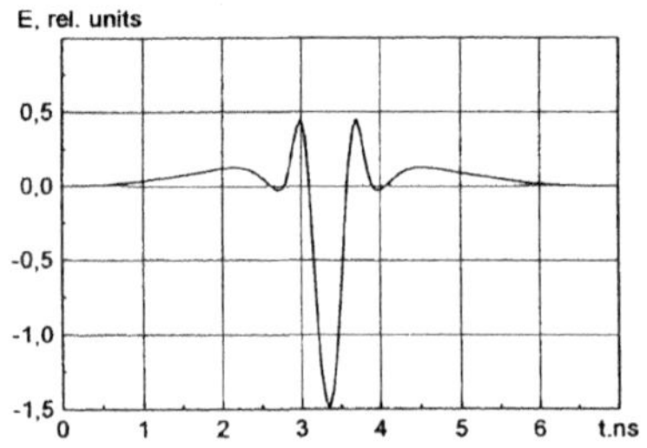

Figure 1. Electromagnetic radiation pulse.

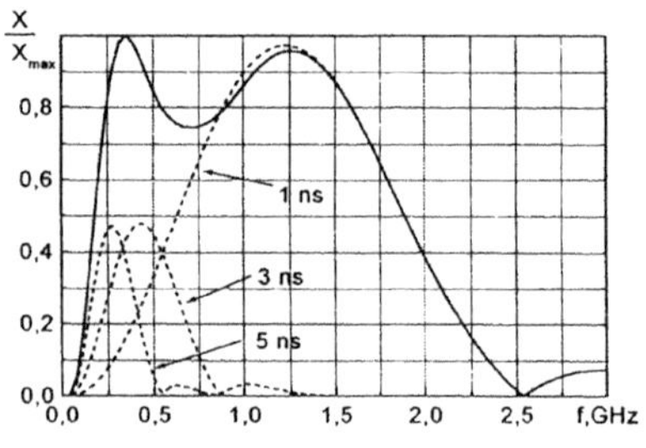

Figure 2. Synthesized pulse spectrum.

SINGLE COMBINED UWB ANTENNA

To specify a magnetic dipole affect on the matching of a combined antenna with feeder, the following experiments have been carried out. Voltage standing wave ratio (VSWR) of the combined antenna (Fig. 3a) and a plane electrical monopole of this antenna placed above the ground plate center (Fig. 3b) were compared. A plate with the dimensions equal to the unrolled case of a combined antenna was used as a ground plate. The results are presented in Fig. 4 where the curve 1 corresponds to VSWR of an electrical monopole and curve 2 corresponds to VSWR of a combined antenna. As it is obvious from the curves

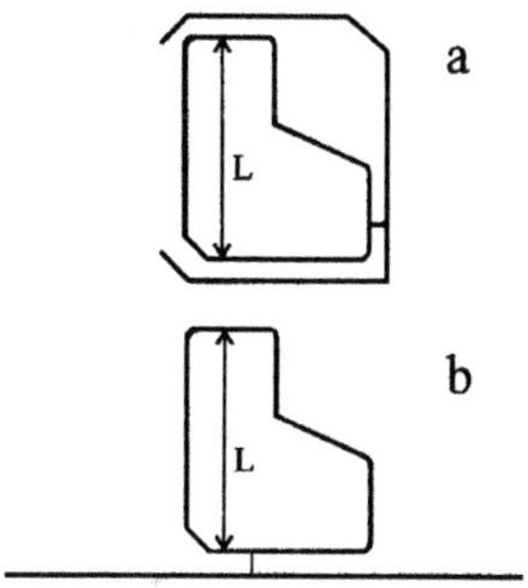

Figure 3. Geometry of a combined antenna (a) and plane electrical monopole (b).

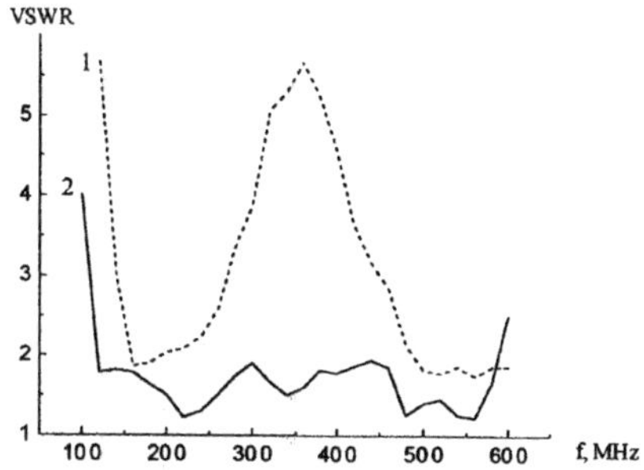

Figure 4. VSWR of a plane monopole (1) and combined antenna (2).

presented, a magnetic dipole addition allows to expand antenna matching into the region of low frequencies approximately by 30% and to match the antenna with a 50-Ohm feeder in the frequency band higher than two octaves at the VSWR less than 2. These results have been obtained for the antenna designed for excitation by a 3-ns long bipolar pulse. Analogous results have been also obtained for the antenna designed for excitation by a 2-ns bipolar pulse. The obtained experimental data agree with the theory[1].

INTERACTION OF COMBINED RADIATORS IN A LINEAR ARRAY

Investigations of interaction of combined antennas were carried out for a two-element and four-element linear arrays. In the experiments, the antenna[3] applied previously to radiate 100-MW pulses have been used.

Let's start with considering the investigation results of a two-element array. Here and further, to excite antennas low-voltage bipolar 3-ns long pulses were used. Measured array patterns by the peak field amplitude for different distances between antennas phase centers d were compared with the calculated array patterns not taking into account interaction of radiators in it. In the calculations, a cardioid pattern for the model array elements was used. The results are presented in Fig. 5a, b, c for d=0.5τc; 0.75τc and 1τc, respectively. Here and further measured patterns are shown by a solid line while the calculated ones by a dotted line. As it is seen from the Figure, measured patterns differ from the calculated ones by less marked side-lobes and insignificant widening of the main lobe. The differences of the patterns are the greater the less is the distance between the antennas.

Besides, in the experiments the energy of the pulse reflected from the array element was measured. Fig. 6 presents the dependence of changing the energy reflected from the array element expressed in percentage on the energy in the input pulse on the distance. As it

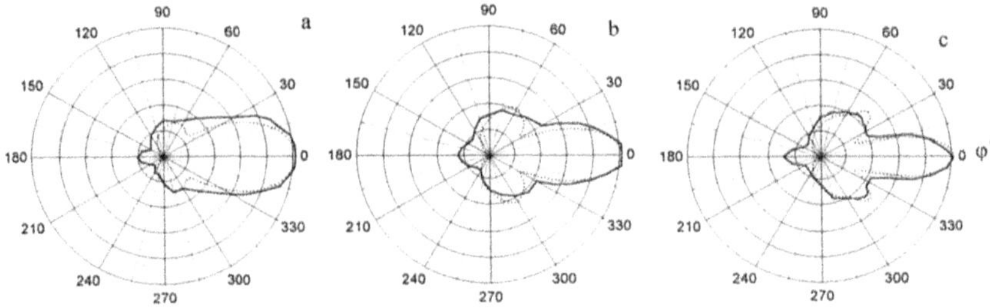

Figure 5. Patterns of a two-element array for d= 0.5τc (a), 0.75τc (b), 1τc (c).

is seen from the Figure, the energy reflected from array elements decreases with the increase of the distance between the antennas and at the distance between antennas phase centers equal to the spatial length of the input pulse becomes close to the energy reflected from a single antenna (13%).

In the direction of the main maximum a coherent summation of signals from two array elements occurs. Fig. 7 presents the oscillograms of radiated pulses at a 3.5-m distance from the transmitting antenna. The oscillogram 1 presents the pulse from a single antenna and oscillogram 2 presents the pulse from a two-element array at d=0.5τc attenuated by a factor of 2.

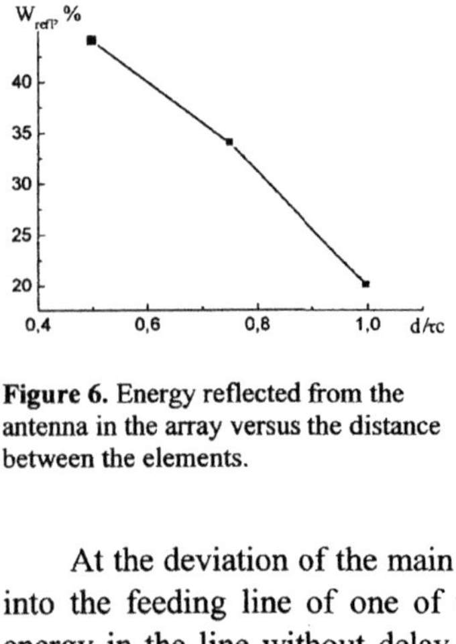

Figure 6. Energy reflected from the antenna in the array versus the distance between the elements.

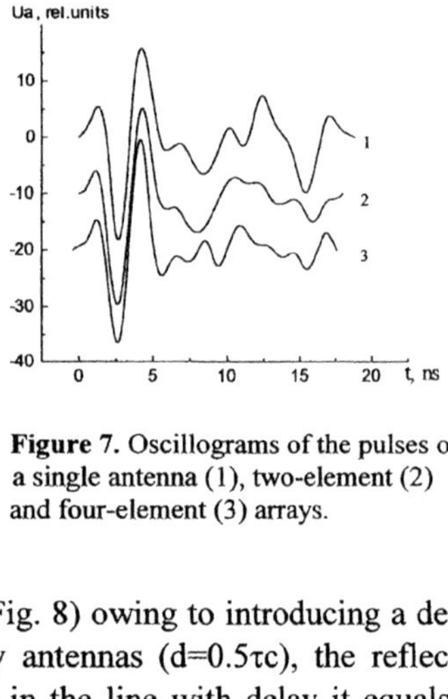

Figure 7. Oscillograms of the pulses of a single antenna (1), two-element (2) and four-element (3) arrays.

At the deviation of the main pattern lobe by 40° (Fig. 8) owing to introducing a delay into the feeding line of one of the two-element array antennas (d=0.5τc), the reflected energy in the line without delay contains ≅15% while in the line with delay it equals to ≅25%. The reflected energy being less than the one without pattern steering is, obviously, related to the change of conditions of radiators interaction.

Comparison of the patterns and determination of the reflected energies has been also carried out for four-element arrays with d=0.5τc and 0.75τc. Fig. 9 presents the calculated and measured patterns for d=0.5τc. Mutual influence in four-element arrays is still stronger than in two-element ones and the pattern measured differs from the calculated one more essentially. The reflected energy in a four-element array for d=0.5τc is equal to 75% and 70% for internal and external antennas, respectively. For d=0.75τc they compose, respectively, 70% and 55%. In spite of the strong mutual influence of the antennas in the

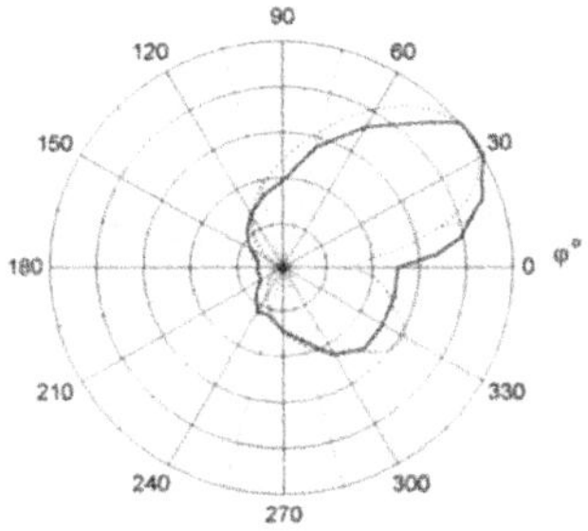

Figure 8. Patterns of a two-element array with d=0.5τc at a steering of pattern.

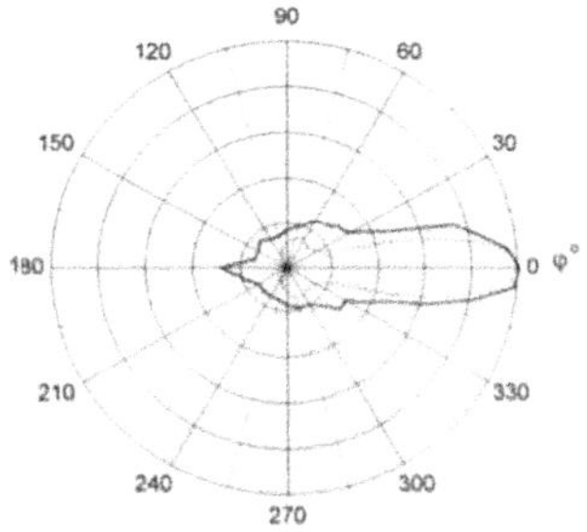

Figure 9. Pattern of a four-element array with d=0.5τc.

direction of the main pattern maximum, coherent signal summation occurs and the field amplitude increases by a factor of 4. The oscillogram 3 presented in Fig. 7 corresponds to the pulse radiated by a four-element array with d=0.5τc and decreased by a factor of 4.

INTERACTION OF PLANE RADIATORS IN A LINEAR ARRAY

A 180x60-cm in dimension conductive plate with 45-cm height, 22.5-cm wide and 2-cm thick radiators installed at it was used for experimental investigation of a linear array with plane radiators (electrical monopole). Fig. 10 presents the pattern of a single radiator. The energy of the pulse reflected from it made up 20% of the incident one.

Investigation for a two element array with the distance between the elements d=(0.25÷1)τc have been carried out. Energy dependence of the pulse reflected from the antenna on the distance between them is presented in Fig. 11. It is seen that minimum reflected energy (28%) is achieved at the distance between the elements d=0.5τc. At the deviation of the pattern main lobe by 45° in a two–element array with d=0.5τc energy of the reflected pulses in the line without delay made up 30% and in the line with delay it was 23%. Fig. 12 presents theoretical and experimental patterns for a 0.5-τc distance between the radiators. Here and further, a real measured pattern of a single radiator (Fig. 10) was taken into account in calculations. However, interaction was not taken into account. The difference between the measured and calculated patterns is analogous to the array of combined antennas.

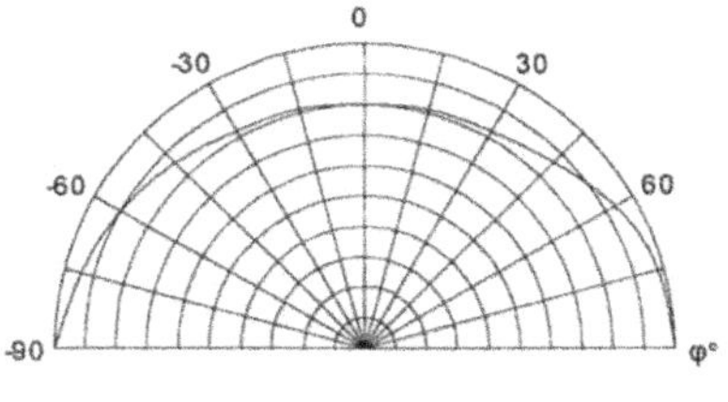

Figure 10. Radiator pattern.

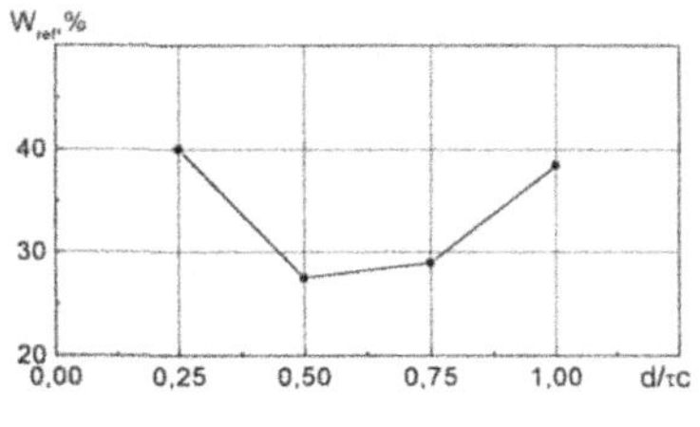

Figure 11. Energy of reflected pulses for a two- element array.

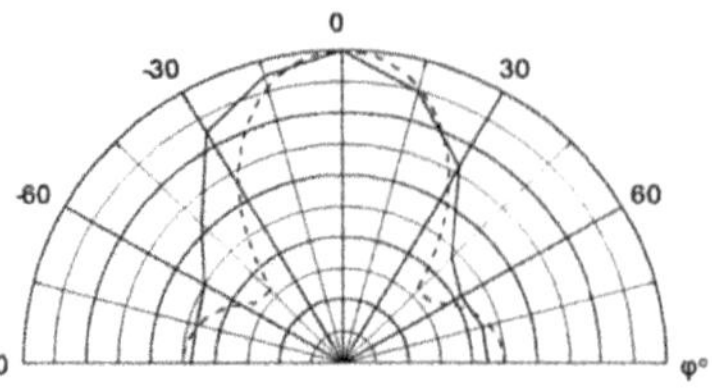

Figure12. Pattern of a two-element array.

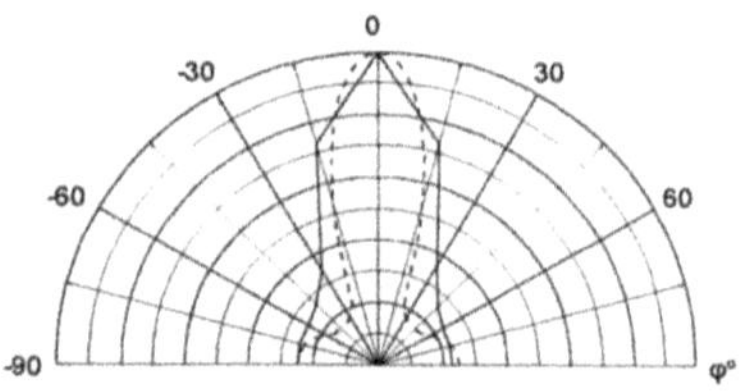

Figure 13. Pattern of a four-element array.

Fig. 13 presents the experimental and theoretical patterns for the four-element array with the distance between the radiators d=0.5τc. Field amplitude in the main maximum direction has increased by a factor of two as compared with a two-element array. Energy of reflected pulses for an external and internal radiators are practically equal and make up approximately 40%. It is obvious that this is higher than for a two-element array (d/τc=0.5, Fig. 11).

CONCLUSION

Interaction of radiators in the array results in the energetic efficiency decrease of a multielement antenna system due to the increase of the energy reflected from the array, widening of the pattern main lobe, and smearing of the side-lobes. Electromagnetic field strength in the main lobe increases proportionally to the number of radiators in spite of the reflected energy increase.

The greater value of the reflected energy for the array of combined radiators in comparison with the array of electrical monopoles is related to the difference in a cross-polarized radiation of these antennas. Special investigations with combined antennas confirmed this assumption. Application of the screens between the radiators will allow to decrease the interaction of the radiators and, respectively, the reflected energy. However, we associate the essential increase of energetic efficiency of antenna arrays with the development of a new variant of ultrawideband antenna which has no cross-polarized radiation.

For expansion of possibilities to control the ultrawideband radiation parameters, multichannel antenna systems with different radiators are promising. Possibility to expand essentially radiated pulse spectrum for a concrete variant of linear array is shown.

REFERENCES

1. V.I.Koshelev, Yu.I.Buyanov, B.M.Kovalchuk, Yu.A.Andreev, V.P.Belichenko, A.M.Efremov, V.V.Plisko, K.N.Sukhushin, V.A.Vizir, and V.B.Zorin, High-Power Ultrawideband Electromagnetic Pulse Radiation, SPIE Proc. 3158: 209 (1997).
2. V.P.Belichenko, Yu.I.Buyanov, V.I.Koshelev, and V.V.Plisko, Short Electromagnetic Pulse Formation by Plane Antenna Array, Proc. of Conf. on Direct and Inverse Problems of Electromagnetic and Acoustic Wave Theory, Lviv: Pidstryhach Institute of Applied Problems of Mechanics and Mathematics of the Ukrain National Academy of Sciences. 43 (1997).
3. Yu.A.Andreev, Yu.I.Buyanov, V.A.Vizir, A.M.Efremov, V.B.Zorin, B.M.Kovalchuk, V.I.Koshelev, and K.N.Sukhushin, A high-Power Ultrawideband Electromagnetic Pulse Generator, *Instruments and Experimental Techniques*. 40: 651 (1997).

ULTRA-WIDEBAND DUPLEXERS FOR LOW AND HIGH PEAK POWER APPLICATIONS

Marc Piette

Royal Military Academy
30, Avenue de la Renaissance
B-1000 Brussels Belgium

INTRODUCTION

In the frame of a research project called HUDEM (HUmanitarian DEMining) involving eight Belgian universities and funded by the MoD and the State Secretariat for Cooperation, the Royal Military Academy (RMA) is developing a lightweight and compact Ultra-Wideband Impulse Radar (400MHz-4GHz) for detecting antipersonnel landmines.

This type of sensor is complementary to the metal detector (MD) because it can detect plastic mines. It yields also a better depth resolution and offers more target recognition capability than a conventional ground penetrating radar (GPR) thanks to the ultra-short pulses (<200 ps) radiated into the ground.

In view of a hand-held detection system, the radar must be lightweight, compact and easy to use. Therefore, using one antenna for transmitting as for receiving is better than two. Moreover because of the short distance between the radar and the ground, a system with two antennas (bistatic radar) would need some focusing of the antennas in order to maximize the signal to noise ratio. Indeed, without focusing the two antennas to the target a few decibels would be lost because the mine is not exactly in the direction of the antenna main beam. This problem is avoided with a monostatic setup, but this one can only be realized with a duplexer routing the signal from the transmitter to the antenna and the echo signal from the antenna to the receiver.

Unfortunately, conventional duplexers like circulators, directive or hybrid couplers are conceived for systems using CW signals or modulated signals with narrow bandwidth (from a few percents to maximum 15 percents depending upon the model). Hence they are unable to transmit short time impulses without waveform distortion.

In this paper we describe the design and the operation of two ultra-wideband duplexers conceived for subnanosecond pulse applications. Their performances are outlined in terms of waveform fidelity, speed, isolation, insertion loss, size, weight and cost. The first duplexer, designed for low power radars, is based on a fast GaAs switch combined with a

Ultra-Wideband, Short-Pulse Electromagnetics 4
Edited by Heyman *et al.*, Kluwer Academic / Plenum Publishers, New York, 1999.

power splitter, a coaxial matched termination and a broadband amplifier. The duplexer is triggered by the radar transmitter so that the power signal is routed to the antenna and that the echo signal, received a few nanoseconds later by the same antenna from the target, is routed to the receiver. The second duplexer can be used in high peak power radars (up to 50 kW). It is based on power splitters and a pulse inverting transformer mounted in a suicide transmission line. A broadband amplifier is integrated in the duplexer to compensate for the loss in the power splitters. Since the price of the duplexers proposed here is lower than that of common ultra-wideband antennas, their use is cost-effective.

THE LOW POWER DUPLEXER

Description

This duplexer, shown in Fig.1, is based on a fast GaAs Single Pole Double Throw switch (SPDT) which exhibits a switching time of 3 ns and a frequency range from dc to 5 GHz. The basic function of a SPDT is to route the RF signal applied to the input port toward one of the two output ports alternatively at the rythm of a clock signal (typically a TTL signal). As shown in Fig.1, one output port of the switch is connected to the radar receiver through a low noise ultra-wideband amplifier, while the other output port is terminated on a 50 Ohm resistance. An ultra-wideband power splitter is used to connect the pulse generator and the antenna to the input port of the SPDT.

Operation

The pulse coming from the generator is splitted in two equal signal in the power divider. The first one is coupled to the antenna and radiated in free space to the target, while the second one is coupled to the SPDT which routes it to the 50 Ohm termination in order to protect the sensitive receiver from this relatively high amplitude pulse. Taking into account the expected round trip time of the radiated wave in free space (about 3 ns for an antenna height above ground of 40 cm), it is possible to drive the switch at the right time in such a way that the echo signal coming back from the target, is routed by the SPDT toward the receiver. Thanks to the wide bandwidth of the switch and the power splitter, the short pulse response of the target can be transmitted to the receiver without significant waveform distortion (Fig.2). This point is particularly important in view of the target identification capability of an UWB GPR. Besides, a high speed switch is absolutely a must in such a duplexing circuit because of the very short round trip time (<5 ns) due to the short distance between the radar antenna and the burried target. Fortunately the GaAs technology offers the opportunity to commute low level signals with switching time of only a few nanoseconds and a very low insertion loss (less than 2 dB). The in-out isolation of commercially available GaAs switch is about 20 dB. This means that when the input signal is routed to output port A, the residual power at output B is 20 dB lower than the power level at output A.

The signal undergoes also a loss in the power splitter (6 dB from branch A to branch B when transmitting and 6 dB from branch B to branch C when receiving) but this is partially compensated by the 8 dB amplifier inserted between the SPDT output and the receiver. All the components of the circuit having coaxial connectors, they can be connected to each other with short rigid cables which make the whole system robust and compact in size.Two key performances of a duplexer are the transmit-receive isolation and the insertion loss. In the circuit shown in Fig.1 the transmit-receive isolation is equal to the sum of the loss in the power splitter and the SPDT isolation minus the amplifier gain, i.e. 6 + 20 - 8 = 18 dB. This

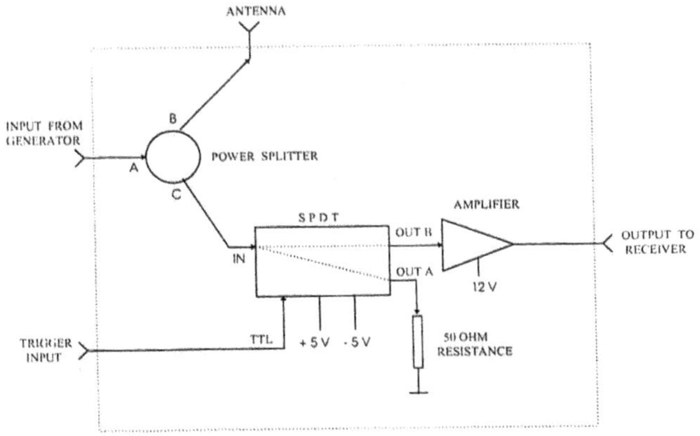

Figure 1. Low peak power duplexer.

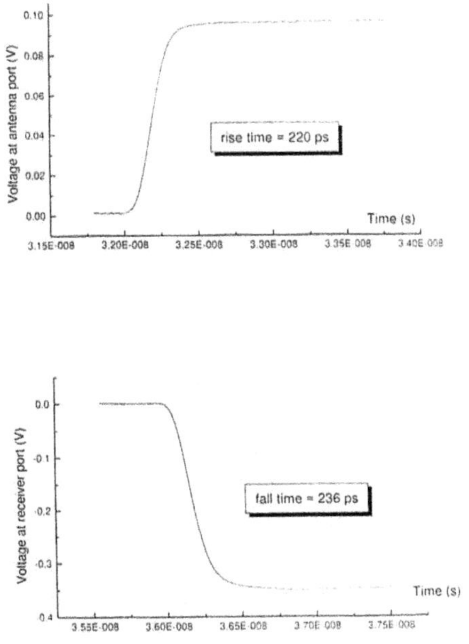

Figure 2a. Transmitting step response of the low power duplexer to a 200 ps Gaussian step voltage applied at the generator port. 2b. Receiving step response of the low power duplexer to a 200 ps Gaussian step voltage applied at the antenna port.

means that if the digital storage oscilloscope used as receiver can withstand a maximum peak voltage of 5 V for example, the maximum output voltage of the pulse generator usable with the duplexer would be 40 V. In practice however, the maximum voltage that can be applied to the input of the duplexer is limited by the linearity of the SPDT, which exhibits typically a 1 dB compression point of about 20 dBm. With the fastest SPDT that are commercially available for the moment [1], the duplexer shown in Fig.1 has been realised and tested. The performances are as follows :

- frequency range : 20 MHz - 4,2 GHz
- transmit-receive isolation : 18 dB
- maximum input power : 100 mW
- cost of the components < 300 US$
- insertion loss from antenna to receiver : 0 dB
- insertion loss from transmitter to antenna : 6 dB
- dimensions : 1" x 2" x 4"
- weight : 130 g

These figures are encouraging if one compares them with those achieved by conventional low power circulators offering typically a relative bandwidth of 5 to 10 % in a frequency range that starts from 500 MHz only, with an isolation of 20 dB and an insertion loss of 0.5 dB.

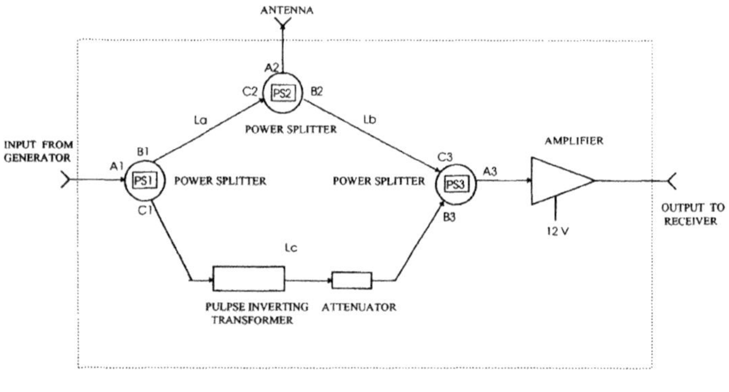

Figure 3. High peak power duplexer.

A question that could arise is how the duplexer would function in case the antenna is not well matched to the 50 Ohm cable connecting it to the power splitter. In general, ultra-wide band antennas are indeed poorly matched to 50 Ohm transmission line so that a significant fraction of power travels back from the antenna. In the case of this duplexer, the reflected power is first reduced by 6 dB in the power splitter before being routed to the 50 Ohm termination by the SPDT that has not yet had the time to switch from port B to port A, at least if the cable between the power splitter and the antenna is not too long.

THE HIGH POWER DUPLEXER

Description

This duplexer is conceived for high peak power radars (up to 50 kW with pulse duration < 300 ns) with moderate average power (4 W). It is based on the use of three power splitters and a pulse inverting transformer configurated to form a suicide transmission line. A broadband amplifier is integrated in the duplexer to compensate for the loss in the power splitters, as shown in Fig.3. Because all the components of the circuit have coaxial connectors, they are connected to each other with semi-rigid cables which make also the circuit robust and compact in size.

Operation

The principle of operation of this circuit is to divide the pulse coming from the generator in two equal signals and to let a part of it interfere destructively before they have reached the receiver, while the other part is applied to the antenna for radiation. As shown in Fig.3, the pulse coming from the generator is splitted in two equal parts in the power splitter PS1. The first part is sent to another power splitter PS2, while the second part is sent to the lower arm C, consisting of a pulse inverting transformer and an attenuator. From power splitter PS2 come two equal signals. The first one is sent to the antenna for radiation and the second one is sent to a third power splitter PS3, along with the signal that is travelling down in arm C.

The function of the pulse inverting transformer is to inverse the polarity of the pulse while the attenuator is used for adjusting the amplitude of the signal in arm C to that of the signal coming from PS2. If the lengths of the two arms are equal (La+Lb=Lc), then the pulses entering the power splitter PS3 interfere destructively and no signal at all appears at the output A3 so that the receiver is completely safe. This means that the transmit-receive isolation of this duplexer is infinite. On the other hand, the target echo signal travelling back from the antenna reaches the receiver through PS2 and PS3. Because the signal undergoes 12 dB attenuation along this path, a broadband amplifier is inserted as output stage of the duplexer to compensate for that loss. Thanks to the wide bandwidth of the power splitters and the amplifier, the short pulse response of the target can be transmitted to the receiver without significant waveform distortion (Fig.4). With the coaxial components that are commercially available for the moment, a duplexer like those shown in Fig.3 has been realised and tested. The performances are as follows :

- frequency range : 60 kHz - 4.2 GHz
- transmit-receive isolation : ∞dB
- cost of the components < 1200 US$
- dimensions : 2" x 4" x 8"
- insertion loss from antenna to receiver : 1dB
- insertion loss from transmitter to antenna : 12dB
- maximum input power: 50kW peak, 4W average
- weight : 1450 g

These figures are quite encouraging if one compares them with those achieved by conventional high power circulators which offer typically a relative bandwidth of 5 to 10 % in a frequency range that start from 500 MHz only, with an isolation of 20 dB and a peak power rating of about 1 kW.

In case the duplexer is used with a poorly matched antenna, the reflection from the antenna will of course reach the receiver, like the echo signal from the target, but the time interval between these two pulses will be about 3 ns for a typical antenna height above ground of 40 cm. By time gating this unwanted reflection on the receiver, it is possible to visualize the uncontaminated response of the target, unless port A2 of the power splitter is also badly matched, which would cause some ringing in this arm of the circuit. But the power splitters used in the circuit exhibit good impedance matching so that this problem does not occur.

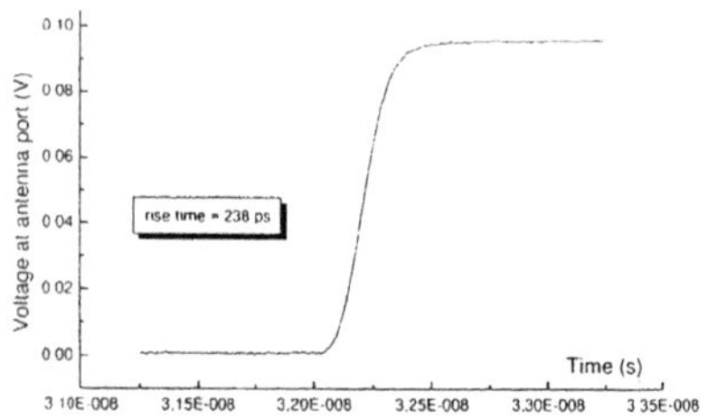

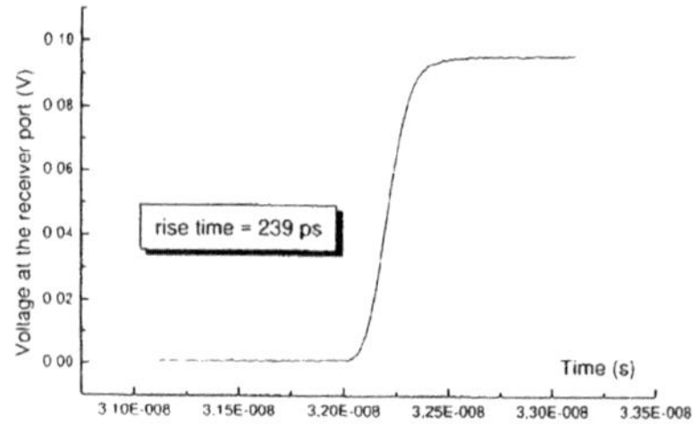

Figure 4a. Transmitting step response of the high peak power duplexer (without amplifier) to a 200 ps Gaussian step voltage applied at the generator port. 4b. Receiving step response of the high peak power duplexer (without amplifier) to a 200 ps Gaussian step voltage applied at the antenna port.

CONCLUSIONS

Two different duplexers have been conceived, designed and tested that allow for using an Impulse UWB Radar in a monostatic configuration. The first one, which is based on a fast GaAs SPDT switch, is usable with a low peak power transmitter (< 100 mW), while the second one, completely passive, is intended for high peak power applications (up to 50 kW). Both exhibit an instantaneous bandwidth covering the frequency range (400 MHz - 4 GHz) which is the most suited frequency band for detecting burried antipersonel landmines by means of UWB radar. These duplexers contribute to a gain of space, weight and money, three important factors in the design of portable detection systems for humanitarian operations.

REFERENCES

1. RF Designer's Guide Minicircuits, pp 118-119, Edition 1998

UWB SHORT PULSE SENSOR FOR TARGET ELECTROMAGNETIC BACKSCATTERING CHARACTERIZATION

M. Le Goff [1], P. Pouliguen [1], Y. Chevalier [2], Y. Imbs [2],
B. Beillard [2], J. Andrieu [2], B. Jecko [2], G. Bouillon [3], B. Juhel [4]

[1]CELAR (DGA), Bruz, France
[2]IRCOM, Brive, France
[3]Dassault Electronique, Saint Cloud, France
[4]Université de Rennes, France

ABSTRACT

The Electronic Armament Centre (CELAR : Centre d'Electronique de l'Armement) has evaluated the use of a relatively low frequency ultra-wide band short pulse measurement facility to characterize target electromagnetic signatures. Specific measurement instrumentation has been developed and implemented. The initial measurement trial began at the beginning of 1998 and led to the establishment of a database on various targets. These measurements made it possible to interpret, in the time domain, the signatures and to work out holograms. Images were calculated either in the time or the frequency domain.

INTRODUCTION

In its mission of evaluation to the benefit of the program services of the Délégation Générale pour l'Armenent (General Armament Delegation), the CELAR measures the radar electromagnetic signatures of military targets. To do this, it uses a number of different facilities. The principle is that of illuminating a target with a frequency ramp then analysing the signal return. Passage through the time domain then allows, by Fourier transform, a target pulse response to be obtained. Using an UWB measurement facility is a means of directly measuring the pulse response of the target under test. In this case, it involves the radiation of a very narrow pulse whose spectral content corresponds to that of the frequency ramps currently being transmitted. Acquisition is directly in the time

Ultra-Wideband, Short-Pulse Electromagnetics 4
Edited by Heyman *et al.*, Kluwer Academic / Plenum Publishers, New York, 1999.

domain. The transient nature of UWB signals is also particularly interesting for the analysis of the electromagnetic phenomenon involved. To this aim, and confronted by the increasing interest in pulse radars and relatively low frequencies in radar, CELAR desires to equip its chambers with a facility to measure radar signatures by ultra-wide band pulse signals.

UWB MEASUREMENT SYSTEM

The measurement system used is relatively simple compared to more conventional facility of measuring Radar Cross Section (RCS) by the harmonic technique. It consists of a pulse generator, two wide-band antennas, a receiver, a digital oscilloscope, a data recording and processing unit and a sequencer. The overhall arrangement of the device is shown in figure 1.

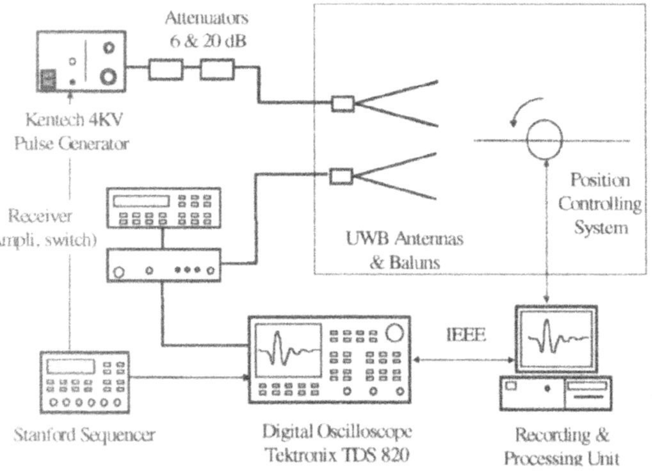

Figure 1 – Measurement system

The generator[2] supplies a pulse with a peak amplitude of 4 kiloVolts, adjustable between 60 and 100 %. High voltage attenuators from 6 to 20 dB are used to bring the level to 2 kiloVolts or 200 Volts depending on the measurement chamber being used (long chamber or short chamber). The supplied signal is highly repetitive and has a rise time of 220 ps with a pulse duration of about 1 ns, covering the 0 to 1.5 GHz frequency band (figure 2). The repetition rate is 1 kHz with 10 ps jitter. The generator is triggered by an external control signal.

The antennas[1] have been designed to radiate and receive short pulses with the least possible distorsion and to illuminate the target with an homogeneous consistent electromagnetic field. The antennas are striplines and based on the principle of TEM horns. They consist of four conducting wires, the upper-half lengths of which are resistively loaded. These loads gradually attenuate the current and avoid reflection at the ends of the wires. The antenna dimensions are optimised so that the gain is maximum in the frequency band in question, from 200 MHz to 1 GHz. A balun ensures antenna matching (200 Ω) on the transmission circuit (50 Ω) and drives the bifilar line with pulses in opposite phases.

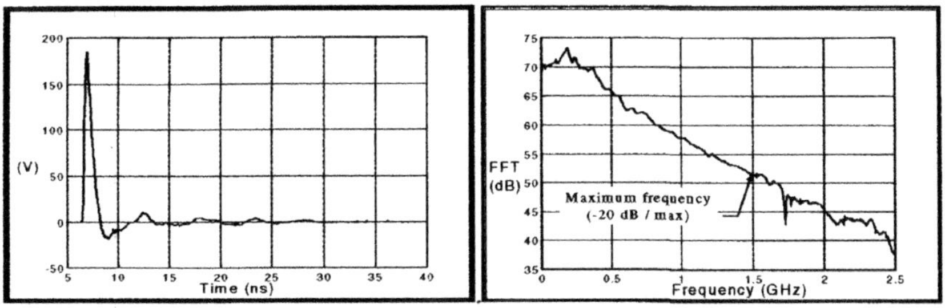

Figure 2 – Pulse supplied by generator and signal spectrum

The purpose of the receiver[2] is not only to amplify low amplitude signals (gain adjustable from 0 to 46 dB) but also to insulate, by time windows, the effective signal from unwanted echoes. To do this, the reception channel has three amplifiers with 18 dB gain, two attenuators programmable from 0 to 31 dB in 1 dB steps and two 30 dB insulation modulators with fast switching times. A limiter at the beginning of the chain protects the receiver from direct leaks between the transmission antenna and the reception antenna.

The oscilloscope used for the acquisition chain is a Tektronix TDS820 6 GHz passband digital unit. Digitizing is carried out in sequence. Using this method, the input signal must be very repetitive. The oscilloscope acquires a single sample per period and gradually delays the moment of sampling of the following points until the entire signal is digitized.

A conventionnal PC is used for initialising, recording and processing of data.

Sequencing is carried out from a Stanford Research Systems[2] delay generator. It triggers the transmitter, generates the reception windows and synchronises the oscilloscope. This delay generator has one reference channel and two independent channels whose delays and widths can be programmed in steps of 50 ps.

DATA ACQUISITION

The first measurement trial was carried out in early 98 in the CHEOPS anechoic chamber (see figure 3) at CELAR.

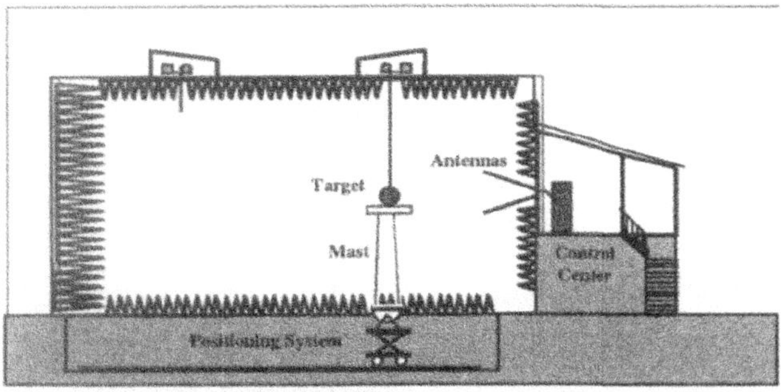

Figure 3 – CHEOPS anechoïc chamber.

The chamber mesures 25 m long, 12 m wide and 12 m high. Targets can be placed at 7 m or 18 m. The positionner, coupled with the measurement instrumentation, puts the target into rotation through 360°. The walls of the chamber are entirely lined with absorbent pyramids with a length of 3 m at the back of the chamber. The measurements are made in quasi-monostatic mode because the transmit and receive antennas are close to each other. The four terms of the polarimetric matrix can be measured per quarter-turn rotation of the antennas.

The advantage of these measurements is to extract the target response and only this response. However, when measurement is made a number of unwanted signals are added : coupling between transmission antenna and reception antenna, reflection on the chamber walls and multipath interference between target and chamber. One of the major advantages of the pulse mode is to be able to use techniques overcoming unwanted echoes. An initial technique is that of simple subtraction in the time domain between target measurement and empty chamber measurement. This method is very efficient (see figure 4) but calls for instrumentation that is very stable in the course of time and for good jitter compared to the digitizer sampling pitch. It is capable of eliminating echoes due to coupling between antennas and to reflections on the walls of the chamber. However, it does not overcome multipath interference between target and chamber. A second step consists in using the fact that measurements are made in terms of time over a short duration and that the target pulse response precedes the various multipath interference. This means that these echoes and those due to coupling between the antennas can be eliminated by carefully centring the reception window. This technique is efficient as long as the target responds in a sufficiently short time compared to the delay between the direct antenna-target-antenna path and the multipath. It means optimum knowledge of the tranquil area of the chamber.

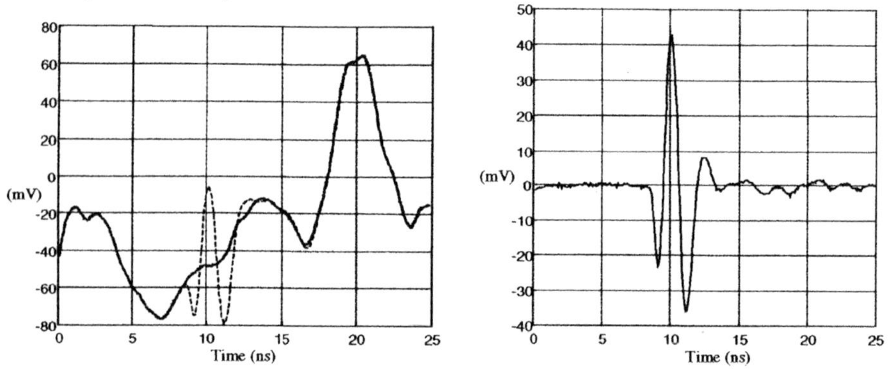

a – Signal with the target installed (dotted line) and removed (solid line)

b – Impulse response after subtraction

Figure 4 – Subtraction in the time domain.

EXAMPLE RESULTS

Determination of the RCS

The measurement of time responses consists in acquiring the signal captured by the reception antenna for each angular position of the target. This generates a time-azimuth chart (see figure 9). The RCS of the various targets is then determined by the application of the following relation[3] :

$$RCS_{\text{measured target}}(\theta, f) = \frac{\left|FFT\left(E_{\text{measured target}}(\theta,t)\right)\right|^2}{\left|FFT\left(E_{\text{measured calibrator}}(\theta,t)\right)\right|^2} * RCS_{\text{theoretical calibrator}}(\theta, f)$$

where $E_{measured\ target}$ represents the time response of the target of interest for various azimuth angles θ and versus time t, minus the empty chamber response. $E_{measured\ calibrator}$ represents, under these conditions, the time response of a standard target whose $RCS_{theoretical\ calibrator}$ is accurately known either analytically or using a computation code. This technique is therefore a way of deducing the RCS of a target without, a priori, knowing anything about the incident signal.

Various targets of variable sizes and materials have been measured to establish a database : spheres, disks, plates, dihedrons, trihedrons, cylinders, dipoles, cavity, cone-spheres, mines, missiles. Figures 5 to 8 shows, for example, the time responses and the RCS of four canonical targets :

- a 80 mm diameter metal sphere
- a 200 mm diameter metal sphere
- a 60 cm sided metal plate
- a 62 cm long, 25 cm diameter metal cylinder seen from one face

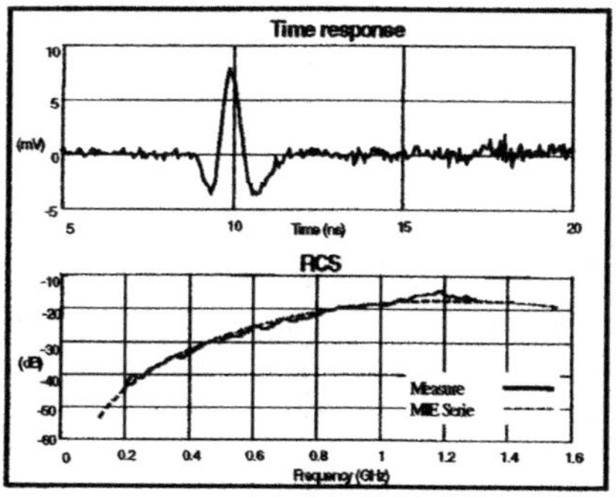

Figure 5 : Sphere (80 mm).

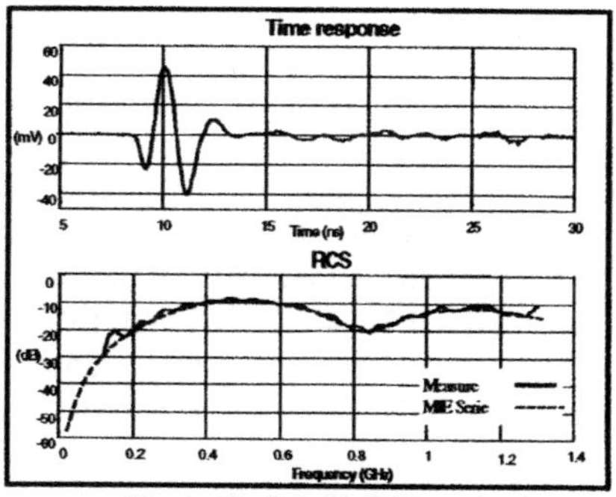

Figure 6 : Sphere (200 mm).

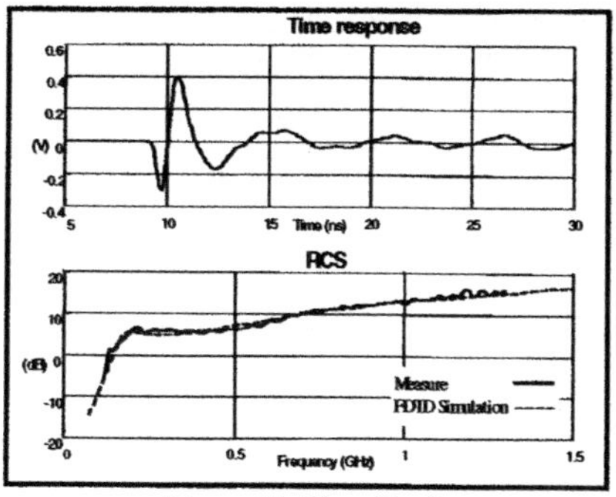

Figure 7 : Plate (1 m).

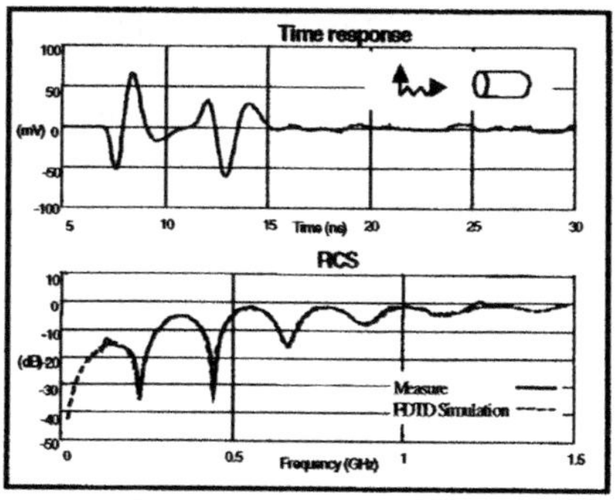

Figure 8 : Cylinder (62 cm).

The RCSs are thus compared either with RCSs derived from modelling code or from analytical RCS curves (MIE series in the case of spheres). The biggest RCS deviations observed are about 1 dB for a frequency band from 200 MHz to 1,2 GHz.

Construction of electromagnetic images

The time method[4] developed here consists in coherently summing pulse responses measured in a given angular sector corresponding to the rotation of the positioning device. This method is used for narrow band holography (tomography) in which pulse responses are calculated by FFT of the hologram. With the UWB technique, pulse responses derive directly from measurement. Calculation can be summarised through the following formula :

$$I(x_i, y_j) = \sum_{k=1}^{n} Rx(d(x_i, y_j), \theta_k)$$

The calculation of the image is made pixel by pixel. For each $I(x_i, y_j)$, the algorithm, by means of each pulse response R_x and thanks to antenna-pixel distance $d(x_i, y_j)$, identify the corresponding range cell. This point has little chance of corresponding accurately to a measured range cell and its value is calculated by linear interpolation between the values of the two closest cells. The final result is obtained by summing these values, obtained from measured pulse responses that correspond to n different positions θ_k of the target in a clearly determined angular sector. This angular sector must be large enough to have a sufficiency of measurements so as to increase the resolution of the characteristic points of the target. But it must not be excessive otherwise the points that appear will no longer be characteristic of the radar direction of incidence. A careful choice is needed. In the following examples an angular sector of 100°, centred on various incident directions, has been chosen.

To illustrate these words, we have chosen to present the results obtained on a totally conducting model of a missile. Figure 9 shows the result chart in the time domain as measured between -90° and +90°. Figure 10 shows the hologram in the frequency domain calculated by Fourier transform of previous data, and as it could have been measured by conventional instrumentation using harmonic technique. Lastly, figure 11 shows several images obtained for different sighting directions. These images reveal the points of the target returning the signal most strongly depending on the direction of incidence of the radar. We can see on this example that the missile fins contribute considerably to the target response whatever the incidence direction of the radar, but differently in level and distribution according to the different directions. This example illustrate the interest of such representation for the analysis of the electromagnetic phenomena involved. Such images can also be used for target identification. However, the concept of "bright spots" disappears little by little as we move down through the frequency band and is rather replaced by the concept of target resonance. If there is a big ratio between the size of the target and the wavelength, the concept of bright spot will be used. Otherwise, the concept of resonance and the technique of poles extraction are more suited to the situation.

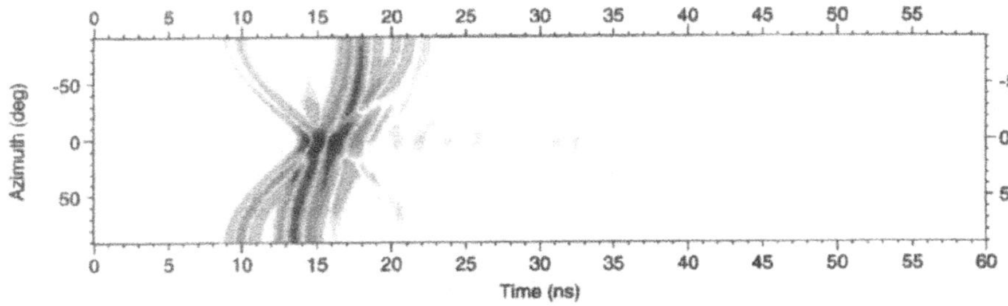

Figure 9 - Measured results in time domain

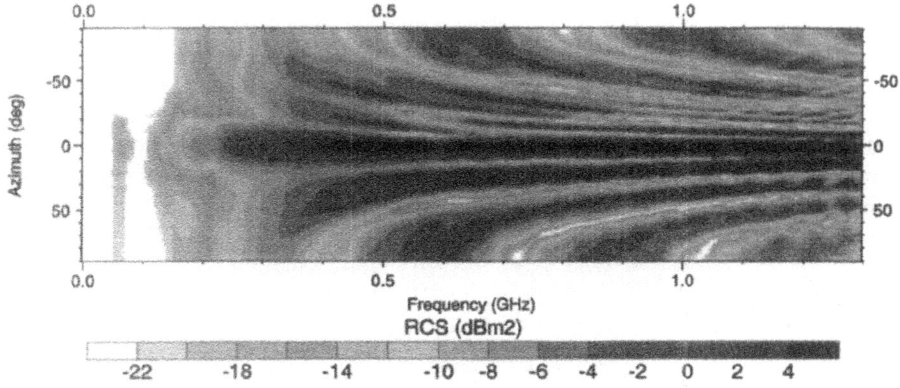

Figure 10 - Hologram in frequency domain

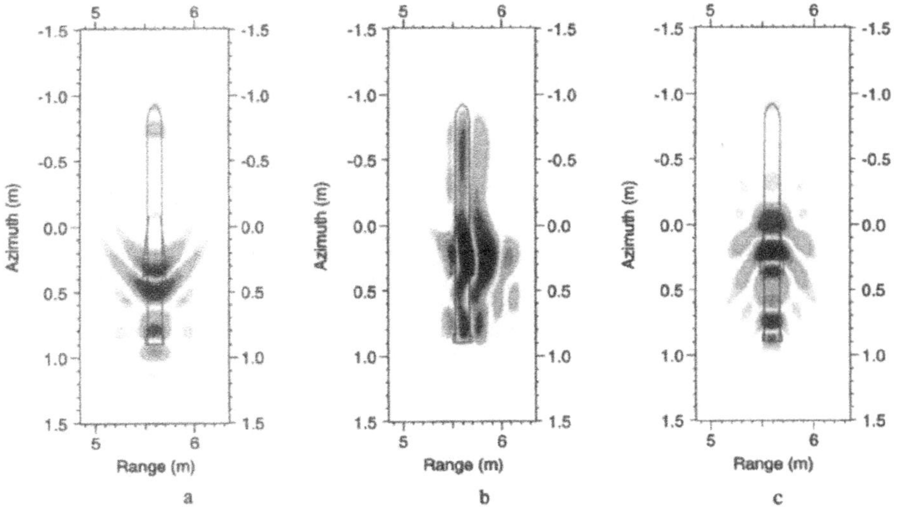

Figure 11 - Missile images for several incident directions
a - in front of the target, b - on the left side, c - from the back.

FUTURE WORKS

An additional measurement trial will be held in June and July 98 in the CELAR SOLANGE facility where large targets will be measured, in particular real missiles and aircraft (full scale). This campaign will be an opportunity to complete the relatively low frequency RCS database already compiled. For measurements, the targets will be positioned at 20 m at the least from the antennas thus entirely overcoming the effects of the near field. Some targets already measured previously will be reused for the very purpose of studying the near field - far field effect.

At present, CELAR is working on the generation of a target resonance pole extraction tool based on the Singularity Expansion Method, introduced by C. Baum[5]. The resonance poles of a structure are specific to this structure independently of the illumination. In theory, it is a way of characterizing the target by a set of identifying parameters specific to it. This tool will be used to characterize targets measured by the UWB instrumentation. Time-frequency analysis methods will also be studied.

CELAR is also developing an external platform to understand the capacities of an UWB synthetic aperture radar. This new instrumentation will be dedicated to pulse measurements on outdoor targets (buried mines, hidden vehicles) and relatively low frequency clutter. Once processed, these measurements will produce high resolution SAR images.

CONCLUSION

We have described and outlined the performance of an Ultra Wide Band short pulse RCS measurement device. This device is capable of determining the RCS within a 200 MHz to 1,2 GHz frequency band. The use of an UWB short pulse sensor makes it possible to measure directly the impulse response from the target. The transient nature of UWB signals is also very useful for data processing and for the analysis of electromagnetic scattering mechanisms. This transient technique should soon be used for applications concerning the detection and identification of targets located on the surface or in the ground.

BIBLIOGRAPHY

1. Y. Chevalier, Y. Imbs, B. Beillard, J. Andrieu, M. Jouvet, B. Jecko, M. Le Goff, P. Pouliguen, E. Legros, "A new broad band wire antenna for ultra-wide-band applications", Ultra Wide Band Short Pulse Electromagnetics, vol 4, in press.
2. G. Bouillon, J. C. Vagneux, , "Chaine minimale radar ULB", Guide de conception, nov 97
3. Y. Chevalier, Y. Imbs, B. Beillard, J. Andrieu, B. Jecko, M. Le Goff, P. Pouliguen, E. Legros, "Mesures de SER basses fréquences via le domaine temporel", JINA 98
4. B. Juhel, "Application de l'Ultra Large Bande à l'imagerie SAR", Rapport n° 5, mai 98
5. C. E. Baum, "On the singularity expansion method for the solution of electromagnetic interaction problems", Air Force Weapons Lab., Tech. Rep. 88, Dec. 1971.

SYNTHESIS OF PURELY DIELECTRIC TRANSIENT LENSES

Carl E. Baum

Air Force Research Laboratory
DEHP, Kirtland AFB
Albuquerque, NM 87117
U.S.A.

Alexander P. Stone

Department of Mathematics and Statistics
University of New Mexico
Albuquerque, NM 87131
U.S.A.

1. INTRODUCTION

The study of inhomogeneous TEM plane waves which propagate on idealized transmission lines with two or more independent perfectly conducting boundaries leads to the study of lens transition regions. These types of inhomogeneous media can be used to synthesize dielectric lenses for transitioning TEM waves, without dispersion between various types (i.e., conical and/or cylindrical) of transmission lines. There are two approaches that have been used rather extensively in transient lens design. One method, which we term the differential-impedance matching and transit-time conservation approach, requires that differential impedances be matched at all waveguide and lens boundaries. This condition ensures that a TEM wave may be transmitted from one region to another without reflections. In addition, it is required that the wave be transmitted undistorted and so the travel time for waves following different paths must be conserved. This synthesis method generally gives rise to a system of ordinary nonlinear differential equations which specify the lens shape and material. The second approach, which transforms an EM problem by a change of coordinates, is called the differential geometric scaling method. In this approach to lens design the basic idea is the creation of a class of electromagnetic problems, each having a complicated geometry and medium, which are equivalent under a scaling to an electromagnetic problem having a simple geometry and medium. The latter problem is the formal problem while the former problem is our lens or real-world problem. Solutions to Maxwell's equations can then be used to specify transient lenses.

One type of problem on which scaling has been used relates to a waveguide filled with a homogeneous isotropic medium. Introduction of a scaling results in an orthogonal curvilinear coordinate system for which the transformed constitutive parameters become inhomogeneous and anisotropic. Thus the boundaries have been simplified but the medium has become more complicated. Many examples have been given of isotropic inhomogeneous media for which certain types of electromagnetic wave propagation could be supported.

The two basic approaches to lens synthesis are equivalent, in the sense that the scaling technique leads to the differential-impedance matching and transit-time conser-

Ultra-Wideband, Short-Pulse Electromagnetics 4
Edited by Heyman *et al.*, Kluwer Academic / Plenum Publishers, New York, 1999.

vation approach and vice-versa. A discussion of this equivalence appears in Baum and Stone[1] .

In this paper purely dielectric lenses which are synthesized by the scaling approach are considered. In this approach we begin with an orthogonal curvilinear coordinate system (u_1, u_2, u_3) with a line element

$$(d\ell)^2 = h_1^2(du_1)^2 + h_2^2(du_2)^2 + h_3^2(du_3)^2 \tag{1}$$

where the scale factors h_i are given by

$$h_i^2 = \left(\frac{\partial x}{\partial u_i}\right)^2 + \left(\frac{\partial y}{\partial u_i}\right)^2 + \left(\frac{\partial z}{\partial u_i}\right)^2 \tag{2}$$

and (x, y, z) are rectangular Cartesian coordinates. The Maxwell equations and constitutive relations are

$$\begin{gathered} \nabla \times \vec{E} = -\frac{\partial \vec{B}}{\partial t}, \; \nabla \times \vec{H} = \vec{J} + \frac{\partial \vec{D}}{\partial t}, \; \nabla \cdot \vec{D} = \rho, \; \nabla \cdot \vec{B} = 0 \\ \vec{D} = (\epsilon_{ij}) \cdot \vec{E}, \; \vec{B} = (\mu_{i,j}) \cdot \vec{H} \end{gathered} \tag{3}$$

In our case $\rho = 0$ and $\vec{J} = \vec{0}$ except on boundaries. The quantities (ϵ_{ij}) and (μ_{ij}) are real constant matrices independent of frequency, but can be functions of position. Matrices $(\alpha_{i,j})$ and $(\beta_{i,j})$, which depend on the scale factor h_i are defined by

$$(\alpha_{i,j}) = (1_{i,j}h_i) \, , \; (\beta_{i,j}) = (1_{i,j}\Delta/h_i) \, , \; \Delta = h_1 h_2 h_3. \tag{4}$$

The formal fields are $\vec{E}', \vec{H}', \vec{B}'$, and $\vec{D}'$ and these are related to the real fields $\vec{E}, \vec{H}, \vec{B}$, and $\vec{D}$ through

$$\vec{E}' = (\alpha_{i,j}) \cdot \vec{E} \, , \; \vec{H}' = (\alpha_{i,j}) \cdot \vec{H} \, , \; \vec{B}' = (\beta_{i,j}) \cdot \vec{B} \, , \; \vec{D}' = (\beta_{i,j}) \cdot \vec{D} \, . \tag{5}$$

The formal Maxwell equations are then obtained by setting

$$(\epsilon'_{i,j}) = (\beta_{i,j}) \cdot (\epsilon_{i,j}) \cdot (\alpha_{i,j})^{-1}, \; (\mu'_{i,j}) = (\beta_{i,j}) \cdot (\mu_{i,j}) \cdot (\alpha_{i,j})^{-1} \tag{6}$$

and replacing all quantities by primed quantities. The restriction to inhomogeneous isotropic media results in the diagonal matrices

$$(\epsilon_{i,j}) = \epsilon(1_{i,j}), \; (\mu_{i,j}) = \mu(1_{i,j}), \; (\epsilon'_{i,j}) = \epsilon(\beta_{i,j})(\alpha_{i,j})^{-1} \text{ and } (\mu'_{i,j}) = \mu(\beta_{i,j})(\alpha_{i,j})^{-1}. \tag{7}$$

Early examples given in Baum and Stone[1] of the scaling technique include the design of converging, diverging, and bending lenses. Special two dimensional lenses, based on conformal transformations, are also described in Baum and Stone[1] . For these lenses one is required to vary only one of the parameters, ϵ or μ, but not both.

In section 2, two dimensional geometries are considered and synthesis procedures for variable permittivity, constant permeability lenses are developed.[2,3] Discrete bends in the propagation direction of plane TEM waves are obtained by utilizing the Brewster angle at a planar interface between media of different uniform isotropic permittivities.

A very practical case is that of a purely dielectric, but inhomogeneous, bending lens as a portion of a body of revolution (BOR). This type of exact lens is considered in section 3. Nondispersive TEM waves propagate through a lens region, which can be used to bend the direction of propagation.[4] Section 4 considers a special kind of dielectric lens as a jacket[5] with two closely spaced (but generally curved) guiding perfectly conducting surfaces.

The case of generalized inhomogeneous TEM plane waves in inhomogeneous, but isotropic, media[6] is the focus of section 5. The use of differential geometric scaling, coupled with the condition that the permeability be that of free space, gives an extra degree of freedom in transient lens synthesis.

2. TWO DIMENSIONAL LENSES

Previously[7–9] some special two-dimensional lenses, suitable for E-plane bends in parallel-plane waveguides were considered. In one case, a uniform-dielectric wedge is inserted in the waveguide with transmission of the wave through the two wedge surfaces at the Brewster angle, thereby producing no reflections. In another case additional conducting sheets are introduced to create an anisotropic lens region which can be thought of as a set of parallel-plane (closely spaced) waveguides which properly match transit times and differential impedances on the wavefront passing through the lens. In Baum[2] the design of more general two-dimensional lenses for E-plane bends in parallel-plane waveguides was considered. The synthesis procedure allowed the permittivity ϵ to be non-uniform while the permeability was constrained to be uniform. Both parameters were to be positive, frequency independent scalars. The solution technique utilized a conformal transformation for generating the (u_1, u_2, u_3) orthogonal curvilinear coordinate system, and an example of an E-plane bend was given.

The problem can be approached from both discrete and continuous viewpoints. The discrete approach relies on the Brewster-angle phenomenon of no reflection at a planar surface between regions of different permittivities. Repetition of this at a set of interfaces will result in various bend angles for the direction of propagation. The continuous approach can be used to search for solutions of the differential geometry equations to synthesize the (u_1, u_3) coordinates, with $u_2 = x$. An example is a class of log-spiral solutions[3] which can give a special form in which constant ϵ occurs on planes of constant ϕ (which can be interpreted as Brewster angle interfaces), giving a continuous variation of ϵ. The procedure allows a lens design which includes both uniform and variable ϵ portions. The log-spiral solutions also lead to coordinates based on the error function and Dawson's integral[10] , and hence additional types of two-dimensional transient lenses may result.

3. BODY-OF-REVOLUTION LENSES

A three dimensional class of exact dielectric lenses with nonuniform (but isotropic) permittivity and constant permeability are those which may be thought of as portions of a dielectric body of revolution (BOR) with propagation in the azimuthal (ϕ) direction. Examples of these bending lenses include an E-plane lens with circular cylindrical conducting boundaries, an H-plane lens with planar conducting boundaries, and new examples[4] with circular conical or spherical conducting boundaries. We start with a homogeneous TEM wave as expressed in terms of the formal fields as

$$E_1' = h_1 E_1 = E_0' f(t - u_3/c') \ , \quad H_2' = h_2 H_2 = \frac{E_0'}{Z_0'} f(t - u_3/c'). \tag{8}$$

Since we have only E_1 and H_2 components, we obtain

$$\epsilon' = \frac{h_2 h_3}{h_1} \epsilon \ , \quad u' = \frac{h_3 h_1}{h_2} \mu \tag{9}$$

Also, since μ is homogeneous we have

$$\mu = \mu' \geq \mu_0 \ , \quad \frac{h_3 h_1}{h_2} = 1 \ , \quad \epsilon = \frac{h_1}{h_2 h_3} \epsilon' = h_3^{-2} \epsilon' \ , \quad \epsilon' \geq \epsilon_0 \tag{10}$$

with μ' and ϵ' real and frequency independent. The BOR coordinates are introduced through the usual cylindrical coordinates (Ψ, ϕ, z) and spherical coordinates (r, θ, ϕ) via

$$x = \Psi \cos(\phi) = r \sin(\theta) \cos(\phi) \ , \quad y = \Psi \sin(\phi) = r \sin(\theta) \sin(\phi) \ ,$$
$$z = \Psi \cos(\theta) \ , \quad \Psi = r \sin(\theta). \tag{11}$$

The dielectric BOR has the z axis as the symmetry axis and we define an orthogonal curvilinear coordinate system (ξ_1, ξ_2, ξ_3) where $\xi_3 = \phi$ is chosen as the propagation coordinate, and ξ_1 and ξ_2 will be used to orient the electric and magnetic fields. Surfaces of constant ξ_3 are planes and surfaces of constant ξ_1 and ξ_2 are BORs. In the (u_1, u_2, u_3) orthogonal system each u_n is only a function of ξ_n. On a plane of constant ξ_3, the ξ_1 and ξ_2 are orthogonal curvilinear coordinates which can be used to generate a conformal transformation $\xi(\xi)$ via complex variables by taking $\xi = \xi_1 \pm j\xi_2 = z + j\Psi$. Then for some initial choice of the ξ_n (scaled as $u_n(\xi_n)$) u_3 will be a function of ϕ alone, and

$$h_\phi = \Psi = r \sin(\theta) \ , \quad h_3 = \Psi \left| \frac{d\phi}{du_3} \right|. \tag{12}$$

Hence u_3 is proportional to ϕ (as ϵ is independent of ϕ) and so we may take

$$u_3 = \Psi_{\max} \phi \ , \quad h_3 = \frac{\Psi}{\Psi_{\max}} = \frac{h_2}{h_1} \tag{13}$$

where $\Psi_{\max}$ is the largest Ψ of interest for the lens region. Since the propagation speeds are

$$c' = (\mu' \epsilon')^{-1/2}, \ c_\omega = (\mu \epsilon)^{1/2} = h_3 c', \text{ local wave speed} \tag{14}$$

we have

$$\epsilon = \left(\frac{\Psi_{\max}}{\Psi} \right)^2 \epsilon_{\min} \ , \ \epsilon_{\min} = \epsilon' \ , \ c_\omega = \left(\frac{\Psi}{\Psi_{\max}} \right) c' \leq c' \leq c \ , \text{ for } \Psi \leq \Psi_{\max} \ . \tag{15}$$

Thus we have enough information to specify ϵ and a range of acceptable Ψ. Choices of u_1 and u_2 will yield various lens geometries. For example, if $u_1 = z, \ h_1 = 1$ we can obtain an H-plane bend, based on cylindrical coordinates as in Baum[4] and shown in Figure 1. Similarly, an E-plane bend based on cylindrical coordinates can be obtained by taking $u_2 = z, \ h_2 = 1$. (See Baum[4] .)

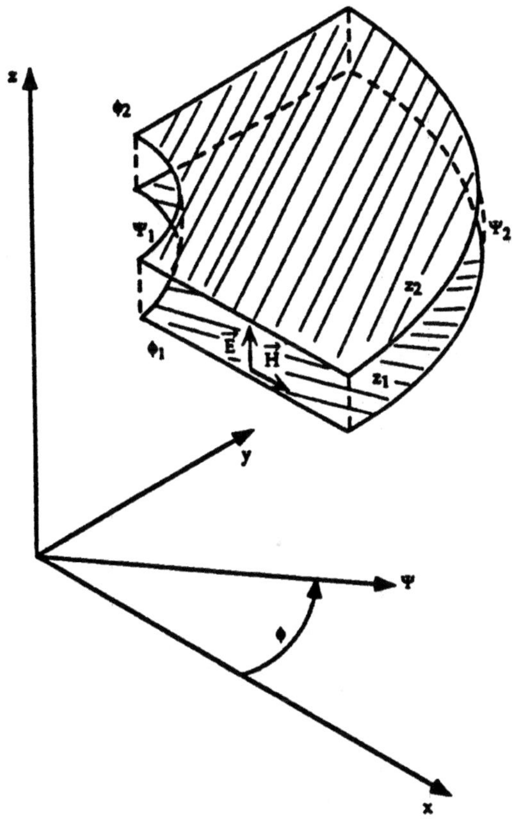

Figure 1: H-Plane Bend

New solutions based on spherical coordinates are also possible using either circular conical or spherical conducting boundaries. For example with the choices of field orientations

$$\vec{E} = -E_1 \vec{1}_\theta \; , \; \vec{H} = H_2 \vec{1}_r \; , \; \vec{1}_1 = -\vec{1}_\theta \; , \; \vec{1}_2 = \vec{1}_r \; , \; u_1 = u_1(\theta) \; , \; u_2 = u_2(r) \tag{16}$$

and

$$\frac{h_2}{h_1} = \frac{\Psi}{\Psi_{\max}} = \frac{r\sin(\theta)}{r_{\max}\sin(\theta_{\max})} \tag{17}$$

we can select

$$h_1 = \left|\frac{rd\theta}{du_1}\right| = \frac{r\sin(\theta_{\max})}{r_{\max}\sin(\theta)}, \quad h_2 = \left|\frac{dr}{du_2}\right| = \left(\frac{r}{r_{\max}}\right)^2 \tag{18}$$

which then yields

$$u_1 = r_{\max}\frac{\cos(\theta)}{\sin(\theta_{\max})}, \quad u_2 = \frac{-r_{\max}^2}{r} . \tag{19}$$

The fields then take the form

$$E_1 = \frac{E_1'}{h_1} = \frac{r_{\max}}{r} \frac{\sin(\theta)}{\sin(\theta_{\max})} E_0' f(t - u_3/c')$$
$$H_2 = \frac{H_2'}{h_2} = \left(\frac{r_{\max}}{r}\right)^2 \frac{E_0'}{Z_0'} f(t - u_3/c'). \tag{20}$$

The electric field is proportional to r^{-1} consistent with increasing conductor spacing for larger r, and to $\sin(\theta)$ due to decreasing ϵ for larger Ψ. The magnetic field is proportional to r^{-2}, since one factor of r^{-1} accounts for increasing conductor spacing with larger r^{-1} and the second factor of r^{-1} accounts for spreading of the magnetic flux proportional to $rd\phi$. This is necessary to give zero divergence for $\vec{H}$ which is like the r^{-2} field from a magnetic monopole at $\vec{r} = \vec{0}$, except for the variation in the ϕ (propagation) direction. Figure 2 shows this bending lens with circular conical conducting boundaries.

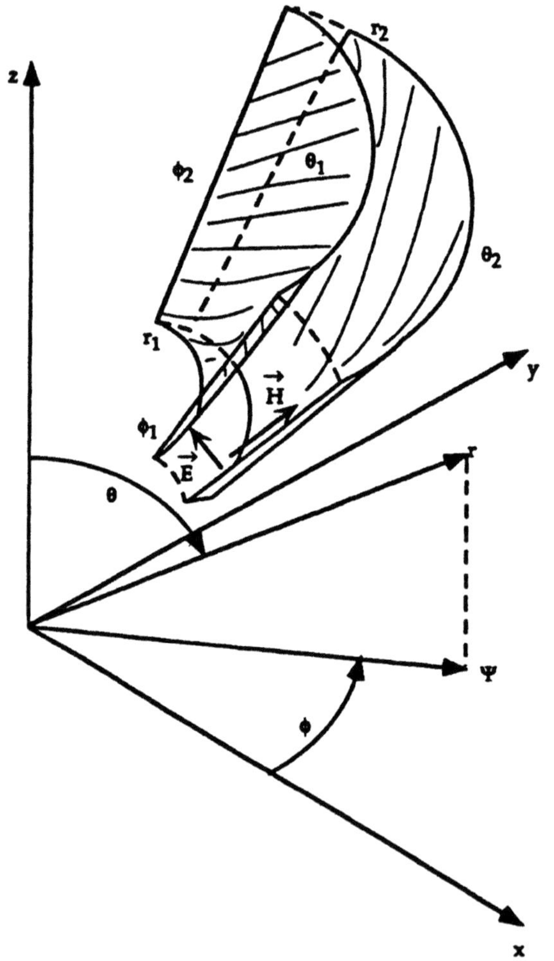

Figure 2: Mixed Bend with Circular Conical Conducting Boundaries

4. JACKET APPROXIMATION

The concept of a jacket[1] is introduced as a special kind of two-dimensional space in which the wave is guided between two closely spaced, perfectly conducting sheets. One may also consider, as in Baum[5] , dielectric jackets as lenses. The jacket consists of two closely spaced, but curved, conducting plates separated by a medium of uniform isotropic permeability and nonuniform isotropic permittivity for the propagation of dispersionless TEM waves.

The fields take the form

$$\vec{E} = \vec{E}_1 \vec{1}_1 = \frac{E'_0}{h_1} f\left(t - \frac{u_3}{c'}\right) \vec{1}_1 \ , \ \vec{H} = \vec{H}_2 \vec{1}_2 = \frac{E'_0}{h_2 Z'_0} f\left(t - \frac{u_3}{c'}\right) \vec{1}_2$$
$$c' = [\mu' \epsilon']^{-\frac{1}{2}} \ , \ Z'_0 = \left[\frac{\mu'}{\epsilon'}\right]^{\frac{1}{2}} \ , \ \mu' = \mu \ , \ \epsilon' = \epsilon_{\min} \tag{21}$$

where μ' and ϵ' are real, positive constants and propagation is in the u_3 (generalized z) direction. There are resulting constraints

$$\frac{h_3 h_1}{h_2} = 1 \ , \ \epsilon = h_3^{-2} \epsilon' \tag{22}$$

The small spacing h_Δ between the two walls given by two values of u_1, namely $u_1^{(1)}$ and $u_1^{(2)}$, is given by

$$h_\Delta(u_2, u_3) \simeq h_1 \left[u_1^{(2)} - u_1^{(1)}\right] . \tag{23}$$

This offers lots of flexibility in the form such a jacket may take.

As an example, nonuniform permittivity can be used for a bend in a coaxial cable, as shown in Figure 3. In this example $\Psi'_3(\phi')$ and $\Psi'_4(\phi')$ are adjusted to match differential impedances (at each ϕ') to straight circular coaxes with uniform dielectric at ϕ_1 and ϕ_2. The permittivity in the lens region $\phi_1 < \phi < \phi_2$ then has the form $\epsilon = (\Psi_{\max}/\Psi)^2 \epsilon_{\min}$.

5. GENERALIZED INHOMOGENEOUS TEM PLANE WAVES

If we allow ϵ' and μ' to be inhomogeneous we then look for formal fields, depending on the (u_1, u_2, u_3) coordinates, in the form

$$\vec{E}' = \vec{E}'_0(u_1, u_2) f(t - \tau(u_3)) \ , \quad \vec{H}' = \vec{H}'_0(u_1, u_2) f(t - \tau(u_3)) \tag{24}$$

with $\vec{E}'_0(u_1, u_2) \cdot \vec{1}_3 = \vec{H}'_0(u_1, u_2) \cdot \vec{1}_3 = 0$, $\vec{1}_3 \times \vec{E}'_0 = Z' \vec{H}'_0$ and consequently the formal parameters ϵ'_3 and μ'_3 will play no role. Maxwell's equations then lead to the result in Baum[6] that

$$\tau = \int_0^{u_3} \frac{du'_3}{v'(u'_3)} = \tau(u_3) \tag{25}$$

and the formal wave speed is

$$v' = [\mu' \epsilon']^{-1/2} = v'(u_3). \tag{26}$$

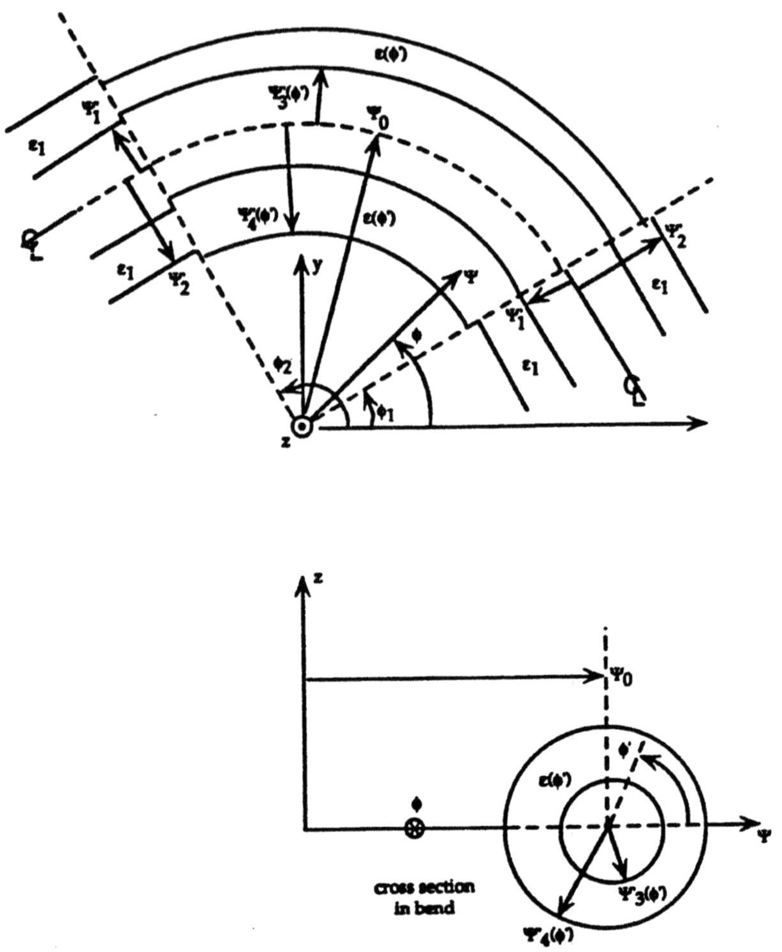

Figure 3: Bend in Coax

Since $\vec{E}'_0$ and $\vec{H}'_0$ are not functions of u_3 we also have

$$Z' = \left[\frac{\mu'}{\epsilon'}\right]^{1/2} = Z'(u_1, u_2). \tag{27}$$

These results suggest that we try solutions of the form

$$\vec{E}'_0(u_1, u_2) = \nabla'_t \Phi_e(u_1, u_2)\,, \quad \vec{H}'_0(u_1, u_2) = \nabla'_t \Phi_h(u_1, u_2) \tag{28}$$

where the transverse gradients in the formal coordinates makes this a two-dimensional potential problem. The electric potential is solved subject to specified potentials on the conductor passing through the lens, while the magnetic potential has to allow for a discontinuity in passing around a conductor. One finds that requiring Φ'_e to satisfy a Laplace equation for an inhomogeneous medium (variable ϵ') yields a two-dimensional potential which satisfies all of the necessary requirements. An analogous statement also holds for Φ'_h. In the context of a bend in a coaxial cable this is discussed in Baum[11] .

If differential geometric scaling is introduced then the formal and real media are inhomogeneous but isotropic. The first two diagonal elements of the constitutive parameter dyadics are equal and we have

$$\mu' = \frac{h_2 h_3}{h_1}\mu = \frac{h_3 h_1}{h_2}\mu\,, \quad \epsilon' = \frac{h_2 h_3}{h_1}\epsilon = \frac{h_3 h_1}{h_2}\epsilon \tag{29}$$

and hence $h_1 = h_2$, which implies that surfaces of constant u_3 can only be spheres or planes[1] . Furthermore we have

$$\begin{aligned} &\mu' = h_3\mu\,, \quad \epsilon' = h_3\epsilon\,, \quad \frac{\mu'}{\epsilon'} = \frac{\mu}{\epsilon} \\ &Z = \left[\frac{\mu}{\epsilon}\right]^{1/2} = Z' \ \text{(a function of } u_1 \text{ and } u_2 \text{ only)} \\ &v = [\mu\epsilon]^{1/2} = h_3[\mu'\epsilon']^{1/2} = h_3 v'\,. \end{aligned} \tag{30}$$

While v' is a function of u_3 only, v can be a function of the various u_i.

If we now impose the requirement that $\mu = \mu_0$, a restriction on the scale factor h_3 results. One finds, in Baum[6] , that h_3 has the form

$$h_3 = \frac{v(u_1, u_2)}{v'(u_3)}. \tag{31}$$

Modification of the usual cylindrical coordinates (Ψ, ϕ, z) so that (u_1, u_2, u_3) become $u_1 = z$, $u_2 = \Psi$, $u_3 = \Psi_{max}\phi$ enables us to specify a bending lens, in which surfaces of constant u_3 are planes. The scale factors become $h_1 = h_2 = 1$ and $h_3 = \Psi/\Psi_{max}$ and we obtain $\vec{E}_0 = \vec{E}'_0$ and $\vec{H}_0 = \vec{H}'_0$. The equation for the electric potential becomes

$$\frac{\partial^2 \Phi'_e}{\partial z^2} + \Psi \frac{\partial}{\partial \Psi}\left[\Psi^{-1}\frac{\partial \Phi'_e}{\partial \Psi}\right] = 0\,. \tag{32}$$

From this Laplace-like equation we can find the solution for the dispersionless TEM wave (inhomogeneous) propagating through the bend for arbitrary cross section (conductor) shapes in an arbitrary (Ψ, z) plane. A coaxial type of cross section is but one example. This solution is quite general.

A further step in the lens synthesis process is a consideration of unipolarized generalized inhomogeneous TEM plane waves in inhomogeneous media. In this case the formal fields would each depend on a single coordinate and the formal media parameters are expressible as products of single variables. The introduction of scaling can then be coupled with the condition that the permeability be that of free space. Further specialization to particular coordinate systems (u_1, u_2, u_3) can then be made. Work on this is in progress.

6. CONCLUSIONS

The generalized form of a TEM plane wave involving inhomogeneous constitutive parameters can be used for the formal fields for synthesizing lens designs based on differential-geometric scaling. This will allow one to have purely dielectric lenses (with free space permeability), and will provide an extra degree of freedom in lens design. We have also seen that guided TEM waves could be propagated in inhomogeneous isotropic media with permittivity proportional to Ψ^{-2}, where Ψ is the cylindrical radius. Also, very general conductor cross sections can be specified, which allows for the construction of dispersionless bends in high-voltage transmission systems in which the pulse rise time is small compared to transit times across the transmission line (e.g. coax) cross section, a matter of importance for applications in high-power pulse generation and antenna design and for low-loss transmission lines.

REFERENCES

1. C. E. Baum and A. P. Stone, *Transient Lens Synthesis: Differential Geometry in Electromagnetic Theory*, Hemisphere (Taylor and Francis), New York, 1991.
2. C. E. Baum, Two-dimensional inhomogeneous dielectric lenses for E-plane bends of TEM waves guided between perfectly conducting sheets, *Sensor and Simulation Note* 388, (October 1995).
3. C. E. Baum, Discrete and continuous E-plane bends in parallel-plate waveguides, *Sensor and Simulation Note* 399, (May 1996).
4. C. E. Baum, Dielectric body-of-revolution lenses with azimuthal propagation, *Sensor and Simulation Note* 393, (March 1996).
5. C. E. Baum, Dielectric jackets as lenses and applications to generalized coaxes and bends in coaxial cables, *Sensor and Simulation Note* 394, (March 1996).
6. C. E. Baum, Use of generalized inhomogeneous TEM plane waves in differential geometric lens synthesis, *Sensor and Simulation Note*, 405, (December 1996), and also in Proceedings of the 1998 International Symposium on Electromagnetic Theory, Thessaloniki, Greece, May 1998.
7. C. E. Baum, Dielectric lenses for TEM waves between parallel plates, *Sensor and Simulation Note* 332, (September 1991).
8. C. E. Baum, Arrays of parallel conducting sheets for two-dimensional E-plane bending lenses, *Sensor and Simulation Note* 341, (April 1992).
9. C. E. Baum and A. P. Stone, Transient lenses for transmission systems and antennas, in *Ultra-Wideband, Short-Pulse Electromagnetics*, H. L. Bertoni, L. Carin, L. B. Felsen, eds., Plenum Press, New York, (1993).
10. M. Abramowitz and I. A. Stegun, *Handbook of mathematical functions*, AMS 55, National Bureau of Standards, 1964 (p. 298).
11. C. E. Baum, Azimuthal TEM waveguides in dielectric media, *Sensor and Simulation Note* 397, (March 1996).

MINIMIZING DISPERSION IN A TEM WAVEGUIDE BEND BY A LAYERED APPROXIMATION OF A GRADED DIELECTRIC LENS

W. Scott Bigelow and Everett G. Farr

Farr Research, Inc.
614 Paseo Del Mar NE
Albuquerque, NM 87123

INTRODUCTION

Waveguide bends pose a problem for high-voltage UWB systems or for any transmission-line system with low-loss/fast-risetime requirements. The difficulty arises because only a straight section of conventional waveguide can support the pure TEM mode necessary to preserve the risetime of a transmitted pulse. When there is a bend in the waveguide, especially when the cross-section of the waveguide is large, as it typically must be for high-voltage or low-loss systems, the risetime of the transmitted signal is lengthened. This can severely limit the system bandwidth.

In several papers, [1, 2, 3, 4, 5, and 6], Dr. Carl E. Baum has described an approach that addresses this problem. We described our initial implementation of some of those ideas in [7]. This paper is an abbreviated description of the work originally reported there, and the reader is referred to that document for additional detail.

Conventional waveguide is filled with a *homogeneous* dielectric material, sometimes air, or is evacuated. Here we consider compensation of a waveguide bend by a purely dielectric lens, characterized by an *inhomogeneous*, frequency-independent, isotropic permittivity. If the relative index of refraction of the lens varies inversely as the radius of curvature, then the optical path length through the bend will be independent of the radius, and the bend will support pure TEM waves.

Although waveguide structures employing lens materials having the required inhomogeneous permittivity profile are easily conceptualized, constructing them is more problematic. In this paper, we describe an approximation employing coarsely graded layers of dielectric materials of various uniform permittivities. We used this approach to compensate a 90° H–plane bend in a strip transmission line.

Here, we first summarize the relevant theory. Then, we describe our hardware implementation of the graded dielectric strip line bend. We analyzed the structure numerically, and we determined its performance experimentally. The measured pulse risetime was 75 ps for a straight air section of strip line. For a 90° air bend, it was 255 ps.

Ultra-Wideband, Short-Pulse Electromagnetics 4
Edited by Heyman *et al.*, Kluwer Academic / Plenum Publishers, New York, 1999.

With our layered compensating lens installed, the risetime of the transmitted pulse was reduced to 185 ps. By thus reducing the dispersion of a transmitted pulse, the strip line bend provided a proof-of-principle demonstration of compensation of a waveguide bend by a graded dielectric lens.

CONCEPT

Consider the waveguide bend shown in Figure 1A, which consists of a strip transmission line above a ground plane, all in air. The straight sections of the waveguide are known to support the non-dispersive propagation of a pure TEM wave, provided that the conductors

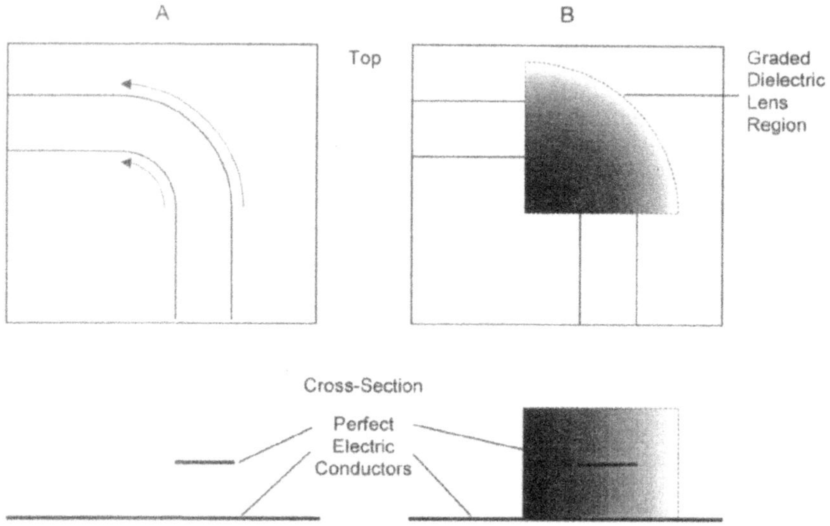

Figure 1. A conventional waveguide bend is dispersive because the path length around the bend is proportional to the radius of curvature (A). A bend may be compensated with a graded dielectric material (B). If the index of refraction within the bend varies inversely as the radius of curvature, optical path lengths through the bend are equalized.

are perfectly conducting. However, the curved section cannot support a pure TEM wave, since the path length through the bend is proportional to the radius of curvature. Rays entering the bend at its outer edge must travel farther than those entering at the inside edge. As a result, they lag behind those at the inside edge. What is needed is a way to equalize the transit times for all rays propagating through the bend. Equivalently, we require that the optical (or electrical) path length through the bend be independent of the radius of curvature. Since the wave speed is inversely proportional to the index of refraction, a conceptually simple solution is to embed the bend in a dielectric medium, the index of which varies inversely with the radius of curvature, as

$$n(\Psi) = \Psi_{max} / \Psi \tag{1}$$

where Ψ is the cylindrical radius of curvature. It is assumed that at some outer radius, Ψ_{max}, the fields no longer contribute significantly, and the index reaches a minimum of unity. Since the square of the index of refraction is the relative permittivity, we can also express (1) as

$$\varepsilon_r(\Psi) = \varepsilon(\Psi)/\varepsilon_0 = \left(\frac{\Psi_{max}}{\Psi}\right)^2 \tag{2}$$

where ε_0 is the permittivity of free space. This approach to solving the waveguide bend problem is indicated conceptually in Figure 1B.

IMPLEMENTATION

We considered several alternative hardware implementations of non-dispersive waveguide bends. Ultimately, we selected a curved strip line embedded within graded dielectric layers machined from sheets of low-loss uniform-permittivity plastics. Use of this layered approximation of a graded dielectric material represented a compromise between modeling fidelity and ease of construction. Other possible implementations are discussed in [7].

Our test fixture layout is depicted in Figure 2. Details of the dielectric material layout in the bend region are shown in Figure 3. With the straight section in place, measurements were made between Port 1 and Port 2. With the curved section, measurements were made between Port 1 and Port 3. The curved section was used both with and without the layered graded dielectric lens.

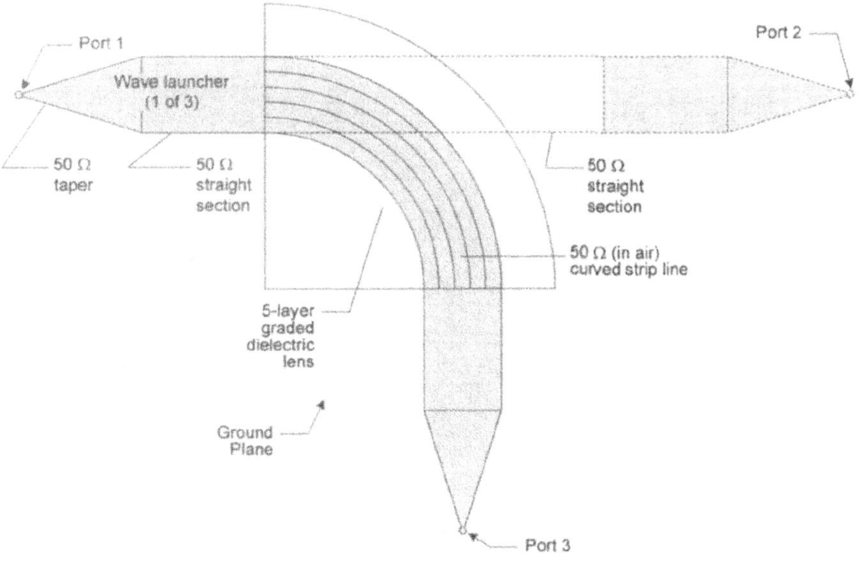

Figure 2. Strip line test fixture layout. The wave launchers and straight strip section were always air-filled. The curved strip section was used either with or without the graded dielectric lens in place.

To find the characteristic impedance of the graded dielectric structure, one first solves a two-dimensional equation in cylindrical coordinates, in a plane of constant azimuth, ϕ. In

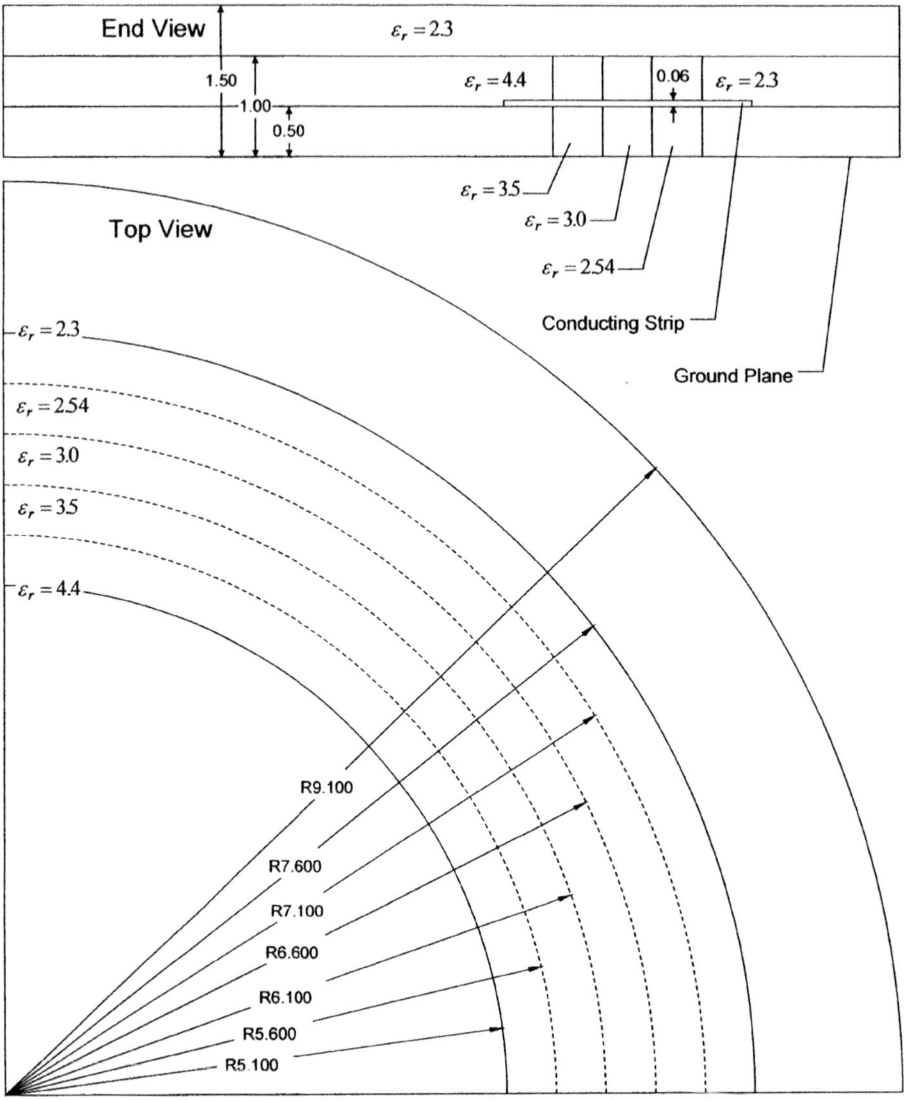

Figure 3. Five-region graded dielectric bend assembly. Straight sections at the entrance and exit of the bend are not shown. The conducting strip is 1.27 cm (0.5 in.) above the ground plane. Its width in both straight and curved regions is a constant 6.35 cm (2.5 in.). Within the strip boundaries, the relative permittivities approximate an inverse square law relationship with the radius of curvature. The relative permittivity is assumed to be 1.0 at a radius of 27.94 cm (11.0 in.). All drawing dimensions are inches.

terms of the electric potential, $V(\Psi, z)$, the equation to be solved is [7]

$$\Psi\nabla \bullet \left(\frac{1}{\Psi}\nabla V\right) = 0 \tag{3}$$

where the boundary conditions are $V = V_0$ on the conducting strip and $V = 0$ on the ground plane. The permittivity, ε, varies inversely as the square of the radius of curvature in the bend region, as specified previously in (2).

As an approximation, we used the finite element method to solve a simpler problem. We assumed that the change in azimuth through the bend could be ignored—a straight strip approximation. This allowed Ψ and z to be treated as Cartesian coordinates, and reduced the problem to solving Laplace's equation,

$$\nabla \bullet \left(\varepsilon_i \; \nabla V\right) = 0 \tag{4}$$

over a rectangular (Ψ, z) domain consisting of sub-domains having piecewise-uniform-permittivities. As described in [7], where the straight strip approximation is justified, this method produced a characteristic impedance of 28 Ω, in agreement with a TDR measurement of the bend (Figure 4).

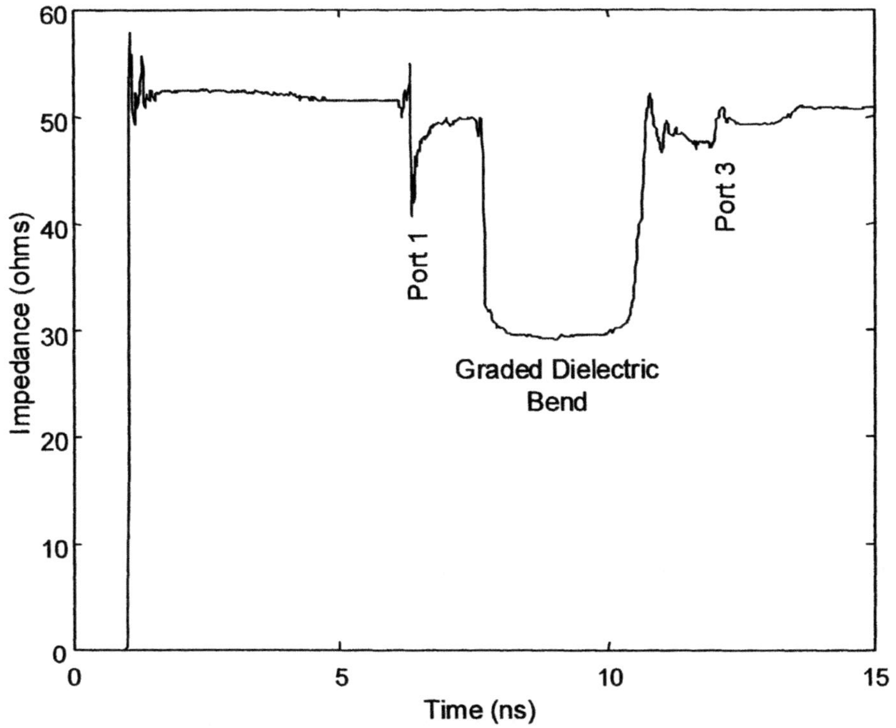

Figure 4. TDR impedance measurement of the strip line test fixture with a 90° bend embedded within a five-layer approximation of a graded dielectric lens. Within the bend region, the impedance is approximately 28 Ω. Note that the discontinuities between the feed sections and the bend could be reduced by use of a dielectric with the same permittivity as the average permittivity within the bend region.

In Figure 5, transmission of a voltage step pulse through the straight, air-filled strip line section (between Port 1 and Port 2), is compared with transmission of the same pulse through the curved section (between Port 1 and Port 3), both air-filled and compensated by the five-layer graded dielectric lens. Transmission through the straight strip line section (and the feed sections) degraded the fall time to 75 ps. This, then, is the best we can expect for a perfectly compensated curved section. The air-filled bend degraded the fall time to 255 ps. The graded dielectric layers improved the transmission through the bend, reducing the fall time to 185 ps.

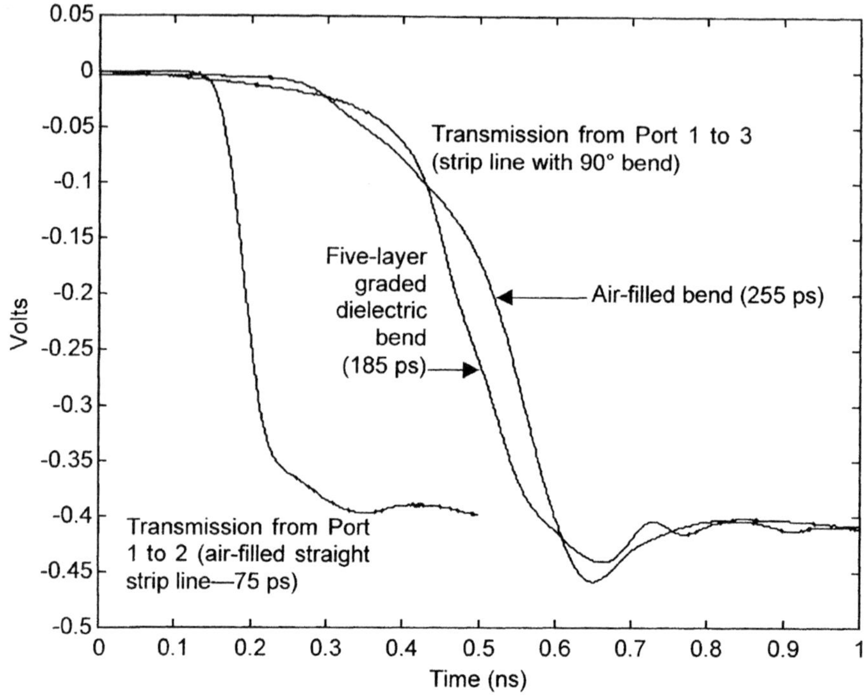

Figure 5. Transmission of a voltage step by straight and bent strip lines. The fall time for the straight line is about 75 ps. With the air-filled, 90° bend, the fall time increases to about 255 ps. For the same bend with a five-layer radially graded dielectric lens, the fall time is reduced to about 185 ps. The data are normalized to correct for reflection losses at the dielectric-air interfaces.

The primary causes of the imperfect compensation of the bend appear to lie in the finite thickness of the dielectric layers, and in the permittivities of the materials filling the fringe field regions above and on each edge of the strip. The difference in transit time around the bend at the strip edges is 330 ps for the air bend; for the layered dielectric bend, the difference is only 110 ps. However, the transit time difference for the dielectric bend climbs to 350 ps when fringe regions 1.3 cm on each side of the strip are considered. Fringe fields propagating above the graded region that covers the strip may also contribute. If we calculate the transit times along each dielectric layer interface, we find that the differences in transit times within the dielectric layers immediately adjacent to each interface range from a low of 74 ps to a high of 170 ps. The aggregate effect of these transit time differences is to stretch the transmitted pulse.

CONCLUSIONS

Although the five-layer graded dielectric bend substantially improved the transmission of a stepped voltage pulse through the curved section of our strip line test fixture, residual variations in bend transit time limited the improvement that could be achieved. Thinner dielectric layers, with well-characterized indices of refraction, as well as use of a closed structure, instead of an open strip line, will be required if the potential of this technique for minimizing dispersion in waveguide bends is to be fully realized.

REFERENCES

1. Baum, Carl E., *Two-Dimensional Inhomogeneous Dielectric Lenses for E-Plane Bends of TEM Waves Guided Between Perfectly Conducting Sheets*, Sensor and Simulation Notes, Note 388, 14 October 1995.
2. Baum, Carl E., *Dielectric Body-of-Revolution Lenses with Azimuthal Propagation*, Sensor and Simulation Notes, Note 393, 9 March 1996.
3. Baum, Carl E., *Dielectric Jackets as Lenses and Application to Generalized Coaxes and Bends in Coaxial Cables*, Sensor and Simulation Notes, Note 394, 23 March 1996.
4. Baum, Carl E., *Azimuthal TEM Waveguides in Dielectric Media*, Sensor and Simulation Notes, Note 397, 31 March 1996.
5. Baum, Carl E., *Discrete and Continuous E-Plane Bends in Parallel-Plate Waveguide*, Sensor and Simulation Notes, Note 399, 1 May 1996.
6. Baum, Carl E., *Use of Generalized Inhomogeneous TEM Plane Waves in Differential Geometric Lens Synthesis*, Sensor and Simulation Notes, Note 405, 5 December 1996.
7. Bigelow, W. Scott, and Everett G. Farr, *Minimizing Dispersion in a TEM Waveguide Bend by a Layered Approximation of a Graded Dielectric Material*, Sensor and Simulation Notes, Note 416, 5 January 1998.

SIMULATION, MODELING, AND EXPERIMENTAL STUDIES OF HIGH-GAIN GALLIUM ARSENIDE PHOTOCONDUCTIVE SWITCHES FOR ULTRA-WIDEBAND APPLICATIONS

E. Schamiloglu, N.E. Islam, C.B. Fleddermann, and B. Shipley

Department of Electrical and Computer Engineering
University of New Mexico, Albuquerque, NM 87131, USA

R.P. Joshi and L. Zheng

Department of Electrical and Computer Engineering
Old Dominion University
Norfolk, VA 23529, USA

INTRODUCTION

The U.S. Air Force Research Laboratory (AFRL, Phillips Site) has recently demonstrated the application of a compact, high-gain photoconductive semiconductor switch (PCSS) for high power ultra-wideband microwave (HPM) generation.[1] The switch is made from commercially available LEC grown semi-insulating (SI) GaAs material, with a resistivity of 2-5 x 10^7 Ω-cm. The switch dimensions are 2.0 cm by 1.3 cm with a thickness of 0.063 cm. The "opposed" contacts, made from layered Pd/Ge/Ti/Pt of 200/400/400/300 μm thicknesses, respectively, are on the front and back of the wafer, about 0.25 cm apart (Fig. 1). The device requires less trigger energy when operated in the "lock-on" mode, and has also shown improved hold-off and risetime characteristics compared with contacts on the same surface of the device. Currently, these switches have a lifetime of 10^4-10^6 shots with a risetime of 350 ps. The minimum trigger energy occurs at 880 nm, near the zero field bandgap of 873 nm.

Improvements in trigger energy, risetime, jitter, and lifetime of the device would make the PCSS a suitable switch component for ultra-wideband HPM applications. A detailed study of the physics associated with the device's operation is therefore essential to the understanding of such phenomena as premature breakdown, the "lock-on" mechanism and high failure rates. Premature breakdown limits high bias device operation and thus affects risetime. Similarly, better knowledge of the "lock-on" phenomenon would contribute to better trigger energy economy.

This paper describes recent results from a collaborative research program between the University of New Mexico (UNM) and Old Dominion University (ODU) to better understand the physics of operation of PCSS's. Computer simulation is a viable tool in the study of the physics of device operation. Such a study of the PCSS will lead to a better

Ultra-Wideband, Short-Pulse Electromagnetics 4
Edited by Heyman *et al.*, Kluwer Academic / Plenum Publishers, New York, 1999.

understanding of the mechanisms described above and may result in the diagnosis of problems and design of a better switch for ultra-wideband HPM applications. We describe simulation results for SI-GaAs PCSS switches used in the AFRL experiments for testing ultra-wideband (UWB) radiation sources. Different simulation tools are being used at the two univerities in this study. In addition, prototype switches are being fabricated at UNM to both validate the simulations and improve existing designs.

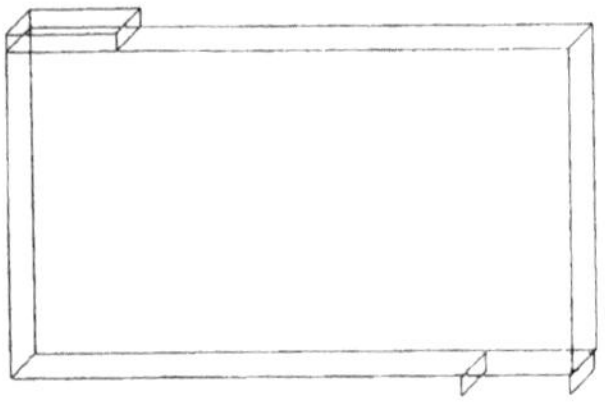

Figure 1. Opposed contact SI-GaAs switch used in simulations at UNM.

BACKGROUND

Carrier transport in a PCSS can be characterized in terms of a two-phase response. Phase one describes the conditions prior to the switching operation when a bias is applied across it, while phase two covers the "on" state of the device, when a laser or other light source creates a conduction path through it. Conditions in phase one thus determine the device performance during phase two. Premature breakdown is of interest during the first phase of the switch operation, the mechanism of which is not fully understood and is an important area of study.[2,3]

PCSS switches utilizing SI-GaAs have also been known to exhibit characteristic "lock-on" behavior under high biasing conditions.[4,5] "Lock-on" is a persistent photoconductivity effect. Persistent photoconductivity (PPC) has been widely observed in a variety of materials including the III-V and II-VI compounds,[6-8] SiC,[9] and GaN.[10,11] The phenomenon is not limited to bulk devices, but also occurs in heterojunctions and modulation-doped structures. Carrier density increases of up to four orders of magnitude have also been associated with this phenomenon.[12] In all cases, PPC has been linked to the presence of deep traps. This is especially relevant to high resistivity devices such as photoconductive switches, since the underlying material always contains a high level of intentional traps and deep level impurities. Even though the "lock-on" mechanism has been utilized to advantage for high power microwave generation requiring lesser trigger energy, the mechanism is still not fully understood and requires further study. Finally, the failure mechanism of the switches also needs to be investigated in order to make the devices more reliable for long-term use. Local heating due to sustained use, the role of junctions, external resistance, or residual carriers need to be investigated as well.

High power operating conditions bring about a number of changes inside a device and at the electrodes. The fact that the PCSS under investigation is made from SI-GaAs (trap-dominated, highly resistive) material may introduce new parameters and must be looked into and characterized for analysis. The device's response to non-equilibrium conditions (minority carrier injection) will depend to a large extent on how the value of $R\tau_{DC}$, (the ratio of the dielectric relaxation time to the carrier lifetime), compares with unity; *i.e.* the response is based on one of the two following conditions:

$$R\tau_{DC} >> 1 \quad (1)$$

$$R\tau_{DC} << 1. \quad (2)$$

Condition (1) is the lifetime regime while condition (2) is the relaxation regime.[13] The two regimes can also be classified by comparing the Debye to Diffusion length ratio, RL_{D0} where $RL_{D0} >> 1$ defines the lifetime regime and $RL_{D0} << 1$ defines the relaxation regime. It has been demonstrated[14,15] that the injection of minority carriers in relaxation regime semiconductors leads to majority carrier *depletion*, whereas injection in lifetime regime semiconductors leads to majority carrier *enhancement*. The depletion of majority carriers in relaxation results in an enhancement of the diffusion component of the current. It has also been demonstrated that the depletion of majority carriers following minority carrier injection is reduced in the relaxation regime if the material is a trap-dominated semiconductor.[16,17] This is because the traps reduce the Debye Length, and in that case it is replaced by the "Screening Length" L_S given by [18]

$$L_S = \frac{L_D}{\sqrt{1 + N_t / n_0}}, \tag{3}$$

where N_t is the trap concentration and n_0 is the equilibrium electron concentration. Since in SI-GaAs material the concentration of EL2 traps is high, the screening length is much shorter than the diffusion length, and the highly resistive material may exhibit properties of lifetime semiconductor material. Our simulations support this trend.

SIMULATION RESULTS

Simulations have been performed at UNM and at ODU, where various aspects of switch operation are being investigated in order to define critical parameters that affect the device's performance. Of particular concern is the breakdown mechanism at high power, the role of contacts on the so-called "lock-on" mechanism, and the device failure rate after prolonged high power operation. Simulations at ODU specifically centered on the "lock-on" mechanism, while at UNM the physics of premature breakdown were studied. At UNM a commercially available simulation code was used in this study.[19] This code has 2-3D capabilities and performs device-circuit-process simulation using a single interface. Flexibility in the simulation is possible through the user defined C-interpreter function where new equations for any model parameter can be tested and verified. For simulation at UNM, the structure has the same dimension as the actual device, but the contact length is reduced (to 100 μm) to save simulation time. Also, only the material that is in intimate contact with the device (Pd) is defined as the contact material for the simulations. This is not expected to affect the results since the conduction properties are essentially dominated by the material characteristics that are in intimate contact with the semiconductor. Manufacturer-supplied data and published reports were the sources for the input.[20]

The code at ODU was developed in-house, written specifically for the device and material under simulation. At ODU, two dimensional (2D) simulations were carried out for a simple square geometry with 10 μm sides. The surface area was chosen to be $1x10^{-4}$ cm^2. The contacts were assumed to be on the top and bottom, diagonally across from each other. A simple 50 Ω resistor was taken to be in series with the photoconductive switch. The SI-GaAs material was assumed to have a marginal n-type donor density of $1x10^{15}$ cm^3 and a trap concentration of $1x10^{16}$/cm^3. The trap level was taken to be 0.6 eV with electron and hole capture cross sections of $1x10^{15}$ cm^2 and $1x10^{17}$cm^2, respectively. This is consistent with observations of predominantly electron trapping with such SI-GaAs material.[21] The semiconductor and metal work functions were assumed to be 4.1 eV and 4.3 eV, while all other transport data were taken from the literature.[22]

RESULTS AND DISCUSSION

Figure 2 presents an I-V characteristic, obtained at UNM, of the device with bias changing by a few orders of magnitude.

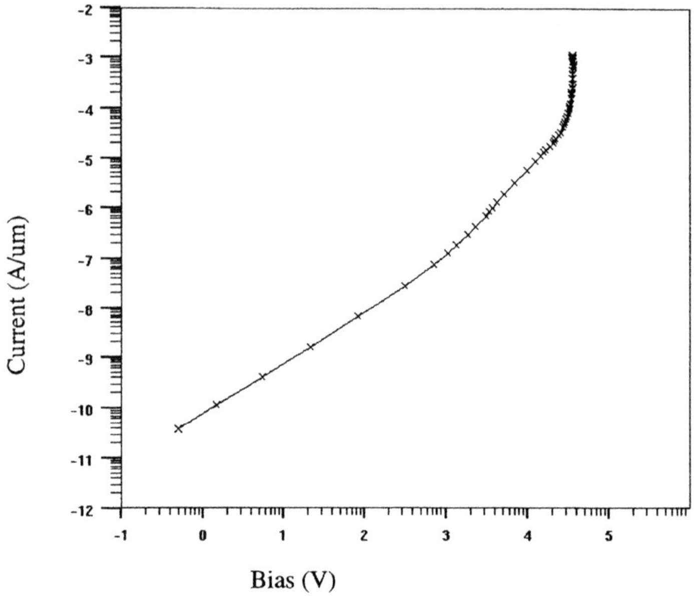

Figure 2. Simulated I-V characteristics of the switch showing super-linear effects.

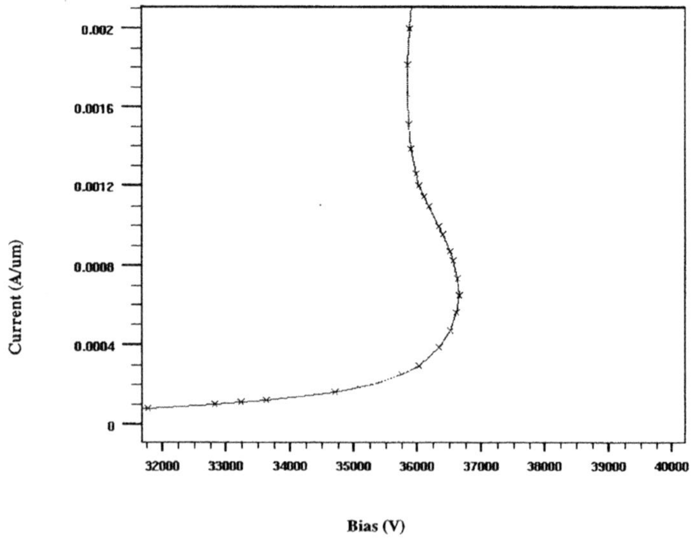

Figure 3. Negative resistivity response of the device at high bias.

This plot was generated with input parameters that gave a reasonable match to the measured device I-V at lower bias.[23] As can be seen from the figure, the response mechanism is similar to experimentally determined I-V for a trap-filled SI-GaAs (Ref. 18, Fig. 3). It shows an extended linear region followed by a super-linear region. A sub-linear region between the two regions showing increased resistivity was not seen. This is perhaps due to the fact that there existed no mechanism to vary lifetime/mobility models.[18] At higher bias the device switches to a higher conducting state and current goes up several orders of magnitude, similar to the experimental plot shown in Ref. 18.

Figure 3 is an I-V plot generated at UNM for the device at high bias. The current controlled negative resistivity (CCNR) characteristics, as shown, may result in open circuit stable behavior, and the current flow between the electrodes will be filamentary in characteristic.[23] Conduction is mainly along microchannels called microplasmas, which consist of regions with high concentration of electrons and holes. Filamentary conduction, however, does not remain stable since the device characteristic changes as bias is increased further.

This instability is attributed to the formation of trap-filled regions near the electrodes at high bias.[23] The response of the device during the "on-state" following the application of a light beam of wavelength 880 nm is shown in Fig. 4. This plot has similar model parameters that generated matching I-V characteristics for the device.

The beam, with its center located at the center of the device, is incident normal to the anode surface with a power of 1525 W/cm^2. The risetime, time for maximum current, and maximum current depend to some extent on the current component running parallel to the surface at high bias conditions. As shown in Fig. 5 at 0.4 ps, the hole current at 10 μm parallel to the surface is much higher when compared to the device center (at 300 μm) or along the line running between the center of the electrodes. This current component also

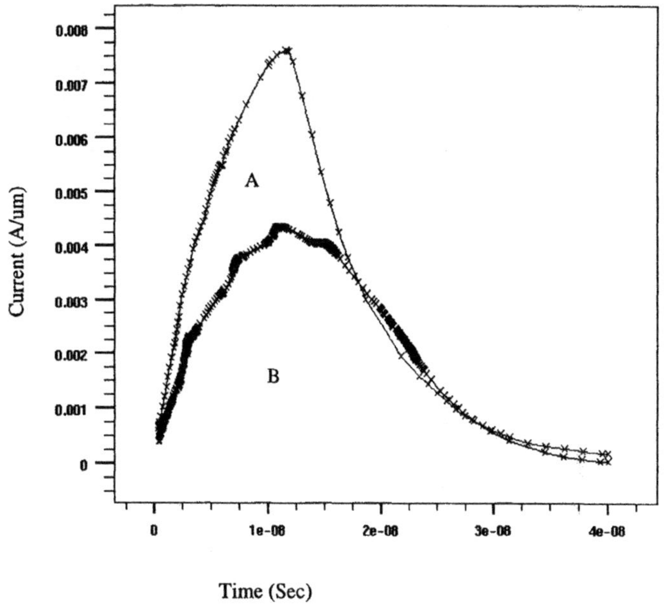

Figure 4. The "on state" response with contact resistivity and (A) with surface effects, (B) without surface effects.

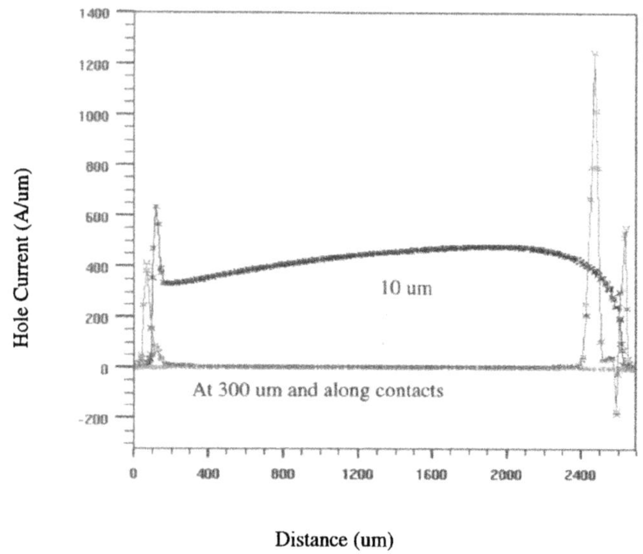

Figure 5. Hole current component at various depths in the device.

determines the maximum current and the rate of collection. The maximum current for the 1.3 cm wide device is about 100 A compared to an experimental value of 133 A.[1] Figure 4A shows the difference over 4B, which has no sub-surface current or contact parameter changes.

Simulation results from ODU for the photocurrent with and without the contact injection process are shown in Fig. 6. An applied voltage waveform consisting of a 1.0 ns ramp followed by a constant 100 V value was assumed. Contact injection due to thermionic emission and the electron tunneling processes were taken into account. The thermionic emission model used an effective Richardson coefficient of 80 $A/cm^2/K^2$. The tunneling currents were calculated based on the Wentzel-Kramers-Brillouin (WKB) approximation within the effective mass theory [24] for a triangular barrier potential. A photoexcitation pulse was assumed to illuminate the semiconductor continually between 10 ns to 15 ns at a fixed generation rate of 2×10^{31} m^{-3} s^{-1}. The initial conduction in Fig. 6 lasting up to 1.0 ns for

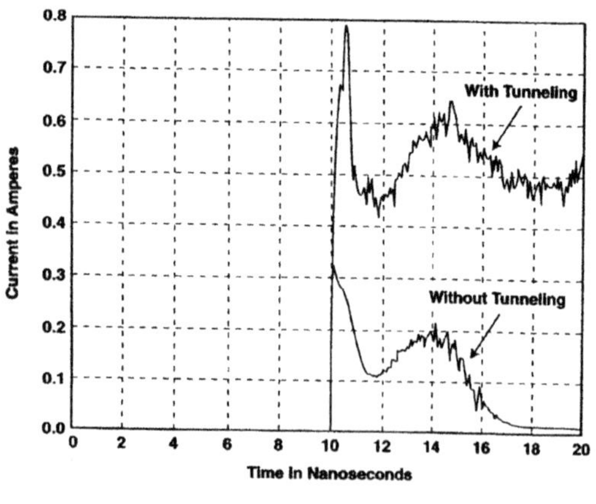

Figure 6. Simulation results from ODU for photocurrent both with and without contact injection.

both cases is the displacement current associated with the external voltage ramp. The rise at about 10 ns follows the optical pulse initiation. The dip at about 12 ns results from internal polarization effects, and the gradual decrease in the device voltage due to the 50 Ω resistive voltage drop. The subsequent increase is simply the result of enhanced electron velocity at the lower fields characteristic of GaAs material. Finally, after the termination of the optical input at 15 ns, the photocurrent is seen to decay if the contact tunneling process is not included. However, *with* tunnel injection included in the model, two primary differences can be seen: (i) first, the current magnitude is greater, and (ii) the photocurrent does *not* appear to decay upon the termination of the photoexcitation pulse at 15 ns. This clearly demonstrates the "lock-on" phenomenon and its link with strong field dependent contact injection. It is also very plausible that lateral inhomogeneities of the barrier height across the semiconductor-contact interface would lead to strong spatial variations in charge injection. This might then lead to detrimental filamentary currents at the contacts. The results also demonstrate that the fabrication of good "blocking contacts" would be a key element in improving the switch stability and its turn-off characteristics.

CONCLUSION

Simulations at UNM show that the I-V characteristic, maximum current, risetime, and time-to-peak depend on the properties of the contacts and the current components that come into play at high bias conditions. Better match with experiments was possible with a model that included contributions from subsurface channels below the GaAs-passivation layer interface. The model assumes that mobility near the surface is affected by surface roughness and phonon scattering, and is field-dependent. A field-dependent band-to-band tunneling model was also used as input. Finally, contact injection and the field-assisted tunneling process were shown by ODU to have an important effect on the photocurrent response under high voltages. It was demonstrated that such injection could potentially lead to the "lock-on" state. Future work will include prototype fabrication at UNM, as well as a comparison of the simulation results obtained at UNM and ODU.

ACKNOWLEDGMENTS

This research was supported by an AFOSR/New World Vistas grant. Useful discussions with the AFRL research group are appreciated.

REFERENCES

1. J.S.H. Schoenberg, J.W. Burger, J.S. Tyo, M.D. Abdalla, M.C. Skipper, and W.R. Buchwald, "Ultra-Wideband Source Using Gallium Arsenide Photoconductive Semiconductor Switches," IEEE Trans. Plasma Sci. 25:327 (1997).
2. W. Van Roosbroeck and H.C. Casey, Jr., "Transport in Relaxation Semiconductors," Phys. Rev. B 5:2154 (1972).
3. K. Zdansky, B.K. Jones, J. Santana and T. Sloan, "Numerical Analysis of Charge Transport in Semi-Insulating GaAs with two Contacts," J. Appl. Phys. 79:3611 (1996).
4. G.M. Loubriel, M.W. O'Malley, and F.J. Zutavern, "Recent Developments in Photoconductive Semiconductor Switches," in "Proc. 6th IEEE Pulsed Power Conf.," IEEE, Piscataway (1987), p. 145 – 148; F.J. Zutavern, G.M. Loubriel, M.W. O'Malley, L.P. Shanwald, W.D. Helgerson, D.L. McLaughlin, and B.B. McKenzie, "Photoconductive Semiconductor Switch Experiments For Pulsed Power Applications," IEEE Trans. Elec. Dev. 37:2472 (1990).
5. R.P. Brinkmann, K.H. Schoenbach, D.C. Stoudt, V. Lakdawala, G.A. Gerdin, and M. Kennedy, "The Lock-On Effect in GaAs Switches," IEEE Trans. Elec. Dev. 38:701 (1991).

6. H.J. Queisser and D.E. Theodorou, "Hall Effect Analysis of Persistent Photocurrents in GaAs Layers," Phys. Rev. Lett. 43:401 (1979).
7. J. Klem, W.T. Masselink, D. Arnold, R. Fischer, T.J. Drummond, H. Morkoc, K. Lee, and M.S. Shur, "Persistent Photoconductivity in Modulation Doped Structures," J. Appl. Phys. 54:5214 (1983).
8. R. Fletcher, E. Zaremba, M.D'Iorio, C.T. Foxon, and J.J. Harris, "Persistent Photoconductivity and Two-Band Effects in GaAs Heterojunctions," Phys. Rev. B 41:10649 (1990).
9. T. Frello, E. Veje, and O. Leistiko, "Observation of Time-Varying and Persistent Photoconductivity in Porous Silicon," J. Appl. Phys. 79:1027 (1996).
10. H.M. Chen, Y.F. Chen, M. Lee, and M.S. Feng, "Persistent Photoconductivity in n-Type GaN," J. Appl. Phys. 82:899 (1997).
11. C.H. Qiu and J.I. Pankove, "Deep Levels and Persistent Photoconductivity," Appl. Phys. Lett. 70:1983 (1997).
12. D.E. Lacklison, J.J. Harris, C.T. Foxon, J. Hewett, D. Hilton, and C. Roberts, "A Comparison of Photocurrent Effects in AlGaAs Heterostructures," Semicond. Sci. Technol. 3:633 (1988).
13. K.A. G. Yakovenko, E.A. Shelonin, and V.I. Fistul, "Dielectric Relaxation in Copper-Doped Germanium," Sov. Phys. Semiconductors 17:218 (1983).
14. B.T. Cavicchi and N.M. Haegel, "Experimental Evidence for Relaxation Phenomena in High Purity Silicon," Phys. Rev. Lett. 63:195 (1989).
15. N.M. Haegel, "Relaxation Semiconductors: In Theory and in Practice," Appl. Phys. A 53:1 (1991).
16. C. Popescu and H.K. Henisch, "Minority Carrier Injections in Insulators," Phys. Rev. B 14:517 (1976).
17. J.C. Mancifer and H.K. Henisch, "The Interpretation of Ohmic Behavior in Semi-insulating Gallium Arsenide Systems," J. Appl. Phys. 52:5195 (1981).
18. N. Derhocobian and N.M. Haegel, "Experimental Study of Transport in Trap-dominated Relaxation Semiconductors," Phys. Rev. B, 44:12754 (1991).
19. SILVACO International, "ATLAS User's Manual" (www.silvaco.com).
20. M. Rogalla, *et al.*, "Characterization of SI-GaAs bulk material for detector application," in "Gallium Arsenide and Related Compounds, Proceedings of the Third International Workshop," P.G. Pelfer, *et al.*, eds., World Scientific, New Jersey (1996), p. 69.
21. W.T. White III, C.G. Dease, M.D. Pocha, and G.H. Khanaka, "Modeling GaAs High-Voltage, Subnanosecond Photoconductive Switch in One Spatial Dimension," IEEE Trans. Elec. Dev. 37:2532 (1990).
22. D.C. Stoudt, R.P. Brinkmann, R.A. Roush, M.S. Mazzola, F.J. Zutavern, and G.M. Loubriel, "Effects of 1-MeV Neutron Irradiation on the Operation of a Bistable Optically Controlled Semiconductor Switch," IEEE Trans. Elec. Dev. 41:913 (1994).
23. N.E. Islam, E. Schamiloglu and C.B. Fleddermann, "Characterization of a Semi-insulating GaAs Photoconductive Semiconductor Switch (PCSS) for High Power Ultra-Wideband Microwave Applications," submitted to Appl. Phys. Lett. (1998).
24. L. Schiff, "Quantum Mechanics," McGraw Hill, New York (1955).

ANALYSIS OF NONLINEAR OPTICAL-ELECTROMAGNETIC INTERACTIONS IN ULTRAFAST MODE-LOCKED LASER DIODES.

J.M. Arnold[(a)], E.A. Avrutin[(a)], J.H. Marsh[(a)], and E.L. Portnoi[(ab)]

[(a)]Department. of Electronics and Electrical. Engineering
University of Glasgow
Glasgow G12 8LT, Scotland, UK
[(b)]A.F. Ioffe Physico-Technical Institute
Polytechnicheskaya ul. 26, St.Petersburg 194021, Russia

I INTRODUCTION

The operation of mode-locked (ML) semiconductor laser diodes (LDs)[1,2], in common with lasers of other types, is a nonlinear process in which a balance of dispersion and various nonlinearities results in the generation of an ultrashort (~1 ps or less) optical pulse, repeated at the cavity round-trip frequency or its harmonic and with parameters, averaged over a round-trip, constant, similar to an optical soliton. However, there are some important features in ML that are specific for semiconductor lasers, e.g.:

- Depending on laser parameters, the ML repetition frequency can cover the extremely broad range, from GHz to sub-THz and even THz range.
- The same nonlinearities that define the ML properties can also be used to achieve various types of (useful) interaction and conversion between optical and electromagnetic (microwave and mm-wave) signals - sinusoidal or pulsed- either within the laser cavity itself or in a monolithically integrated scheme, thus broadening the range of possible applications.

Here, we theoretically analyse the fundamentals of high-frequency mode-locked laser diodes (Section II) and their application to generating and/or frequency shifting ultrafast electrical signals (Section III)

II. NONLINEAR PROCESSES IN ULTRAFAST MODE-LOCKED DIODE LASERS: PHYSICS AND THEORETICAL APPROACHES

General considerations

Mode-locking (ML) is a regime of laser generation whereby the laser emits light in several modes with a constant phase relation, i.e. with precisely equidistant frequencies. Usually, the term is used more specifically, assuming *amplitude-modulation* (AM) mode

Ultra-Wideband, Short-Pulse Electromagnetics 4
Edited by Heyman *et al.*, Kluwer Academic / Plenum Publishers, New York, 1999.

locking, meaning that the phases of the (longitudinal) modes are not only constant but may be considered approximately equal. This picture corresponds to an *optical pulse* that is repetitively reshaped as it circulates in the laser cavity. The laser thus emits a train of pulses with a repetition frequency approximately equal to an integer multiple (a harmonic) of the cavity round-trip frequency:

$$F \approx \frac{Mv_g}{2L} \tag{1}$$

where $v_g = c/\eta_g$ (velocity of light divided by the group refractive index) is the group velocity of light in the laser waveguide (average over cavity length), L the cavity length and M the number of pulses coexisting in the cavity (in the simplest ML configurations, M=1). The standard methods of achieving ML are either modulation of one of the laser parameters at a frequency F (*active* ML) or exploiting nonlinear properties of the medium, usually by introducing a *saturable absorber* (SA) into the laser cavity (*passive* ML). The combination of these methods is known as *hybrid* ML; in particular, ML by external modulation is in the form of short pulses is known as *synchronous* ML.

During the ML cycle, the parameters of the pulse experience periodic evolutions, but on the average remain constant (as in an optical soliton) due to balanced action of pulse-shortening (saturable absorption saturation) and pulse-broadening (gain saturation and dispersion) effects. These effects are common for lasers of all kinds. However, there are some important properties that are peculiar to the semiconductor active media (amplifying sections and SA) used in LD's, mainly because of the very high density of population inversion, of the order 10^{18} excited particles (electron-hole pairs) per cm^3, and the broad energy spectrum of these particles. One such property is the strong coupling between gain/loss and refractive index changes, as a result of which changes of carrier density in the gain and SA sections are accompanied by strong self-phase modulation (SPM) leading to ML pulses often being chirped. Another important point specific to semiconductor active media is that saturation of both gain and saturable absorption contains both "slow" (with characteristic times longer than the ML pulse duration) components due to changes in the total carrier population in the gain or SA, and "fast" components associated with "fast" gain and saturable absorption nonlinearities such as spectral hole burning and dynamic carrier heating. Similar to the "slow" gain saturation leading to "slow" SPM, gain/SA nonlinearities may be accompanied by fast *refractive index* nonlinearities. Finally, the short cavity lengths in diode bring the values of the duty cycle, or the ratio of the laser length to the "spatial duration" of a single optical pulse, down to the order of units to several tens (compared with hundreds to thousands in gas and most solid-state lasers).

The considerations above lead to some differences in the theoretical approaches to ML in diode lasers as compared to modelling other lasers. For example, peculiarities of semiconductor lasers render some traditional techniques of ML analysis (lumped time-domain models) less accurate and make other approaches (see below) both more necessary and more computationally viable. This is particularly true for ML lasers, that is, laser constructions capable, in principle, of generating a stream of pulses that are not only subpicosecond in duration but are repeated at very high (subterahertz) frequencies. Below, we describe two approaches we use for analysis of such lasers.

Distributed time-domain modelling of mode-locked laser diodes.

In time domain, the ML is described as propagation of an optical pulse through a waveguide medium. A distributed time-domain model (DTDM) then starts with decomposing the optical field in the laser cavity into components propagating right and left in the longitudinal direction (say z).

$$E(\mathbf{r},t) = \Phi(x,y)\left(E_R \exp(-i\beta_0 z) + E_L \exp(i\beta_0 z)\right)\exp(i\omega_0 t) \quad (2)$$

with Φ being the transverse/lateral waveguide mode profile and ω_0 and $\beta_0=n(\omega_0)\omega_0/c$ being the central optical frequency and corresponding wavevector respectively. Then, for slowly varying amplitudes $E_{R,L}$, one obtains the reduced equations :

$$\pm\frac{1}{v_g}\frac{\partial E_{R,L}}{dz} + \frac{\partial E_{R,L}}{dt} = \Gamma(\hat{g} + i\Delta\hat{\beta} - a_i)E_{R,L} + F_L(z,t) \quad (3)$$

Here, the operators $\boldsymbol{g}$ and $\boldsymbol{\Delta\beta}$ describe the gain and variable part of propagation constant (refractive index) respectively and include both the nonlinearity (saturation) and frequency dependence (dispersion; hence the operator nature of these parameters), a_i is the unsaturable loss. At z values within the saturable absorber (SA), the gain value g in (3) is changed to $-a$, a being the saturable absorption coefficient. The model is completed by standard rate equations for carrier densities in the gain section (4a) and the SA (4b):

$$\frac{\partial N(z,t)}{\partial t} = \frac{J(z,t)}{ed} - \frac{N}{\tau_G(N)} - v_g g S(z,t), \quad (4a)$$

$$S(z,t) = |E_R(z,t)|^2 + |E_L(z,t)|^2$$

$$\frac{\partial N(z,t)}{\partial t} = -\frac{N}{\tau_A} + v_g a S(z,t) \quad (4b).$$

For carrier density dependence of gain and saturable absorption, our version of the model may use either simple linear expansions:

$$g_L(N)=A_G(N-N_{0G})\ ;\ a_L(N)=A_A(N_{0A}-N) \quad (5a)$$

with the subscript L standing for "linear", or logarithmic approximations. Fast nonlinearities included using phenomenological nonlinearity coefficients $\varepsilon_{A,G}$:

$$g = g_L(N)/(1+\varepsilon_G S), \qquad a = a_L(N)/(1+\varepsilon_A S) \quad (5b)$$

For subterahertz high frequency ML, where both the pulse width and the repetition period are (sub)picosecond, a finite relaxation time of the nonlinearities is also important. In our work, a phenomenological relaxation equation

$$\frac{da}{dt} = \frac{a_L - a(1+\varepsilon_A S)}{\tau_{nl}^{(a)}} \quad (6)$$

is used to account for this finite relaxation time.

To achieve self-starting of the model and to model the noise/jitter properties of the LD's, the r.h.s. of Eq.(3) is completed with a Langevin noise source $F_L(z,t)$. This source describes spontaneous emission and is implemented numerically by adding complex random numbers to the propagating fields.

To account for gain dispersion, the gain spectrum profile is approximated as a Lorentzian, which in time domain is Fourier-transformed as

$$\hat{g}E = g\cdot\omega_L\cdot\int_0^\infty E(t-\tau)\exp(-\omega_L\cdot\tau)d\tau \quad (7)$$

(which in practice is efficiently calculated using a simple recursive formula).

Group-velocity dispersion is implemented as a second time derivative, so that the total refractive operator, including SPM, is

$$\Delta\hat{\beta}E = \alpha_G(g - g_{th})E - \frac{1}{2v_g^2}\frac{dv_g}{d\omega}\frac{\partial^2 E}{\partial t^2} \quad (8)$$

We find that, like gain dispersion, it is important for modelling subpicosecond pulses.

Frequency-domain approaches to mode-locking in laser diodes.

Model formulation. For a frequency-domain approach, instead of (2), one uses an expansion of the optical field in the form

$$E(\mathbf{r},t) = \Phi(x,y)\sum_k E_k(t)u_k(z)\exp(i(\omega_k t + \varphi_k(t))) \tag{9}$$

where E_k, u_k, ω_k and φ_k are respectively the amplitude, longitudinal mode profile, frequency and phase of the mode k. The functions $u_k(z)$ are defined by requiring that these functions satisfy the wave equation plus reflection boundary conditions at all the reflectors present in the cavity. The time dependence of E_k and φ_k is considered slow in the sense that the variables do not change significantly on a timescale comparable to the round-trip time ($d/dt << v_g/L$). All the fast (faster than round-trip) time dependence of lasing field is included in the expansion (9). The gain g and saturable absorption a in a frequency-domain model are also treated as "slow" variables; so are the carrier densities in gain and absorber sections; their dynamics are therefore described by (4) with no z dependence and a "slow" photon density $S(t) = \sum_k E_k^2$. The dynamics of the modal amplitudes and phases are described by a system of (ordinary) differential equations for complex modal amplitudes $\mathbf{E}_k = E_k(t)\exp(i\varphi_k(t))$ in the form:

$$\frac{\partial}{\partial t}\mathbf{E}_k = [\gamma_k + i(\Omega_k - \omega_k)]\mathbf{E}_k + \sum_m G_m \mathbf{E}_{k+m} + F_k^{(L)}(t) \tag{10}$$

In (10),

$$\gamma_k = \frac{1}{2}v_g[\Gamma(f_g g - f_a a) - \Delta g_k - a_{ck}] \tag{11}$$

and Ω_k are the net amplitude gain and eigenfrequency of the k-th mode, Δg_k and a_{ck} being the net gain discrimination (due to gain and SA dispersion) and cavity loss for the mode k and $f_{g,a} = L_{g,a}/L$ fractions of the cavity occupied by gain and SA sections. The parameters a_{ck} and Ω_k are defined simultaneously with $u_k(z)$. The second term in (10) describes the mode interactions that lead to ML;

$$\begin{aligned} G_m = (\sum_k \mathrm{E}_k \mathrm{E}_{k+m}^*)\cdot\Bigg\{ & \xi_m^{(a)} a \left[\frac{v_g(1+i\alpha_A)\cdot A_G \cdot \tau_{\Sigma a}}{1 - im\Delta\Omega\tau_{\Sigma a}} + \frac{\varepsilon_A}{1 - im\Delta\Omega\tau_{nl}^{(a)}}\right] \\ & - \xi_m^{(g)} g \left[\frac{v_g(1+i\alpha_G)\cdot A_A \tau_{\Sigma g}}{1 - im\Delta\Omega\tau_{\Sigma g}} + \frac{\varepsilon_G}{1 - im\Delta\Omega\tau_{nl}^{(g)}}\right]\Bigg\} \end{aligned} \tag{12}$$

In (12), the first factor is the component of the light intensity oscillating at the frequency $m\Delta\Omega$ due to beating between modes; the second factor (the braces) is the response of the active medium of the SA section (the first term in the braces) and of the gain section (the second term) to these light intensity oscillations. In each set of square brackets, the first term is the response of the active medium due to the oscillations of total carrier density, the second is due to fast nonlinearities. The parameters $1/\tau_{\Sigma a,g} = 1/\tau_{a,g} + 1/\tau_{a,gs}$ are total (non-stimulated plus stimulated) recombination rates, $\tau_{a,gs} = 1/(v_g A_{a,g} S)$ being the stimulated recombination times. Finally,

$$\xi_m^{(g,a)} = \int_{g,a} dz \cdot (u_k u_{k+m}^*) \cos\left(\frac{m\pi}{L}(z + \frac{L}{2})\right) \tag{13}$$

are the coefficients of overlap between the light intensity oscillations and the gain/SA sections (e.g. for a tandem structure with $f_a << f_g$, $\xi_m^{(g)} \approx 0.5 f_g$, $\xi_m^{(a)} \approx f_a$; for a CPM structure, $\xi_{2m}^{(a)} \approx 2f_a$, $\xi_{2m-1}^{(a)} \approx 0$; m=1,2,3... More accurate expressions including the case of multiple absorbers are given in).

Mode locking criterion. The frequency-domain approach to ML was first applied to monolithic LD's by Lau [3] using, in effect, a simplified version of (3) with only three modes, no SPM and no fast nonlinearities taken into account. In particular, this three-mode version of the FD model was used [3] to derive the criterion defining the parameter range of the ML regime. This is done by analysing the (in)stability of the single-mode solution of (10) with respect to the excitation of weak side modes. In our work [1,2,4,5], the ML criterion was generalised, using the model in the form presented here, for an arbitrary passive ML configuration ; the result has the form:

$$\xi_m^{(a)} a\left[\frac{v_g A_a \tau_{\Sigma a}}{1+(m\Delta\Omega\tau_{\Sigma a})^2}+\frac{\varepsilon_A}{1+(m\Delta\Omega\tau_{nl}^{(a)})^2}\right]- \xi_m^{(g)} g\left[\frac{v_g A_g \cdot \tau_{\Sigma g}}{1+(m\Delta\Omega\tau_{\Sigma g})^2}+\frac{\varepsilon_G}{1+(m\Delta\Omega\tau_{nl}^{(g)})^2}\right] > \frac{(a_{c,\pm m}-a_{c0})+\Delta g_{\pm m}}{S} \quad (15)$$

for the case of no SPM ($\alpha_{G,A}=0$) (he criterion for the case with nonzero $\alpha_{G,A}$ is given elsewhere [5]). Although this is a fairly approximate criterion [2], it allows some useful conclusions to be made, in particular, when applied to ultrafast mode-locked lasers.

Ultrafast (subterahertz) mode-locked lasers.

As follows from the formula (1) for the ML freqiency, the two distinct ways of achieving high ML rates are either using a shorter laser cavity or introducing an advanced LD construction capable of ML operation at a higher (Mth, $M>1$) harmonic of the laser round-trip frequency. By mode-locking short lasers, the ML frequency was increased from ~100 GHz achieved initially [6] up to 240 GHz[7]. Higher repetition rates have been achieved only with ML at a harmonic of the round-trip frequency. Positioning one or more saturable absorber(s) (SA's) at multiples of an integer fraction $1/M$ of the cavity length has allowed ML to be achieved at the Mth harmonic of the round-trip frequency, with M increasing from $M=2$ (colliding pulse mode-locking, CPM [8]) to $M=3$ or 4 (multiple CPM, MCPM [4]) and $M=5,7$ (asymmetric CPM, ACPM [9]). These techniques have produced HML at frequencies from 240 to 860 GHz. The highest repetition rates of 800 GHz ($M=20$) or1.6 THz ($M=40$) have been achieved with the technique of harmonic ML (HML) using a DBR LD[10]. In a theoretical paper [5], we have proposed a modification of this structure, with an additional Fabry-Perot subcavity formed by an intermediate reflector (e.g. an etched groove) in the cavity.

From the frequency-domain theory point of view, harmonic ML takes place if the ML condition (15) is fulfilled for $m>1$ rather than $m=1$.

Our analysis shows that for *multiple / asymmetric CPM*, the factors that provide such modal selectivity are the mode coupling parameters G_m or, more accurately, the overlap parameters ξ_m that scale the coupling strengths. Harmonic mode-locking in these devices is therefore an essentially nonlinear effect. Both simulations end experiment reveal that it is highly sensitive to external influences with switching possible between both different repetition rates and different supermodes within the same repetition rate. On the other hand, since pulse collisions only happen in the SA sections, short pulses may be achieved.

In the case of a *compound-cavity*, the modal selectivity is ensured by the values of the cavity losses a_c which within the first approximation may be estimated as

$$a_{ck}=\frac{1}{L}\ln\frac{1}{r_1 r_2}+\Delta a_{ck};\quad \Delta a_{ck}\approx-\frac{r_i}{L}\left(\frac{r_1^{f_2}}{r_2^{f_1}}\cos(2\pi k f_1)+\frac{r_2^{f_1}}{r_1^{f_2}}\cos(2\pi k f_2)\right) \quad (16)$$

with L the cavity length and $f_{1,2}\approx L_{1,2}/L$ sub-cavity length fractions, $L_{1,2}$ being the lengths of right/left sub-cavities respectively, and r_1, r_2, r_i being left, right and intemediate reflectances. Harmonic nature of mode-locking is therefore a linear effect and may be expected to be more stable than in MCPM lasers. Indeed, in our preliminary experiments with ring constructions [11] we did not observe any fundamental ML under any conditions studied. On the other hand, the presence of pulse collisions in the gain medium in these lasers may lead to pulse broadening, particularly for relatively low repetition frequencies (100-200 GHz). Figures 1a,b illustrate DTDM-simulated pulse trains for an MCPM laser and for a compound-cavity laser.

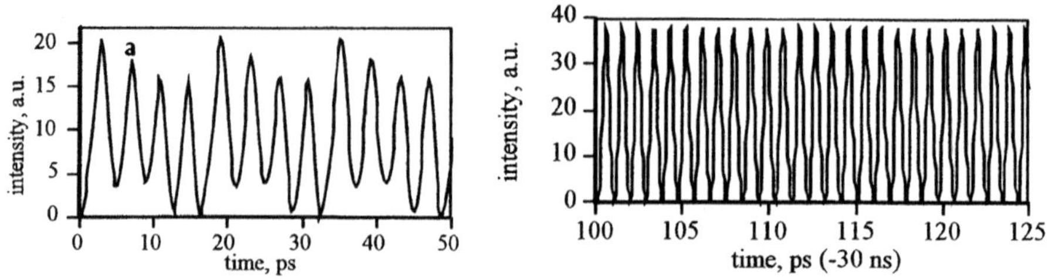

Fig. 1. Simulated pulse profile from a multiple-colliding-pulse mode-locked laser at 240 GHz (a) and a compound-cavity mode-locked laser diode at 1.06 THz (b).

III. OPTOELECTRONIC INTERACTIONS IN MODE-LOCKED LASER DIODES AND THEIR APPLICATION FOR ELECTROMAGNETIC SIGNAL GENERATION AND FREQUENCY CONVERSION.

As an electromagnetic signal, light emitted by a by a ML laser (Eq.9) consists of a microwave signal (pulse train at the repetition frequency) superimposed on the optical carrier. Then, the most straightforward way of generating electrical signals from the ML light is by detecting it and thus removing the optical carrier. For lasers mode-locked at tens, or low hundreds, of GHz this may be achieved using the relatively standard photodetectors such as ultrafast photodiodes. For the case of (sub-) terahertz lasers, the detection scheme is discussed below. The disadvantage of this simplest approach is that the fundamental frequency of the signal obtained is fixed at the ML repetition frequency and the possibilities of tuning it are very limited. Generation of more broadband – and tunable- electrical signals by mode-locked lasers requires a more complex approach using *optoelectronic frequency conversion* [1,2,12]

Optoelectronic mixing and frequency conversion in mode-locked laser diodes.

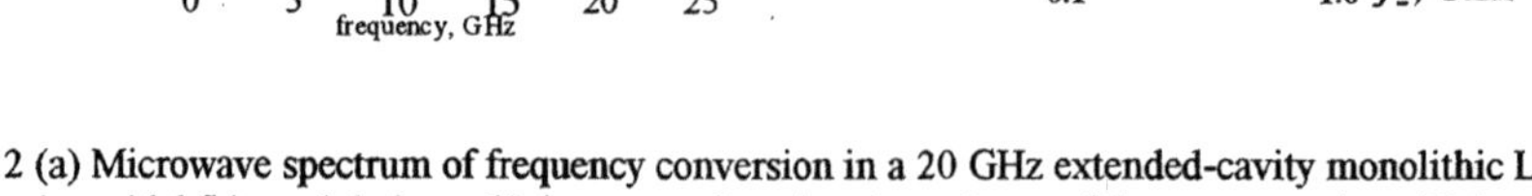

Fig. 2 (a) Microwave spectrum of frequency conversion in a 20 GHz extended-cavity monolithic LD under sinusoidal SA modulation; (b) frequency detuning dependence of down-conversion efficiency in the same laser

This could be achieved by using a ML LD as a mixing element which would provide its output in the form of an amplitude-modulated *optical* signal. A possible way of realising such a function would be to apply a voltage modulation at the signal frequency f_s to the SA section. The *local oscillator* (LO) source would be provided by oscillations of light intensity, and therefore of the optical properties of the active medium, at the fundamental ML frequency F (or one of its harmonics). Mixing of the LO and the signal occurs due to parametric effects since the induced recombination rate depends on the product of the net modal gain (which is modulated at f_s) and photon density (which has a component at F according to Eq.(9)). The microwave spectrum of the output optical intensity would then contain, apart from components at $f_{LO}=F$, its harmonics and f_s, also some modulation at the beating frequencies $f_+=f_{LO}+f_s$, $f_-=|f_S-f_{LO}|$. Since the laser performs the functions of a LO generator and a mixing element simultaneously, we may speak of an *integrated frequency down- or up- converter,* depending of which of the two frequencies f_- ,f_+ is treated as the useful output frequency.

The set-up for optoelectronic conversion is exactly the same as that for hybrid ML but mixing is, obviously, realised outwith the parameter range for successful locking of the laser to the signal. Using DTDM modelling[13], we mapped the regions corresponding to different regimes and found that microwave mixing only occurred for sufficiently large microwave frequency detunings (f_-) and/or small signal intensities.

However, even with small signal amplitudes, the microwave mixing is still a complicated non-linear process, with the modulation at $f_\pm$ reacting back at the signal modulation, if the LO modulation itself is not small, which is usually the case for well-developed ML. For instance, this effect leads to the presence of some additional lines such as $f_{LO} - f_- = 2f_S - f_{LO}$, in the output spectra (Fig. 2a). Therefore, rather than trying to obtain a small-signal estimate for the conversion efficiency, we performed full numerical (DTDM) simulations with subsequent Fourier analysis of the output intensity spectrum. The results are plotted in Fig.2b as a function of the frequency detuning $f_-=f_s-f_{LO}$. The curve has similar features to a standard laser modulation curve, with a sharp peak seen at $f_-=f_{RES}$~1GHz the electron-photon resonance frequency, but with a relatively flat region at $f_- < f_{RES}$. We believe this shows that, despite the complexity of the mixing process, this mixing may still be used to down-convert weak signals detuned from the ML frequency by less than f_{RES} without significant distortion. For down-converting very narrowband signals, it may, on the other hand, be advantageous to adjust the (current-dependent) resonance frequency so as to perform resonant conversion ($f_- \sim f_{RES}$) to maximise conversion efficiency.
The numerical examples above are at typical "communications" repetition frequencies (tens of GHz); however, our simulations have also shown that, given the possibility of modulating the SA with the signal frequency, optoelectronic conversion would be possible at (sub)-terahertz frequencies (Fig.3). Thus, in principle, a mode-locked laser diode in an integrated scheme may be used for (sub)-THz down conversion, for example, for atmospheric/astrophysical/plasma measurements. This, however, requires solving a technical problem of applying the signal to the SA section.

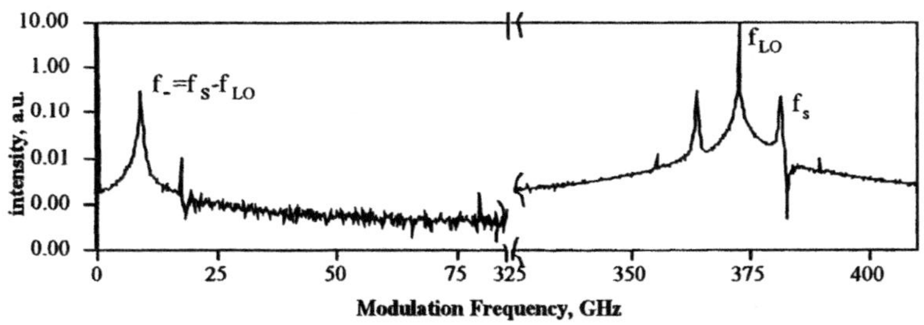

Fig. 3: Part of the simulated RF spectrum of the optical output from a CPM laser diode under direct modulation of the saturable absorber.

Mode-locked laser diodes for terahertz electromagnetic signals?

In a THz integrated optoelectronic down-converter, the THz local oscillator exists only in the form of oscillations of net gain inside the laser. It may be also attractive to use an utlrafast ML LD as part of an optoelectronic terahertz source to provide an *electromagnetic* THz signal. Since standard photodetectors at THz range have a very low efficiency, some alternative ways of detection may be worthwhile analysing. One possible way of such detection is *direct optical rectification* in a structure including an optical waveduide for the THz-modulated light from the laser monolithically integrated with a THz (e.g. strip-line) waveguide for the detected THz signal. The conversion in this case would happen through the virtually instantaneous optical nonlinearity in the semiconductor material, being in a way "zeroth-harmonic generation" as opposed to the wider known effect of second-harmonic generation. As usual in nonlinear optic, the structure needs to be engineered in such a way that the group velocity of light in the light waveguide coincides with the phase velocity of the signal in the THz waveguide to achieve phase synchronism. Our initial estimates show the importance of obtaining good overlap factors between the THz and optical guided waves for achieving good conversion efficiency. Another way of enhancing this efficiency, as well as improving the purity of the THz signal, is to move from a scheme with a travelling-wave light-terahertz interaction to a resonator structure, seen by either the light, or the THz wave, or ideally both. Finally, it may be possible to take this idea one step further and analyse of a structure in which the non-linear element (the rectifier) is integrated *within the THz laser cavity*. Further studies are now in progress.

1 E.A.Avrutin, J.H.Marsh, E.L.Portnoi, "Mode-locked diode lasers for microwave optoelectronic applications", Internat. J. of High-Speed Electron. and Systems, 1997, V.8, No.3, pp. 495-523

2 E.A.Avrutin, J.H.Marsh, E.L.Portnoi, (invited review), in preparation for *IEE Proceedings Pt. J (Optoelectronics)*

3 K. Y. Lau, "Narrow-band modulation of semiconductor lasers at millimetre wave frequencies (>100 GHz) by mode locking", *IEEE J. Quantum Electron.*, **26** (1990), 250-261

4 J. F. Martins-Filho , E.A.Avrutin, C.N. Ironside, J.S. Roberts, "Monolithic multiple colliding pulse mode locked quantum well lasers: experiment and theory", *IEEE J. Select. Topics in Quantum Electron.*,**1**(1995), 539-52

5 E.A.Avrutin, J.H.Marsh,J.M.Arnold, "Modelling of semiconductor laser structures for passive harmonic mode locking at terahertz frequencies", *International J. Optoelecronics*, **10** (1996), pp. 427-432

6 P.P. Vasil'ev, A.B. Sergeev, "Generation of bandwidth-limited 2 ps pulses with 100 GHz repetition rate from multi-segmented injection laser", *Electron. Lett.*, **25** (1989), 1049-1050

7 E. L. Portnoi, A.V. Chelnokov, "Passive mode-locking in a short-cavity laser", *in: Digest, IEEE 12th International Semiconductor Laser Conference, Davos, Switzerland, Sept. 1990,* pp. 140-141

8 Y.-K. Chen, M.C. Wu, "Monolithic colliding-pulse mode locked quantum-well lasers", *IEEE J. Quantum Electron.*, **28** (1992), 2176-2184

9 Shimizu T, Ogura I, Yokoyama H, "860 GHz rate asymmetric colliding pulse modelocked diode lasers", *Electron. Lett.*, , 1997, Vol.33, No.22, pp.1868-1869

10 S.Arahira, Y.Matsui, Y.Ogawa,"Mode locking at very high repetition rates more than terahertz in passively mode locked distributed-Bragg-reflector laser diodes", *IEEE J.Quantum Electron.*, **32** (1996), 1782-1790

11 11 E.A.Avrutin, J.H.Marsh, J.M.Arnold, T.F. Krauss, H. Pottinger, R.M. De La Rue, "Analysis of Harmonic (Sub-)THz Passive Mode Locking in Monolithic Compound Cavity Fabry-Perot and Ring Laser Diodes",in: Proc. of the International Meeting on Microwave Photonics, Duisburg, Germany, 3-5 Sep 1997, pp. 47-50; *journal paper submitted*

12 E.L. Portnoi, V.B. Gorfinkel, E.A. Avrutin, I.G. Thayne, D.A. Barrow, J.H. Marsh, and S. Luryi, "Optoelectronic Microwave-Range Frequency Mixing in Semiconductor Lasers", *IEEE J. Select.Topics in Quantum Electron.*, **1**(1995), 451-460

13 E.A.Avrutin, J.M.Arnold, J.H.Marsh, "Analysis of dynamics of monolithic passively mode-locked laser diodes under external periodic excitation", *IEE Proc. Pt.J.*, **143** (1996), 81-8

ULTRA-WIDEBAND ANALOGUE CHANNELS USING SOLITONS

J. M. Arnold

Department of Electronics and Electrical Engineering
University of Glasgow
Glasgow G12 8LT
Scotland, UK

1. Introduction

Proposals for the use of the nonlinear phenomenon of the soliton have been made for a number of different wave propagation regimes, including optical fibres and transmission lines with signal-dependent wave velocity. The principal advantage of soliton pulse propagation is that the effects of dispersion in the medium can be compensated by nonlinearity, so that the soliton propagates without the temporal broadening that would occur in the linear regime. In this way, pulses of the order of picoseconds duration have been demonstrated experimentally in both optical fibre and nonlinear transmission line environments.

Because the soliton is specifically a nonlinear phenomenon, the behaviour of a channel in which solitons are the carriers of information is different to the conventional type of channel where the wave propagation is linear. In addition, the soliton is an idealisation which can only exist under specific conditions which are not generic, namely that the underlying PDE which describes the wave propagation is integrable, meaning that it has an infinite number of conserved integrals. In the generic case, certain pulse waveforms known as solitary waves share some of the properties of solitons such as invariant shape with propagation, but lack an infinite number of conserved integrals.

The conventional theory of linear channels makes extensive use of the Fourier transform in conjunction with the defining property of linear systems, the superposition principle. There is no direct superposition principle for nonlinear systems, even for integrable ones, and the question of formulating an adequate theory of the information capacity of the nonlinear channel becomes problematic. For integrable systems the Inverse Scattering Transform (IST) is a powerful tool of mathematical analysis which resembles the linear Fourier Transform in many respects, but it can be difficult to apply the exact theory in all but the simplest of cases, and in any case the theory does not apply to generic nonintegrable systems.

Recently a theoretical method has begun to be developed which has the potential to describe the information-theoretical behaviour of discrete-time-encoded nonlinear channels quite effectively, while being much more tractable than the standard IST, and which is applicable to nonintegrable systems as well as integrable ones. In this theory, each soliton or solitary wave is regarded as a distinct and indestructible entity, or quasi-particle, which has its own small number of degrees of freedom such as position and velocity. These quasi-particles interact through a nonlinear potential in the manner of a dynamical chain, and the interacting chain provides a rather accurate description of the underlying multi-soliton dynamics in the circumstances typically encountered in a communication system or periodic oscillator. This theory shows that decodable information channels can be created

Ultra-Wideband, Short-Pulse Electromagnetics 4
Edited by Heyman *et al.*, Kluwer Academic / Plenum Publishers, New York, 1999.

using trains of solitary waves even when the individual solitary waves interact quite strongly, a situation which was previously thought to give rise to unacceptable errors in optical fibre systems.

2. Nonlinear wave equations

The two principle types of integrable nonlinear wave equations encountered in potential applications to communication channels are the Nonlinear Schrödinger Equation (NSE)

$$i\partial_z\psi + \tfrac{1}{2}\partial_t^2\psi + |\psi|^2\psi = 0 \tag{1}$$

for a complex field ψ, and the Korteweg-de Vries (KdV) Equation

$$\partial_z v + \partial_t^3 v + 3\partial_t v^2 = 0 \tag{2}$$

for a real field v. Here z is the propagation coordinate and t is the local time in a coordinate system moving with the wave velocity. Equation (1) describes the complex envelope of an optical wave on a fibre, and equation (2) describes the variation of line voltage on a dispersive transmission line with nonlinear capacitance. Also related to (2) is the Boussinesq equation

$$\partial_z^2 v - \partial_t^2 v - \varepsilon(\partial_t^4 v + \partial_t^2 v^2) = 0 \tag{3}$$

with ε a small parameter and t the absolute time, from which (2) is obtained by rescaling the variables, replacing $t \to t + z$ and neglecting backward-travelling waves for $\varepsilon \to 0$. In all cases the variables are appropriately scaled and normalised to remove as many physical parameters as possible. The equations (1)-(3) are all examples of integrable PDEs, meaning that they possess an infinite number of conserved integrals and are exactly soluble by the Inverse Scattering Transform for arbitrary initial conditions at $z = 0$.

Since the equations (1)-(3) are integrable they possess soliton solutions. These are

$$\psi = \eta\exp[i(\kappa z - \omega t)]\operatorname{sech}[\eta(t + \omega z)] \tag{4}$$

for NSE, with $\kappa = (\eta^2 - \omega^2)/2$,

$$v = 2\eta^2\operatorname{sech}^2[\eta(t - 4\eta^2 z)] \tag{5}$$

for KdV, and

$$v = 6\eta^2\operatorname{sech}^2[\eta(t - \sqrt{1 + 4\varepsilon\eta^2}\, z)] \tag{6}$$

for Boussinesq, with η and ω arbitrary real constants. All solitons have the character of a localised pulse whose duration varies inversely with amplitude, moving with some velocity relative to the coordinate frame.

3. Soliton lattices

In the remainder of this paper we consider the case of KdV solitons (2), appropriate to the propagation of pulses on a nonlinear transmission line. The case of NSE solitons (1) has been completely described in a number of recent papers [1-4]. Although the theory for KdV solitons is quite similar to that for NSE solitons, there are a number of differences in the construction of the two theories.

Suppose that initial conditions of (2) are set up at $z = 0$ consisting of a superposition of solitons

$$v_{|z=0} = \sum_{k\in\mathbf{Z}} 2\eta_k^2 \operatorname{sech}^2[\eta_k(t-t_k)]. \tag{7}$$

Suppose further that the solitons are separated by large times $t_{k+1} - t_k = O(T)$, $T >> 1$, and introduce the small parameter $\varepsilon = \exp(-\eta T)$, where η is the average amplitude of the solitons. Suppose finally that the soliton amplitudes η_k fluctuate around the mean by $O(\varepsilon)$, so that $p_k = \eta_k - \eta = O(\varepsilon)$. It can then be shown that the evolution of the solitons from $z = 0$ is given to a good approximation by allowing the parameters t_k to be z-dependent in (7), and requiring $t_k(z)$ to follow the Toda Lattice Equations

$$d_z^2 t_k = -K\{\exp[-2\eta(t_{k+1} - t_k)] - \exp[-2\eta(t_k - t_{k-1})]\}, \tag{8}$$

where K is a real positive constant. For a finite number of solitons equal to N, it is assumed that $t_0 = -\infty$ and $t_{N+1} = \infty$. By rescaling $2\eta t_k \to t_k, 2\eta K \to K, z\sqrt{K} \to z$, the system (8) can be brought to a canonical normalised form for the Toda Lattice,

$$d_z^2 t_k = -\exp[-(t_{k+1} - t_k)] + \exp[-(t_k - t_{k-1})]. \tag{9}$$

This result was first obtained by Gorshkov and Papko [5], and derived systematically for the case of $N = 2$ solitons by Gorshkov and Ostrovski [6]. The system (8) applies equally well in the limit $N \to \infty$ under suitable boundary conditions such as

$$\lim_{k\to\pm\infty} (t_{k+1} - t_k) = T \tag{10}$$

for fixed $T >> 1$.

The Toda Lattice equation (9) was also obtained as the underlying dynamical system for solitons of the Nonlinear Schrödinger Equation [1] when the phase difference between alternate solitons is exactly π. When the phase difference α is different from π, the Toda Lattice becomes complex, with the variable t_k in (9) replaced by a complex variable $s_k = t_k + i(\theta_k - \pi)$, where θ_k is the phase of the k-th soliton [1, 3, 4].

The existence of an approximating discrete dynamical system which describes the evolution of trains of nearly-identical solitons is a great advantage in the study of various types of communication channels, in which the carrier of information is a periodic pulse train deformed by modulations on the soliton parameters such as their position relative to the periodic points. We give an example of this analysis in the next section.

An additional interesting consequence of the fact that both NSE and KdV systems are approximated by the same discrete dynamical system is that it is therefore possible to simulate optical fibre systems (NSE) by equivalent millimetre-wave systems (KdV), even though the solitary waves (4) and (5) of the two systems are of different forms. Experimental devices have been made by a number of research groups in which a millimetre-wave planar transmission line supports KdV solitons, and the design of the devices is derived originally from an understanding of fibre solitons [7-10].

4. Linear channel and pulse-position modulation

Stationary states of the infinite Toda Lattice ($N = \infty$) are formed by periodic distributions $t_k = kT$. A *linear* channel is created by small fluctuations of the soliton positions t_k about the periodic points kT. Let $t_k = kT + q_k$. The parameters q_k can be encoded with information, a type of modulation that is known as pulse-position modulation (PPM). The linearised Toda Lattice Equations for small q_k are

$$d_z^2 q_k = -\tfrac{1}{4}\lambda_0^2 (q_{k+1} - 2q_k + q_{k-1}), \tag{11}$$

where

$$\lambda_0^2 = 4\exp(-T),$$

and the general solutions of the linearised Toda Lattice Equation are

$$q_k(z) = \sum_l q_l(0) J_{2(k-l)}(\lambda_0 z) + d_z q_l(0) \int_0^z J_{2(k-l)}(\lambda_0 z')dz', \tag{12}$$

where $q_k(0)$ and $d_x q_k(0)$ are the initial values of the soliton positions and velocities respectively at $z = 0$ and $J_n(.)$ are Bessel functions of the first kind of order n.. An alternative representation to (12) is the Fourier Transform

$$q_k(z) = (2\pi)^{-1} \int_{-\pi}^{\pi} \{Q(\beta)\cos(\lambda(\beta)z) + P(\beta)\sin(\lambda(\beta)z)\} e^{ik\beta} d\beta \tag{13}$$

where

$$Q(\beta) = \sum_k q_k(0) e^{-ik\beta}, \; P(\beta)\lambda(\beta) = \sum_k d_z q_k(0) e^{-ik\beta} \tag{14}$$

and it follows from (11) that the function $\lambda(\beta)$ is given by

$$\lambda(\beta) = \lambda_0 \sin\beta/2. \tag{15}$$

The Fourier transform variable β is essentially a normalised frequency parameterising the spectrum of the discrete-time function q_k. If all the solitons are started at $z = 0$ with equal velocities, so that $P(\beta) = 0$, then the transfer function for soliton displacements is

$$H(\beta, z) = \cos[\lambda(\beta)z], \tag{16}$$

which has zeroes at $\lambda(\beta) = (2r+1)\pi/2z$ for $r \in \mathbf{Z}$; these frequencies in the Fourier spectrum of the soliton displacements cannot be transmitted by the soliton channel. Conversely, the transfer function is equal to 1 at the frequencies $-\pi < \beta < \pi$ for which $\lambda(\beta) = 2r\pi/z$.

5. Invariant encoding

The one distinguishing characterstic of soliton pulse trains, as contrasted with any other kind of pulse train, is that they are solutions of *integrable* equations, and hence possess an infinite number of conserved integrals, or invariants. It follows that a communication channel must exploit this property if a soliton channel is to have any advantages over nonsoliton channels. It turns out that an N-soliton solution of the KdV equation has exactly N invariants, or one *per* soliton. These invariants are, essentially, the discrete eigenvalues of the Lax operator defining the integrable equation, which for KdV is the Schrödinger operator

$$L = -\partial_t^2 + v \tag{17}$$

where v is an N-soliton solution of the KdV equation. These eigenvalues remain fixed during the evolution of the KdV wavefunction v. The Toda Lattice equation (9) for N particles also has the same number of eigenvalues, one *per* particle, of its Lax operator

$$L = \begin{pmatrix} d_z t_1 & e^{-(t_2-t_1)} & & \\ e^{-(t_2-t_1)} & d_z t_2 & e^{-(t_3-t_2)} & \\ & e^{-(t_3-t_2)} & \ddots & \\ & & & d_z t_N \end{pmatrix}. \tag{18}$$

It can be shown that there is a direct mapping between the eigenvalues of the two systems. Thus it is possible to determine the eigenvalues of the KdV Lax operator, and hence the evolution of the KdV wavetrain, by fixing the Lax eigenvalues of the approximating Toda Lattice. These eigenvalues are fixed by initial conditions and remain invariant under the evolution of the dynamical system.

The existence of the N invariants for an N-soliton (N-particle) train suggests that these should be chosen as the modulation parameters, since they retain an exact memory of the initial conditions and are the same after propagation as before, despite other evolutions that might take place in the pulse train as it propagates. This concept was suggested in the context of optical fibre solitons by Hasegawa and Nyu [11], but for the case of noninteracting pulses. Here we see that the concept extends to the interacting case, and to KdV-type wave equations.

It is particularly important in the Inverse Scattering Transform to be able to express invariants in terms of the dynamical variables t_k and p_k ($= d_z t_k$) as well as in terms of the eigenvalues ζ_k of the Lax operator L. For the Toda Lattice the following set of invariants is known from the IST:

$$J_r = \mathrm{Tr}\left[L^r\right] = \sum_{k=1}^{N} (\zeta_k)^r,\ r \in \{1, 2, \ldots\}. \tag{19}$$

The trace values can also be computed directly from (17) in terms of the dynamical variables, so that

$$\begin{aligned} J_1 &= \sum_{k=1}^{N} p_k \\ J_2 &= \sum_{k=1}^{N} \left(p_k^2 + 2\exp[-(t_{k+1} - t_k)]\right) \\ J_r &= \cdots \end{aligned} \tag{20}$$

The concept extends to infinite pulse trains by taking limits as $N \to \infty$, with a mathematical technicality to ensure the convergence of the sums in (20); this consists of an initial renormalisation of the invariants (20) by subtracting from each an amount corresponding to an exactly stationary periodic lattice with period T, and requiring the dynamical variables to satisfy boundary conditions $p_k \to$constant and t_{k+1}-$t_k \to T$ as $k \to \pm\infty$. In the infinite pulse train the discrete eigenvalues ζ_k degenerate to a continuum $\zeta(\theta)$, and the variable θ plays a role similar to the frequency β in the linearised PPM theory of Section 4. A full family of invariants for the infinite Toda Lattice is the set of functionals $J_r = \int \rho(\theta) \sin^r\theta d\theta$, where $\rho(\theta)$ is a function fixed by the intial conditions and which therefore encodes the initial conditions into the pulse stream as data [12].

If the N initial values $\{t_k(0)\}$ are prescribed as data, and $p_k(0)$ = constant, then the eigenvalues ζ_k and the invariants J_r are determined. Conversely, if the $2N$ data $\{p_k, t_k\}$ are detected after a propagation distance z , then the invariants are similarly determined. It is in principle possible to determine the original values at $z = 0$ from the detected values at $z \neq 0$, by running the IST algorithm backwards. However, if the invariants themselves are regarded as the communicated data, then the backward IST algorithm is dispensed with, and only the detection needs to be done. This is called an *invariant-encoded* dynamical system.

An invariant-encoded pulse train consists of Nsolitons, giving rise to N eigenvalues ζ_k each of which can be encoded with one analogue value. An obvious disadvantage with this method of coding is that all N values have to be fixed at $z = 0$ for all time, whereas in practice one would like a dynamical encoding where shorter sequences of pulses can be encoded in a temporally localised manner. That this is possible is actually demonstrated by the example of small-deviation PPM, where the individual frequency components of the PPM are the separate analogue channels belonging to the invariant frequencies β; nevertheless, PPM can be impressed as a time-local modulation at the initial distance $z = 0$, although the time-local property is gradually lost as the pulses propagate due to the dynamical interaction between the pulses. The loss of time-locality is a form of *dispersion*.

A compromise can be reached between the requirements of absolute invariance of the eigenvalues and the need for some time-localisation of a practical modulation scheme by the concept of 'windowing' the channels, as is done in the short-time Fourier transform for the classical linear case. Suppose we divide the eigenvalue continuum of an infinite pulse train into narrow intervals of width $\Delta\theta$; suppose further that we allow the data $\rho(\theta)$ carried in each of these intervals to be changed slowly, over a large time-scale of Δt. It would be intuitively expected that if $\Delta\theta\Delta t>h$, where h is some fixed number, then there is a minimal error involved in reconstructing the initial data from the final received signal, which may even be zero in some classes of data.

6. Conclusions

The dynamics of N-solitons of the KdV equation can be reduced to the Toda Lattice equations for the temporal positions of the solitons when the solitons are nearly identical and widely separated in time. Invariants of the KdV equation map to corresponding invariants of the Toda Lattice, and the map exists in the limit $N\rightarrow\infty$. These invariants form analogue channels for the communication of data in which an exact memory of the initial conditions is preserved, even though the KdV wavetrain itself is evolving dynamically. It follows that there exist channels which are immune to the effects of the interpulse dynamics in the KdV wavetrain.

7. References

1. J. M. Arnold, Complex Toda Lattice and its application to the theory of interacting optical solitons, *J. Opt. Soc. Am. (A)*, **15**, 1450-1458, 1998.
2. J. M. Arnold, Varieties of solitons and solitary waves, *Opt. Quant. Elect.*, to appear in 1998.
3. V. S. Gerdjikov, I. M. Uzunov, E. G. Evstatiev and G. L. Diankov, Nonlinear Schrödinger equation and N-soliton interactions: Generalised Karpman-Solov'ev approach and the complex Toda chain, *Phys. Rev. E.*, **55**, 6039-6060, 1997.
4. V. S. Gerdjikov, D. J. Kaup, I. M. Uzunov and E. G. Evstatiev, Asymptotic behavior of N-soliton trains of the nonlinear Schrödinger equation, *Phys. Rev. Lett.*, **77**, 3943-3946, 1996.
5. K. A. Gorshkov and L. A. Ostrovski, Interactions of solitons in nonintegrable systems: direct perturbation method and applications, *Physica (D)*, **3**, 428-438, 1981.
6. K. A. Gorshkov and V. V. Papko, Dynamic and stochastic oscillations of soliton lattices,*Sov. Phys. JETP*, **46**, 92-97, 1977.
7. A. Gasch, T. Berning and D. Jager, Generation and parametric amplification of solitons in a nonlinear resonator with a Korteweg-de Vries medium, *Phys. Rev. A*, **34**, 4528-4531, 1986.
8. M. J. W. Rodwell, S. T. Allen, R. Y. Yu, M. G. Case, U. Bhattacharya, M. Reddy, E. Carman, M. Kegawa, Y. Konishi, J. Pusl and R. Pullela, Active and nonlinear wave propagation devices in ultrafast electronics and optoelectronics, *Proc. IEEE*, **82**, 1037-1058, 1994.
9. G. J. Ballantyne, P. T. Gough and D. P. Taylor, Baseband soliton oscillator, *Chaos, solitons and fractals*, **5**, 1013-1029, 1995.
10. G. J. Ballantyne, P. T. Gough and D. P. Taylor, Periodic solutions of the Toda Lattice in a loop nonlinear transmission line, *Elect. Lett.*, **29**, 607-609, 1993.
11. A. Hasegawa and T. Nyu, Eigenvalue communication, *J. Light. Tech.*, **11**, 395-399, 1993.
12. L. D. Fadde'ev and L. A. Takhtajan, *Hamiltonian methods in the theory of solitons*, Springer, 1987.

ELECTROMAGNETIC IMPULSE RESPONSE OF TRIPLY-DISTILLED WATER

Judith E. K. Laurens and Kurt E. Oughstun

College of Engineering and Mathematics
University of Vermont
Burlington, VT 05405-0156

INTRODUCTION

As a pulse propagates through a linear dispersive, lossy medium, each spectral component propagates at its own characteristic phase velocity $v_p(\omega) = \omega/\beta(\omega)$ and is attenuated at a rate characterized by its attenuation coefficient $\alpha(\omega) = \Im\{\tilde{k}(\omega)\}$, where $\tilde{k}(\omega) = \beta(\omega) + i\alpha(\omega)$ is the complex wavenumber of a time-harmonic plane wave of angular frequency ω. The relative phases and amplitudes of the spectral components of the initial pulse then change by differing amounts as the propagation distance increases, giving rise to the observed distortion of the propagated pulse. The usual depiction of the dynamical pulse evolution relies upon the approximations associated with the group velocity description[1-4]; however, the accuracy of that approach breaks down as the pulse becomes either ultrashort or ultrawideband and so is inappropriate in this regime[5,6].

The correct description of ultrawideband pulse propagation in a linear, causally dispersive dielectric originated in 1914 with the classical analysis of Sommerfeld[7] and Brillouin[8,9] who used the then newly developed asymptotic method of steepest descents[10] to describe the dynamical evolution of the propagated field due to an input unit step-function modulated signal of constant carrier frequency in a single resonance Lorentz model dielectric. The modern asymptotic description[11-14] has shown the critical role that the precursor fields, originally introduced by Sommerfeld[7] and Brillouin[8,9], play in both the signal arrival and the resultant pulse distortion for ultrawideband fields as they propagate in a single resonance Lorentz medium. This modern asymptotic description is based on Olver's saddle point method[15], a generalization of the method of steepest descents, together with appropriate uniform asymptotic expansion techniques[14], and provides an accurate description of the entire dynamical field evolution in the mature dispersion regime. This modern description has shown that the precursor fields dominate the propagated field structure in dispersive, lossy dielectrics[16-20] in the mature dispersion regime and that they carry significant energy deeper into the lossy, dispersive medium[21] than that predicted by traditional analyses. These precursor fields persist for noninstantaneous rise-time input pulses[5,6,14,22] provided that the initial rise-time (or fall-time) is on the order of or less than the characteristic medium relaxation time. Their potentially profound impact on health and safety studies is just beginning to be appreciated[23].

This asymptotic description is a canonical theory, the most fundamental problems involving the impulse response of the dispersive medium and the propagation of a unit step-function modulated signal through the medium. The theory has primarily considered the single resonance Lorentz model dielectric[5-9,11-14,21] and has only been extended to the double resonance Lorentz model in the special case when the two resonances have similar loss[24]; this extension has shown the formation of an additional precursor field that evolves between the Sommerfeld and Brillouin precursors[6,24]. In addition, numerical results for a single relaxation time Debye model dielectric[16-18] have shown the presence of a Brillouin precursor, but the absence of any Sommerfeld precursor in the propagated field structure due to an input ultrawideband microwave pulse. The present analysis provides an extension of these results to a causal model of the dielectric dispersion of triply-distilled H_2O whose orientational polarization dispersion is described by several relaxation times and whose electronic polarization dispersion contains several resonance lines with dissimilar loss coefficients[25].

The exact integral representation of the propagated plane wave field in the half-space $z \geq 0$ is given by[14]

$$A(z,t) = \frac{1}{2\pi}\int_C \tilde{f}(\omega)\exp\left\{\frac{z}{c}\phi(\omega,\theta)\right\}d\omega \quad , \tag{1}$$

where $\tilde{f}(\omega)$ is the temporal Fourier spectrum of the initial pulse $f(t) = A(0,t)$ at the plane $z = 0$. Here $A(z,t)$ represents any scalar component of the electric or magnetic field vector whose spectrum $\tilde{A}(z,\omega)$ satisfies the scalar Helmholtz equation

$$\left(\nabla^2 + \tilde{k}^2(\omega)\right)\tilde{A}(z,\omega) = 0 \quad . \tag{2}$$

The complex wavenumber appearing here is given by

$$\tilde{k}(\omega) = \omega n(\omega)/c \quad , \tag{3}$$

with $\beta(\omega) = \Re\{\tilde{k}(\omega)\}$ the plane wave propagation factor and $\alpha(\omega) = \Im\{\tilde{k}(\omega)\}$ the plane wave attenuation factor, where c denotes the speed of light in vacuum, and where $n(\omega) = [\mu\varepsilon(\omega)]^{1/2}$ is the complex index of refraction of the dispersive medium occupying the half-space $z \geq 0$ with relative magnetic permeability μ and complex-valued, relative dielectric permittivity $\varepsilon(\omega) = \varepsilon_r(\omega) + i\varepsilon_i(\omega)$. If $f(t) = 0$ for $t < 0$, then Eq. (1) is taken to be a Laplace representation for which the contour of integration C in the complex ω-plane is the Bromwich contour[14] that is given by the straight line $\omega = \omega' + ia$ with a being a real-valued constant that is greater than the abscissa of absolute convergence for the function $f(t)$, and where $\omega' \equiv \Re\{\omega\}$ varies from negative to positive infinity. The complex phase function $\phi(\omega,\theta)$ appearing in Eq. (1) is given by

$$\phi(\omega,\theta) = i\frac{c}{z}\left[\tilde{k}(\omega)z - \omega t\right] = i\omega\left[n(\omega) - \theta\right] \quad , \tag{4}$$

where

$$\theta = ct/z \tag{5}$$

is a dimensionless parameter that characterizes any given space-time point in the propagated field. The asymptotic theory provides a uniformly valid description of the temporal evolution of the field at any sufficiently large, fixed value of the propagation distance z .

A dielectric material of central importance to bioelectromagnetic effects is triply-distilled water[16-18,23], as it forms the base for many biological systems. Representative experimental data points for the real frequency dispersion of both the real index of refraction $n_r(\omega) = \Re\{n(\omega)\}$ and the absorption coefficient $\alpha(\omega) = (\omega/c)n_i(\omega)$ with $n_i(\omega) = \Im\{n(\omega)\}$ of triply-distilled water are illustrated in Fig. 1. The measured low frequency dependence of $n_r(\omega)$ for $0 \le \omega/2\pi \le 10^{13}\,Hz$ is characteristic of rotational polarization in polar dielectrics while the high frequency dependence observed for $\omega/2\pi > 10^{13}\,Hz$ is characteristic of resonance polarization in dielectrics[26]. The frequency dispersion due to rotational polarization is described here using the Rocard-Powles model[27], which is a first-order correction to the classic Debye model[28,29] of polar media, while the frequency dispersion due to resonance polarization is described here using the classical Lorentz oscillator model[30,31]. Both of these physical models are causal.

A composite Rocard-Powles-Lorentz model[25] with two relaxation times and four separate resonance frequencies that is appropriate for the observed frequency dispersion of the dielectric permittivity of triply-distilled water is given by

$$\varepsilon(\omega) = \varepsilon_\infty + \sum_{j=0,2} \frac{a_j}{\left(1 - i\omega\tau_j\right)\left(1 - i\omega\tau_{fj}\right)} - \sum_{j=11,13,15,17} \frac{b_j^2}{\omega^2 - \omega_j^2 + 2i\delta_j\omega} \quad . \tag{6}$$

Here τ_j is the macroscopic Debye relaxation time and τ_{fj} is the associated friction time in the Rocard-Powles model. In the infinitely large friction limit, each friction time vanishes and the first summation in Eq. (6) reduces to the expression for the dielectric permittivity in the Debye theory[25,28,29]. In the second summation over resonance polarization terms, each Lorentz model resonance line is characterized by the parameter triple $(\omega_j, b_j, \delta_j)$, where ω_j is the undamped resonance frequency, b_j the plasma frequency, and δ_j the phenomenological damping constant associated with the jth Lorentz oscillator type[25,30,31]. Finally, the quantity ε_∞ denotes the value of the dielectric permittivity in the limit as $\omega \to \infty$.

The medium parameters for the Rocard-Powles model component of Eq. (6) were fit to the data points first over the frequency domain $0 \le \omega/2\pi \le 10^{13}\,Hz$ using just the first three terms on the right-hand side of Eq. (6) with ε_∞ replaced by the parameter $\varepsilon_{\infty rot}$ which is appropriate for this model component[25]. The rms best fit result is depicted by the solid curves in Fig. 2 along with the associated parameter values. The dashed curves in the figure shows the rms best fit Debye model result $(\tau_{fj} = 0, j = 0,2)$ for comparison. Notice that when $\omega > 1/\tau_2 \approx 6\pi \times 10^{12}\,Hz$, the theoretical Debye model absorption curve rapidly approaches a constant limiting value that is known as the Debye plateau[27]; the inadequacy of this model for all $\omega > 1/\tau_2$ is clearly evident.

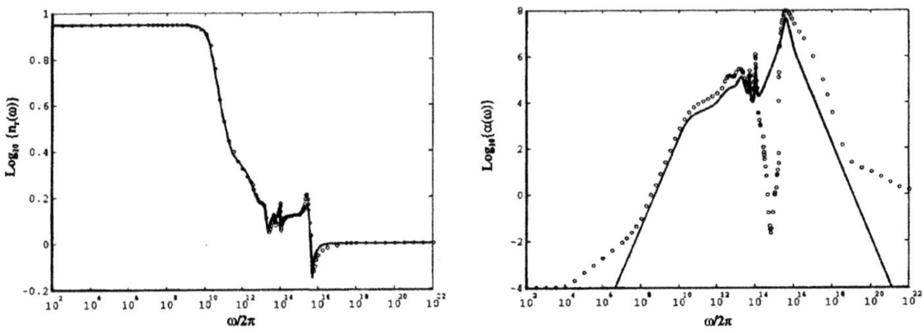

Figure 1. Experimental values (open circles) and optimized theoretical behavior (solid curves) of the frequency dispersion of the real index of refraction $n_r(\omega)$ and the absorption coefficient $\alpha(\omega)$, in units of m^{-1}, of triply-distilled water over the frequency domain from $\omega/2\pi = 1\times10^2\,Hz$ to $\omega/2\pi = 1\times10^{22}\,Hz$. (Data supplied by the School of Aerospace Medicine, Armstrong Laboratory, Brooks Air Force Base).

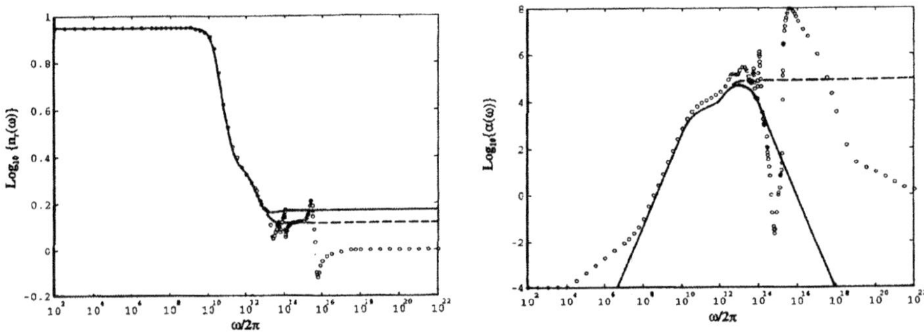

Figure 2. Experimental values (open circles) and rms best fit behavior of the frequency dispersion of the real index of refraction $n_r(\omega)$ and the absorption coefficient $\alpha(\omega)$ of triply-distilled water for the double relaxation time Rocard-Powles model (solid curves) with $\varepsilon_{\infty_{rot}} = 2.1$, $a_0 = 74.1$, $\tau_0 = (5.3/2\pi)\times10^{-11}\,\text{sec}$, $\tau_{f0} = (2.9/2\pi)\times10^{-13}\,\text{sec}$, $a_2 = 3.0$, $\tau_2 = (4.1/2\pi)\times10^{-13}\,\text{sec}$, and $\tau_{f2} = (9.0/2\pi)\times10^{-15}\,\text{sec}$, and for the double relaxation time Debye polarization model (dashed curves) with parameters $\varepsilon_{\infty_{rot}} = 1.7$, $a_0 = 74.7$, $\tau_0 = (5.3/2\pi)\times10^{-11}\,\text{sec}$, $a_2 = 2.8$, and $\tau_2 = (3.0/2\pi)\times10^{-13}\,\text{sec}$.

The medium parameters for the Lorentz model component of Eq.(6) were then fit to the data points by introducing each separate Lorentz model resonance line and determining its rms best fit parameters over a narrow frequency domain about that particular resonance frequency. Because the frequency dependence of each individual component of the full model extends over the entire frequency domain, the addition of each new resonance feature alters the rms fit of previously introduced features. As a consequence, the optimization procedure must be repeatedly applied so as to obtain the optimum rms fit for the entire model given in Eq.(6).

This was done for the Rocard-Powles model parameters whose values were used in Fig. 2. The final optimized result in the infrared to ultraviolet frequency domain is illustrated in Fig.3 where the location of each Lorentz model resonance frequency is indicated. Notice that the absorption due to the ultraviolet Lorentz resonance line at ω_{17} fills up the remaining half of the visible window in water. The optimized values of the Rocard-Powles-Lorentz model medium parameters appearing in Eq.(6) are given in Table 1. These parameter values are in good agreement with published results[32] for a simpler composite Debye-Lorentz model of the complex permittivity of water below $1THz$.

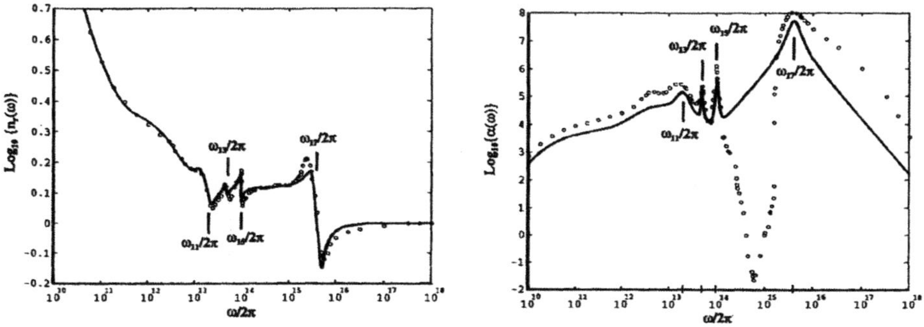

Figure 3. Experimental values (open circles) and optimized theoretical behavior (solid curves) of the frequency dispersion of the real index of refraction $n_r(\omega)$ and the absorption coefficient $\alpha(\omega)$ of triply-distilled water for the full Rocard-Powles-Lorentz polarization model with medium parameters as given in Table 1. The locations of the four Lorentz resonance frequencies are indicated in each diagram.

A comparison of the theoretical behavior and the experimental data of the real index of refraction and absorption for triply-distilled water over the entire frequency domain of interest is illustrated in Fig. 1. Notice that the experimental absorption data forms an upper bound to the theoretical curve over all but the visible region of the electromagnetic spectrum. Since the Rocard-Powles-Lorentz model (6) of the dielectric dispersion is strictly causal, it is then seen that the experimental absorption data for triply-distilled water is overestimated in both the low $(0 \le \omega < 2\pi \times 10^8 Hz)$ and high $(\omega > 2\pi \times 10^{18} Hz)$ frequency domains. In its present form, Eq.(6) is seen to provide an accurate model of the dielectric dispersion of triply-distilled water from the static $(\omega = 0)$ through the infrared $(\omega \approx 2\pi \times 10^{14} Hz)$ region of the electromagnetic spectrum. The most notable shortfall of this model is the near complete loss of the visible window, which is nearly equally due to both the rotational and resonance portions of the model.

IMPULSE RESPONSE OF THE ROCARD-POWLES-LORENTZ MODEL OF THE DIELECTRIC DISPERSION OF TRIPLY-DISTILLED WATER

The integral representation of the impulse response of the model dielectric is obtained from Eq.(1) with $\tilde{f}(\omega) = 1$ as

$$A(z,t) = \frac{1}{2\pi}\int_C \exp\left\{\frac{z}{c}\phi(\omega,\theta)\right\}d\omega \quad . \tag{7}$$

An exact evaluation of Eq. (7) may be achieved by application of Cauchy's residue theorem to that integral taken over the closed contour $C_O = C + C_R$ if the integral over the semi-circular arc C_R at infinity vanishes. From the limiting behavior

$$\lim_{|\omega|\to\infty}\{n(\omega)\} = \sqrt{\varepsilon_\infty} \quad , \tag{8}$$

obtained from Eq. (6) for any value of $\arg(\omega)$, the integrand appearing in Eq.(7) takes on the limiting form

$$\exp\left\{\frac{z}{c}\phi(\omega,\theta)\right\} \to \exp\left\{\frac{z}{c}|\omega|\left[(-\sin\xi + i\cos\xi)\left(\varepsilon_\infty^{1/2} - \theta\right)\right]\right\} \tag{9}$$

as $|\omega| \to \infty$, where $\omega = |\omega|\exp(i\xi)$. If the inequality $\varepsilon_\infty^{1/2} - \theta > 0$ is satisfied, then the contour may be completed by taking C_R as the semi-circular arc in the upper-half plane since the integrand (9) goes to zero exponentially along C_R as its radius $|\omega| \to \infty$. Since the integrand in Eq.(7) has no poles in the upper-half plane, then the residue theorem yields the exact result

$$A(z,t) = 0 \quad , \theta < \varepsilon_\infty^{1/2} \quad . \tag{10}$$

For $\theta > \varepsilon_\infty^{1/2}$ the integral representation (7) of the propagated field may be evaluated either numerically or by using asymptotic methods of analysis.

The complete precursor field structure that characterizes the electromagnetic impulse response of the Rocard-Powles-Lorentz model of the frequency dispersion of the dielectric permittivity of triply-distilled water, as given by Eq.(6) with the medium parameters listed in Table 1, is illustrated in the sequence of plots given in Fig. 4. These results were obtained using a numerical inverse Laplace transform algorithm[33] applied to the integral representation (7) at the fixed propagation distance $z = 6\pi \times 10^{-5} m$, which is sufficiently large to place the field evolution in the mature dispersion regime[11-14], and are displayed as a function of the dimensionless space-time parameter $\theta = ct/z$ which, at fixed z, is a dimensionless time parameter.

The overall structure of the propagated field evolution illustrated in Fig. 4(a) is quite similar to that obtained for either a single[11-14] or a double[13,24] resonance Lorentz model dielectric. The front of the field arrives at the speed of light in vacuum with an infinite instantaneous oscillation frequency that chirps down towards the upper end of the upper absorption band modeled in Eq.(6), and is referred to as a Sommerfeld precursor[20]. This is followed by a quasi-static field whose amplitude rapidly builds to a near-peak value that decays with the propagation distance only as $z^{-1/2}$, and is referred to as a Brillouin precursor[20]. However, the dynamical field evolution following the main peak in the Brillouin precursor, illustrated in Figs. 4(b,c,d), is quite different from that observed in previous studies for a Lorentz model medium. In addition, the peak value in the Brillouin precursor should appear[18,20] at or very near to the space-time point $\theta_0 \equiv n(0)$ which has the numerical value $n(0) \cong 8.90$ for triply-distilled water. Notice also in Fig. 4(a) that the oscillatory relaxation of the Brillouin precursor has been quenched. The detailed field evolution depicted in Fig. 4(b) for intermediate values of θ clearly exhibits the presence of interference between two distinct frequency components that decay at essentially the same rate, while Fig. 4(c) shows the gradual buildup of a quasi-static field component. The detailed field evolution depicted in Fig. 4(d) for intermediate to large values of θ clearly shows the long time exponential decay of the impulse response in triply-distilled water. Each of these new and uniquely interesting features is explained in the remainder of this paper using asymptotic methods of analysis applied to the impulse response of the separate rotational and resonance portions of the composite Rocard-Powles-Lorentz model of the dielectric dispersion given in Eq.(6).

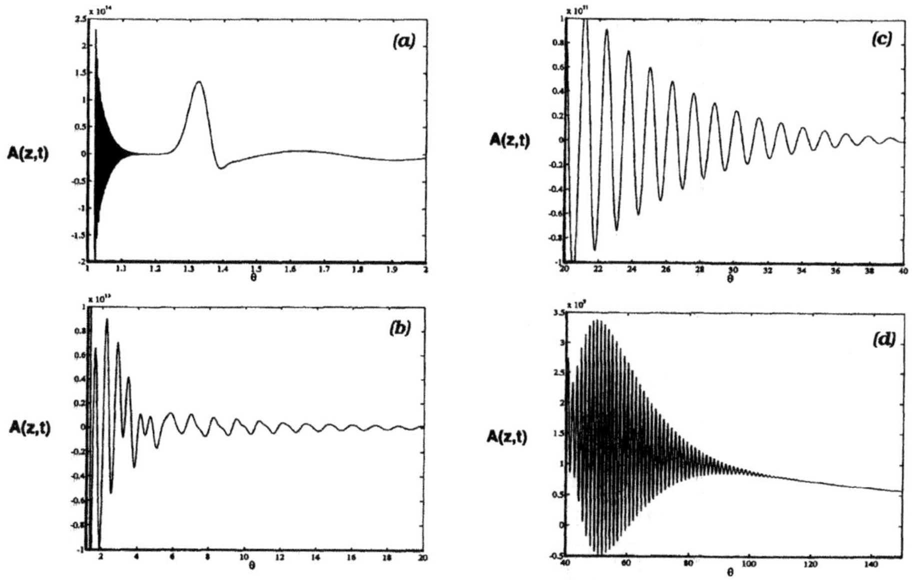

Figure 4. Dynamical evolution of the propagated field due to an input delta function pulse in the full Rocard-Powles-Lorentz model of triply-distilled water at the fixed propagation distance $z = 6\pi \times 10^{-5} m$ for small through large values of the space-time parameter $\theta = ct/z$.

ASYMPTOTIC DESCRIPTION OF THE IMPULSE RESPONSE OF A SINGLE RELAXATION TIME ROCARD-POWLES MODEL DIELECTRIC

Complex Phase Behavior in the Complex ω-Plane

The single relaxation time Rocard-Powles model of the orientational polarization response of a dielectric results in a complex-valued relative dielectric permittivity whose frequency dispersion is described by[25,27]

$$\varepsilon(\omega) = \varepsilon_{\infty_{rot}} + \frac{a_0}{(1 - i\omega\tau_0)(1 - i\omega\tau_{f0})} \quad , \qquad (11)$$

where $\varepsilon_{\infty_{rot}} = \varepsilon(\infty)$, $\varepsilon_s = \varepsilon(0)$, $a_0 = \varepsilon_s - \varepsilon_{\infty_{rot}}$, τ_0 is the macroscopic relaxation time, and τ_{f0} is the friction time with $\tau_{f0} < \tau_0$. The corresponding complex index of refraction is then given by

$$n(\omega)=\left[\frac{\varepsilon_{\infty_{rot}}+a_0-i\varepsilon_{\infty_{rot}}\left(\tau_0+\tau_{f0}\right)\omega-\varepsilon_{\infty_{rot}}\tau_0\tau_{f0}\omega^2}{1-i\left(\tau_0+\tau_{f0}\right)\omega-\tau_0\tau_{f0}\omega^2}\right]^{1/2} , \tag{12}$$

where $\mu/\mu_0=1$. For a single relaxation time Debye model dielectric, $\tau_{f0}=0$.

The set of branch points of $n(\omega)$ are comprised of the pair of branch point singularities and branch point zeroes

$$\omega_{p1}=-\frac{i}{\tau_{f0}} \quad , \quad \omega_{p2}=-\frac{i}{\tau_0} \quad , \tag{13a}$$

$$\omega_{z1,2}=-i\frac{\tau_0+\tau_{f0}}{2\tau_0\tau_{f0}}\mp\frac{1}{2\tau_0\tau_{f0}}\left[4\frac{\varepsilon_{\infty_{rot}}+a_0}{\varepsilon_{\infty_{rot}}}\tau_0\tau_{f0}-\left(\tau_0+\tau_{f0}\right)^2\right]^{1/2} , \tag{13b}$$

respectively. The two branch point zeroes coalesce on the negative imaginary axis when the friction time τ_{f0} equals the critical value

$$\tau_{f0}^c\equiv\tau_0\left\{\left(2\frac{\varepsilon_{\infty_{rot}}+a_0}{\varepsilon_{\infty_{rot}}}-1\right)-\sqrt{\left(2\frac{\varepsilon_{\infty_{rot}}+a_0}{\varepsilon_{\infty_{rot}}}-1\right)^2-1}\right\} . \tag{14}$$

The branch cut $\left[\omega_{z2},\omega_{p2}\right]$ associated with the relaxation time τ_0 is situated along the negative imaginary axis for $\tau_{f0}\le\tau_{f0}^c$. The branch cut $\left[\omega_{p1},\omega_{z1}\right]$ that is associated with the friction time is located on the imaginary axis at $-i\infty$ when $\tau_{f0}=0$ (in which limit the Debye model is obtained) and moves up the imaginary axis and approaches the lower end of the branch cut $\left[\omega_{z2},\omega_{p2}\right]$ as τ_{f0} increases toward the critical value τ_{f0}^c. When $\tau_{f0}=\tau_{f0}^c$, the two branch point zeroes coalesce at the point $\omega_{z1}=\omega_{z2}=-i\left(\tau_0+\tau_{f0}\right)/\left(\tau_0\tau_{f0}\right)$. Finally, when $\tau_{f0}>\tau_{f0}^c$ the branch point zeros move off the imaginary axis and are located at

$$\omega_{z1,2}=\frac{1}{2\tau_0\tau_{f0}}\left\{\mp\sqrt{4\frac{\varepsilon_{\infty_{rot}}+a_0}{\varepsilon_{\infty_{rot}}}\tau_0\tau_{f0}-\left(\tau_0+\tau_{f0}\right)^2}-i\left(\tau_0+\tau_{f0}\right)\right\} . \tag{15}$$

The branch cuts are now the line segments $\left[\omega_{p1},\omega_{p2}\right]$ and $\left[\omega_{z1},\omega_{z2}\right]$.

For the particular case of water with the Rocard-Powles model parameter values given in Fig.2, the branch points and their associated branch cuts are found to be situated along the imaginary axis. For notational simplicity the expressions for the branch point singularities and zeros in this case may be respectively expressed as

$$\omega_{p1,2}=-i\frac{\tau_p\pm c_p}{2\tau_m^2} \quad , \quad \omega_{z1,2}=-i\frac{\tau_p\mp c_z}{2\tau_m^2} \quad , \tag{16}$$

where $\tau_p\equiv\tau_0+\tau_{f0}$, $\tau_m^2\equiv\tau_0\tau_{f0}$, $c_p\equiv\sqrt{\tau_p^2-4\tau_m^2}$, and $c_z\equiv\sqrt{\tau_p^2-4\tau_m^2\left(1+a_0/\varepsilon_{\infty_{rot}}\right)}$ are all real-valued constants of the medium. This particular case is assumed to hold throughout the remaining analysis. Other cases may be treated in a similar manner when the medium parameters warrant.

Saddle Point Dynamics of the Complex Phase Function

The saddle points of the complex phase function are located in the complex ω -plane at those values of ω that satisfy the saddle point equation $\phi'(\omega,\theta)=0$, which results in the general expression

$$n(\omega)+\omega\frac{\partial n(\omega)}{\partial\omega}-\theta=0 \quad , \tag{17}$$

so that the saddle point locations evolve with the space-time parameter θ. For the single relaxation time Rocard-Powles model dielectric, the saddle point equation becomes intractable and approximate solutions must be obtained in specific regions of the complex ω -plane[25].

For large $|\omega| >> 1/\tau_m$, the complex index of refraction (12) may be approximated as

$$n(\omega)\approx\varepsilon_{\infty_{rot}}^{1/2}-\frac{a_0}{2\varepsilon_{\infty_{rot}}^{1/2}\tau_m^2\omega\left(\omega+i\tau_p/\tau_m^2\right)} \quad . \tag{18}$$

With this substitution the saddle point equation (17) yields the pair of approximate solutions

$$\omega_{SP_D}^{\pm}(\theta)\approx\pm\left[\frac{a_0}{2\varepsilon_{\infty_{rot}}^{1/2}\tau_m^2\left(\theta-\varepsilon_{\infty_{rot}}^{1/2}\right)}\right]^{1/2}-i\frac{\tau_p}{\tau_m^2} \quad , \tag{19}$$

which are found[25] to be accurate for all $\theta\geq\sqrt{\varepsilon_{\infty_{rot}}}$. Because this pair of distant saddle points remain in the lower-half plane below the branch cut $[\omega_{z2},\omega_{p2}]$, they are inaccessible and provide no contribution to the asymptotic behavior of the propagated field.

For small $|\omega| << |\omega_{p2}|$, the complex index of refraction (12) may be approximated by the quadratic expression[25]

$$n(\omega)\approx\frac{a_0\tau_m^2}{2\varepsilon_s^{1/2}}\left[1-\frac{\tau_p^2}{4\varepsilon_s\tau_m^2}\left(\varepsilon_{\infty_{rot}}+3\varepsilon_s\right)\right]\omega^2+i\frac{a_0\tau_p}{2\varepsilon_s^{1/2}}\omega+\varepsilon_s^{1/2} \quad . \tag{20}$$

With this substitution the saddle point equation (17) yields the roots

$$\omega^{\pm}(\theta)\approx\frac{2\varepsilon_s^{1/2}}{3a_0\tau_m^2\left[1-\frac{\tau_p^2}{4\varepsilon_s\tau_m^2}\left(\varepsilon_{\infty_{rot}}+3\varepsilon_s\right)\right]}$$

$$\left\{-i\frac{a_0\tau_p}{2\varepsilon_s^{1/2}}\pm\sqrt{\frac{3a_0\tau_m^2}{2\varepsilon_s^{1/2}}\left[1-\frac{\tau_p^2}{4\varepsilon_s\tau_m^2}\left(\varepsilon_{\infty_{rot}}+3\varepsilon_s\right)\right]\left(\theta-\theta_{RP}\right)}\right\} \quad , \tag{21}$$

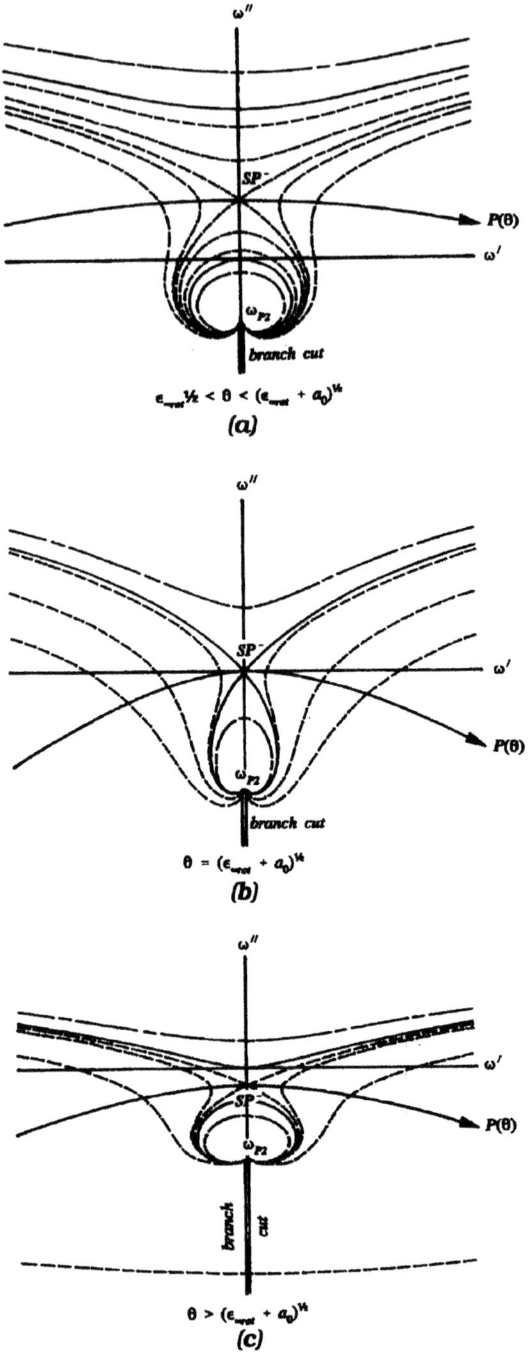

Figure 5. Topography of the isotimic contours of $X(\omega,\theta) = \Re\{\phi(\omega,\theta)\}$ in the complex ω-plane for a single relaxation time Rocard-Powles model dielectric with $\tau_{f0} < \tau_{f0}^{c}$ for several fixed, increasing values of the space-time parameter $\theta = ct/z$. The contour $P(\theta)$ is an Olver-type path with respect to the dominant saddle point at each value of θ depicted.

where

$$\theta_{RP} = \varepsilon_s^{1/2}\left[1 + \frac{a_0}{6\varepsilon_s \frac{\tau_m^2}{\tau_p^2} - \frac{3}{2}\left(\varepsilon_{\infty_{rot}} + \varepsilon_s\right)}\right] \quad . \tag{22}$$

The required inequality $|\omega^{\pm}(\theta)| << |\omega_{p2}|$ is not satisfied for $\theta \leq \theta_{RP}$ and so neither root given in Eq.(21) yields a saddle point in that space-time domain. For $\theta > \theta_{RP}$ only the downward moving root $\omega^{-}(\theta)$ satisfies the required inequality, and so one obtains a single saddle point solution $\omega_{SP}^{-}(\theta) = \omega^{-}(\theta)$ for which the approximation given in Eq.(21) is valid over the space-time domain

$$\theta_{RP} < \theta < \varepsilon_s^{1/2} + \frac{a_0 \tau_p^2}{6\varepsilon_s^{1/2}\tau_m^2\left[1 - \frac{\tau_p^2}{4\varepsilon_s \tau_m^2}\left(\varepsilon_{\infty_{rot}} + 3\varepsilon_s\right)\right]} \quad . \tag{23}$$

For the Rocard-Powles medium parameters given in Fig.2, this inequality becomes $7 < \theta < 10.77$, which agrees with the results of numerical studies[25]. The dynamical evolution of the saddle point location $\omega_{SP}^{-}(\theta)$ is illustrated in Fig. 5 over the space-time domain $\theta \geq \sqrt{\varepsilon_{\infty_{rot}}}$. At $\theta = \sqrt{\varepsilon_{\infty_{rot}}}$ this first-order saddle point is located in the upper-half plane along the imaginary axis and moves down the imaginary axis as θ increases, as seen in Fig. 5(a). At $\theta = n(0) = \sqrt{\varepsilon_{\infty_{rot}} + a_0}$ this saddle point crosses the origin, as seen in Fig. 5(b), and it approaches the branch point ω_{p2} in the limit as $\theta \to \infty$, as seen in Fig. 5(c). Similar behavior for this near saddle point is obtained for the single relaxation time Debye model dielectric[25].

Impulse Response of the Rocard-Powles Model Dielectric

The asymptotic description of the propagated field due to a delta function input in a single relaxation time Rocard-Powles model dielectric over the space-time domain $\theta > \varepsilon_{\infty_{rot}}^{1/2}$ is obtained through application of Olver's theorem[15] to the integral representation (7) when the contour C is deformed to an Olver-type path $P(\theta)$ with respect to the dominant, accessible saddle point SP^{-} with location $\omega_{SP}^{-}(\theta)$, as depicted in Fig. 5, with the result[14]

$$A(z,t) \sim \Re\left\{\left[-\frac{c}{2\pi z \phi''\left(\omega_{SP}^{-},\theta\right)}\right]^{1/2} e^{\frac{z}{c}\phi\left(\omega_{SP}^{-},\theta\right)}\right\} \quad , \ \theta > \varepsilon_{\infty_{rot}}^{1/2} \quad , \tag{24}$$

as $z \to \infty$. Numerical and asymptotic calculations of the impulse response of the Rocard-Powles model dielectric whose medium parameters are given in Fig.2 at a fixed propagation distance of $z = 0.25m$ are essentially identical over the entire θ-domain of interest and are both illustrated by the solid curve in Fig. 6. The observed dynamical field evolution is completely explained by the saddle point dynamics illustrated in Fig. 5 which describe a nonoscillatory field whose amplitude first increases over the space-time domain $1.77 \leq \theta < 8.89$ and reaches a maximum value at $\theta = n(0) = \sqrt{\varepsilon_s} = 8.89$ when the saddle point crosses the origin, at which space-time point the propagated field has no exponential decay and only decreases algebraically as $z^{-1/2}$, and whose amplitude then decreases for increasing $\theta > 8.89$ as the saddle point moves into the lower-half plane and approaches the branch point singularity at ω_{p2}. For comparison, the numerically determined impulse response of the corresponding single relaxation time Debye model dielectric is depicted by the dashed curve in Fig.6. The observed close agreement is a simple consequence of the similarity of the dominant saddle point dynamics for each medium model[25]. The identification of this temporally transient field structure as a Brillouin precursor follows from the analysis presented in Ref. 20. Although the Debye and Rocard-Powles models predict nearly identical Brillouin precursors, both of which are in approximate qualitative agreement with the Brillouin precursor exhibited in the impulse response of the complete dielectric permittivity model of water given in Fig. 4(a), they occur at a space-time point θ that differs substantially from that obtained with the complete model.

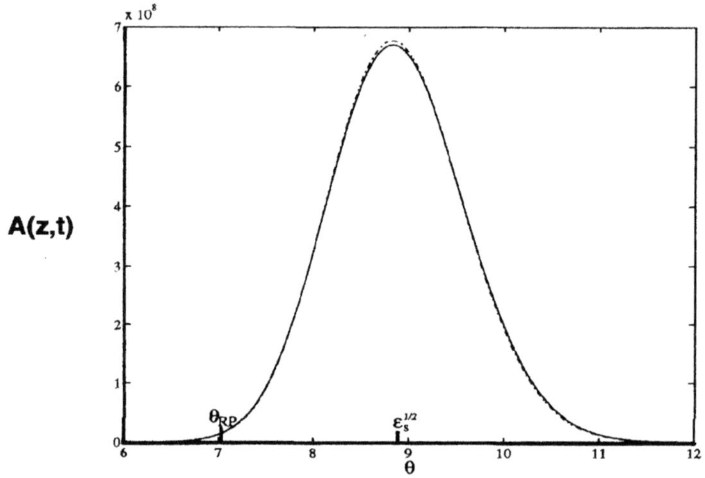

Figure 6. Dynamical field evolution due to an input delta function pulse at the fixed propagation distance $z = 0.25m$ over the space-time domain $\theta > \varepsilon_{\infty_{rot}}^{1/2}$ in the single relaxation time Debye (dashed curve) and Rocard-Powles (solid curve) models of the dielectric dispersion of triply-distilled water, where $\varepsilon_s^{1/2} = 8.89$ for both models.

ASYMPTOTIC DESCRIPTION OF THE IMPULSE RESPONSE OF A MULTIPLE RESONANCE LORENTZ MODEL DIELECTRIC

Complex Phase Behavior in the Complex ω-Plane

The Lorentz model for the resonance polarization response of a dielectric whose frequency dispersion is given by the second summation appearing in Eq.(2.1) results in the complex index of refraction

$$n(\omega)=\left[\varepsilon_\infty-\frac{b_{11}^2}{\omega^2-\omega_{11}^2+i2\delta_{11}\omega}-\frac{b_{13}^2}{\omega^2-\omega_{13}^2+i2\delta_{13}\omega}-\frac{b_{15}^2}{\omega^2-\omega_{15}^2+i2\delta_{15}\omega}-\frac{b_{17}^2}{\omega^2-\omega_{17}^2+i2\delta_{17}\omega}\right]^{1/2}, \tag{25}$$

where $\mu/\mu_0=1$, and where ε_∞ is the limiting value of the dielectric permittivity as $\omega\to\infty$. These four Lorentz lines with the medium parameter values given in Table 1, which include the two resonance lines ω_{11} and ω_{13} in the far infrared, the single resonance line ω_{15} in the near infrared, and the single resonance line ω_{17} in the near ultraviolet regions of the electromagnetic spectrum, provide a fairly complete description of the medium response due to the prominent resonance features appearing in the frequency dispersion of triply-distilled water.

The branch point singularities of $n(\omega)$ are given by the set of relations

$$\omega_{j_p}^{\pm}=\pm\left(\omega_j^2-\delta_j^2\right)^{1/2}-i\delta_j \quad , \ j=11,13,15,17 \quad , \tag{26}$$

while the locations of the branch point zeroes of the multiple resonance Lorentz model medium are given by the zeroes of the argument of the square root expression appearing in Eq.(25) and are denoted by $\omega_{j_z}^{\pm}$, $j=11,13,15,17$. The branch cuts are taken as the set of line segments $\left[\omega_{j_z}^{-},\omega_{j_p}^{-}\right]$ and $\left[\omega_{j_p}^{+},\omega_{j_z}^{+}\right]$, $j=11,13,15,17$, that are pair-wise symmetrically located about the imaginary axis, as seen in Fig. 7.

Saddle Point Dynamics of the Complex Phase Function

The saddle point dynamics for the multiple resonance Lorentz medium considered here are obtained from Eq.(17) with the complex index of refraction (25). This results in a 32nd-order polynomial equation in ω which yields 16 saddle points; the analytical solution of this equation is intractable so that numerical solutions must be relied upon.

The numerically determined paths followed by the 16 saddle points of the four resonance Lorentz model for triply-distilled water in the complex ω-plane over the space-time domain $\theta\geq\sqrt{\varepsilon_\infty}$ are presented in Fig. 7. Part (a) of that figure depicts a region of the plane that is appropriate for the saddle points that evolve in a region removed from the origin, while part (b) depicts a region that is appropriate for the saddle points that evolve in a region about the origin. The direction that each saddle point $\omega_{SP_j}(\theta)$ moves along its path as θ increases is indicated by the arrowhead along that path. Notice that the two weak resonance lines ω_{11} and ω_{13} in the far infrared region create an extremely complicated motion of the saddle points in the region about the origin and that this, in turn, will result in a corresponding intricate behavior in the dynamical evolution of the propagated field. The behavior of $X\left(\omega_{SP_j},\theta\right)=\Re\left\{\phi\left(\omega_{SP_j},\theta\right)\right\}$ at each saddle point for $\theta\geq\sqrt{\varepsilon_\infty}$ is illustrated in Fig. 8. Part (a) presents the overall behavior of each function $X\left(\omega_{SP_j}(\theta),\theta\right)$, while part (b) presents a detailed view of the significant order and dominance changes that occur at various critical values of θ. In order to unambiguously establish both the dominance and the accessibility of each of the saddle points over the entire space-time domain $\theta \geq \sqrt{\varepsilon_\infty}$, it is necessary to consider the general topography of the isotimic contours of $X(\omega,\theta)$ in the complex ω-plane at each critical value of θ that marks a change in the order, the dominance, or the accessibility of any of the relevant saddle points, as determined from Fig. 8.

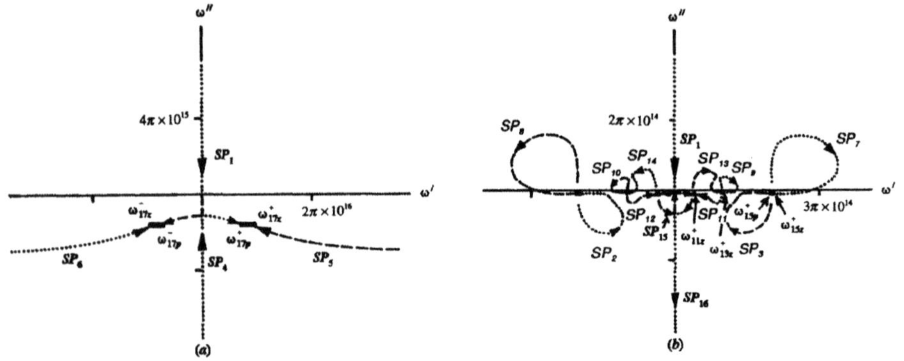

Figure 7. Saddle point trajectories in the complex ω-plane as a function of the space-time parameter $\theta \geq \varepsilon_\infty^{1/2}$ for the four resonance Lorentz model dielectric with parameters appropriate for triply-distilled water as given in Table 1. The branch cuts of $\phi(\omega,\theta)$ are indicated by the double line segments, and the arrowheads indicate the direction that a particular saddle point moves as θ increases.

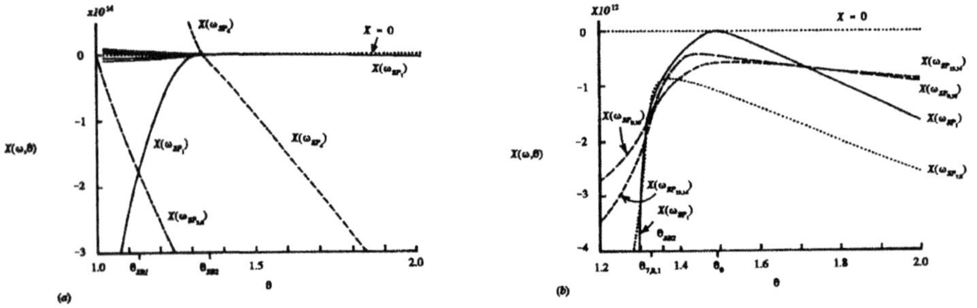

Figure 8. Behavior of $X(\omega_{SP_j},\theta) = \Re\{\phi(\omega_{SP_j},\theta)\}$ as a function of the space-time parameter $\theta \geq \varepsilon_\infty^{1/2}$ at the saddle points of the four resonance Lorentz model dielectric with parameters appropriate for triply-distilled water as given in Table 1.

The topography of the isotimic contours of $X(\omega,\theta)$ illustrated in Fig. 9(a) shows that the distant saddle point pair $SP_{5,6}$ is dominant and accessible by an Olver-type path $P(\theta)$ over the space-time domain $\varepsilon_\infty^{1/2} \leq \theta < \theta_{SB_1}$ and determines the initial behavior of the propagated field. At $\theta = \varepsilon_\infty^{1/2}$ this distant saddle point pair is located at $\pm\infty - i2\delta_{17}$ and from there moves in and approaches the respective branch point zeroes $\omega_{17_z}^{\pm}$ in the limit as $\theta \to \infty$.

When $\theta = \theta_{SB_1} \cong 1.133$, the distant saddle point pair $SP_{5,6}$ and the near saddle point SP_1 are of equal dominance, as seen in Fig. 9(b). The near saddle point SP_1 moves down the positive imaginary axis and remains the dominant accessible saddle point over the space-time interval $\theta_{SB_1} < \theta < \theta_{7,1}$ with $\theta_{7,1} \cong 1.3235$, as illustrated in Fig. 10. This near saddle point continues to move down the imaginary axis as θ increases and crosses the origin at $\theta = \theta_0 = 1.49$. Notice that the pair of saddle points $SP_{7,8}$ become dominant over the near saddle point SP_1 at $\theta = \theta_{1,7} \cong 1.303$ but are inaccessible because there is no Olver-type path with respect to them.

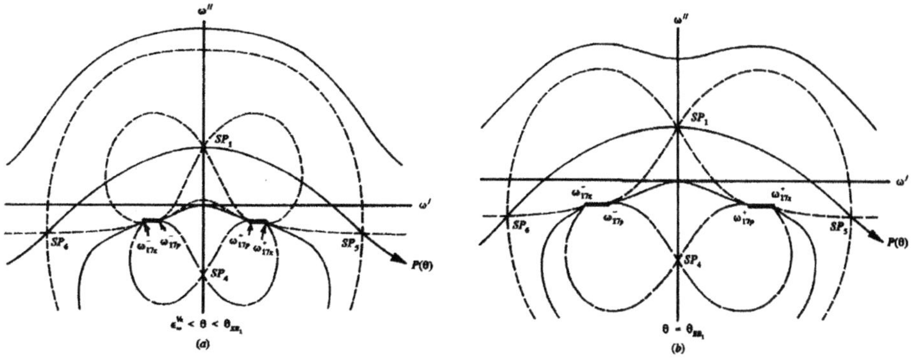

Figure 9. Topography of the isotimic contours of $X(\omega,\theta) = \Re\{\phi(\omega,\theta)\}$ in the complex ω-plane for large $|\omega|$ for the four resonance Lorentz model dielectric with parameters appropriate for triply-distilled water as given in Table 1.

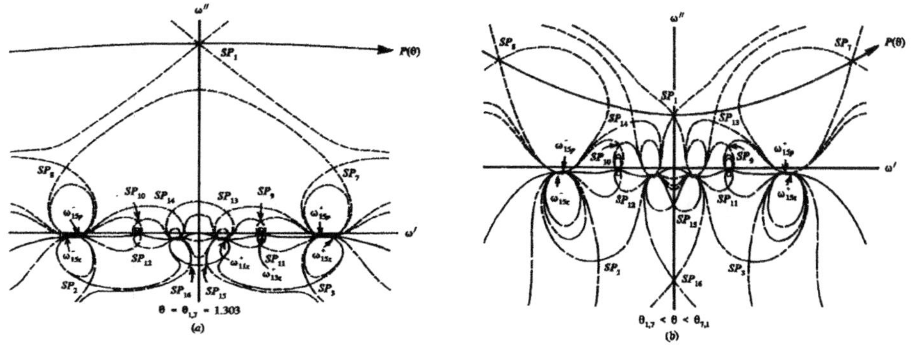

Figure 10. Topography of the isotimic contours of $X(\omega,\theta) = \Re\{\phi(\omega,\theta)\}$ in the complex ω-plane for small $|\omega|$ for the four resonance Lorentz model dielectric with parameters appropriate for triply-distilled water as given in Table 1.

When $\theta = \theta_{7,1} \cong 1.3235$, the saddle point pair $SP_{7,8}$ and the near saddle point SP_1 are again of equal dominance and this saddle point pair $SP_{7,8}$ is now accessible, as seen in Fig. 11(a). Notice that the saddle point pair $SP_{13,14}$ is also accessible at this space-time point, and the Olver-type path $P(\theta)$ with respect to the equally dominant saddle points SP_1 and $SP_{7,8}$ has been chosen so as to pass through these less dominant saddle points. The saddle point pair $SP_{7,8}$ is dominant and accessible over the short space-time interval $\theta_{7,1} < \theta < \theta_{SB_2}$, and the near saddle point SP_1 regains its dominance at $\theta = \theta_{SB_2} \cong 1.359$ and remains dominant and accessible over the space-time interval $\theta_{SB_2} < \theta < \theta_1$, where the two first-order saddle points SP_1 and SP_{15} coalesce into a single second-order saddle point along the negative imaginary axis when $\theta = \theta_1 \cong 1.51$. These two first-order near saddle points then separate and symmetrically move off into the lower-half of the complex ω -plane for $\theta > \theta_1$, as seen in Fig. 11(b), approaching the innermost branch points $\omega_{11_p}^{\pm}$ in the limit as $\theta \to \infty$, and are the dominant, accessible saddle points over the space-time interval $\theta_1 < \theta < \theta_{9,1}$. Changes in dominance from the saddle point pair $SP_{1,15}$ to the saddle point pair $SP_{9,10}$ at $\theta = \theta_{9,1} \cong 1.71$ and then from that saddle point pair to the saddle point pair $SP_{13,14}$ at $\theta = \theta_{13,9} \cong 1.73$ establish the final set of dominance changes that occur. The accessible saddle point pair $SP_{13,14}$ remains dominant for all $\theta > \theta_{13,9}$ and approaches the branch points $\omega_{13_p}^{\pm}$ in the limit as $\theta \to \infty$. Notice that the saddle point pair $SP_{9,10}$ also remains accessible and is just barely dominated by the saddle point pair $SP_{13,14}$ for all $\theta > \theta_{13,9}$, that is, $X(\omega_{13,14},\theta) \cong X(\omega_{9,10},\theta)$ with $X(\omega_{13,14},\theta) < X(\omega_{9,10},\theta)$ over this entire space-time interval, and that this saddle point pair approaches the branch points $\omega_{13_z}^{\pm}$ in the limit as $\theta \to \infty$, as seen in Fig. 11(b).

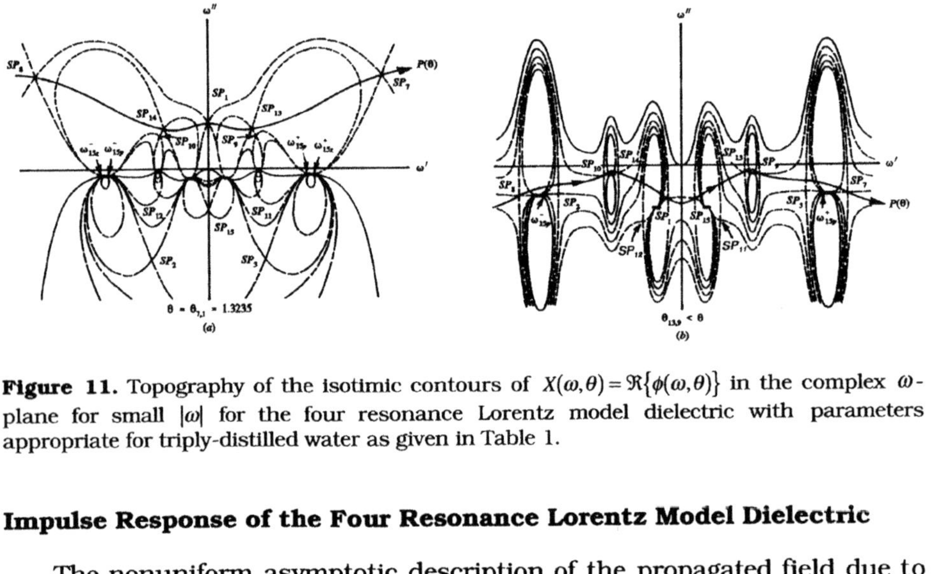

Figure 11. Topography of the isotimic contours of $X(\omega,\theta) = \Re\{\phi(\omega,\theta)\}$ in the complex ω-plane for small $|\omega|$ for the four resonance Lorentz model dielectric with parameters appropriate for triply-distilled water as given in Table 1.

Impulse Response of the Four Resonance Lorentz Model Dielectric

The nonuniform asymptotic description of the propagated field due to a delta function input in a four resonance Lorentz model dielectric over the space-time domain $\theta > \varepsilon_\infty^{1/2}$ is obtained through a straightforward application of Olver's theorem[15] to the integral representation given in Eq.(7) when the contour C is deformed to an Olver-type path $P(\theta)$ with respect to the dominant, accessible saddle points of $\phi(\omega,\theta) = i\omega(n(\omega) - \theta)$, as depicted in Figs. 9-11. The complete, numerically determined precursor structure that characterizes the electromagnetic impulse response of the four resonance Lorentz model of the frequency dispersion of the complex index of refraction of triply-distilled water, as given by Eq. (25) with the medium parameters listed in Table 1, is illustrated in the sequence of plots given in Fig. 12 at the fixed propagation distance $z = 6\pi \times 10^{-5} m$. The detailed analysis of the saddle point dynamics given in the preceding subsection is now used to explain each important feature observed in this computed impulse response. Because of the exceptional quantitative agreement between the field structures depicted in Figs. 12(a)-(c) with their counterparts depicted in Fig. 4(a)-(c), this description also applies to the full model results obtained from Eq. (6) through intermediate values of the space-time parameter θ.

Over the initial space-time interval $\varepsilon_\infty^{1/2} < \theta < \theta_{SB_1}$, where $\theta_{SB_1} \cong 1.133$, the distant saddle point pair $SP_{5,6}$ is dominant and results in the asymptotic contribution

$$A(z,t) \sim \left(\frac{2c}{\pi z}\right)^{1/2} \Re\left\{\left[-\phi''(\omega_5,\theta)\right]^{-1/2} e^{\frac{z}{c}\phi(\omega_5,\theta)} + \left[-\phi''(\omega_6,\theta)\right]^{-1/2} e^{\frac{z}{c}\phi(\omega_6,\theta)}\right\} \quad , \tag{27}$$

for $\varepsilon_\infty^{1/2} < \theta < \theta_{SB_1}$ as $z \to \infty$. This contribution yields the dynamical evolution of the Sommerfeld precursor field $A_S(z,t)$, whose front has an infinite instantaneous oscillation frequency and travels at the velocity $c/\varepsilon_\infty^{1/2}$, which is equal to the speed of light in vacuum for most (if not all) materials, including water[20]. As these distant saddle points move in from infinity and approach the branch points $\omega_{17_z}^{\pm}$ in the limit as $\theta \to \infty$, the attenuation $X(\omega_5,\theta)/c = X(\omega_6,\theta)/c$ increases so that the Sommerfeld precursor amplitude decreases monotonically, and the instantaneous frequency of oscillation $\omega_S(\theta) \cong \Re\{\omega_5(\theta)\}$ decreases monotonically from infinity and approaches the value $\Re\{\omega_{17_z}^{+}\}$ from above as $\theta \to \infty$, as seen in both Fig. 4(a) and Fig. 12(a).

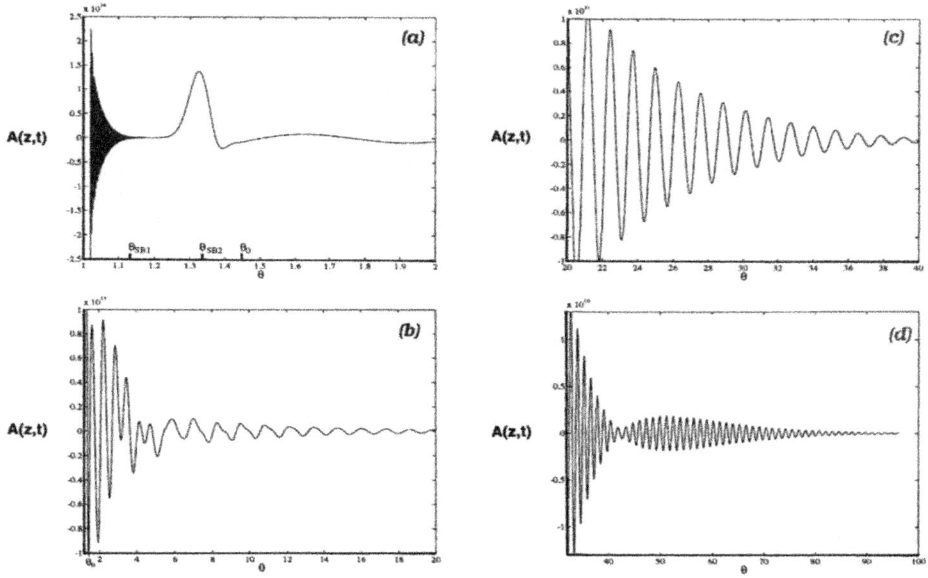

Figure 12. Dynamical evolution of the propagated field due to an input delta function pulse in the four resonance Lorentz model of triply-distilled water at the fixed propagation distance $z = 6\pi \times 10^{-5} m$ for small through large values of the space-time parameter $\theta = ct/z$.

Over the space-time interval $\theta_{SB_1} < \theta < \theta_{7,1}$, where $\theta_{7,1} \cong 1.3235$, the near saddle point SP_1 is dominant and results in the asymptotic contribution

$$A(z,t) \sim \left(\frac{c}{2\pi z}\right)^{1/2} \Re\left\{\left[-\phi''(\omega_1,\theta)\right]^{-1/2} e^{\frac{z}{c}\phi(\omega_1,\theta)}\right\} \quad , \theta_{SB_1} < \theta < \theta_{7,1} \quad , \tag{28}$$

as $z \to \infty$, during which the propagated field is nonoscillatory (or quasi-static) and increases in amplitude as the dominant saddle point SP_1 moves down the positive imaginary axis with a resultant decreasing attenuation factor. Over the short space-time interval $\theta_{7,1} < \theta < \theta_{SB_2}$, where $\theta_{SB_2} \cong 1.359$, the pair of intermediate saddle points $SP_{7,8}$ become dominant and result in the asymptotic contribution

$$A(z,t) \sim \left(\frac{2c}{\pi z}\right)^{1/2} \Re\left\{\left[-\phi''(\omega_7,\theta)\right]^{-1/2} e^{\frac{z}{c}\phi(\omega_7,\theta)} + \left[-\phi''(\omega_8,\theta)\right]^{-1/2} e^{\frac{z}{c}\phi(\omega_8,\theta)}\right\} \quad , \tag{29}$$

for $\theta_{7,1} < \theta < \theta_{SB_2}$ as $z \to \infty$, which introduces a very brief oscillatory contribution with instantaneous frequency $\omega_I(\theta) \cong \Re\{\omega_7(\theta)\}$ to the propagated field. The first-order near saddle point SP_1 is again dominant over the space-time interval $\theta_{SB_2} < \theta < \theta_1$, where $\theta_1 \cong 1.51$, and results in the asymptotic contribution

$$A(z,t) \sim \left(\frac{c}{2\pi z}\right)^{1/2} \Re\left\{\left[-\phi''(\omega_1,\theta)\right]^{-1/2} e^{\frac{z}{c}\phi(\omega_1,\theta)}\right\} \quad , \theta_{SB_2} < \theta < \theta_1 \quad , \tag{30}$$

as $z \to \infty$, during which the propagated field is quasi-static. At $\theta = \theta_0 = 1.49$, where $\theta_{SB_2} < \theta_0 < \theta_1$, $\phi(\omega_1(\theta_0),\theta_0) = 0$ and the asymptotic contribution (30) experiences no exponential decay, varying only as $z^{-1/2}$. As θ increases above θ_0 the exponential attenuation monotonically increases. Over the space-time interval $\theta_1 < \theta < \theta_{9,1}$, where $\theta_{9,1} \cong 1.71$, the near saddle point pair $SP_{1,15}$ is dominant and move apart symmetrically with respect to the imaginary axis, producing the asymptotic contribution

$$A(z,t) \sim \left(\frac{2c}{\pi z}\right)^{1/2} \Re\left\{\left[-\phi''(\omega_{15},\theta)\right]^{-1/2} e^{\frac{z}{c}\phi(\omega_{15},\theta)} + \left[-\phi''(\omega_1,\theta)\right]^{-1/2} e^{\frac{z}{c}\phi(\omega_1,\theta)}\right\} \quad , \tag{31}$$

for $\theta_1 < \theta < \theta_{9,1}$ as $z \to \infty$. The exponential attenuation of this field component increases monotonically with increasing θ, as seen in Fig. 8, while its instantaneous frequency of oscillation $\omega_I \cong \Re\{\omega_{15}(\theta)\}$ chirps upward from its quasi-static value at $\theta = \theta_1 \cong 1.51$ and approaches the limiting value $\Re\{\omega_{11_p}^+\}$ as $\theta \to \infty$. Over the space-time interval $\theta_{9,1} < \theta < \theta_{13,9}$, where $\theta_{9,1} \cong 1.71$ and $\theta_{13,9} \cong 1.73$ for triply-distilled water, the middle saddle point pair $SP_{9,10}$ is dominant and produces the asymptotic contribution

$$A(z,t) \sim \left(\frac{2c}{\pi z}\right)^{1/2} \Re\left\{\left[-\phi''(\omega_9,\theta)\right]^{-1/2} e^{\frac{z}{c}\phi(\omega_9,\theta)} + \left[-\phi''(\omega_{10},\theta)\right]^{-1/2} e^{\frac{z}{c}\phi(\omega_{10},\theta)}\right\} \quad , \tag{32}$$

for $\theta_{9,1} < \theta < \theta_{13,9}$ as $z \to \infty$, whose instantaneous frequency of oscillation $\omega_I \cong \Re\{\omega_9(\theta)\}$ chirps downwards towards the value $\Re\{\omega_{13_z}^+\}$ in the limit as $\theta \to \infty$. Finally, the first-order saddle point pair $SP_{13,14}$ are dominant for all $\theta > \theta_{13,9}$ and produce the asymptotic contribution

$$A(z,t) \sim \left(\frac{2c}{\pi z}\right)^{1/2} \Re\left\{\left[-\phi''(\omega_{13},\theta)\right]^{-1/2} e^{\frac{z}{c}\phi(\omega_{13},\theta)} + \left[-\phi''(\omega_{14},\theta)\right]^{-1/2} e^{\frac{z}{c}\phi(\omega_{14},\theta)}\right\} \quad , \tag{33}$$

for $\theta_{13,9} < \theta$, as $z \to \infty$, whose instantaneous frequency of oscillation $\omega_I \cong \Re\{\omega_{13}(\theta)\}$ chirps upwards and approaches the value $\Re\{\omega_{13_p}^+\}$ as $\theta \to \infty$.

The set of expressions given in Eqs. (28)-(33) yields the dynamical evolution of the Brillouin precursor field $A_B(z,t)$. The initial behavior of this precursor field is dominated by the asymptotic contribution due to the first-order saddle point SP_1 which results in the quasi-static growth of this field, where the field amplitude at the critical space-time point $\theta = \theta_0$ decays only algebraically as $z^{-1/2}$. However, due to interference with the contributions from the two saddle point pairs $SP_{7,8}$ and $SP_{13,14}$, the observed amplitude peak in the Brillouin precursor field is shifted towards an earlier space-time point just prior to the value θ_{SB_2}, as seen in Fig. 12(a). As the propagation distance increases, this amplitude peak shifts upwards and approaches the critical value θ_0 from below in the limit as $z \to \infty$. The velocity of this peak in the Brillouin precursor then slows down from its initial value c/θ_{SB_2} and approaches the value c/θ_0 as the propagation distance increases.

As θ increases above θ_1 the propagated field is initially dominated by the asymptotic contribution due to the saddle point pair $SP_{1,15}$ which results in an increasingly attenuated field with an upwardly chirping instantaneous angular frequency of oscillation. For all $\theta > \theta_{9,1}$ the propagated field is dominated by the asymptotic contributions due to the two pairs of saddle points $SP_{9,10}$ and $SP_{13,14}$, which possess nearly equal attenuation over this entire space-time domain. Because of their different instantaneous oscillation frequencies, their superposition then results in the observed quenching of the oscillatory relaxation of the Brillouin precursor and the resultant beat structure over the space-time domain $\theta > \theta_{9,1}$ that is clearly evident in both Figs. 4 and 12.

DISCUSSION

The electromagnetic impulse response of the full Rocard-Powles-Lorentz model for the frequency dispersion of the dielectric permittivity of triply-distilled water with two relaxation times and four resonance lines, depicted in Fig. 4, displays the complete Sommerfeld precursor field structure followed by a Brillouin precursor field structure whose oscillatory relaxation is quenched. Each of the predominant features observed in this propagated field is completely explained by the four resonance Lorentz model (depicted in Fig. 12) with the single exception of the long-time exponential relaxation observed in Fig. 4 for the full model. It is quite surprising that the effects of rotational polarization, which dominate the observed frequency dispersion depicted in Figs. 1-2, have only a minor effect on the propagated field. In particular, one might expect that the quasi-static peak in the Brillouin precursor would occur at the space-time point $\theta = n(0) \cong 8.902$ that is a characteristic of the rotational polarization component when the full model is used, and not at the space-time point $\theta_0 = n(0) \cong 1.501$ that is predicted by the quartic resonance Lorentz model given in Eq. (25). However, the latter result is correct while the former is not even close. This surprising result, together with the quenched oscillatory and long-time exponential relaxation of the Brillouin precursor field demonstrate intriguing interactions between the rotational and resonance components of the Rocard-Powles-Lorentz model considered here.

Table 1. Optimized parameter values for the Rocard-Powles-Lorentz model for the frequency dispersion of the dielectric permittivity of triply-distilled water.

$\varepsilon_\infty = 1$

Rocard-Powles Model Parameters

(Microwave Region)

$a_0 = 74.1$	$\tau_0 = 8.44 \times 10^{-12} s$	$\tau_{f0} = 4.93 \times 10^{-14} s$
$a_2 = 2.90$	$\tau_2 = 6.05 \times 10^{-14} s$	$\tau_{f2} = 8.59 \times 10^{-15} s$

Lorentz Model Parameters

(Infrared Region)

$\omega_{11}/2\pi = 1.8 \times 10^{13} Hz$	$b_{11}/2\pi = 1.2 \times 10^{13} Hz$	$\delta_{11}/2\pi = 4.3 \times 10^{12} Hz$
$\omega_{13}/2\pi = 4.9 \times 10^{13} Hz$	$b_{13}/2\pi = 6.8 \times 10^{12} Hz$	$\delta_{13}/2\pi = 8.4 \times 10^{11} Hz$
$\omega_{15}/2\pi = 1.0 \times 10^{14} Hz$	$b_{15}/2\pi = 2.0 \times 10^{13} Hz$	$\delta_{15}/2\pi = 2.8 \times 10^{12} Hz$

(Ultraviolet Region)

$\omega_{17}/2\pi = 3.7 \times 10^{15} Hz$	$b_{17}/2\pi = 3.2 \times 10^{15} Hz$	$\delta_{17}/2\pi = 8.0 \times 10^{14} Hz$

ACKNOWLEDGMENTS

The research presented in this paper was supported, in part, by the United States Air Force Office of Scientific Research under Contract F49620-89-C-0057 and Grants F49620-92-J-0419 and F49620-94-1-0430.

REFERENCES

1. J. Jones, On the propagation of a pulse through a dispersive medium, Am. J. Phys. 42:43 (1974).
2. D. G. Anderson and J. I. H. Askne, Wave packets in strongly dispersive media, Proc. IEEE 62:1518 (1974).
3. D. Anderson, J. Askne, and M. Lisak, Wave packets in an absorbing and strongly dispersive medium, Phys. Rev. A 12:1546 (1975).
4. I. P. Christov, Generation and propagation of ultrashort optical pulses, in: "Progress in Optics," vol. XXIX, E. Wolf, ed., Elsevier, Amsterdam (1991), pp. 199-291.
5. K. E. Oughstun and C. M. Balictsis, Gaussian pulse propagation in a dispersive, absorbing dielectric, Phys. Rev. Lett. 77:2210 (1996).
6. K. E. Oughstun and H. Xiao, Failure of the quasimonochromatic approximation for ultrashort pulse propagation in a dispersive, attenuative medium, Phys. Rev. Lett. 78:642 (1997).
7. A. Sommerfeld, Über die fortpflanzung des lichtes in disperdierenden medien, Ann. Phys. 44:177 (1914).
8. L. Brillouin, Über die fortpflanzung des licht in disperdierenden medien, Ann. Phys. 44:203 (1914).
9. L. Brillouin. "Wave Propagation and Group Velocity," Academic, New York (1960).
10. P. Debye, Näherungsformeln für die zylinderfunktionen für grosse werte des arguments und unbeschränkt verander liche werte des index, Math. Anal. 67:535 (1909).
11. K. E. Oughstun and G. C. Sherman, Propagation of electromagnetic pulses in a linear dispersive medium with absorption (the Lorentz medium), J. Opt. Soc. Am. B 5:817 (1988).
12. K. E. Oughstun and G. C. Sherman, Uniform asymptotic description of electromagnetic pulse propagation in a linear dispersive medium with absorption (the Lorentz medium), J. Opt. Soc. Am. A 6:1394 (1989).
13. K. E. Oughstun and G. C. Sherman, Asymptotic theory of pulse propagation in absorbing and dispersive dielectrics, in: "Review of Radio Science, 1990-1992," W. R. Stone, ed., Oxford (1993), pp. 75-105.
14. K. E. Oughstun and G. C. Sherman. "Electromagnetic Pulse Propagation in Causal Dielectrics," Springer-Verlag, Berlin (1994).
15. F. W. J. Olver, Why steepest descents?, SIAM Rev. 12:228 (1970).
16. K. Moten, C. Durney, and T. Stockham, Jr., Electromagnetic pulse propagation in dispersive planar dielectrics, Bioelectromagnetics 10:35 (1989).
17. K. Moten, C. Durney, and T. Stockham, Jr., Electromagnetic pulsed-wave radiation in spherical models of dispersive biological substances, Bioelectromagnetics 12:319 (1991).
18. R. Albanese, J. Penn, and R. Medina, Short rise-time microwave pulse propagation through dispersive biological media, J. Opt. Soc. Am. A 6:1441 (1989).
19. K. E. Oughstun, J. E. K. Laurens, and C. M. Balictsis, Asymptotic description of electromagnetic pulse propagation in a linear dispersive medium, in: "Ultra-Wideband, Short-Pulse Electromagnetics," L. Carin and L. B. Felsen, eds., Plenum, New York (1993), pp. 223-240.
20. K. E. Oughstun, Dynamical structure of the precursor fields in linear dispersive pulse propagation in lossy dielectrics, in: "Ultra-Wideband, Short-Pulse Electromagnetics 2," L. Carin and L. B. Felsen, eds., Plenum, New York (1995), pp. 257-272.
21. P. D. Smith and K. E. Oughstun, Electromagnetic energy dissipation of ultrawideband plane wave pulses in a causal, dispersive dielectric, in:

"Ultra-Wideband, Short-Pulse Electromagnetics 2," L. Carin and L. B. Felsen, eds., Plenum, New York (1995), pp. 285-295.
22.K. E. Oughstun, Noninstantaneous, finite rise-time effects on the precursor field formation in linear dispersive pulse propagation, J. Opt. Soc. Am. A 12:1715 (1995).
23.R. Albanese, J. Blaschak, R. Medina, and J, Penn, Ultrashort electromagnetic signals: biophysical questions, safety issues, and medical opportunities, Aviat. Space Environ. Med. 65:116 (1994).
24.S. Shen and K. E. Oughstun, Dispersive pulse propagation in a double resonance Lorentz medium, J. Opt. Soc. Am. B 6:948 (1989).
25.J. E. K. Laurens. "Plane Wave Pulse Propagation in a Linear, Causally Dispersive Polar Medium," Ph.D. Dissertation, University of Vermont, 1993.
26.C. F. Bohren and D. R. Huffman. "Absorption and Scattering of Light by Small Particles," Wiley, New York (1983), ch. 9.
27.J. McConnell," Rotational Brownian Motion and Dielectric Theory," Academic, London (1980).
28.P. Debye. "Polar Molecules," Dover, New York (1929).
29.C. J. F. Böttcher and A. Bordewijk. "Theory of Electric Polarization," vol. 2, Elsevier, Amsterdam (1980).
30.H. A. Lorentz. "The Theory of Electrons," Dover, New York (1952).
31.L. Rosenfeld. "Theory of Electrons," Dover, New York (1965).
32.H. J. Liebe, G. A. Hufford, and T. Manabe, A model for the complex permittivity of water at frequencies below 1 Thz, Int. J. Infrared and Millimeter Waves 12:659 (1991).
33.P. Wyns, D. P. Foty, and K. E. Oughstun, Numerical analysis of the precursor fields in linear dispersive pulse propagation, J. Opt. Soc. Am. A 6:1421 (1989).

ULTRAWIDEBAND ELECTROMAGNETIC PULSE PROPAGATION IN TRIPLY-DISTILLED WATER

Paul D. Smith and Kurt E. Oughstun

College of Engineering and Mathematics
University of Vermont
Burlington, VT 05405-0156

INTRODUCTION

As a pulse propagates through a linear dispersive, lossy medium, each spectral component propagates at a rate characterized by its phase velocity $v_p(\omega) = \omega/\beta(\omega)$ and is attenuated at a rate characterized by its attenuation coefficient $\alpha(\omega) = \Im\{\tilde{k}(\omega)\}$, where $\tilde{k}(\omega) = \beta(\omega) + i\alpha(\omega)$ is the complex wavenumber for a time-harmonic plane wave of angular frequency ω. The relative phases and amplitudes of the spectral components of a given pulse then change by differing amounts as the propagation distance increases, thereby giving rise to the observed distortion of the propagated wave field.

A precise description of ultrawideband pulse propagation in a linear, causally dispersive medium originated in 1914 with the classical analysis of Sommerfeld[1] and Brillouin[2,3] who used the asymptotic method of steepest descents[4] to describe the dynamical evolution of the propagated field due to an input unit step-function modulated signal of constant carrier frequency in a single resonance Lorentz model dielectric. The modern asymptotic description[5-8] has shown the critical role that the precursor fields, originally introduced by Sommerfeld[1] and Brillouin[2,3], play in both the signal arrival[5-8] and the resultant pulse distortion[8-11] in a single resonance Lorentz medium. This modern description is based on Olver's saddle point method[12], together with the appropriate uniform asymptotic expansion techniques[8], and provides an accurate description of the entire dynamical field evolution in the mature dispersion regime of the medium. This description has clearly shown that the precursor fields dominate the propagated field structure in dispersive, lossy dielectrics when the propagation distance is in the mature dispersion regime and that they carry significant energy much deeper into the lossy, dispersive medium[13,14] than that predicted by traditional analyses. These precursor fields persist for noninstantaneous rise-time input pulses[8,10,11,15] provided that the initial rise-time (or fall-time) is on the order of or less than the characteristic medium relaxation time. Their potentially profound impact on health and safety studies is just beginning to be appreciated[16-19].

Ultra-Wideband, Short-Pulse Electromagnetics 4
Edited by Heyman *et al.*, Kluwer Academic / Plenum Publishers, New York, 1999.

The dynamical evolution of the propagated field due to an input delta function pulse in a particular dispersive medium displays the complete precursor field structure that is characteristic of that dispersive material. For example, the impulse response of a single resonance Lorentz model dielectric is entirely comprised of the leading-edge Sommerfeld precursor and the trailing-edge Brillouin precursor fields, while the impulse response of a multiple resonance dielectric may be enhanced by the appearance of additional middle precursor fields[20-23] that evolve between the Sommerfeld and Brillouin precursors. On the other hand, the impulse response of either a Debye model or the Rocard-Powles extension of the Debye model of orientational polarization phenomena in dielectrics is completely described by the Brillouin precursor field[18,22,23]. Finally, for a composite Rocard-Powles-Lorentz model dielectric, the dynamical field evolution is again comprised of a high-frequency, leading-edge Sommerfeld precursor and a low-frequency, trailing-edge Brillouin precursor with the possibility of additional intermediate-frequency precursor fields that evolve between the Sommerfeld and Brillouin precursors. However, for this more physically realistic model of a dielectric material, the Brillouin precursor is more a characteristic of the low-frequency limiting behavior of the higher-frequency resonance polarization phenomena than it is of the low-frequency orientational polarization phenomena[23].

The detailed dynamics of the individual precursor field components in a complex dispersive medium that is described by both orientational and resonance polarization phenomena are better observed with an ultrawideband pulse whose spectral bandwidth is sufficiently narrow in order to sample individual polarization structures that are a characteristic of the material. An initial pulse shape that is ideally suited for this purpose is the rectangular envelope modulated signal

$$f(t) = \Pi_T(t)\sin(\omega_c t) \quad , \tag{1}$$

where ω_c denotes the carrier frequency of the electromagnetic field and where

$$\Pi_T(t) = \begin{Bmatrix} 0 & , t < 0 \\ 1 & , 0 < t < T \\ 0 & , T < t \end{Bmatrix} \tag{2}$$

denotes the unit amplitude rectangle function of duration $T > 0$. The initial pulse spectrum is then given by

$$\tilde{f}(\omega) = \frac{T}{2i}\exp(i\omega T/2)\cdot \left\{\exp(i\omega_c T/2)\operatorname{sin}c((\omega+\omega_c)T/2) - \exp(-i\omega_c T/2)\operatorname{sin}c((\omega-\omega_c)T/2)\right\} \quad , \tag{3}$$

whose spectral bandwidth along the positive frequency axis is given by $\Delta\omega = 4\pi/T$ and is centered on the carrier frequency ω_c of the field. The bandwidth can then be made as narrow as desired in order to probe any dispersion structure of interest in the host material without any effect upon the ultrawideband character of the pulsed field.

The exact integral representation of the propagated plane wave field in the half-space $z \geq 0$ that is occupied by the dispersive material is given by[8]

$$A(z,t)=\frac{1}{2\pi}\int_C \tilde{f}(\omega)\exp\left\{\frac{z}{c}\phi(\omega,\theta)\right\}d\omega \quad , \tag{4}$$

where $f(t)=A(0,t)$. Here $A(z,t)$ represents any scalar component of the electric or magnetic field vector whose spectrum $\tilde{A}(z,\omega)$ satisfies the scalar Helmholtz equation

$$\left(\nabla^2+\tilde{k}^2(\omega)\right)\tilde{A}(z,\omega)=0 \quad . \tag{5}$$

The complex wavenumber appearing here is given by

$$\tilde{k}(\omega)=\omega n(\omega)/c \quad , \tag{6}$$

with $\beta(\omega)=\Re\{\tilde{k}(\omega)\}$ and $\alpha(\omega)=\Im\{\tilde{k}(\omega)\}$, where c denotes the speed of light in vacuum, and where $n(\omega)=[\mu\varepsilon(\omega)]^{1/2}$ is the complex index of refraction of the dispersive medium occupying the half-space $z\geq 0$ with relative magnetic permeability μ and complex-valued, relative dielectric permittivity $\varepsilon(\omega)=\varepsilon_r(\omega)+i\varepsilon_i(\omega)$. Since $f(t)=0$ for $t<0$, then Eq. (4) is a Laplace representation for which the contour of integration C in the complex ω-plane is the Bromwich contour[8] $\omega=\omega'+ia$, where a is a real-valued constant that is greater than the abscissa of absolute convergence for the function $f(t)$, and where $\omega'\equiv\Re\{\omega\}$ varies from negative to positive infinity. The complex phase function $\phi(\omega,\theta)$ appearing in Eq. (4) is given by

$$\phi(\omega,\theta)=i\frac{c}{z}\left[\tilde{k}(\omega)z-\omega t\right]=i\omega[n(\omega)-\theta] \quad , \tag{7}$$

where

$$\theta=ct/z \tag{8}$$

is a dimensionless parameter that characterizes any given space-time point in the propagated field.

The asymptotic theory[8,9] provides a uniformly valid description of the temporal evolution of the propagated field due to an input rectangular envelope modulated signal at any sufficiently large, fixed value of the propagation distance z. For this purpose it is convenient to express the initial field that is given in Eq. (1) as

$$A(0,t)=f(t)=[U(t)-U(t-T)]\sin(\omega_c t) \quad , \tag{9}$$

where $U(t)$ denotes the Heaviside unit step-function, and where $T>0$ denotes the initial pulse width. The propagated field is then given by

$$A(z,t)=U(z,t)-U(z,t-T) \quad , \tag{10}$$

where $U(z,t)$ denotes the propagated field due to the unit step-function modulated signal $U(0,t)=U(t)\sin(\omega_c t)$. The uniform asymptotic description of this signal in any complex dispersive medium may be represented as[5-8]

$$U(z,t) \sim U_S(z,t) + U_m(z,t) + U_B(z,t) + U_c(z,t) \tag{11}$$

as $z \to \infty$. Here $U_S(z,t)$ denotes the Sommerfeld precursor field component that is due to the distant saddle points of $\phi(\omega,\theta)$, $U_B(z,t)$ denotes the Brillouin precursor field component that is due to the near saddle points of $\phi(\omega,\theta)$, and $U_m(z,t)$ denotes the middle precursor field components that are due to the middle saddle points that may evolve in the intermediate frequency region of the complex ω-plane whose maximum frequency magnitude is bounded above by the upper frequency characteristic of the dispersive medium. Finally, the component $U_c(z,t)$ denotes the propagated signal component that is due to the pole contribution in the spectrum of $U(0,t)$; in the absence of dispersion, this residue contribution would represent the entire propagated field structure so that the precursor fields describe the effects of dispersion on the propagated field. With the asymptotic representation given in Eq. (11), the asymptotic description of the propagated field due to the rectangular envelope pulse given in Eq. (10) then shows that the pulse distortion due to dispersion is due to the leading and trailing edge precursor fields, the pulse distortion becoming severe when these two sets of precursor fields begin to overlap and interfere with each other, this overlap increasing as the propagation distance increases.

ULTRAWIDEBAND PULSE DISPERSION IN TRIPLY-DISTILLED WATER

The dynamical field evolution due to an input rectangular envelope modulated pulse in triply-distilled water is now presented as a function of the propagation distance into the medium for several different values of the carrier frequency of the initial pulse. For each case considered, the initial pulse width T has been chosen to be given by 10 oscillation periods of the specified input carrier frequency ω_c. For each carrier frequency considered, the propagated field has been calculated at propagation distances of *1*, *3*, *5* and *10* absorption depths $\delta_s = \alpha^{-1}(\omega_c)$ at that particular carrier frequency. Each carrier frequency has been selected in order to sample a different relaxation or resonance feature that appears in the dielectric dispersion of triply-distilled water, as described by the composite Rocard-Powles-Lorentz model[23]

$$\varepsilon(\omega) = \varepsilon_\infty + \sum_{j=0,2} \frac{a_j}{(1 - i\omega\tau_j)(1 - i\omega\tau_{fj})} - \sum_{j=11,13,15,17} \frac{b_j^2}{\omega^2 - \omega_j^2 + 2i\delta_j\omega} \quad , \tag{12}$$

where $\varepsilon_\infty = 1$ is the high frequency limit of the dielectric permittivity, $a_0 = 74.1$, $\tau_0 = 8.44\times10^{-12}s$, $\tau_{f0} = 4.93\times10^{-14}s$ and $a_2 = 2.90$, $\tau_2 = 6.05\times10^{-14}s$, $\tau_{f2} = 8.59\times10^{-15}s$ are the Rocard-Powles model parameters for the observed frequency dispersion in the microwave region, $\omega_{11}/2\pi = 1.8\times10^{13}Hz$, $b_{11}/2\pi = 1.2\times10^{13}Hz$, $\delta_{11}/2\pi = 4.3\times10^{12}Hz$, $\omega_{13}/2\pi = 4.9\times10^{13}Hz$, $b_{13}/2\pi = 6.8\times10^{12}Hz$, $\delta_{13}/2\pi = 8.4\times10^{11}Hz$, and $\omega_{15}/2\pi = 1.0\times10^{14}Hz$, $b_{15}/2\pi = 2.0\times10^{13}Hz$, $\delta_{15}/2\pi = 2.8\times10^{12}Hz$ are the Lorentz model parameters for the observed frequency dispersion in the infrared region, and where $\omega_{17}/2\pi = 3.7\times10^{15}Hz$, $b_{17}/2\pi = 3.2\times10^{15}Hz$, $\delta_{17}/2\pi = 8.0\times10^{14}Hz$ are the Lorentz model parameters for the observed frequency dispersion in the ultraviolet region of the electromagnetic spectrum. The frequency dependence of the real part of the complex index of refraction for this Rocard-Powles-Lorentz model of triply-distilled water is presented in Figure 1. The angular frequency locations of the five carrier frequency values $f_1 = 1GHz$, $f_2 = 10Ghz$, $f_3 = 100GHz$, $f_4 = 1THz$, and $f_5 = 10THz$ considered here are illustrated in this figure as is the input pulse bandwidth about each frequency value.

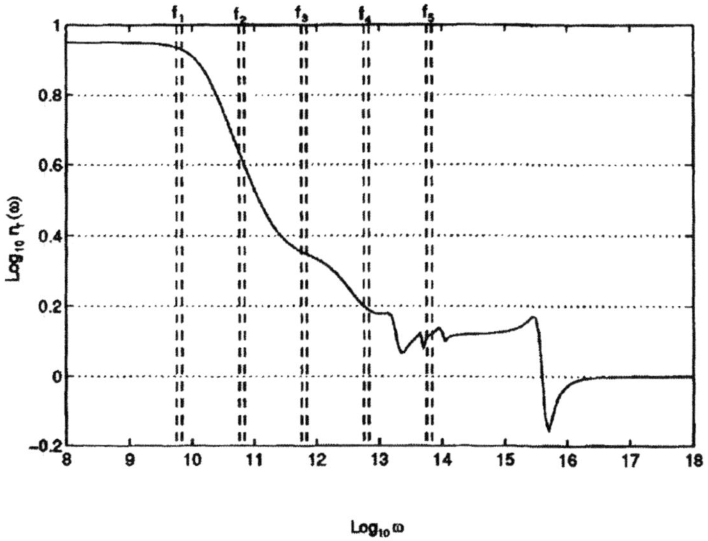

Figure 1. Frequency dependence of the real part of the complex index of refraction for the Rocard-Powles-Lorentz model of the frequency dispersion of triply-distilled water. The vertical dashed lines indicate the frequency bands that are subtended by the main lobes of the input pulse spectra for the five cases of the rectangular pulse envelope modulated sinusoidal fields that are considered in this paper.

The numerically determined dynamical field evolution for each of the five cases considered here is depicted in Figures 2-6. In each case the temporal location of the peak in the leading edge Brillouin precursor is denoted by

$$t_0^{eff} = (z/c)\theta_0^{eff} \quad , \tag{13}$$

where θ_0^{eff} denotes the fixed space-time point whose value is given by the effective zero frequency limit of the index of refraction that is sampled by the initial pulse spectrum. The temporal location of the peak in the trailing edge Brillouin precursor is then given by $t_0^{eff} + T$.

For the $f_1 = 1GHz$ case, $n_r(\omega_1) = 8.574$, $n_i(\omega_1) = 1.303$, and the e^{-1} absorption depth is $\delta_s = 5.831 \times 10^{-3} m$. The effective zero frequency limit of the index of refraction for this case is given by

$$\theta_0^{eff} \cong \left(\varepsilon_\infty + a_0 + a_2 + b_{11}^2/\omega_{11}^2 + b_{13}^2/\omega_{13}^2 + b_{15}^2/\omega_{15}^2\right)^{1/2} = 8.92 \tag{14}$$

whose value is dominated by the Rocard-Powles component of the frequency dispersion. The dynamical field evolution depicted in Figure 2 clearly shows the exponential decay of the carrier signal component of the propagated field while the leading and trailing edge Brillouin precursors, whose peak amplitudes only decay as z^{-1}, dominate the propagated field structure for all $z \geq \delta_s$. The predicted values t_0^{eff} and $t_0^{eff}+T$ are seen to accurately describe the peak field points in the leading and trailing edge Brillouin precursors, which are then seen to be a characteristic of the rotational polarization phenomena in the observed frequency dispersion of the dielectric permittivity of water at this input pulse frequency in the UHF radio frequency band, as seen in Figure 1.

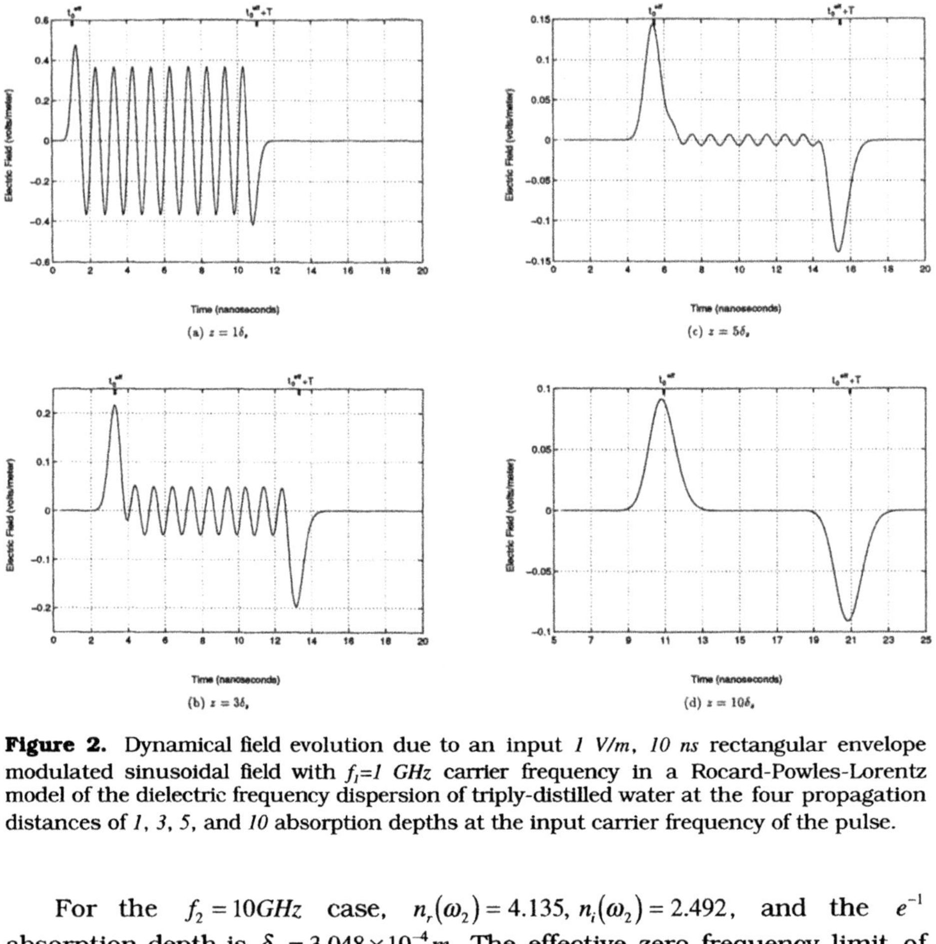

Figure 2. Dynamical field evolution due to an input *1 V/m*, *10 ns* rectangular envelope modulated sinusoidal field with f_1=*1 GHz* carrier frequency in a Rocard-Powles-Lorentz model of the dielectric frequency dispersion of triply-distilled water at the four propagation distances of *1*, *3*, *5*, and *10* absorption depths at the input carrier frequency of the pulse.

For the $f_2 = 10GHz$ case, $n_r(\omega_2) = 4.135$, $n_i(\omega_2) = 2.492$, and the e^{-1} absorption depth is $\delta_s = 3.048 \times 10^{-4} m$. The effective zero frequency limit of the index of refraction for this case is approximately given by Eq. (14). The dynamical field evolution depicted in Figure 3 again shows the exponential decay of the carrier signal component of the propagated field while the leading and trailing edge Brillouin precursors, whose peak amplitudes only decay as z^{-1}, dominate the propagated field structure for all $z \geq \delta_s$. Notice that the accuracy of the predicted values t_0^{eff} and $t_0^{eff}+T$ in describing the peak field points in the leading and trailing edge Brillouin precursors has decreased from that obtained in the previous case, so that they are still seen to primarily be a characteristic of the rotational polarization phenomena in the observed frequency dispersion of the dielectric permittivity of water at this particular input pulse frequency in the SHF radio frequency band, as seen in Figure 1.

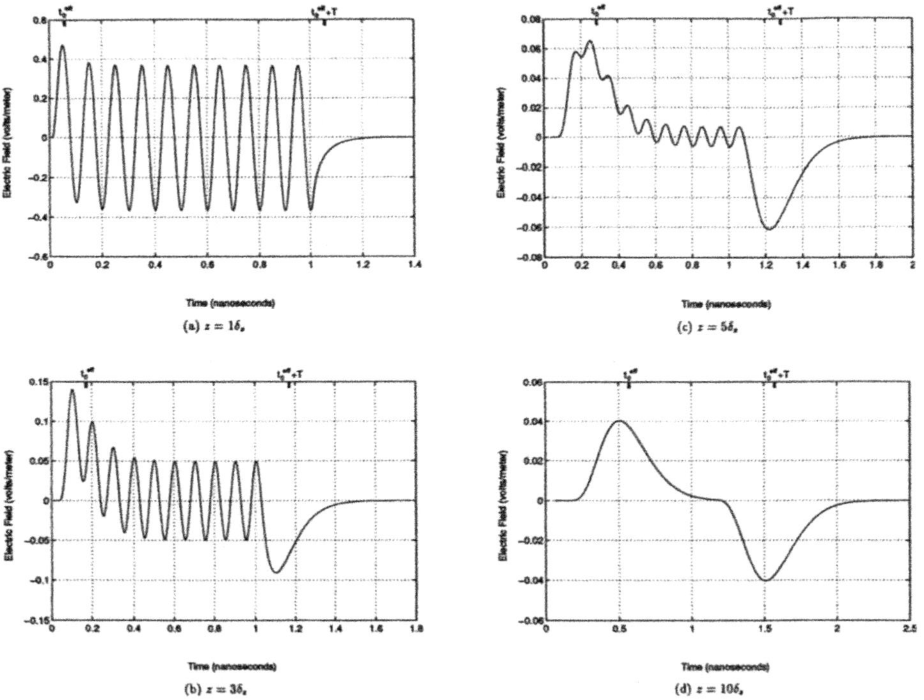

Figure 3. Dynamical field evolution due to an input *1 V/m*, *1 ns* rectangular envelope modulated sinusoidal field with f_1=*10 GHz* carrier frequency in a Rocard-Powles-Lorentz model of the dielectric frequency dispersion of triply-distilled water at the four propagation distances of *1*, *3*, *5*, and *10* absorption depths at the input carrier frequency of the pulse.

For the $f_3 = 100GHz$ case, $n_r(\omega_3) = 2.245$, $n_i(\omega_3) = 0.6483$, and the e^{-1} absorption depth is $\delta_s = 1.172\times 10^{-4} m$. The effective zero frequency limit of the index of refraction for this case is bounded above by the value given in Eq. (14) and is bounded below by the value

$$\theta_0^{eff} \cong \left(\varepsilon_\infty + a_2 + b_{11}^2/\omega_{11}^2 + b_{13}^2/\omega_{13}^2 + b_{15}^2/\omega_{15}^2\right)^{1/2} = 2.55 \quad , \tag{15}$$

which no longer depends upon the first Rocard-Powles component of the model given in Eq. (12). The dynamical field evolution depicted in Figure 4 again shows the exponential decay of the carrier signal component of the propagated field while the leading and trailing edge Brillouin precursors dominate the propagated field structure for all $z > \delta_s$. Notice that only the upper bounds to the values t_0^{eff} and $t_0^{eff} + T$ are indicated in Figure 4. In this EHF radio frequency band case the leading and trailing edge Brillouin precursors are characteristic of both the orientational and resonance polarization phenomena in the observed frequency dispersion of the dielectric permittivity of water, as seen in Figure 1. The observed slow decay of both the leading and trailing edge Brillouin precursors in Figure 4 is due to the slow relaxation of the orientational polarization of the medium.

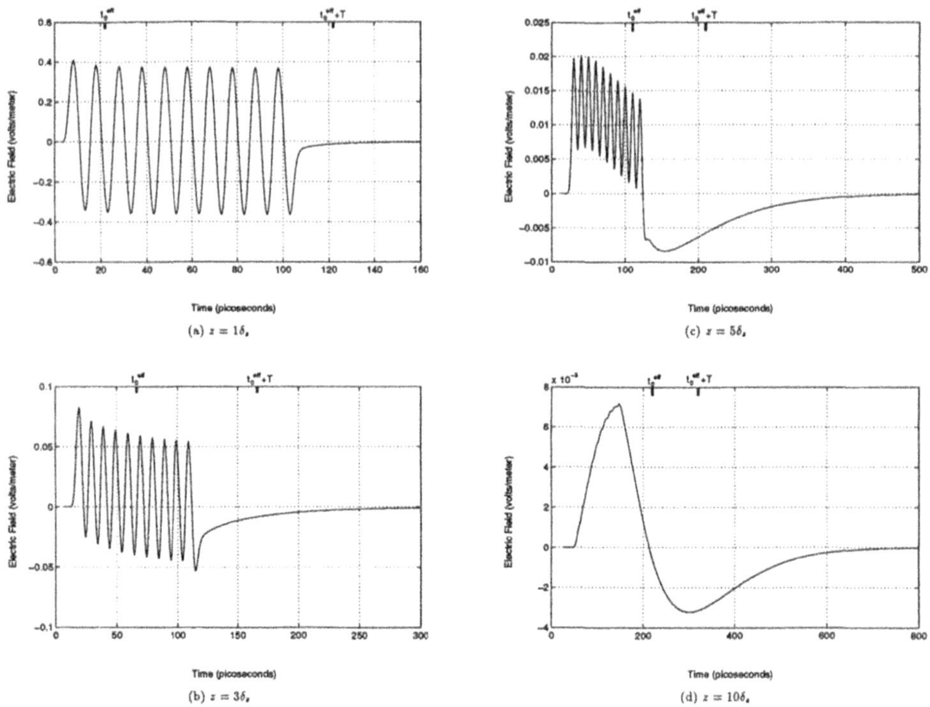

Figure 4. Dynamical field evolution due to an input *1 V/m, 100 ps* rectangular envelope modulated sinusoidal field with f_1*=100 GHz* carrier frequency in a Rocard-Powles-Lorentz model of the dielectric frequency dispersion of triply-distilled water at the four propagation distances of *1, 3, 5,* and *10* absorption depths at the input carrier frequency of the pulse.

For the $f_4 = 1THz$ case, $n_r(\omega_4) = 1.558$, $n_i(\omega_4) = 0.3848$, and the e^{-1} absorption depth is $\delta_s = 1.975\times10^{-5}m$. The effective zero frequency limit of the index of refraction for this case is approximately given by Eq. (15). The dynamical field evolution depicted in Figure 5 once again shows the exponential decay of the carrier signal component of the propagated field while the leading and trailing edge Brillouin precursors dominate the propagated field structure for all $z > \delta_s$. In this low-infrared frequency case the leading and trailing edge Brillouin precursors are now primarily a characteristic of the infrared resonance polarization phenomena in the observed frequency dispersion of the dielectric permittivity of water, as seen in Figure 1. The observed slow decay of both the leading and trailing edge Brillouin precursors in Figure 5 is due to the slow relaxation of the orientational polarization of the medium. Finally, notice the appearance of small amplitude leading and trailing edge Sommerfeld precursors in the dynamical field structure, particularly at the larger propagation distances depicted in the figure.

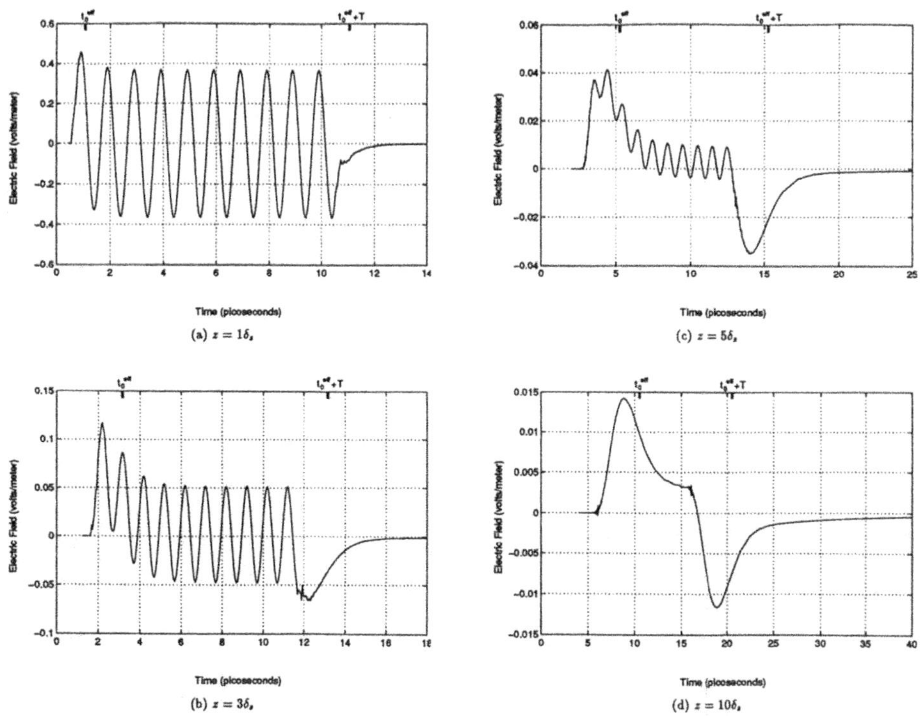

Figure 5. Dynamical field evolution due to an input *1 V/m, 10 ps* rectangular envelope modulated sinusoidal field with f_1=*1 THz* carrier frequency in a Rocard-Powles-Lorentz model of the dielectric frequency dispersion of triply-distilled water at the four propagation distances of *1, 3, 5,* and *10* absorption depths at the input carrier frequency of the pulse.

Finally, for the $f_5 = 10THz$ case, $n_r(\omega_5) = 1.308$, $n_i(\omega_5) = 0.0108$, and the e^{-1} absorption depth is $\delta_s = 7.024\times10^{-5}m$. The infrared carrier frequency $\omega_c = \omega_5$ is now completely removed from the Rocard-Powles model structure and the sampled material dispersion is nearly completely determined by the Lorentz model component. The effective zero frequency limit of the index of refraction for this case is then given by

$$\theta_0^{eff} \cong \left(\varepsilon_\infty + b_{11}^2/\omega_{11}^2 + b_{13}^2/\omega_{13}^2 + b_{15}^2/\omega_{15}^2 + b_{17}^2/\omega_{17}^2\right)^{1/2} = 1.50 \quad . \tag{16}$$

The dynamical field evolution depicted in Figure 6 shows that the observed pulse distortion is no longer primarily due to the leading and trailing edge Brillouin precursor, as it was in each of the previous, lower frequency cases, but is now primarily due to the leading and trailing edge middle precursors that are a characteristic of the infrared resonance lines in the observed frequency dispersion of the dielectric permittivity of triply-distilled water (see reference 23). The observed slowly decaying trailing edge of the propagated field structure depicted in Figure 6 is due to the trailing edge Brillouin precursor that is primarily a characteristic of the zero frequency limit of the infrared Lorentz lines in the composite Rocard-Powles-Lorentz model given in Eq. (12).

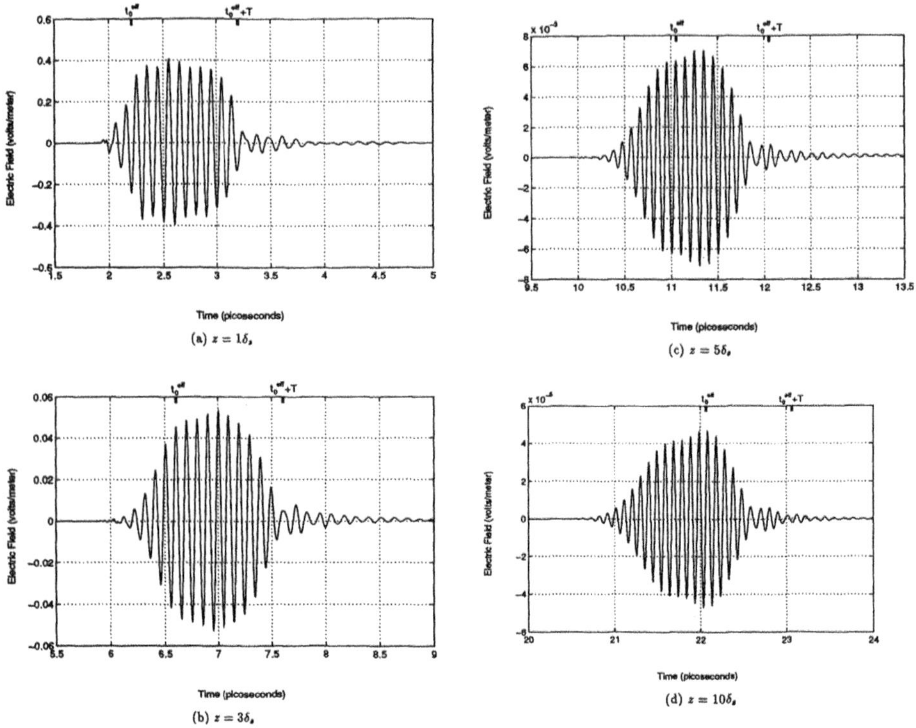

Figure 6. Dynamical field evolution due to an input *1 V/m*, *1 ps* rectangular envelope modulated sinusoidal field with f_l=*10 THz* carrier frequency in a Rocard-Powles-Lorentz model of the dielectric frequency dispersion of triply-distilled water at the four propagation distances of *1*, *3*, *5*, and *10* absorption depths at the input carrier frequency of the pulse.

DISCUSSION

The observed dynamical field evolution due to an input ultrawideband electromagnetic pulse as it propagates through a complex dispersive medium is primarily due to the causal model components of the dielectric permittivity dispersion that is spanned by the bandwidth of the input pulse. The various model components of a given dispersive medium may then be probed individually by a properly designed ultrawideband pulse so that, by varying the carrier frequency of the input pulse, different features may be studied and their associated parameter values may possibly be extracted. An ideal pulse for this purpose is the rectangular envelope modulated signal since it is always ultrawideband even though it may not be ultrashort. The initial pulse width may then be tailored so that its spectrum may sample any desired feature in the dielectric dispersion of the chosen material.

ACKNOWLEDGMENTS

The research presented in this paper was supported, in part, by the United States Air Force Office of Scientific Research under Grant # 49620-92-J-0419.

REFERENCES

1. A. Sommerfeld, Über die fortpflanzung des lichtes in disperdierenden medien, *Ann. Phys.* 44:177 (1914).
2. L. Brillouin, Über die fortpflanzung des licht in disperdierenden medien, *Ann. Phys.* 44:203 (1914).
3. L. Brillouin. "Wave Propagation and Group Velocity," Academic, New York (1960).
4. P. Debye, Näherungsformeln für die zylinderfunktionen für grosse werte des arguments und unbeschränkt verander liche werte des index, *Math. Anal.* 67:535 (1909).
5. K. E. Oughstun and G. C. Sherman, Propagation of electromagnetic pulses in a linear dispersive medium with absorption (the Lorentz medium), *J. Opt. Soc. Am. B* 5:817 (1988).
6. K. E. Oughstun and G. C. Sherman, Uniform asymptotic description of electromagnetic pulse propagation in a linear dispersive medium with absorption (the Lorentz medium), *J. Opt. Soc. Am. A* 6:1394 (1989).
7. K. E. Oughstun and G. C. Sherman, Asymptotic theory of pulse propagation in absorbing and dispersive dielectrics, in: "Review of Radio Science, 1990-1992," W. R. Stone, ed., Oxford (1993), pp. 75-105.
8. K. E. Oughstun and G. C. Sherman. "Electromagnetic Pulse Propagation in Causal Dielectrics," Springer-Verlag, Berlin (1994).
9. K. E. Oughstun and G. C. Sherman, Uniform asymptotic description of ultrashort rectangular optical pulse propagation in a linear, causally dispersive medium, *Phys. Rev. A* 41: 6090 (1990).
10. C. M. Balictsis and K. E. Oughstun, Uniform asymptotic description of ultrashort gaussian pulse propagation in a causal, dispersive dielectric, *Phys. Rev. E* 47: 3645 (1993).
11. C. M. Balictsis and K. E. Oughstun, Generalized asymptotic description of the propagated field dynamics in gaussian pulse propagation in a linear, causally dispersive medium, *Phys. Rev. E* 55: 1910 (1997).
12. F. W. J. Olver, Why steepest descents?, *SIAM Rev.* 12:228 (1970).
13. P. D. Smith and K. E. Oughstun, Electromagnetic energy dissipation of ultrawideband plane wave pulses in a causal, dispersive dielectric, in: "Ultra-Wideband, Short-Pulse Electromagnetics 2," L. Carin and L. B. Felsen, eds., Plenum, New York (1995), pp. 285-295.
14. P. D. Smith and K. E. 'Oughstun, Electromagnetic energy dissipation and propagation of an ultrawideband plane wave pulse in a causally dispersive dielectric, *Radio Science* (to be published).
15. K. E. Oughstun, Noninstantaneous, finite rise-time effects on the precursor field formation in linear dispersive pulse propagation, *J. Opt. Soc. Am. A* 12:1715 (1995).
16. K. Moten, C. Durney, and T. Stockham, Jr., Electromagnetic pulse propagation in dispersive planar dielectrics, Bioelectromagnetics 10:35 (1989).
17. K. Moten, C. Durney, and T. Stockham, Jr., Electromagnetic pulsed-wave radiation in spherical models of dispersive biological substances, Bioelectromagnetics 12:319 (1991).
18. R. Albanese, J. Penn, and R. Medina, Short rise-time microwave pulse propagation through dispersive biological media, J. Opt. Soc. Am. A 6:1441 (1989).
19. R. Albanese, J. Blaschak, R. Medina, and J, Penn, Ultrashort electromagnetic signals: biophysical questions, safety issues, and medical opportunities, Aviat. Space Environ. Med. 65:116 (1994).

20.S. Shen and K. E. Oughstun, Dispersive pulse propagation in a double resonance Lorentz medium, J. Opt. Soc. Am. B 6:948 (1989).
21.K. E. Oughstun, J. E. K. Laurens, and C. M. Balictsis, Asymptotic description of electromagnetic pulse propagation in a linear dispersive medium, in: "Ultra-Wideband, Short-Pulse Electromagnetics," L. Carin and L. B. Felsen, eds., Plenum, New York (1993), pp. 223-240.
22.K. E. Oughstun, Dynamical structure of the precursor fields in linear dispersive pulse propagation in lossy dielectrics, in: "Ultra-Wideband, Short-Pulse Electromagnetics 2," L. Carin and L. B. Felsen, eds., Plenum, New York (1995), pp. 257-272.
23.J. E. K. Laurens and K. E. Oughstun, Electromagnetic impulse response of triply-distilled water, in: "Ultra-Wideband, Short-Pulse Electromagnetics 4."

PULSED BEAM PROPAGATION IN LOSSLESS DISPERSIVE MEDIA

Timor Melamed[1] and Leopold B. Felsen[2]

[1]Odin Technologies, P.O. Box 248, Yokne'am Elit 20698, Israel
[2]Department of Aerospace and Mechanical Engineering
Boston University, 110 Cummington St., Boston MA 02215.

INTRODUCTION

In this paper, we employ general (wavenumber – frequency) spectral analysis and synthesis to asymptotically treat the evolution of the signal excited by pulsed beam (PB) wavepacket [1, 2, 3] initial conditions in a lossless homogeneous dispersive medium described generically by the frequency-dependent wavenumber $k(\omega)$. For the well-explored plane wave case, this requires asymptotic analysis of the Fourier inversion integral from the *frequency* domain. However, unlike the plane wave case, the spatially and temporally localized PB input also requires spectral analysis and synthesis in the wavenumber domain. This has already been performed in previous nondispersive studies [3], with emphasis on paraxial asymptotics for approximating the PB propagators. We shall employ this useful approximation for the present PB studies in the *dispersive* environment and perform the asymptotic paraxial inversion from the spectral domain under these generalized conditions. Our goal is to extract via simple high frequency saddle point asymptotics the effects of dispersion on the paraxial PB propagators. These effects will be characterized and quantified in terms of *critical nondimensional* parameters which capture the relevant dispersive wave physics pertaining to various space-time observation domains for various choices of PB spatial width D and pulse length T. In the well developed (late-time) dispersive regime, this can be done for the generic $k(\omega)$. However, near the first arrival (wavefront regime) at the observer, specific dispersion models must be invoked in order to determine the uniform transition from the nondispersive behavior very near the wavefront to the fully dispersive behavior sufficiently far behind the wavefront [4, 5]. The critical nondimensional descriptors, i.e. the footprints, of various observables of the paraxial

Ultra-Wideband, Short-Pulse Electromagnetics 4
Edited by Heyman *et al.*, Kluwer Academic / Plenum Publishers, New York, 1999.

PB-excited signal are calibrated via higher-order off-axis expansions beyond the second order paraxial range, thereby allowing quantitative limits to be established by examining the conditions for neglecting the higher order terms. A variety of PB observables are parameterized nondimensionally in this fashion. In the text to follow, after stating the problem, we first summarize time domain (TD) results obtained by beam inversion from the frequency domain (FD) as in [4]. We then proceed to the new direct TD treatment in terms of wavenumber spectra of pulsed plane waves, followed by wavenumber synthesis and asymptotics.

STATEMENT OF THE PROBLEM

The problem of PB wavepacket propagation in a lossless, homogeneous, isotropic dispersive medium is first defined by the PB-matched initial distribution in the frequency domain and is Fourier inverted from there into the time domain.

Time-frequency Transforms

Given a TD field, $u(\mathbf{r}, t)$, the corresponding FD field $\widehat{u}(\mathbf{r}; \omega)$ is defined by the Fourier transform relations

$$\widehat{u}(\mathbf{r}, \omega) = \int_{-\infty}^{\infty} u(\mathbf{r}, t) e^{i\omega t} dt, \quad u(\mathbf{r}, t) = \frac{1}{2\pi} \int_{-\infty}^{\infty} \widehat{u}(\mathbf{r}, \omega) e^{-i\omega t} d\omega \tag{1 a,b}$$

where $\mathbf{r} = (x_1, x_2, z)$ are conventional Cartesian coordinates. Here, and henceforth, frequency domain fields are denoted by a caret $\hat{}$. Throughout, we shall utilize the analytic signal formulation for time-dependent fields in order to accommodate wave constituents with evanescent spectra, as encountered in the PB. The analytic field $\overset{+}{u}(\mathbf{r}, t)$ (denoted by the symbol $^+$) corresponding to the FD field $\widehat{u}(\mathbf{r}, \omega)$ is obtained by the one-sided inverse Fourier transform

$$\overset{+}{u}(\mathbf{r}, t) = \frac{1}{\pi} \int_{0}^{\infty} \widehat{u}(\mathbf{r}, \omega)\, e^{-i\omega t} d\omega, \ \ \mathrm{Im}\, t \leq 0, \tag{2}$$

where $\widehat{u}(\mathbf{r}, \omega)$ is defined in (1a). The real field is obtained in this formulation from the real part of the analytic field,

$$u(\mathbf{r}, t) = \mathrm{Re}\, \overset{+}{u}(\mathbf{r}, t). \tag{3}$$

Space-wavenumber Transforms

The time-domain initial distribution $u_o(\mathbf{x}, t)$ for the PB on the $z = 0$ plane will be synthesized via inversion of the corresponding frequency-domain distribution $\widehat{u}_o(\mathbf{x}, \omega)$. The FD wavenumber spectral amplitude on the initial surface is given by

$$\widehat{\widetilde{u}}_o(\boldsymbol{\xi}, \omega) = \int_{-\infty}^{\infty} d^2x\ \widehat{u}_o(\mathbf{x}, \omega)\ e^{-ik(\omega)\boldsymbol{\xi}\cdot\mathbf{x}} \tag{4a}$$

where $\boldsymbol{\xi} = (\xi_1, \xi_2)$ is the normalized spatial wavenumber vector, $\mathbf{x} = (x_1, x_2)$, $k(\omega)$ is the frequency-dependent wavenumber in the ambient medium, and $\tilde{}$ identifies a wavenumber spectral function. The reconstruction of the FD initial field is, accordingly,

$$\widehat{u}_o(\mathbf{x}, \omega) = (k(\omega)/2\pi)^2 \int d^2\xi\ \widehat{\widetilde{u}}_o(\boldsymbol{\xi}, \omega)\ e^{ik(\omega)\,\boldsymbol{\xi}\cdot\mathbf{x}}. \tag{4b}$$

The normalization with respect to the wavenumber $k(\omega)$ anticipates inversion to the time domain, rendering $\boldsymbol{\xi}$ frequency-independent, with direct geometrical interpretation in terms of the spectral plane wave propagation angles. For simplicity, the $\omega-$dependence of $k = k(\omega)$ and of $\widehat{u}(\mathbf{r}) = \widehat{u}(\mathbf{r},\omega)$ will be suppressed unless specifically required for clarity. Also, integration limits are omitted on all integrals extending from $-\infty$ to $+\infty$.

By inclusion of the plane wave spectral propagator $\exp(ik\zeta z)$, the plane-wave representation for the FD field away from the initial plane is

$$\widehat{u}(\mathbf{r},\omega) = (k/2\pi)^2 \int d^2\xi \, \widehat{\widetilde{u}}_o(\boldsymbol{\xi})\, e^{ik\mathring{\boldsymbol{\kappa}}\cdot\mathbf{r}}, \tag{5}$$

where $\mathring{\boldsymbol{\kappa}} = (\boldsymbol{\xi},\zeta)$ is a unit vector in the direction of propagation, with

$$\zeta = \sqrt{1-\boldsymbol{\xi}^2}, \qquad \boldsymbol{\xi}^2 \equiv \boldsymbol{\xi}\cdot\boldsymbol{\xi}, \qquad \operatorname{Im}\zeta \geq 0. \tag{6}$$

Inserting (5) into (2) , we obtain the formal plane wave spectral representation of the TD analytic field at any observation point $\mathbf{r}$,

$$\overset{+}{u}(\mathbf{r},t) = \frac{1}{\pi}\int_0^\infty d\omega \left(\frac{k(\omega)}{2\pi}\right)^2 \int d^2\xi \, \widehat{\widetilde{u}}_o(\boldsymbol{\xi},\omega)\, e^{-i\omega t + ik(\omega)\mathring{\boldsymbol{\kappa}}\cdot\mathbf{r}} \tag{7}$$

The Initial Distribution

The initial field distribution is chosen so that it generates a pulsed beam (PB) wavepacket in the dispersive environment. For purposes of focusing such a wavepacket at a particular space-time point, it is desirable to select the beam parameters in such a way that the focusing distance is independent of frequency. In the FD such initial distributions have the form [3]

$$\widehat{u}_o(\mathbf{x},\omega) = \widehat{f}(\omega)e^{-\frac{1}{2}k(\omega)\,\rho^2/\beta}, \tag{8}$$

where $\rho^2 \equiv \mathbf{x}\cdot\mathbf{x}$, $\widehat{f}$ is some frequency-dependent function and $\beta = \beta_r + i\beta_i$ (with $\beta_r > 0$ for $\omega > 0$) is a frequency-independent parameter. The corresponding plane-wave spectrum $\widehat{\widetilde{u}}_o$ of the above distribution, obtained by inserting (8) into (4a), is

$$\widehat{\widetilde{u}}_o(\boldsymbol{\xi},\omega) = (2\pi\widehat{f}\beta/k)e^{-\frac{1}{2}k\,\beta\boldsymbol{\xi}^2}. \tag{9}$$

The TD initial distribution is obtained via the Fourier inversion in (2), yielding

$$\overset{+}{u}_o(\mathbf{x},t) = \int_0^\infty d\omega \; \widehat{f}(\omega)e^{-i\omega t - \frac{1}{2}k(\omega)\,\mathbf{x}^2/\beta} \tag{10}$$

Note that the integral in (10) depends on the details of the dispersive medium which is so far described by the generic dispersion relation $k = k(\omega)$.

In evaluating the integral in (7), two approaches may be considered: "FD First", in which the frequency domain field is evaluated via $d^2\xi$ integration before transforming the result into the time domain via (1b); and 2) "TD first", by interchanging the order of integration in (7) and evaluating the $d\omega$ integral first, thereby obtaining a *transient plane-wave* spectral representation for the field in the dispersive medium. The first approach, presented in detail in [4, 5], is summarized in next section. The second approach is summarized in the section thereafter; further details and generalizations are in preparation for separate publication [6].

FD SPECTRAL INTEGRAL AND PARAXIAL ASYMPTOTICS

The FD field, corresponding to the FD initial distribution in (8), is given by the plane-wave spectral integral in (5),

$$\widehat{u}(\mathbf{r},\omega) = (\beta k(\omega)\widehat{f}(\omega)/2\pi)\int d^2\xi \; e^{-i\omega t + ik(\omega)\,(\frac{i}{2}\beta\boldsymbol{\xi}^2 + \mathring{\boldsymbol{\kappa}}\cdot\mathbf{r})}. \tag{11}$$

Using the saddle point technique, as well as a paraxial approximation, this integral may be evaluated asymptotically [3]. The result is

$$\widehat{u}(\mathbf{r},\omega) \sim \widehat{f}(\omega)\frac{-i\beta}{z-i\beta}\, e^{ik(\omega)S(\mathbf{r})} \tag{12}$$

where

$$S(\mathbf{r}) = z + \tfrac{1}{2}\rho^2/(z-i\beta). \tag{13}$$

is the normalized paraxial phase.

From (12) with (13), the field is identified as a paraxial Gaussian beam (GB) propagating in the z direction. In order to parameterize the beam field, we write the real and imaginary parts of S as follows

$$S = z + \tfrac{1}{2}\rho^2/(z-Z-iF) = z + \tfrac{1}{2}\rho^2(1/R + i/I), \quad I = kD^2 \tag{14}$$

where

$$Z = -\beta_i, \quad F = \beta_r, \tag{15}$$

and

$$D = \sqrt{F/k}\sqrt{1+(z-Z)^2/F^2}, \quad R = (z-Z) + F^2/(z-Z). \tag{16}$$

By substituting (14) into (12) one readily identifies $2\sqrt{2}D$ and R as the e^{-1} *beam width* and the *phase-front radius of curvature*, respectively. The GB *waist* is located at $z = Z$ and its *collimation length* is F. Note that the waist location Z, the collimation length F, and therefore I as well as the phase S as a whole, are frequency independent, as intended.

TD INVERSION

The TD representation for the PB field, corresponding to the asymptotic (paraxial) FD field in (12), is obtained by the one-sided Fourier inversion in (2), i.e.,

$$\overset{+}{u}(\mathbf{r},t) = \frac{-i\beta}{z-i\beta}\,\frac{1}{\pi}\int_0^\infty d\omega\, \widehat{f}(\omega)e^{-i\Phi(\omega;\mathbf{r},t)}, \quad \Phi(\omega;\mathbf{r},t) = \omega t - k(\omega)S(\mathbf{r}) \tag{17}$$

where S is defined in (13).

Nondispersive Case

For nondispersive media, the integral in (17) can be evaluated in closed form. Inserting $k(\omega) = \omega/c$, we obtain $u(\mathbf{r},t) = \mathrm{Re}\,\overset{+}{u}(\mathbf{r},t)$ where

$$\overset{+}{u}(\mathbf{r},t) = \frac{-i\beta}{z-i\beta}\,\overset{+}{f}\,[t - c^{-1}(z + \frac{1}{2}\rho^2/(z-i\beta))]. \tag{18}$$

A conventional waveform is obtained by choosing [3] $\widehat{f}(\omega) = e^{-\omega T/2}$ which implies that

$$\overset{+}{f}(t) = \overset{+}{\delta}(t - iT/2), \quad f(t) = \mathrm{Re}\,\overset{+}{f} = \frac{1}{\pi}\,\frac{T/2}{t^2+(T/2)^2}, \tag{19}$$

where $\overset{+}{\delta}(t) = 1/\pi i t$ is the analytic δ-function defined in the lower half of the complex t-plane. This waveform peaks at $t = 0$ and its pulse length and peak values are given by T and $2/\pi T$, respectively. Also note that the maximum frequency of this signal may be estimated by

$$\omega_{max} \simeq T^{-1}. \tag{20}$$

Therefore, such a band-limited pulse may be regarded as a model for a physical (sampled) signal. Inserting (19) into (18), we obtain

$$\overset{+}{u}(\mathbf{r},t) = \frac{-i\beta}{z-i\beta}\ \overset{+}{\delta}\,[t - i\frac{T}{2} - c^{-1}(z + \frac{1}{2}\rho^2/(z-i\beta))]. \tag{21}$$

This PB field has been discussed thoroughly in [1, 2, 3]. Here, we shall summarize the main features since they are also relevant for the dispersive case. The PB field in (21) propagates along the z axis. Its space-time structure may be inferred from the elements of the frequency-independent normalized phase S in (14), with $R(z)$ in (16) now interpreted as the wavefront radius of curvature and $T_p(\mathbf{r}) = T + c^{-1}\rho^2/I(z)$ interpreted as the temporal half-amplitude width of the $\overset{+}{\delta}$ pulse, which is inversely proportional to the pulse amplitude (see $f(t)$ in (19)). The field is strongest on the beam axis $\rho = 0$ where $T_p(\mathbf{r}) = T$ is minimal, and it decays as T_p grows away from the beam axis. On the wavefront, $t = c^{-1}[z + \rho^2/2R(z)]$, the field amplitude is proportional to $T_p^{-1}(\rho)$ and the half-amplitude beam width in the ρ direction is found by solving $T_p(\rho) = 2T_p(0)$, giving

$$D(z) = 2\sqrt{cTI(z)} \tag{22}$$

The collimation length is given by F in (15) and the waist is located at $z = Z$ with the width $2\sqrt{cTF}$. From (22) with (14) one notes that in the collimation (Fresnel) zone $|z - Z| < F$, the PB profile is essentially unchanged, whereas outside this zone, the profile broadens and approaches the asymptotic far field diffraction angle $\Theta = 2\sqrt{vT/F}$.

Asymptotic Evaluation of the Dispersive PB Field

When the medium is dispersive, the integral in (17) cannot generally be evaluated in closed form. We shall evaluate (17) asymptotically by continuing the integrand analytically into the complex ω-plane and applying the saddle point method, with the saddle point denoted by ω_s. Our goal is the parameterization of the effect of dispersion on the paraxial PB field. To quantify the effect, it is necessary to retain higher order terms in the transverse coordinate ρ at various stages of the analysis. Accordingly, when necessary, paraxial $O(\rho^2)$ quantities will be identified by the subscript p.

Using $\hat{f}(\omega) = \exp(-\omega T/2)$ in (17) yields

$$\overset{+}{u}(\mathbf{r},t) = \frac{-i\beta}{z-i\beta}\,\frac{1}{\pi}\int_0^{\infty} d\omega\, e^{-i\Phi(\omega;\mathbf{r},t)}, \quad \Phi(\omega;\mathbf{r},t) = \omega(t - i\frac{T}{2}) - k(\omega)S(\mathbf{r}). \tag{23}$$

From (23), we find that the stationary frequency ω_s satisfies

$$\frac{t - iT/2}{S} = k'(\omega_s), \tag{24}$$

with the prime denoting the derivative with respect to the argument. Solving (24) even for on-axis observation points requires analytic continuation of the dispersion surface $ck(\omega)$ into the complex ω-domain. In the following analysis we assume that *both* T and ρ are small parameters with respect to t and z, respectively.

We denote the real on-axis stationary point corresponding to $T = 0$ by $\bar{\omega}_s \equiv \omega_s(\rho = 0, T = 0)$. Using $T = 0$ and $\rho = 0$ in (24), we obtain

$$c\frac{dk}{d\omega}\bigg|_{\omega=\bar{\omega}_s} = \bar{\Omega}, \quad \bar{\Omega} \equiv \frac{ct}{z}. \tag{25}$$

For interpretation of this equation in terms of space-time rays and the (k,ω) dispersion surface, see [7, Sec. 1.6] as well as Fig. 1. For points near the beam axis, an approximate expression for the field in (17) can be obtained by a expanding $\Phi(\mathbf{r},\omega)$ in a Taylor series about the on-axis stationary point $\bar{\omega}_s$ (only the ω-dependence is shown),

$$\Phi(\omega) = \Phi_0 + \Phi_1(\omega - \bar{\omega}_s) + \frac{1}{2}\Phi_2(\omega - \bar{\omega}_s)^2 \tag{26}$$

where

$$\Phi_o \equiv \Phi(\bar{\omega}_s) = \bar{\omega}_s(t - \frac{i}{2}T) - k(\bar{\omega}_s)S(\mathbf{r}) \tag{27}$$

and

$$\Phi_1 \equiv \frac{\partial \Phi}{\partial \omega}\Big|_{\bar{\omega}_s} = t - \frac{i}{2}T - k'(\bar{\omega}_s)S = t(1 - \frac{S}{z}) - \frac{i}{2}T. \tag{28}$$

The last equality in (28) is obtained by using (25). Φ_2 is given by

$$\Phi_2 \equiv \frac{\partial^2 \Phi}{\partial \omega^2}\Big|_{\bar{\omega}_s} = -k''(\bar{\omega}_s)S. \tag{29}$$

Thus, from (26), the stationary point $\Phi'(\omega_s) = 0$ is seen to be

$$\omega_s = \bar{\omega}_s - \Phi_1/\Phi_2. \tag{30}$$

Choosing Re $\bar{\omega}_s > 0$, the field in (17) is asymptotically evaluated by the lowest order saddle point formula [7] to yield

$$\overset{+}{u}(\mathbf{r},t) \sim A(\mathbf{r},t)e^{-i\Psi(\mathbf{r},t)}, \tag{31}$$

where the phase is obtained by inserting (30) into (26)

$$\Psi(\mathbf{r},t) \equiv \Phi(\omega_s;\mathbf{r},t) = \Phi_o - \frac{1}{2}\Phi_1^2/\Phi_2, \tag{32}$$

and the amplitude is given by

$$A(\mathbf{r},t) = \frac{-i\beta}{z - i\beta}\sqrt{\frac{2}{\pi i k''(\bar{\omega}_s)S}}. \tag{33}$$

Recall that ω_s and $\bar{\omega}_s$ are functions of $(\mathbf{r},t)$ and (z,t), respectively, whereas $\Phi_{0,1,2}(\mathbf{r},\bar{\omega}_s)$ depend on $\mathbf{r}$ explicitly and on (z,t) implicitly through $\bar{\omega}_s$. Note that in (33) the amplitude is evaluated using $\bar{\omega}_s$ instead of ω_s. This approximation is based on the reasonable assumption that the deviation of the off-axis stationary point from the on-axis stationary point may be neglected in the *amplitude*. Also note that in (32), the ρ-dependence is not restricted to $O(\rho^2)$ and therefore goes beyond the the paraxial approximation.

Inserting (27–29) into (32), we obtain

$$\Psi = \Psi_p + \Psi_d \tag{34}$$

where

$$\Psi_p(\mathbf{r},t) = \bar{\omega}_s(t - \frac{i}{2}T) - k(\bar{\omega}_s)S(z,\rho) \tag{35}$$

is the conventional paraxial approximation and

$$\Psi_d = -\frac{1}{8}[\frac{t\rho^2}{(z - i\beta)z} - iT]^2/Sk''(\bar{\omega}_s) \tag{36}$$

is the deviation of Ψ from the paraxial approximation. The condition $\Psi_d \ll 2\pi$, under which Ψ_d in (34) may be neglected, is discussed and parameterized in [4, 5]. Subject to this condition, the field is paraxially approximated by

$$\overset{+}{u}_p(\mathbf{r},t) = \frac{-i\beta}{z - i\beta}\sqrt{\frac{2}{\pi i k''(\bar{\omega}_s)S}}e^{-i\Psi_p(\mathbf{r},t)}, \quad \Psi_p(\mathbf{r},t) = \bar{\omega}_s(t - i\frac{T}{2}) - k(\bar{\omega}_s)S. \tag{37}$$

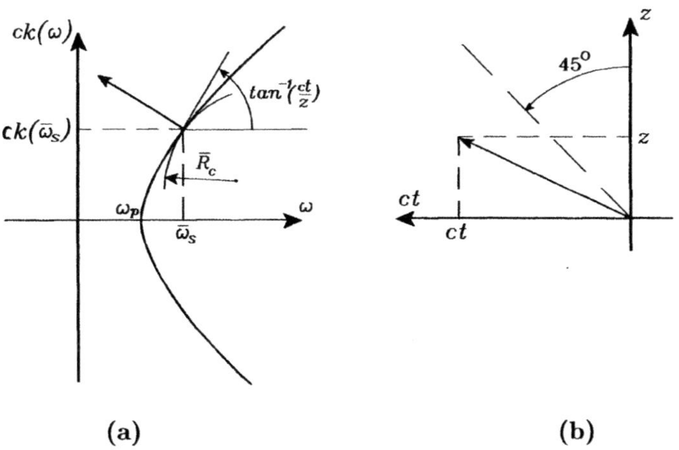

Figure 1. On–axis PB asymptotics, dispersion surface and space–time rays. a) $k(\omega)$ dispersion surface. The normal to the surface (see (25)) is parallel to the space-time ray to the observation point (z, ct). The construction determines the saddle point values $\bar{\omega}_s(z,t)$ and $k[\bar{\omega}_s(z,t)]$. The local on–axis radius of curvature $\bar{R}_c$ of the dispersion curve is also shown (see (54)). b) Space–time ray to the observation point (z, ct).

TRANSIENT SPECTRUM REPRESENTATION OF THE PB FIELD

Transient Dispersive Plane Wave Spectrum

We now pursue the second option after eq. (10) and invert the order of integration in (7), thereby evaluating the ω-integral first, and obtaining the wavenumber spectral representation

$$\overset{+}{u}(\mathbf{r},t) = \int d^2\xi \;\; \overset{+}{\tilde{u}}(\mathbf{r},t;\boldsymbol{\xi}). \tag{38}$$

Here, the spectral wave functions $\overset{+}{\tilde{u}}(\mathbf{r},t;\boldsymbol{\xi})$ are dispersive transient plane waves

$$\overset{+}{\tilde{u}}(\mathbf{r},t;\boldsymbol{\xi}) = \frac{1}{\pi}\int_0^\infty d\omega \;\left[\frac{k(\omega)}{2\pi}\right]^2 \hat{\tilde{u}}_o(\boldsymbol{\xi},\omega)\; e^{-i\omega t + ik(\omega)\mathring{\boldsymbol{\kappa}}\cdot\mathbf{r}} \tag{39}$$

propagating in the direction of the unit vector $\mathring{\boldsymbol{\kappa}} = (\boldsymbol{\xi},\zeta)$. The plane wave spectrum $\hat{\tilde{u}}_o$ for the PB initial conditions is given by (9).

Formal solution and asymptotics

The TD spectral constituents excited by the frequency domain PB initial distribution in (8), with (9), are given formally by inserting (9) into (39),

$$\overset{+}{\tilde{u}}(\mathbf{r},t;\boldsymbol{\xi}) = \frac{1}{\pi}\int_0^\infty d\omega\; \beta k(\omega)\; e^{-i\Phi(\mathbf{r},t;\boldsymbol{\xi},\omega)}, \quad \operatorname{Im} t \le 0 \tag{40}$$

where $k(\omega)$ is the generic bulk wavenumber characterizing the frequency-dependent propagation properties in the lossless homogeneous dispersive medium, and

$$\Phi(\mathbf{r},t;\boldsymbol{\xi};\omega,T) = \omega(t - \frac{i}{2}T) - k(\omega)\widetilde{S}, \quad \widetilde{S}(\boldsymbol{\xi},\mathbf{r}) = \mathring{\boldsymbol{\kappa}}\cdot\mathbf{r} + \frac{i}{2}\beta\boldsymbol{\xi}^2. \tag{41}$$

Except when required for clarity, we shall not include in the notation all of the parameters in the argument of Φ. The integral in (40) cannot generally be evaluated in closed form. We shall evaluate (40) asymptotically by continuing the integrand analytically into the complex ω-plane and applying the saddle point method, with the stationary frequency denoted by $\widetilde{\omega}_s(\mathbf{r}, t; \boldsymbol{\xi})$. The stationary frequency satisfies the relation $(d\Phi/d\omega) = 0$ at $\widetilde{\omega}_s$, so that, using (41),

$$\frac{t - \frac{i}{2}T}{\widetilde{S}} = k'(\widetilde{\omega}_s), \quad \widetilde{\omega}_s = \widetilde{\omega}_s(\mathbf{r}, t; \boldsymbol{\xi}; T). \tag{42}$$

The lowest order saddle point formula [7] yields for the field in (40) the asymptotic approximation

$$\overset{+}{\widetilde{u}}(\mathbf{r}, t; \boldsymbol{\xi}) \sim \overset{+}{\widetilde{A}}(\mathbf{r}, t; \boldsymbol{\xi})\, e^{-i\widetilde{\Psi}(\mathbf{r},t,\boldsymbol{\xi})} \tag{43}$$

with the spectral phase

$$\widetilde{\Psi}(\mathbf{r}, t; \boldsymbol{\xi}) \equiv \Phi(\mathbf{r}, t; \boldsymbol{\xi}, \widetilde{\omega}_s, T) = (t - \frac{i}{2}T)\widetilde{\omega}_s - k(\widetilde{\omega}_s)\widetilde{S}, \tag{44}$$

and amplitude

$$\overset{+}{\widetilde{A}}(\mathbf{r}, t, \boldsymbol{\xi}) = \beta k(\widetilde{\omega}_s)\left[-2\pi i\widetilde{S}k''(\widetilde{\omega}_s)\right]^{-1/2}. \tag{45}$$

For $T = 0$ (temporal impulse) and $\boldsymbol{\xi} = 0$ (propagation along the beam axis), the resulting $\widetilde{S} = z$ is real, and eq. (42) has a real solution, $\bar{\omega}_s$. For more general solutions of (42), the spatial localization, expressed by $\exp[-\frac{1}{2}k(\omega)\beta\boldsymbol{\xi}^2]$ in (9), confines the main contribution to the $\boldsymbol{\xi} \approx 0$ spectral constituents. With T "small" (quantified later on (see (57) and (58)) and $\boldsymbol{\xi} \approx 0$, we may perform perturbation around $\bar{\omega}_s(\mathbf{r}, t)$ in (25) and $\widetilde{\Psi}(\mathbf{r}, t; \boldsymbol{\xi})$ in (44). Expanding $\widetilde{\Psi}(\mathbf{r}, t; \boldsymbol{\xi})$ in a Taylor series about $T = 0$ and $\boldsymbol{\xi} = 0$, and truncating the expression after the quadratic term yields (only the $\boldsymbol{\xi}$ and T-dependence is shown)

$$\widetilde{\Psi}(T, \boldsymbol{\xi}) = \widetilde{\Psi}_{00} + \widetilde{\Psi}_{01}T + \widetilde{\boldsymbol{\Psi}}_{01} \cdot \boldsymbol{\xi} + \frac{1}{2}\widetilde{\Psi}_{20}T^2 + \widetilde{\boldsymbol{\Psi}}_{11} \cdot \boldsymbol{\xi}T + \frac{1}{2}\boldsymbol{\xi} \cdot \widetilde{\boldsymbol{\Psi}}_{02} \cdot \boldsymbol{\xi} + \ldots \tag{46}$$

with coefficients

$$\widetilde{\Psi}_{00} \equiv \widetilde{\Psi}|_{T=0,\boldsymbol{\xi}=0} = \bar{\omega}_s t - k(\bar{\omega}_s)z, \tag{46a}$$

$$\widetilde{\Psi}_{10} \equiv \partial_T\widetilde{\Psi}|_{T=0,\boldsymbol{\xi}=0} = -\frac{i}{2}\bar{\omega}_s, \tag{46b}$$

$$\widetilde{\Psi}_{20} \equiv \partial_T^2\widetilde{\Psi}|_{T=0,\boldsymbol{\xi}=0} = [-4k''(\bar{\omega}_s)z]^{-1}, \tag{46c}$$

$$\widetilde{\boldsymbol{\Psi}}_{01} \equiv \nabla_\xi\widetilde{\Psi}|_{T=0,\boldsymbol{\xi}=0} = -k(\bar{\omega}_s)\mathbf{x}, \tag{46d}$$

$$\widetilde{\boldsymbol{\Psi}}_{11} \equiv \partial_T\nabla_\xi\widetilde{\Psi}|_{T=0,\boldsymbol{\xi}=0} = it\mathbf{x}/(2k''(\bar{\omega}_s)z^2) \tag{46e}$$

$$\widetilde{\boldsymbol{\Psi}}_{02} \equiv \nabla_\xi\nabla_\xi\widetilde{\Psi}|_{T=0,\boldsymbol{\xi}=0} = -k(\bar{\omega}_s)\mathbf{M} + \frac{t^2}{k''(\bar{\omega}_s)z^3}\mathbf{L} \tag{46f}$$

where

$$\mathbf{M} = \begin{bmatrix} i\beta - z & 0 \\ 0 & i\beta - z \end{bmatrix}, \quad \mathbf{L} = \begin{bmatrix} x_1^2 & x_1x_2 \\ x_1x_2 & x_2^2 \end{bmatrix}. \tag{47}$$

Details of the straightforward but somewhat lengthy derivations are given in [6].

Paraxial approximation and its validation

Using (46), the phase $\widetilde{\Psi}$ may be written in the form

$$\widetilde{\Psi} = \widetilde{\Psi}_p + \widetilde{\Psi}_d \tag{48}$$

where

$$\widetilde{\Psi}_p = (t - \frac{i}{2}T)\bar{\omega}_s - k(\bar{\omega}_s)\left[z + \mathbf{x}\cdot\boldsymbol{\xi} + \frac{1}{2}\boldsymbol{\xi}\cdot\mathbf{M}\cdot\boldsymbol{\xi}\right] \tag{49}$$

is the *spectral paraxial* phase, and

$$\widetilde{\Psi}_d = \frac{1}{k''(\bar{\omega}_s)z}\left[-\frac{1}{8}T^2 + \frac{t^2}{2z^2}\boldsymbol{\xi}\cdot\mathbf{L}\cdot\boldsymbol{\xi} + \frac{itT\mathbf{x}\cdot\boldsymbol{\xi}}{2z}\right] \tag{50}$$

is the non-paraxial term. Inserting (47) into (50) yields

$$\widetilde{\Psi}_d = [2k''(\bar{\omega}_s)z]^{-1}\left[\frac{iT}{2} + \frac{t}{z}\boldsymbol{\xi}\cdot\mathbf{x}\right]^2. \tag{51}$$

With $\widetilde{\Psi}_p$ from (49) and the amplitude term from (45), we obtain the paraxial approximation for the plane wave field in (43),

$$\overset{+}{\tilde{u}}(\mathbf{r},t;\boldsymbol{\xi}) = \beta k(\bar{\omega}_s)\left[-2\pi izk''(\bar{\omega}_s)\right]^{-1/2}\exp[-i\widetilde{\Psi}_p(\mathbf{r},t;\boldsymbol{\xi})]. \tag{52}$$

To justify the paraxial approximation we shall examine the parametric regimes for which $\widetilde{\Psi}_d$ in (51) may be neglected. We shall anchor the parameterization to the (ck,ω) dispersion surface. The location of the plane wave stationary point $\bar{\omega}_s$ on the dispersion surface is found from the condition (25). The radius of curvature R_c of the dispersion surface $ck(\omega)$ is given by

$$R_c(\omega) = \left\{1 + \left[ck'(\omega)\right]^2\right\}^{3/2}/ck''(\omega). \tag{53}$$

Using $\bar{\omega}_s$ in (25), its value $\bar{R}_c$ at the space-time *on-axis* observation point (z,t) is

$$\bar{R}_c \equiv R_c(\bar{\omega}_s) = \frac{(1+\bar{\Omega}^2)^{3/2}}{ck''(\bar{\omega}_s)}, \tag{54}$$

which has the dimensionality $\omega \sim t^{-1}$ (see Fig. 1). Then from (51) and (54), the phase $\widetilde{\Psi}_d$ has the form

$$\widetilde{\Psi}_d = \frac{c}{-2z}\left(1+\bar{\Omega}^2\right)^{-3/2}|\bar{R}_c|\left[\frac{T}{2} - \frac{it}{z}\boldsymbol{\xi}\cdot\mathbf{x}\right]^2, \tag{55}$$

and may be neglected provided that $|\widetilde{\Psi}_d| \ll 2\pi$. Accordingly, we define a critical on-axis $(\mathbf{x}=0)$ *nondimensional* parameter based on the equality $|\widetilde{\Psi}_d| = 2\pi$,

$$Q_T = |\widetilde{\Psi}_d(\boldsymbol{\xi}=0)|/(2\pi) = \frac{c}{16\pi z}\left(1+\bar{\Omega}^2\right)^{-3/2}|\bar{R}_c|T^2. \tag{56}$$

The condition $|\widetilde{\Psi}_d| \ll 2\pi$, for legitimacy of the paraxial approximation in (49), thus becomes

$$Q_T \ll 1. \tag{57}$$

Within the paraxial approximation, the inequality (57) with (56), estimates the smallest on-axis observation point z, for a given pulse length T, or maximum pulse length T for specified z on the beam axis. The estimates fail near the first arrival of the signal ($\bar{\Omega} \to 1$), where dispersion is not yet fully developed ($k(\omega) \to \omega/c$, $k''(\bar{\omega}_s) \to 0$ and $\bar{R}_c \to \infty$) (see [4][eq. 28]), and also near inflection points on the dispersion surface for complex materials. Better asymptotics than those based on first order saddle point methods are required in these transition regions [5]. The result in (57) agrees with that in [4, 5] which was derived via the conventional "FD first" route summarized in the first portion of this paper.

For *off-axis* observation points, we define from (51) and (53) an additional critical nondimensional estimator, $Q_\xi \equiv |\Psi_d(T=0)|/(2\pi)$, i.e.,

$$Q_\xi = \frac{ct^2}{4\pi z^3}\left(1+\bar{\Omega}^2\right)^{-3/2}|\bar{R}_c|\,|\boldsymbol{\xi}\cdot\mathbf{x}|^2 \ll 1, \tag{58}$$

which validates in terms of the on-axis parameters $\bar{\Omega}$, $\bar{R}_c$ and z, the paraxial approximation (49) off-axis at $\mathbf{x}$ with respect to the tilted plane wave direction $\boldsymbol{\xi}$. Thus, for a given off-axis observation point $(z, \mathbf{x}, t)$, (58) bounds the maximum excursion of $\boldsymbol{\xi}$ from 0 (the allowable angular spread) that can be included in the transient plane wave superposition in (38), using the spectral paraxial phase in (49). In (52), for $\boldsymbol{\xi}$ constant, (58) bounds the maximum excursion $\mathbf{x}$ from the beam axis. For on-axis observation points $\mathbf{x}=0$, the estimator $Q_\xi = 0$; off axis, the allowable maximum angular spread $\boldsymbol{\xi}$ decreases with the distance from the beam axis. As in (56), Q_ξ is parameterized by the *on-axis* observation point (z, t), which determines the location behind the wavefront, $\bar{\Omega} = ct/z$, and the local radius of curvature of the dispersion surface $\bar{R}_c$.

The paraxial spectral plane wave $\overset{+}{\tilde{u}}$ in (52) can be regarded either as a configuration-space object, where $\boldsymbol{\xi}$ is kept constant and the space-time observation point $(\mathbf{r}, t)$ is allowed to vary, or as a spectral distribution, where $(\mathbf{r}, t)$ is kept constant and $\boldsymbol{\xi}$ is allowed to vary. Details pertaining to these alternative interpretations are given in [6].

Asymptotic evaluation of TD spectral integral

Formal spectral synthesis in (38) over the plane wave constituents in (43) yields the TD field

$$\overset{+}{u}(\mathbf{r},t) \sim \int d^2\xi\ \overset{+}{\tilde{A}}(\mathbf{r},t;\boldsymbol{\xi})\ e^{-i\tilde{\Psi}}, \tag{59}$$

which will be evaluated for the the *paraxial* approximation in (52). The integrand in (59) has a stationary point $\boldsymbol{\xi}_s$ in the complex $\boldsymbol{\xi}$ domain that satisfies

$$\nabla_\xi \tilde{\Psi}_p(\boldsymbol{\xi})_{\boldsymbol{\xi}=\boldsymbol{\xi}_s} = 0, \quad \boldsymbol{\xi}_s = \boldsymbol{\xi}_s(\mathbf{r};T). \tag{60}$$

Lowest order two-dimensional saddle point asymptotics [7][eq. 4.7.3] yields

$$u(\mathbf{r},t) \sim \overset{+}{A}(\mathbf{r},t;T)e^{-i\Psi(\mathbf{r},t;\bar{\boldsymbol{\xi}})} \tag{61}$$

with

$$\Psi(\mathbf{r},t;T) = \tilde{\Psi}_p(\mathbf{r},t;\boldsymbol{\xi};T)_{\bar{\boldsymbol{\xi}}=\boldsymbol{\xi}_s}, \tag{62}$$

and

$$\overset{+}{A}(\mathbf{r},t;T) = \overset{+}{\tilde{A}}(\mathbf{r},t;\boldsymbol{\xi};T)|_{\boldsymbol{\xi}=0}\ \frac{-2\pi i}{\left[\mathrm{Det}\nabla_\xi\nabla_\xi\tilde{\Psi}_p|\boldsymbol{\xi}_s\right]^{1/2}}. \tag{63}$$

Since the fields in the integrand are approximated by their paraxial (quadratic) form, the integration can be performed exactly. Inserting (49) into (60), one obtains

$$\nabla_\xi\left[z+\mathbf{x}\cdot\boldsymbol{\xi}+\frac{1}{2}\boldsymbol{\xi}\cdot\mathbf{M}\cdot\boldsymbol{\xi}\right]_{\boldsymbol{\xi}=\boldsymbol{\xi}_s} = 0, \tag{64}$$

and therefore

$$\boldsymbol{\xi}_s = -\mathbf{M}^{-1}\cdot\mathbf{x}. \tag{65}$$

From (65) we note that the displacement of $\boldsymbol{\xi}_s$ from the on-axis real value 0 is proportional to $\mathbf{x}$. The phase Ψ is obtained by inserting (65) into (62),

$$\Psi(\mathbf{r},t) = (t - \frac{i}{2}T)\bar{\omega}_s - k(\bar{\omega}_s)\left[z - \frac{1}{2}\mathbf{x}\cdot\mathbf{M}^{-1}\cdot\mathbf{x}\right]. \tag{66}$$

Using (47) we obtain

$$-\mathbf{x}\cdot\mathbf{M}^{-1}\cdot\mathbf{x} = \rho^2/(z - i\beta), \tag{67}$$

and the paraxial phase is

$$\Psi(\mathbf{r},t) = \bar{\omega}_s(t - \frac{i}{2}T) - k(\bar{\omega}_s)\left[z + \frac{1}{2}\rho^2/(z - i\beta)\right]. \tag{68}$$

The phase in (68) consists of the on-axis phase $\bar{\omega}_s(t - \frac{i}{2}T) - k(\bar{\omega}_s)z$ and an additional paraxial off-axis term $\frac{1}{2}\rho^2/(z - i\beta)$. The result in (68) is valid as long as T satisfies condition (57) and as long as the stationary point $\boldsymbol{\xi}_s$ satisfies condition (58). Substituting (65) into (58), we obtain

$$Q_\rho \equiv Q_{\boldsymbol{\xi}}|_{\boldsymbol{\xi}=\boldsymbol{\xi}_s} = \rho^4 \frac{\bar{\Omega}^2}{(1+\bar{\Omega}^2)^{3/2}} \frac{|\bar{R}_c|}{4\pi|z - i\beta|^2\, cz} \ll 1. \tag{69}$$

Here Q_ρ is the critical nondimensional estimator that parameterizes the maximum off-axis excursion $\mathbf{x}$ for which the paraxial phase in (68) is valid, i.e. the maximum off-axis excursion $\mathbf{x}_{max}$ is obtained for $Q_\rho \simeq 1$. Recall that for a non-dispersive field, this deviation is parameterized by $|\mathbf{x}| << z$ alone. When dispersion is introduced, the parameters $\bar{R}_c$ and $\bar{\Omega}$ play a role, in addition to z. The estimate in (69) becomes invalid when $|\bar{R}_c| \to \infty$ because the amplitude in (52) shows that the asymptotic field diverges in that limit (see also the amplitude in (70); for a uniform correction when $\bar{\boldsymbol{\xi}} = 0$, see [5]).

Inserting (46f) together with (45) into (63) yields

$$\overset{+}{A}(\mathbf{r},t) = \sqrt{\frac{2}{i\pi z k''(\bar{\omega}_s)}\,\frac{-i\beta}{z - i\beta}}. \tag{70}$$

Comparing this result with the one in (33) we find that by setting $S = z + \rho^2/(z - i\beta) \approx z$ in (33), the two results are identical. Therefore, the generalized result in (70) is the zero order off-axis approximation for the *amplitude*. Substitution of (68) and (70) into (61) completes the asymptotic evaluation of the TD pulsed beam field.

CONCLUSIONS

In this paper, we have been concerned with the parameterization of the effects of frequency dispersion on the propagation characteristics of a paraxially approximated pulsed beam (PB) wavepacket in a lossless medium with generic wavenumber profile $k(\omega)$. Various nondimensional measures – critical parameters – have been defined in order to systematically assess and quantify a) the effect of dispersion on various observables associated with the PB field and thereby on the resolution of these observables, and b) the range of validity of the paraxial approximation under these conditions.

After first summarizing the "conventional" spectral synthesis via the previously performed sequence in [4, 5], which begins with the wavenumber superposition in the time harmonic frequency domain (FD), followed by the FD inversion to the time domain (TD), the remaining portion of this paper has dealt with the "unconventional" synthesis whereby the frequency inversion is performed first to yield a wavenumber spectral distribution of pulsed plane waves which are assembled subsequently to constitute the TD PB field. While the final outcomes from these alternative approaches for a given set of problem conditions coincide, each spectral sequence elucidates its own wave physics. For understanding short pulse (i.e., PB) phenomenology at its most fundamental level, the TD spectral plane wave decomposition is the preferred approach. The (k, ω) dispersion profile plays a crucial role in parameterizing all space time observables self-consistently in terms of space-time dependent frequencies and wavenumbers: $\omega[k(\mathbf{r}, t)]$ or $k[\omega(\mathbf{r}, t)]$. Within the condensed format of this paper, details of derivations and extensions to parametric regimes beyond those presented here could not be included. Such material is contained in an article on the "direct TD" approach which is being submitted separately for publication [6].

ACKNOWLEDGMENT

This research was supported in part by the Naval Coastal Systems Center, Panama City, FL, under ONR Grant N61331-96-K-0028. T. Melamed also acknowledges partial support from a 1996-97 Dean's Postdoctoral Fellowship at Boston University as well as partial support from Odin Technologies, Yokne'am Elit, Israel.

References

[1] E. Heyman and L.B. Felsen, "Complex source pulsed beam fields," *J. Opt. Soc. Am. A*, 6:806–817, 1989.

[2] E. Heyman, "Pulsed beam propagation in an inhomogeneous medium," *IEEE Trans. Antennas Propagat.*, AP-42:311–319, 1994.

[3] T. Melamed, "Phase-space beam summation: A local spectrum analysis for time-dependent radiation," *J. Electromag. Waves Appl.*, 11:739–773, 1997.

[4] T. Melamed and L.B. Felsen, "Pulsed beam propagation in lossless dispersive media, Part I: Theory," *J. Opt. Soc. Am. A*, 15:1268–1276, 1998.

[5] T. Melamed and L.B. Felsen, "Pulsed beam propagation in lossless dispersive media, Part II: A numerical example," *J. Opt. Soc. Am. A*, 15:1277–1284, 1998.

[6] T. Melamed L.B. Felsen, "Pulsed beam propagation in dispersive media via pulsed plane wave spectral decomposition,", Submitted to *IEEE Trans. Antennas Propagat.*

[7] L.B. Felsen and N. Marcuvitz, *Radiation and Scattering of Waves*, IEEE Press, Piscataway, N.J., 1994, Classic reissue.

SPECTRAL ALTERNATIVES FOR THE SYNTHESIS OF SHORT-PULSE WAVEFIELDS IN WAVEGUIDES

B.P. de Hon[1], E. Heyman[2] and L. B. Felsen[3]

[1] Department of Electrical Engineering, Eindhoven University of Technology, Eindhoven, The Netherlands
[2] Department of Electrical Engineering – Physical Electronics, Tel-Aviv University, Tel Aviv, Israel
[3] Department of Aerospace and Mechanical Engineering and Department of Electrical and Computer Engineering, Boston University, Boston, MA

I ABSTRACT

For the synthesis of short-pulse 2D wavefields in simple waveguiding configurations, various alternative spectral field representations can be constructed. In the Spectral Theory of Transients (STT), a wavefield is expanded in terms of non-dispersive time-domain plane wave constituents. The STT furnishes an appropriate setting for the description of short-pulse fields within the Fresnel zone in a waveguide. In the far zone a ray-field expansion is not very effective, since the constituents cease to be well resolved in space and time. There, a global field expansion in terms of a frequency spectrum of time-harmonic modes is more appropriate, but does not yield sufficient information about the time behavior of the field. Fortunately, the scope of the STT can be extended through the construction of an angular spectrum of global dispersive time-domain spectral-mode (TDSM) constituents, which exploits the advantages of both synthesis options. The field representation in terms of a spectral integral of a series of TDSM constituents proves to be equivalent to the representation in which the field is assembled with contributions from a periodic infinite array of image sources.

The use of analytic signals allows for the complexification of the space-time source coordinates in the resulting expressions for the field, thus yielding the 2D complex-source pulsed-beam (CSPB) solutions, associated with a class of directive sources. The asymptotic evaluation of the remaining spectral integral leads to an expression for the modal field in terms of instantaneous frequencies and instantaneous angles of propagation, through which a cogent description of the space-time-domain wave phenomena is obtained.

II INTRODUCTION

Many applications in physics and engineering involve complex layered environments. With the advent of ultra-wideband measurement techniques, the time-domain interrogation of these environments, based on the interaction of electromagnetic, acoustic and elastodynamic waves with contrasting materials has come to the fore. Hence, the development of efficient modelling methods to describe these phenomena in a physically cogent fashion will become increasingly important.

For short-pulse fields — especially for well-collimated fields — an expansion in terms of time-domain wave objects is most appropriate, both phenomenologically and numerically. The Spectral Theory of Transients (STT), in which the time-domain wave objects are *non-dispersive* time-domain plane waves, constitutes a well-established framework for the description of ray-type short-pulse fields, i.e., fields which can be represented by a limited number of individual well-resolved wave constituents that have interacted a distinct number of times with interfaces (cf. Heyman and Felsen [1–3]). An elegant alternative technique to evaluate ray-type short-pulse fields that preceded the STT is the Cagniard-de Hoop method (cf. De Hoop [4]).

Considering short-pulse fields in waveguides, a ray-field expansion constitutes an appropriate description of the field relatively close to the source (within the Fresnel zone). That is to say: the time evolution of the total field may be decomposed into localized events which are well resolved in space and time, i.e., on a phenomenological level, the properties of the total field can be identified in terms of the individual ray constituents. In the far zone however, a ray expansion is not very effective: many ray fields have to be incorporated before convergence is reached, and the local features of a ray constituent can no longer be discerned within the total field. Hence, one would like to synthesize the global features of the total short-pulse field in terms of mode-like constituents.

In the present paper we extend the scope of the STT so as to encompass time-domain global wave objects. Specifically, we shall construct an angular spectrum of *dispersive* time-domain spectral-mode (TDSM) constituents in terms of which the field in the far zone can be adequately expanded. We illustrate this novel extension of the STT through the treatment of a simple canonical problem, in which a 2D wavefield is excited by a current line source within a parallel-plate waveguide environment with perfectly conducting boundaries. The use of analytic signals allows for the complexification of the space-time source coordinates in the resulting expressions for the field, thus yielding the corresponding 2D complex-source pulsed-beam (CSPB) solutions, associated with a class of highly directive sources (cf. Heyman [5]). The analysis has been inspired by the research conducted by Felsen and Niu [6–8] on the short-pulse leaky-mode synthesis of the radiation field emanating from a grounded dielectric layer.

III FORMULATION

The parallel-plate waveguide configuration under consideration is sketched in Figure 1. A current line source parallel to the y-axis excites an electromagnetic field, which is readily

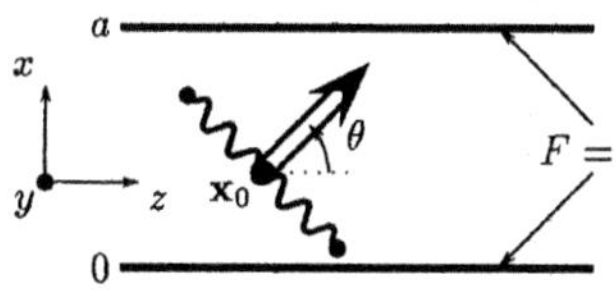

Figure 1. Schematic depiction of the configuration.

derived from the scalar field $F(\mathbf{x}, t)$ that for an isotropic line source satisfies the scalar wave equation

$$\left(\partial_x^2 + \partial_z^2 - c^{-2}\partial_t^2\right) F = -\delta(\mathbf{x} - \mathbf{x}')f(t - t') \tag{1}$$

where c denotes the wavespeed, f denotes the source signature, and $\{\mathbf{x}', t'\}$ denote the space-time source coordinates, respectively. The parallel plates are perfect conductors, implying that $F|_{x=0} = F|_{x=a} = 0$.

We shall synthesize the field from its spectral representation. Once the field has been determined for an isotropic line source, the complex-source pulsed-beam solutions follow upon complexification of the source coordinates according to (cf. Heyman [5])

$$\mathbf{x}' \to \mathbf{x}_0 + i\mathbf{b}, \quad \mathrm{t}' \to ib/c, \tag{2a}$$

$$f(t) \to \overset{+}{f}(t), \quad F(\mathbf{x}, t) \to \overset{+}{F}(\mathbf{x}, t), \tag{2b}$$

in which the real point $\mathbf{x}_0$ denotes the center of beam waist, while the direction θ and the magnitude b of the real vector $\mathbf{b}$ respectively describe the beam direction and the collimation distance of the CSPB, with $b_x = b\sin(\theta)$ and $b_z = b\cos(\theta)$. In view of the complex source coordinates we employ the analytic signal extension $\overset{+}{f}(t)$ of $f(t)$, defined as

$$\overset{+}{f}(t) = \begin{cases} \frac{1}{\pi}\int_0^\infty d\omega\ \hat{f}(\omega)e^{-i\omega t}, & \mathrm{Im}(t) < 0, \\ f(t) - i\mathcal{H}\{f(t)\}, & \mathrm{Im}(t) = 0, \end{cases} \tag{3}$$

in which $\hat{f}(\omega)$ and $\mathcal{H}\{f(t)\}$ denote the frequency spectrum and the Hilbert transform of $f(t)$, respectively. Similarly, $\overset{+}{F}(\mathbf{x}, t)$ is the analytic field solution. The real PB field for a real observation point $(\mathbf{x}, t)$ is obtained via $F(\mathbf{x}, t) = \mathrm{Re}(\overset{+}{F}(\mathbf{x}, t))$.

IV SPECTRAL REPRESENTATIONS

Our starting point is the spectral-frequency-domain representation of the field. Recall that in the chosen system of coordinates, the x-axis points in the direction transverse to the waveguide, while the z-axis is directed along the waveguide. We employ a transverse spectral synthesis, in which the spectral variable is the transverse slowness. This is the natural synthesis option for the description of propagation along the waveguide. The pertaining representation is also termed z-propagation representation [9]. It may be derived, for example, from the more conventional longitudinal Fourier transform representation along the z-axis via contour deformation in the complex k_z plane.

The field can be written as the linear superposition of four distinct wave species, i.e.,

$$\overset{+}{F} = \sum_{j=1}^{4} \overset{+}{F}_j. \tag{4}$$

Each of the space-time-domain wave species may be synthesized in terms of its spectral-frequency-domain constituents according to

$$\overset{+}{F}_j = \int_{\xi=-\infty}^{\infty} \int_{\omega=0}^{\infty} \mathrm{d}\omega \, \mathrm{d}\xi \; e^{-i\omega[(t-t')-\zeta(z-z')/c]} \hat{f}(\omega) \hat{\tilde{G}}_j(\xi, \omega) \tag{5}$$

in which $\hat{\tilde{G}}_j$ denotes a wave-species Green's function constituent, while ξ and $\zeta = \zeta(\xi) = \sqrt{1-\xi^2}$, with $\mathrm{Im}(\zeta) \geq 0$, denote the transverse and longitudinal normalized slownesses, respectively. They are related to the Cartesian components of the wavevector, via $k_x = \omega\xi/c$ and $k_z = \omega\zeta/c$. The exponential term in Eq. (5) accounts for the propagation along the waveguide, while the wave-species Green's function constituents $\hat{\tilde{G}}_j$ account for the amplitude and the propagation in the transverse direction. For wideband applications, the slowness representation is preferable to the more common wavenumber representation, since it describes the field in terms of a frequency-independent geometrical quantity, namely the plane wave angle φ, defined through

$$\xi = \sin(\varphi), \qquad \zeta = \cos(\varphi). \tag{6}$$

For the interpretation of the final results it is convenient to define φ (and thereby ξ) to be the plane wave departure angle at the source, rather than the plane wave direction at the observer.

The wave-species Green's function constituents $\hat{\tilde{G}}_j$ can be constructed in several alternative fashions. One option, which has already been mentioned, is to start with the Fourier transform along the z-axis. Instead, we employ the method of images. Assuming, without loss of generality, that $\mathrm{Re}(x - x') > 0$, $\mathrm{Re}(z - z') > 0$, we have

$$\hat{\tilde{G}}_1 = \sum_{n=0}^{\infty} \tilde{A}(\xi) e^{i\omega\xi(x-x'+2an)/c} \quad \text{for} \quad \omega > 0, \mathrm{Im}(\xi) \downarrow 0, \tag{7a}$$

$$\hat{\tilde{G}}_2 = \sum_{n=1}^{\infty} -\tilde{A}(\xi) e^{i\omega\xi(-x-x'+2an)/c} \quad \text{for} \quad \omega > 0, \mathrm{Im}(\xi) \downarrow 0, \tag{7b}$$

$$\hat{\tilde{G}}_3 = \sum_{n=0}^{\infty} -\tilde{A}(\xi) e^{i\omega\xi(-x-x'-2an)/c} \quad \text{for} \quad \omega > 0, \mathrm{Im}(\xi) \uparrow 0, \tag{7c}$$

$$\hat{\tilde{G}}_4 = \sum_{n=1}^{\infty} \tilde{A}(\xi) e^{i\omega\xi(x-x'-2an)/c} \quad \text{for} \quad \omega > 0, \mathrm{Im}(\xi) \uparrow 0. \tag{7d}$$

in which $\tilde{A}(\xi) = i/4\pi^2\zeta(\xi)$ denotes the wave amplitude. The infinitesimal imaginary parts of ξ in Eq. (7) render the series convergent. Note that the definition of the signs of $\mathrm{Im}(\xi)$

is consistent with the fact that φ represents the plane wave departure angle at the source. A wave species is characterized by a frequency-independent amplitude term that accounts for the source excitation, followed by a series of propagation terms from the source level to the receiver level that include all multiple reflections at the boundaries. Evaluation of these series into expressions with a resonant denominator yields

$$\hat{\tilde{G}}_1 = \frac{\tilde{A}(\xi)e^{i\omega\xi(x-x')/c}}{1-e^{2i\omega\xi a/c}} \quad \text{for} \quad \omega > 0, \mathrm{Im}(\xi) \downarrow 0, \tag{8a}$$

$$\hat{\tilde{G}}_2 = \frac{-\tilde{A}(\xi)e^{i\omega\xi(2a-x-x')/c}}{1-e^{2i\omega\xi a/c}} \quad \text{for} \quad \omega > 0, \mathrm{Im}(\xi) \downarrow 0, \tag{8b}$$

$$\hat{\tilde{G}}_3 = -\hat{\tilde{G}}_2 \quad \text{for} \quad \omega > 0, \mathrm{Im}(\xi) \uparrow 0, \tag{8c}$$

$$\hat{\tilde{G}}_4 = -\hat{\tilde{G}}_1 \quad \text{for} \quad \omega > 0, \mathrm{Im}(\xi) \uparrow 0. \tag{8d}$$

IV.1 Alternative spectral synthesis options

Basically, there are four ways of synthesizing the space-time-domain wavefield. First, upon performing a double integral with dissimilar variables of integration, the order in which the integrals are performed remains to be chosen. In the conventional approach the spectral integral is performed first, thus leaving a frequency integral of space-frequency-domain field constituents. In the (non-conventional) STT, the frequency integral is performed first, thus leaving an integral over the angular spectrum of spectral-time-domain field constituents.

Second, for each of the options above one may choose to expand the spectral-frequency-domain wave species in terms of ray-type constituents, as described by the ray series in Eq. (7), or in terms of mode-type constituents, described by the global expressions in Eq. (8). In Figure 2 we have schematically depicted the alternative synthesis options. In the near-zone, the field is most appropriately described in terms of the local ray-type constituents. Here exact closed form inversion of ultra wideband wavefields into the time domain may be performed

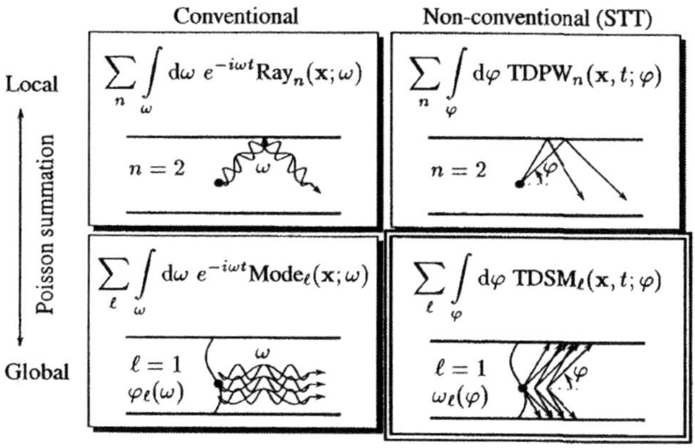

Figure 2. The alternative synthesis options.

by either the Cagniard-de Hoop method or the STT approach (see e.g., [1–4]). The STT approach, in which the field is expanded as the superposition with respect to the angle of propagation φ of time-dependent plane waves (TDPW), is depicted in the upper-right corner of Figure 2.

However, for long-distance propagation — specifically, for distances beyond the Fresnel distance a^2/cT of the waveguide, in which T denotes the pulse length — the ray events are no longer separable in time, and hence a representation in terms of global (i.e., modal) constituents is prefered. In this case, the conventional frequency-domain global synthesis option in which the field is expanded in a frequency spectrum of mode constituents (lower-left corner of Figure 2) is the obvious setting for the analysis of narrow-band signals. However, for short-pulse wavefields, this frequency-domain representation does not reveal sufficient information about the time behavior of the pertaining wavefields. In that space-time regime the STT modal synthesis option depicted in the lower-right corner of Figure 2 is the most suitable choice, i.e., the modal fields are the appropriate building blocks of the far-zone field, and the fact that the spectral constituents are time-domain wave objects renders the time behavior explicit. Here, the field is described as an angular superposition of time-dependent spectral modes (TDSM), to be defined in Section IV.2. Moreover, for a wavefield that is generated by a highly directive source, e.g., a complex-source pulsed-beam, the dominant contribution to the spectral integral arises from a localized portion of the spectrum, which renders the STT modal synthesis even more appropriate.

IV.2 Global field expansions

The resonance denominator in Eq. (8) gives rise to a countably infinite set of poles, for which the resonance equation

$$e^{2i\omega\xi a/c} = 1 = e^{2i\pi\ell} \tag{9}$$

is satisfied.

Figure 3. Conventional field-synthesis contours of integration in the complex ξ- and ω-planes. The symbols X and ● denote simple poles and algebraic branch points, respectively.

For the *conventional* slowness before frequency integral field synthesis, we keep ω fixed and real positive, implying that the poles are located at

$$\xi_\ell = \frac{\pi\ell c}{\omega a} \quad \text{with} \quad \ell \in \mathbb{Z} \tag{10}$$

on the real axis of the complex ξ-plane (cf. Figure 3(a)). For species 1 we have $\mathrm{Im}(\xi) > 0$ (cf. Eq. (8a)), and hence the poles are located below the contour of integration. For species 4 we have $\mathrm{Im}(\xi) < 0$ (cf. Eq. (8d)), and hence the poles are located above the contour of integration. Closing the contours of integration in Figure 3(a) we may express the field in terms of pole contributions that describe the frequency dependent modes, supplemented with integrals along branch cuts. However, since species 4 is the opposite of species 1 on the real ξ axis, it follows that the frequency-domain field associated with the combined species 1 and 4 can be expressed in terms of pole contributions only. As regards the combination of species 2 and 3 the situation is similar. This leaves the time-domain field described through frequency integration of the modal fields as depicted in the lower-left corner of Figure 2. Note that for each mode the frequency domain is separated into evanescent ($\omega < \pi\ell c/a$) and propagation ($\omega > \pi\ell c/a$) regimes, as depicted in Fig. 3(b).

For the *non-conventional* frequency before slowness-integral field synthesis, we keep ξ fixed, which gives rise to poles in the complex ω-plane. These poles describe the frequencies for which a given plane wave satisfies a transverse resonance condition. In the relevant right-half of the complex ω-plane these poles are located at

$$\omega_\ell = \frac{\pi\ell c}{\pm\xi a} \quad \text{for} \quad \mathrm{Re}(\xi) \gtrless 0,\ \ell \in \mathbb{Z}^+. \tag{11}$$

For species 1 we have $\mathrm{Im}(\xi) > 0$, and hence the poles in the complex ω-plane are located below and above the positive real ω-axis, depending on whether $\mathrm{Re}(\xi) > 0$ or $\mathrm{Re}(\xi) < 0$, respectively (cf. Figures 4(a)–4(b)). For species 4 we have $\mathrm{Im}(\xi) < 0$, and hence the situa-

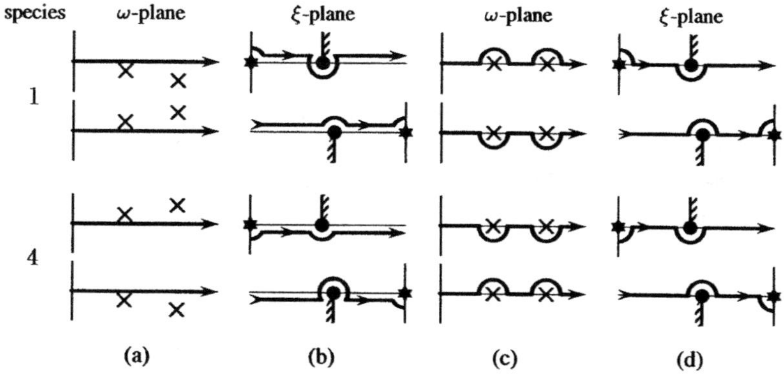

Figure 4. Non-conventional field-synthesis contours of integration in the complex ω- and ξ-planes, and the corresponding limiting contours. The symbols X, ● and ✱ denote simple poles, algebraic branch points and essential singularities, respectively.

tion is reversed. In the limiting case of $\mathrm{Im}(\xi) \to 0$, the contours of integration in the complex ω-plane must be deformed such that the poles are encircled in the correct fashion (cf. Figure 4(c)). Evaluating the frequency integral in Eq. (5) for species 1, we obtain the following sum of a series of half-residue contributions and a diffraction-like principal value integral

$$\int_{i0}^{\infty+i0} \mathrm{d}\omega\, \hat{\tilde{G}}_1 = \frac{1}{2}\sum_{\ell=1}^{\infty} \overset{+}{\tilde{G}}{}^{\mathrm{M}}_{1;\ell} + \mathrm{PV}\int_0^{\infty} \mathrm{d}\omega\, \hat{\tilde{G}}_1, \tag{12}$$

in which a single residue contribution

$$\overset{+}{G}{}^{\mathrm{M}}_{1;\ell} = \frac{\pm ic}{4\pi\xi\zeta a} e^{-i\omega_\ell[t-t'-\zeta(z-z')/c]} e^{\pm i\pi\ell(x-x')/a} \quad \text{for} \quad \xi \gtrless 0 \tag{13}$$

defines a time-dependent spectral mode (TDSM).

The evaluation of the frequency integral for species 4 leads to a similar result except for an opposite sign of the principal value integral i.e.,

$$\int_{-i0}^{\infty-i0} \mathrm{d}\omega\, \hat{\tilde{G}}_4 = \frac{1}{2}\sum_{\ell=1}^{\infty} \overset{+}{G}{}^{\mathrm{M}}_{1;\ell} - \mathrm{PV}\int_0^{\infty} \mathrm{d}\omega\, \hat{\tilde{G}}_1. \tag{14}$$

Upon combining the species, only the TDSM constituents remain, i.e.,

$$\int_0^{\infty} \mathrm{d}\omega\, (\hat{\tilde{G}}_1 + \hat{\tilde{G}}_4) = \sum_{\ell=1}^{\infty} \overset{+}{G}{}^{\mathrm{M}}_{1;\ell} \tag{15}$$

The cancellation of the remainder integrals is not surprising, if one recalls that the sum of the species 1 and 4 comprises all the multiple reflections of the plane wave that, for ξ positive, emanates upward from the source and arrives at the observer from below. For ξ negative, the sum comprises the downward emitted plane-wave constituents that arrive at the observer from above. Stated in a different fashion, the species 1 and 4 describe the collective contributions of two semi-infinite arrays of image sources at the source locations $x'_n = x' + 2an$ for the integers $n \leq 0$ and $n > 0$, respectively.

A similar result is obtained for species 2 and 3 that, for ξ positive, describe all multiple reflections of plane-wave constituents that emanate upward from the source and arrive at the observer from above, and that, for ξ negative, describe all multiple reflections of plane-wave constituents that emanate downward from the source and arrive at the observer from below. Alternatively, the species 2 and 3 may be interpreted as describing the collective contributions of two semi-infinite arrays of image sources at the source locations $x'_n = -x' + 2an$ for the integers $n > 0$ and $n \leq 0$, respectively.

V SPECTRAL EVALUATION OF THE TIME-DOMAIN MODAL FIELDS

An explicit expression for a TDSM comprising species 1 and 4 is given in Eq. (13). To explore the properties of the corresponding time-domain modal field, we apply the saddle point method to evaluate the dominant contribution to the following generic spectral-integral representation (see Eqs. (5) and (6))

$$\overset{+}{G}{}^{\mathrm{M}}_{1;\ell} = \left(\int_0^{\pi/2} + \int_{\pi/2}^{\pi/2-i\infty}\right) \mathrm{d}\varphi\ \Phi_\ell(\varphi) e^{-i\phi_\ell(\varphi)}, \tag{16}$$

in which

$$-i\phi_\ell(\varphi) = -i\omega_\ell[t-t'-\zeta(z-z')/c], \quad \text{with} \quad \omega_\ell = \frac{\pi\ell c}{a\xi} \tag{17}$$

denotes the phase, and

$$\Phi_\ell(\varphi) = \frac{ic\,\cos[\pi\ell(x-x')/a]}{2\pi\xi a} \tag{18}$$

represents an amplitude term. Note that the essential singularity at $\xi = 0$ is circumnavigated as depicted in Figure 4(d). This essential singularity indicates that for spectral directions about $\varphi = 0$ the associated frequency must tend to infinity, so as to retain sufficient variation in the transverse field distribution to satisfy the boundary conditions. Observing that the slownesses depend on the spectral angle via Eq. (6), we have derived Eq. (16) from Eq. (13) by replacing the variable of integration ξ by φ for ξ positive, and by $-\varphi$ for ξ negative. As regards the TDSM associated with species 2 and 3, we note that the spectral-integral representation for $\overset{+}{G}{}^{\mathrm{M}}_{2;\ell}$ can be retrieved from Eqs. (16)–(18) through the substitutions $x' \to -x'$ and $\Phi_\ell \to -\Phi_\ell$.

As the source signature, we now choose the mth derivative of the Rayleigh pulse, defined by $\partial_t^m \overset{+}{f}(t) = \partial_t^m \overset{+}{\delta}(t - iT)$, in which $T > 0$ is a measure of the pulsewidth. The result in Eq. (16) can be generalized to describe response to this specific pulse history upon performing the substitution

$$\mathrm{t}' \to i(b/c+T), \qquad \Phi_\ell \to (-i\omega_\ell)^m \Phi_\ell \tag{19}$$

This procedure enables us to extend the scope of the analysis to encompass a class of pulse excitations, without having to carry out a convolution with respect to time. Furthermore, for $b > 0$, the resulting integral representation describes the response to a collimated pulsed-beam source as explained below Eq. (2).

Without loss of generality we choose the origin of our system of coordinates such that $z_0 = 0$. It is readily shown that the real part of the modal impulse response to an isotropic line source excitation (i.e., $m = 0$, $T = 0$ and $b = 0$) vanishes for $t < z/c$. Addition of all the modal constituents renders the resulting signal causal, i.e., the real part of the signal vanishes for $t < \sqrt{(x-x_0)^2 + z^2}/c$. As a result of the complexification the strict causality is lost, but the longitudinal travel time z/c is still an appropriate measure for the time delay.

The dominant contribution to the spectral integral is furnished by TDSM constituents at spectral angles in the neighborhood of the stationary point. At the stationary point $\varphi = \varphi_{\mathrm{s}}(t)$, which is defined through $\partial_\varphi \phi|_{\varphi=\varphi_{\mathrm{s}}} = 0$, we have

$$\zeta_{\mathrm{s}}(t) = \cos(\varphi_{\mathrm{s}}) = \frac{z - ib\cos(\theta)}{ct - i(b+cT)} \tag{20a}$$

$$\xi_{\mathrm{s}}(t) = \sin(\varphi_{\mathrm{s}}) = \sqrt{1-\zeta_{\mathrm{s}}^2}, \tag{20b}$$

The instantaneous longitudinal and transverse slownesses in Eq. (20) define the instantaneous direction of propagation. The associated instantaneous angle of propagation $\varphi_{\mathrm{s}}(t)$ and the instantaneous frequency $\omega_{\mathrm{s}}(t)$ are most conveniently expressed as (cf. Eq. (17))

$$\varphi_{\mathrm{s}}(t) = \frac{1}{2i}\ln\left(\frac{\zeta_{\mathrm{s}} + i\xi_{\mathrm{s}}}{\zeta_{\mathrm{s}} - i\xi_{\mathrm{s}}}\right) \quad \text{and} \quad \omega_{\ell;\mathrm{s}}(t) = \frac{\pi\ell c}{\xi_{\mathrm{s}} a}, \tag{21}$$

respectively. For $T = b = 0$, the instantaneous frequency approaches infinity for $t \downarrow z/c$ and decreases monotonically as time progresses, implying that the signal chirps down in time. On average, this behavior is retained for signals generated by complex sources.

The phase at the stationary point and its derivative with respect to time are found to be

$$-i\phi_{\ell;\mathrm{s}} = \frac{-i\pi\ell}{a}[ct - i(b+cT)]\xi_{\mathrm{s}} \quad \text{and} \quad -i\partial_t\phi_{\ell;\mathrm{s}} = \frac{-i\pi c\ell}{a\xi_{\mathrm{s}}}. \tag{22}$$

A close examination of Eq. (20) reveals that

$$\mathrm{Im}(\zeta_s) \gtrless 0 \text{ for } t \lessgtr t_d = \frac{z(b+cT)}{cb\cos(\theta)} = \frac{z}{c\cos(\theta')}, \quad \text{with} \quad \theta' = \arccos\left[\frac{b\cos(\theta)}{b+cT}\right]. \tag{23}$$

From Eqs. (20b), (22) and (23), we infer that $\mathrm{Re}(-i\partial_t \phi_{\ell;s})|_{t=t_d} = 0$. Hence, $|e^{-i\phi_{\ell;s}}|$ attains its maximum at the instant $t = t_d$. As regards the impulse response, for which $T = 0$, this result implies that upon arrival the spectral direction in which the field is excited coincides with the beam direction. For $T > 0$ we infer from Eq. (23) that the dominant spectral angle of propagation θ' exceeds the pulsed-beam angle θ, which indicates the effect of destructive interference at times prior to $t = t_d$.

Application of the saddle-point method to evaluate Eq. (16) leads to

$$\overset{+}{G}{}^{M}_{1;\ell} \sim -\frac{(-i\omega_{\ell;s})^{m+1}}{\pi^2 \ell}\sqrt{\frac{\pi}{i\omega_{\ell;s}t}}\cos[\pi\ell(x-x')/a]e^{-i\phi_{\ell;s}}, \tag{24}$$

which is accurate if

$$\left|\frac{\Phi''_{\ell;s}}{2i\Phi_{\ell;s}\phi''_{\ell;s}}\right| \ll 1. \tag{25}$$

To determine the condition for the corresponding region of validity, we distinguish between the pulsed-beam modal field excited by a directive source, and the modal field excited by a line source. We have seen above that for pulsed-beam modal fields, the signal is strong at the instant $t = t_d$. Confining the analysis of the region of validity of Eq. (24) to the instant $t = t_d$, it can be shown that Eq. (25) leads to

$$\sqrt{\frac{z^2 + b^2\cos^2(\theta)}{a^2}} \gg \frac{m+1}{2\pi\ell}\cot(\theta')\left[1 + (m+1)\cos^2(\theta')\right]. \tag{26}$$

For the isotropic line-source excitation the angles θ and θ' are not defined. From Eq. (20) we infer that as time progresses, the stationary point in the complex ξ-plane moves away from the singularity at $\xi = 0$, which in view of Eq. (18) implies that the variation of Φ_ℓ in the neighborhood of the stationary point is large at early times. Confining the analysis of the region of validity of Eq. (24) to the early instant $t = z/c$, it can be shown that Eq. (25) leads to

$$\frac{z}{a} \gg \frac{F}{a}\,\frac{(m+1)^2(m+2)^2}{8\pi^2\ell^2}, \tag{27}$$

in which $F = \frac{a^2}{cT}$ denotes the Fresnel distance.

VI CONCLUDING REMARKS

We have constructed an angular spectrum of time-domain spectral-mode (TDSM) constituents, which furnishes an appropriate setting for the expansion of the field in the far zone in a waveguiding environment. Thus, we have incorporated novel time-domain *dispersive* modal wave objects in the STT framework.

In the spectral-time domain, a single wave species associated with a semi-infinite image array of sources or receivers has been shown to correspond to a series of half-amplitude

TDSM constituents, supplemented with a remainder diffraction-like integral. The remainder diffraction integrals cancel upon combining wave species associated with complementary semi-infinite image arrays.

Through the application of the saddle-point method, we have extracted physical parameters associated with a time-domain modal field constituent, e.g., instantaneous frequencies and angles of propagation, which characterize the chirped signals. Owing to the use of analytic signals the space-time source coordinates could be complexified in the resulting expressions for the field, yielding the corresponding 2D CSPB solutions associated with a class of highly directive sources. These pulsed-beams fields are modelled efficiently and in a physically cogent fashion in terms of the time-domain spectral modes.

Having established the properties of the STT TDSM expansion, the prospects for future research in this area seem promising. Among the interesting topics, we mention the exploration of TDSM constituents in complex waveguiding environments with penetrable boundaries, the Brewster-angle phenomenon, and the classification of properties of such environment in terms of the characteristic parameters of the chirped signals.

VII ACKNOWLEDGEMENTS

The research presented in this paper has been carried out at Tel Aviv University, and has been made possible through a fellowship awarded to the first author by the Royal Netherlands Academy of Arts and Sciences. The other authors have received partial support by the US-Israel Binational Science Foundation, Jerusalem, Israel, under grants No. 92-00273 and 95-00399, which is gratefully acknowledged.

REFERENCES

1. E. Heyman and L. B. Felsen. Weakly dispersive spectral theory of transients (STT). Part I: Formulation and interpretation. *IEEE Trans. Antennas Propagat.*, 35(1):80–86, 1987.

2. E. Heyman and L. B. Felsen. Weakly dispersive spectral theory of transients (STT). Part II: Evaluation of the spectral integral. *IEEE Trans. Antennas Propagat.*, 35(5):574–580, 1987.

3. E. Heyman. Weakly dispersive spectral theory of transients (STT). Part III: Applications. *IEEE Trans. Antennas Propagat.*, 35(11):1258–1266, 1987.

4. A. T. de Hoop and H. J. Frankena. Radiation of pulses generated by a vertical electric dipole above a plane, non-conducting earth. *Appl. Sci. Res., Sec. B*, 8:369–377, 1960.

5. E. Heyman. Complex source pulsed beam representation of transient radiation. *Wave Motion*, 11:337–349, 1989.

6. L. B. Felsen and F. Niu. Spectral analysis and synthesis options for short pulse radiation from a point dipole in a grounded dielectric layer. *IEEE Trans. Antennas Propagat.*, 41(6):747–754, 1993.

7. F. Niu and L. B. Felsen. Time-domain leaky modes on layered media: Dispersion characteristics and synthesis of pulsed radiation. *IEEE Trans. Antennas Propagat.*, 41(6):755–761, 1993.

8. F. Niu and L. B. Felsen. Asymptotic analysis and numerical evaluation of short pulse radiation from a point dipole in a grounded dielectric layer. *IEEE Trans. Antennas Propagat.*, 41(6):762–769, 1993.

9. L. B. Felsen and N. Marcuvitz. *Radiation and Scattering of Waves*. IEEE PRESS, Piscataway, NJ, 1994. Originally published by Prentice-Hall, Englewood Cliffs, NJ, 1973.

SPACE-TIME GREEN FUNCTION AND SHORT PULSE PROPAGATION IN DIFFERENT MEDIA

A. L. Gutman

Voronezh State Forestry Engineering Academy
Department of Physics
Timiryazev St., 8, Voronezh 394613, Russia
E-mail: olc@ns.vglta.ac.ru

INTRODUCTION

Space-time Green function (GF) was suggested by L. Weinstein in 1976. He used GF to prove some general properties of the pulse propagation in different media. And then he had passed to study the propagation of a long time pulse body. It is natural that here he had crossed from time domain analyses to a frequency one, where a band width is a small parameter of the problem. This analyses is the main contents of the Weinstains's paper [1]. May be this is the reason of an inadequate attention to the first part of the paper. We shall try to employ it in a time domain consideration, where an initial pulse time duration will be a small parameter of the problem.

SPACE-TIME GREEN FUNCTION AND ITS QUALITIES

The general definition of GF is

$$g(z,t)=\frac{1}{2\pi}\int_{-\infty}^{\infty}e^{i[h(\omega)z-\omega t]}d\omega\,, \tag{1}$$

where $h(\omega)$ is a law of a wave dispersion and attenuation in a given medium or line. $h(\omega)$ has the following form for different media:

- for conducting media with conductance $\sigma = \alpha \;/\; 2\pi$, which is independent of a frequency, and a line in a vacuum with conductance, also independent of a frequency

$$h(\omega)=\frac{\omega}{c}\sqrt{1+\frac{2i\alpha}{\omega}},\qquad \alpha = const>0,\quad c=const>0\,; \tag{2}$$

–for a long line with losses in wires and in the medium between them

$$h(\omega)=\frac{\omega+i\alpha}{c}\sqrt{1+\frac{\gamma^2}{(\omega+i\alpha)^2}},\qquad(\alpha\geq\gamma\geq 0);\tag{3}$$

– for a cold plasma with a plasma frequency ω_p and a collisions frequency - ν

$$h(\omega)=\frac{\omega}{c}\sqrt{1-\frac{\omega_p^2}{\omega(\omega+i\nu)}},\qquad(\omega_p=const>0,\quad \nu=const>0);\tag{4}$$

– for dielectric medium with elastically bound charges

$$h(\omega)=\frac{\omega}{c}\sqrt{1-\frac{\omega_p^2}{\omega^2+i\nu\omega-\omega_r^2}},\qquad(\omega_r=const>0),\tag{5}$$

– for active medium (system)

$$h(\omega)=\frac{\omega}{c}\sqrt{1+\frac{\gamma^2}{\omega^2}}.\tag{6}$$

$h(\omega)$ has a positive imaginary part by all real ω in all passive system (2), (3), (4) and (5). $h(\omega)$ allows analytic continuation to complex-valued ω and in upper half plate (Im$\omega\geq 0$) is a holomorphic function of ω. Its asymptotic expansion (by $|\omega|\to\infty$) has the form

$$h(\omega)=\frac{\omega}{c}+b_0-\frac{b_1}{\omega+i\nu}+\ldots,\tag{7}$$

where b_0 and b are constants and in (2) and (6) $\nu=0$.

It is possible to prove important qualities of GF and pulse propagation using these general properties of $h(\omega)$.

Thus we get

$$g(z,t)=0,\quad\text{if}\quad t<\frac{z}{c}.\tag{8}$$

Then if $f(z,t)$ is the pulse, we get

$$f(z,t)=\int_{-\infty}^{\infty}g(z,t-\bar{t})f(0,\bar{t})d\bar{t},\tag{9}$$

$$\text{and} \quad \left.\begin{array}{ll} \text{if} \quad f(0,t) = 0 & \text{with} \quad t < 0, \\ \text{then} \quad f(z,t) = 0 & \text{with } t < \dfrac{z}{c}. \end{array}\right\} \tag{10}$$

i.e. the pulse front moves with a limiting velocity - c.

The equation

$$g(0,t) = \delta(t), \tag{11}$$

where $\delta(t)$ is delta-function, follows from the definition of GF (1).

The integral presentation (1) for the dispersion laws (2), (3), (4) and (6) (by $\nu = 0$) can be expressed in an explicit form through the known functions:

$$g(z,t) = \left[\delta\left(t - \frac{z}{c}\right) + \frac{\gamma z}{c\sqrt{t^2 - z^2/c^2}} I_1\left(\gamma\sqrt{t^2 - z^2/c^2}\right)\right] e^{-\alpha t}, t \geq \frac{z}{c}. \tag{12}$$

Thence if $\gamma = \alpha$, we get $g(z,t)$ for the law (2), if $\alpha = 0$, we get it for (6). If $\gamma = i\omega_p$, $\alpha = 0$, $\nu = 0$, we get the following expression for a cold plasma without collisions:

$$g(z,t) = \delta\left(t - \frac{z}{c}\right) - \frac{\omega_p z}{c} \frac{J_1\left(\omega_p\sqrt{t^2 - z^2/c^2}\right)}{\sqrt{t^2 - z^2/c^2}}, \quad t \geq \frac{z}{c}. \tag{13}$$

The integral presentation $g(z, t)$ for the dispersion law (5) cannot be expressed through the known functions. But it can be calculated approximately for different ratio $z/t \leq c$ and for great z. Thus, from asymptotic expression (7) it is possible to get an approximate correlation

$$g(t,z) = \left[\delta\left(t - \frac{z}{c}\right) - \sqrt{\frac{b_1 z}{t - z/c}} J_1\left(2\sqrt{b_1 z(t - z/c)}\right) e^{-\nu(t - z/c)}\right] e^{ib_0 z}, \quad t \geq \frac{z}{c}. \tag{14}$$

It may be used with a small positive $t - z/c$, which provides small or finite argument for Bessel function.

GENERAL CORRELATIONS FOR THE PULSE PROPAGATION

All expressions for $g(z, t)$ ((12), (13) and (14)) have two parts. The first part contains delta-function. Therefore the corresponding part of the integral presentation (9) for $g(z, t)$ is calculated simply for any pulse form:

$$\int_{-\infty}^{\infty} \delta(t - \bar{t} - z/c) e^{-\alpha(t - \bar{t})} f(0, \bar{t}) d\bar{t} = f(0, t - z/c) e^{-\alpha\frac{z}{c}}, \tag{15}$$

$$\int_{-\infty}^{\infty} \delta(t - \bar{t} - z/c) f(0, \bar{t}) d\bar{t} = f(0, t - z/c), \tag{16}$$

$$\int_{-\infty}^{\infty} \delta(t-\bar{t}-z/c)e^{-\nu(t-\bar{t}-z/c)}e^{ib_0 z}f(0,\bar{t})d\bar{t} = f(0,t-z/c)e^{ib_0 z}. \tag{17}$$

In order to calculate the second part of (9), let us correct the integral's limits, using correlations (8) and (9). Then general correlation (9) takes three forms corresponding to equation pairs (12) and (15), (13) and (16), (14) and (17):

$$\begin{aligned} f(z,t) &= f(0,t-\frac{z}{c})e^{-\alpha\frac{z}{c}} + \\ &+ \int_0^{t-\frac{z}{c}} \frac{\gamma z}{c\sqrt{(t-\bar{t})^2 - z^2/c^2}} I_1\left(\gamma\sqrt{(t-\bar{t})^2 - z^2/c^2}\right)e^{-\alpha(t-\bar{t})} f(0,\bar{t})d\bar{t}, \end{aligned} \tag{18}$$

$$f(z,t) = f(0,t-\frac{z}{c}) - \frac{\omega_p z}{c}\int_0^{t-\frac{z}{c}} \frac{J_1\left(\omega_p\sqrt{(t-\bar{t})^2 - z^2/c^2}\right)}{\sqrt{(t-\bar{t})^2 - z^2/c^2}} f(0,\bar{t})d\bar{t}, \tag{19}$$

$$\begin{aligned} f(z,t) &= f(0,t-\frac{z}{c})e^{-b_0 z} - \\ &- \int_0^{t-\frac{z}{c}} \sqrt{\frac{b_1 z}{t-\bar{t}-z/c}} J_1\left(2\sqrt{b_1 z(t-\bar{t}-z/c)}\right)e^{-\nu(t-\bar{t}-z/c)}e^{ib_0 z} f(0,\bar{t})d\bar{t}. \end{aligned} \tag{20}$$

The integrals calculation in (18), (19) and (20) depends on the pulse form $f(0,t)$ essentially. We have chosen two forms:

$$f_1(0,t) = \begin{cases} 0, & t<0 \\ \sigma(t)-\sigma(t-T), & t \ge 0, \end{cases} \qquad \sigma(t) = \begin{cases} 0, & t<0 \\ 1, & t \ge 0 \end{cases} \tag{21}$$

$$f_2(0,t) = \begin{cases} 0, & t<0 \\ e^{-\beta t^2}, & t \ge 0. \end{cases} \tag{22}$$

In this paper we shall study pulse propagation in the media, which are discribed by equations (18) and (19).

THE RECTANGULAR PULSE PROPAGATION

By substituting $\sigma(t)$ from (21) into (19) we obtain

$$f_1(z,t) = \sigma(t-\frac{z}{c}) - \frac{\omega_p z}{c}\int_0^{t-\frac{z}{c}} \frac{J_1(\omega_p\sqrt{(t-\bar{t})^2 - z^2/c^2})}{\sqrt{(t-\bar{t})^2 - z^2/c^2}}d\bar{t}, \qquad t \ge \frac{z}{c}.$$

If we introduce dimensionless coordinates

$$\theta = \omega_p t, \qquad \xi = \frac{\omega_p z}{c}, \qquad \bar{\theta} = \omega_p \bar{t}\ , \tag{23}$$

this integral takes the form

$$f_1(\xi,\theta) = \sigma(\theta - \xi) - \xi \int_0^{\theta-\xi} \frac{J_1\left(\sqrt{(\theta-\bar{\theta})^2 - \xi^2}\right)}{\sqrt{(\theta-\bar{\theta})^2 - \xi^2}} d\bar{\theta}\ . \tag{24}$$

Let us introduce a new variable of integration

$$x = \frac{\theta - \bar{\theta}}{\xi}\ . \tag{25}$$

Then

$$f_1(\xi,\theta) = \sigma(\theta - \xi) - \xi \int_1^{\frac{\theta}{\xi}} \frac{J_1\left(\xi\sqrt{x^2-1}\right)}{\sqrt{x^2-1}} dx, \qquad \theta \geq \xi\ . \tag{26}$$

By $\theta/\xi \to \infty$

$$\xi \int_1^{\frac{\theta}{\xi}} \frac{J_1\left(\xi\sqrt{x^2-1}\right)}{\sqrt{x^2-1}} dx \to \xi\, I_{1/2}(\xi/2) K_{1/2}(\xi/2) = 1 - e^{-\xi}\ . \tag{27}$$

The results of computer calculation of $f_1(\xi,\theta)$ accoding to (26) are given in Fig 1. We see the head front of pulse, which moves with light velocity, and the oscillating trace behind it. The oscillation dies as the front goes away. According to (27) the trace level is $e^{-\xi}$, where ξ is the head front coordinate.

By substituting $f_1(0,t)$ from (21) into (19) and repetition transformation (23)-(26) we obtain the correlation for a rectangular pulse:

$$f_1(\xi,\theta) = \sigma(\theta - \xi) - \sigma(\theta - \tau - \xi) - \xi \int_1^{\frac{\theta}{\xi}} \frac{J_1\left(\xi\sqrt{x^2-1}\right)}{\sqrt{x^2-1}} dx + \\ + \xi \int_1^{\frac{\theta-\tau}{\xi}} \frac{J_1\left(\xi\sqrt{x^2-1}\right)}{\sqrt{x^2-1}} dx, \qquad \theta \geq \xi, \tag{28}$$

where

$$\tau = \omega_p T\ . \tag{29}$$

The results of computer calculation of $f_1(\theta, \xi)$ according to (28) are given in Fig. 2 and Fig. 3. Here one can see that if the pulse duration time is commensurable with a period of the oscillations in trace then an essential change of the pulse form is possible not far from the pulse movement beginning.

By analogy with (26) and (28) it is possible to get the correlations for rectangular pulse propagation in medium, which is described by equation (18):

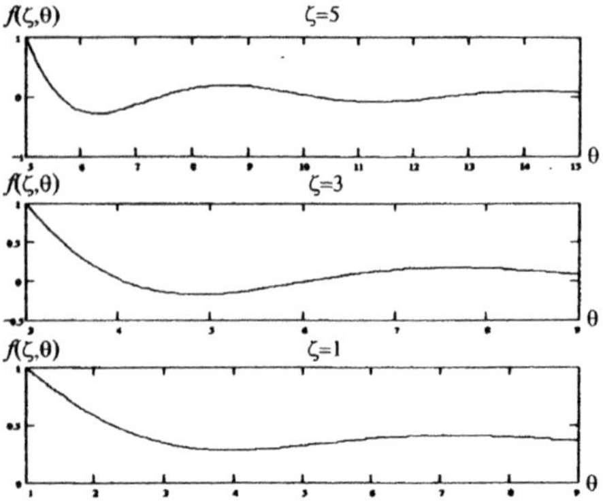

Figure 1.

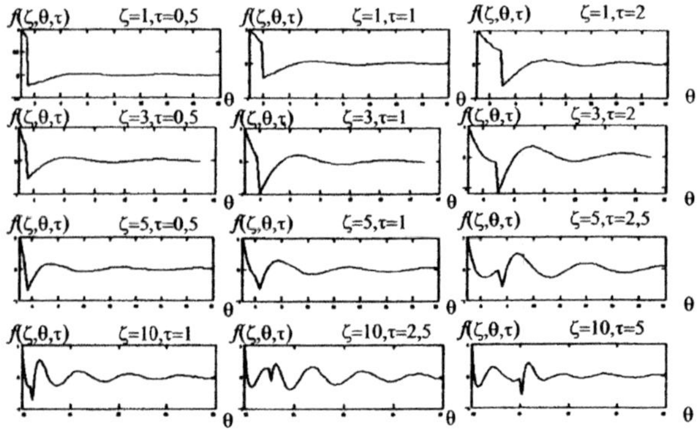

Figure 2.

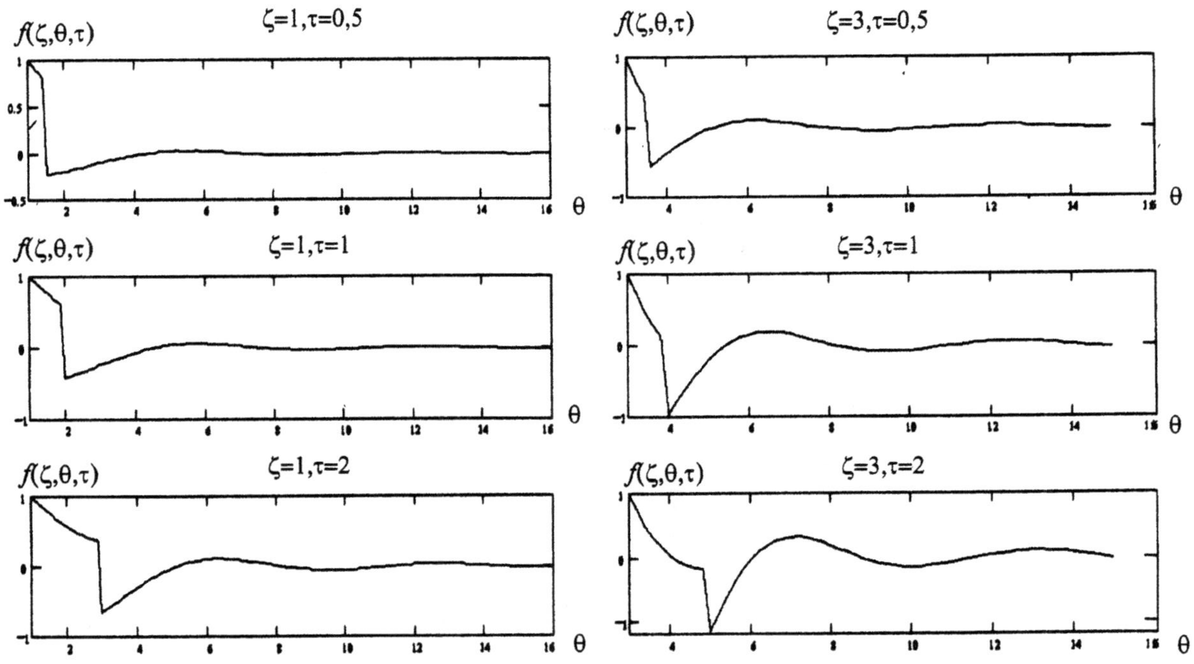
f(ζ,θ,τ)
ζ=1,τ=0,5
θ
f(ζ,θ,τ)
ζ=3,τ=0,5
θ
f(ζ,θ,τ)
ζ=1,τ=1
θ
f(ζ,θ,τ)
ζ=3,τ=1
θ
f(ζ,θ,τ)
ζ=1,τ=2
θ
f(ζ,θ,τ)
ζ=3,τ=2
θ

Figure 3.

$$f_1(\xi,t)=\sigma(\theta-\xi)e^{-\eta\xi}-\sigma(\theta-\tau-\xi)e^{-\eta(\xi-\tau)}+$$
$$+\xi\int_1^{\frac{\theta}{\xi}} e^{-\eta\xi x}\frac{I_1\left(\xi\sqrt{x^2-1}\right)}{\sqrt{x^2-1}}dx-\xi\int_1^{\frac{\theta-\tau}{\xi}} e^{-\eta\xi x}\frac{I_1\left(\xi\sqrt{x^2-1}\right)}{\sqrt{x^2-1}}dx, \quad (30)$$

where $\theta=\gamma t$, $\xi=\gamma\frac{z}{c}$, $\eta=\frac{\alpha}{\gamma}$.

FACILITIES OF AN ASYMPTOTIC PRESENTATION OF THE SHORT PULSE PROPAGATION

We shall consider these facilities using the second pulse form (22):

$$f_2(0,t)=\begin{cases}0, & t<0,\\ e^{-\beta t^2}, & t\geq 0.\end{cases} \quad (22)$$

Here we intend to use a small parameter – effective time pulse duration. In (22) this parameter is

$$T^*=\frac{1}{(\mathrm{Re}\,\beta)^{1/2}}. \quad (31)$$

Correlations (18) and (19) now take the form

$$f_2(z,t)=e^{-\beta\left(t-\frac{z}{c}\right)^2}e^{-\alpha\frac{z}{c}}+\frac{\gamma z}{c}\int_0^{t-\frac{z}{c}}\frac{e^{-\beta \bar{t}^2}I_1\left(\gamma\sqrt{(t-\bar{t})^2-z^2/c^2}\right)}{\sqrt{(t-\bar{t})^2-z^2/c^2}}e^{-\alpha(t-\bar{t})}d\bar{t}, \quad (32)$$

$$f_2(z,t)=e^{-\beta\left(t-\frac{z}{c}\right)^2}-\frac{\omega_p z}{c}\int_0^{t-\frac{z}{c}}\frac{e^{-\beta \bar{t}^2}J_1\left(\omega_p\sqrt{(t-\bar{t})^2-z^2/c^2}\right)}{\sqrt{(t-\bar{t})^2-z^2/c^2}}d\bar{t}. \quad (33)$$

Reβ is a large parameter in accordance with (31). Therefore the application of Laplace method is possible [2] for calculating the integrals in (32) and (33). After standard computations according to Watson lemma [2] we obtain the expressions for the asymptotic presentation $f_2(z, t)$ for (32):

$$f_2(z,t)=e^{-\beta(t-z/c)^2}e^{-\alpha z/c}+\frac{\gamma z}{c}e^{-\alpha t}\left\{\sqrt{\pi}\frac{I_1\left(\gamma\sqrt{t^2-z^2/c^2}\right)}{\sqrt{t^2-z^2/c^2}}+\right.$$
$$\left.+\frac{\gamma^2 z}{2c}\frac{I_2\left(\gamma\sqrt{t^2-z^2/c^2}\right)}{t^2-z^2/c^2}tT^2+\ldots\right\}, \quad (34)$$

where $T=\beta^{-1/2}$, if β is real value, then $T=T^*$. (35)

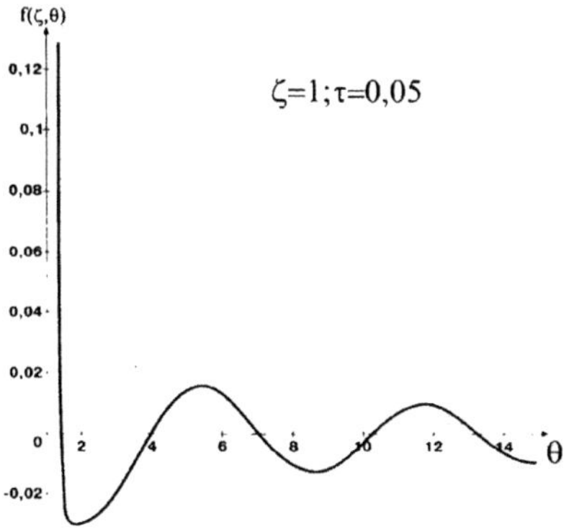

Figure 4.

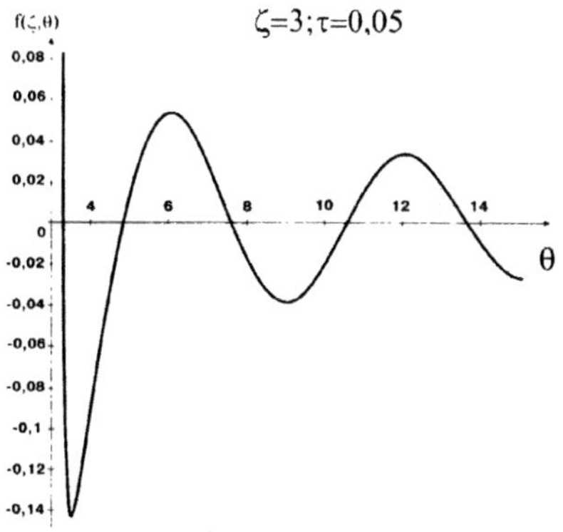

Figure 5.

For (33)

$$f_2(z,t) = e^{-\beta\left(t-\frac{z}{c}\right)^2} - \frac{\omega_p z}{c}\left\{\sqrt{\pi}\,\frac{J_1\left(\omega_p\sqrt{t^2 - z^2/c^2}\right)}{\sqrt{t^2 - z^2/c^2}}T + \right.$$
$$\left. + \frac{\omega_p^2 z}{2c}\frac{J_2\left(\omega_p\sqrt{t^2 - z^2/c^2}\right)}{t^2 - z^2/c^2}tT^2 + \ldots\right\}. \tag{36}$$

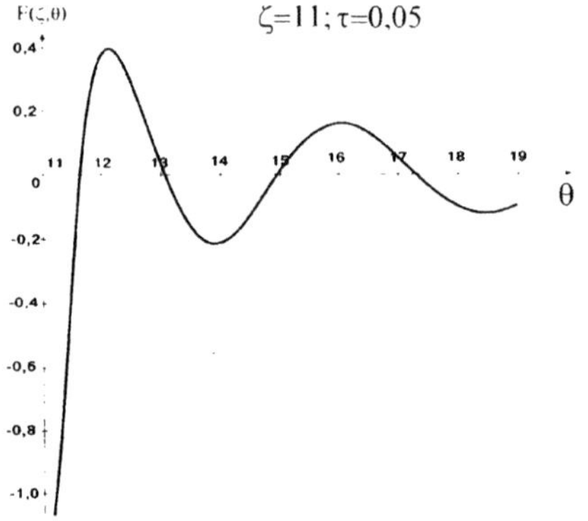

Figure 6.

One can see, that the next terms of this expansion have a similar structure. For instance, the third term in (36) has the form:

$$\frac{\sqrt{\pi}}{8}\frac{\omega_p^2 z}{c}\left[\frac{J_2\left(\omega_p\sqrt{t^2 - z^2/c^2}\right)}{t^2 - z^2/c^2} - \frac{\omega_p t^2 J_3\left(\omega_p\sqrt{t^2 - z^2/c^2}\right)}{t^2 - z^2/c^2}\right]T^3. \tag{36a}$$

It means that the factor t^{n-1} arises side by side with factor T^n.

The influence of the factor t^{n-1} is compensated by the factor $J_n\left(\omega_p\sqrt{t^2 - z^2/c^2}\right)/\left(t^2 - z^2/c^2\right)^{n/2}$ if $t^2 >> z^2/c^2$, because

$$\frac{J_n\left(\omega_p\sqrt{t^2 - z^2/c^2}\right)}{\left(t^2 - z^2/c^2\right)^{n/2}} \xrightarrow[t\to\infty]{} \frac{\sqrt{2}}{\sqrt{\pi\omega_p t}\, t^n} = \sqrt{\frac{2}{\pi\omega_p}} t^{-n-1/2}. \tag{37}$$

Thus the worst convergence of expansions in (33) and (34) will be with $t - z/c \to 0$. Then, for instance, the ratio between the second term and the first one in (36) will be equal to

$$\frac{\omega_p^2 ztT}{2\sqrt{\pi}\, c} \frac{J_2(\omega_p x)}{x^2} \cdot \frac{x}{J_1(\omega_p x)} \xrightarrow[x\to 0]{} \frac{\omega_p^3 ztT}{8\sqrt{\pi}\, c} = \frac{\xi\theta\tau}{8\sqrt{\pi}}, \tag{38}$$

where ξ, θ, τ are the dimensionless coordinates according to (23). The correlation (38) gives us an estimation of applicability of the expansions (34), (36) for a small value $\theta - \xi$:

$$\frac{\xi\theta\tau}{8\sqrt{\pi}} < 1. \tag{39}$$

The curves in Fig. 4,5 and Fig. 6 were calculated for equation (34) without the third term (36). In these figures one can see an efficiency of the estimation (39): the condition (39) is broken in Fig. 6.

CONCLUSIONS

The facilities of using the space-time GF for the short pulse propagation analysis were discussed. The GF application reduces a boundary-value problem to an integral calculation. The usefulness of this way is determined by the facilities of an exact and simple enough GF construction and a convenient calculation of obtained integrals. An explicit form of GF can be expressed exactly through the known functions for the conducting media, for the long line and cold plasma without collisions. In another cases it should be calculated approximately. In this work the use of the exact GF form was considered. The propagation of two forms of the pulse was observed. The integral presentation for the rectangular pulse was calculated numerically. The asymptotic expansion was obtained for the pulse $f(0,t) = e^{-\beta t^2}$, where a smallness of the effective pulse time duration was used.

REFERENCES

1. L. Weinstein. Pulses propagation. Uspekhi Fis. Nauk, Moscow, 1976, v. 118. p.339-367.
2. A. Erdelyi. Asymptotic expansions. Moscow, 1962, p.128.

PROPAGATION OF ULTRAWIDEBAND PULSED RADIATION IN CONDUCTING MEDIUM

Victor P. Belichenko, Vladimir I. Koshelev, and Konstantin A. Pervakov

High Current Electronics Institute RAS,
4, Akademichesky Ave., 634055 Tomsk, Russia

INTRODUCTION

In the papers[1,2] a group theoretical method is introduced for the description of separation of variables in the principal partial differential equations of mathematical physics. Systematic development of ideas of this method contains in the book[3]. A large number of solutions of the above-mentioned equations with nonseparable variables are also presented in it. In a number of recently published papers[4-6] it has been shown that some problems concerning propagation of nonharmonic electromagnetic waves in dispersive media, waveguides as well as in nonlinear media can be reduced to solving the Klein-Gordon equation. For its solving it was suggested to use definite solutions from a set of the ones[3].

In this paper, some formal constructions of papers[5,6] are defined more precisely, the preferential usage of the Klein-Gordon equation solutions with nonseparable variables is explained, different approaches for solving the obtained infinite systems of simultaneous linear algebraic equations are described, illustrative numerical results are presented as an example of a simple problem solving.

STATEMENT OF A PROBLEM

Let us suppose that half-space $z > 0$ is occupied by a motionless stationary nonmagnetic and conducting medium the boundary of which coincides with the plane $z = 0$. We suppose the medium to be linear, uniform and isotropic. Plane electromagnetic wave having the only nonvanishing components $E_x(z,t)$ and $H_y(z,t)$ and characterized by an arbitrary time dependence of these components falls from free half-space $z < 0$ normally to the boundary. It is required to find fields reflected and transmitted through the medium interface. Obviously, owing to the symmetry of the problem these fields will contain only the same field components.

As it is shown,[5,6] components E_x^+ and H_y^+ of the transmitted field can be presented by the following expansions

Ultra-Wideband, Short-Pulse Electromagnetics 4
Edited by Heyman *et al.*, Kluwer Academic / Plenum Publishers, New York, 1999.

$$E_x^+(\eta,\tau) = -\frac{A_0}{T_\sigma}\sum_{q=0}^{\infty} b_q e_q, \ (1) \qquad H_y^+(\eta,\tau) = -\frac{A_0}{WT_\sigma}\sum_{q=0}^{\infty} b_q h_q, \ (2)$$

$$e_q(\eta,\tau) = \frac{1}{2}e^{-\tau}\{\psi_{q-1} + \psi_{q+1} - 2\psi_q\}, \quad h_q(\eta,\tau) = \frac{1}{2}e^{-\tau}\{\psi_{q-1} - \psi_{q+1}\}, \quad W = \sqrt{\frac{\mu}{\varepsilon}}.$$

Here A_0 is the constant with vector potential dimension, T_σ is the characteristic transient period of the field, ε and μ are the absolute permittivity and permeability of the medium, respectively, $\eta = \frac{z}{T_\sigma}\sqrt{\varepsilon\mu}$, $\tau = \frac{t}{T_\sigma}$ and b_q are the unknown coefficients.

Functions

$$\psi_q(\eta,\tau) = \left(\frac{\tau-\eta}{\tau+\eta}\right)^{\frac{q}{2}} I_q\left(\sqrt{\tau^2-\eta^2}\right), \quad \tau \geq \eta$$

are the solutions[5,6] of the Klein-Gordon equation

$$\frac{\partial^2 F}{\partial\eta^2} - \frac{\partial^2 F}{\partial\tau^2} = -F$$

with nonseparable variables η and τ.

SPECTRA OF THE FUNCTIONS

Laplace transformation of the function

$$\psi_q(\eta,\tau) = 0 \ \text{at} \ \tau < \eta; \qquad \psi_q(\eta,\tau) = \left(\frac{\tau-\eta}{\tau+\eta}\right)^{\frac{q}{2}} I_q(\sqrt{\tau^2-\eta^2}) \ \text{at} \ \tau \geq \eta$$

after transition from variables η, τ again to the variables z, t can be represented[7] as

$$L\{\psi_q(z,t);p\} = \frac{\exp\left\{-z[\varepsilon\mu(p^2-\omega_\sigma^2)]^{\frac{1}{2}}\right\}}{(p^2-\omega_\sigma^2)^{\frac{1}{2}} i^q}\left[\frac{\omega_\sigma}{p+(p^2-\omega_\sigma^2)^{\frac{1}{2}}}\right]^q,$$

where $\omega_\sigma = T_\sigma^{-1}$.

Further, using well-known[7] coupling of Fourier and Laplace transformations

$$F\{f(t);\omega\} = L\{f(t);i\omega\} + L\{f(-t);-i\omega\},$$

we'll obtain the sought spectrum of the function $\psi_q(z,t)$

$$F\{\psi_q(z,t);\omega\} = \frac{\exp\left\{iz[\varepsilon\mu(\omega^2+\omega_\sigma^2)]^{\frac{1}{2}}\right\}}{i(\omega^2+\omega_\sigma^2)^{\frac{1}{2}}}(-1)^q\left[\frac{\omega_\sigma}{\omega+(\omega^2-\omega_\sigma^2)^{\frac{1}{2}}}\right]^q.$$

Expressions describing Fourier spectra of the functions $e_q(z,t)$ and $h_q(z,t)$ are easily obtained by applying the formula

$$F\{e^{-\alpha t}f(t);\omega\} = F\{f(t);\omega - i\alpha\}.$$

As it follows from the obtained expressions, function $\psi_0(z,t)$ has the widest and slowly decaying spectrum. Functions $\psi_q(z,t)$ with small q also have wide spectra. The expediency to use basic functions $e_q(z,t)$ and $h_q(z,t)$ at the field representation in a conducting medium (especially for studying propagation of the pulse with very small duration) is justified just by these circumstances .

REDUSING OF THE PROBLEM TO SOLVING AN INFINITE SYSTEM OF SIMULTANEOUS LINEAR ALGEBRAIC EQUATIONS

According to (1), (2) a field in a medium will be completely determined if unknown coefficients b_q are found. We'll state a method of their determination in common features defining more precisely some constructions and statements of the papers.[5,6]

We'll expand the functions $E_x(0,t)$ and $H_y(0,t)$ describing the field of the incident pulsed plane electromagnetic wave at $z = 0$ in the series about some full orthonormalized in the interval $0 \le t \le \infty$ function system. Supposing these functions to be different from zero only at $t \ge 0$ (a signal with a strongly expressed rise time), we'll choose as the desired functions the Lagerre functions.[6] Then

$$E_x(0,t) = \sum_{m=0}^{\infty} a_m l_m\left(\frac{t}{T_0}\right), \quad H_y(0,t) = \frac{1}{W_0}\sum_{m=0}^{\infty} a_m l_m\left(\frac{t}{T_0}\right),$$

$$a_m = \frac{1}{T_0}\int_0^{\infty} E_x(0,t) l_m\left(\frac{t}{T_0}\right)dt,$$

where W_0 - intrinsic impedance of the free space, T_0 - the characteristic temporal scale of the incident pulse.

We'll expand the harmonics $e_q(0,t)$ and $h_q(0,t)$ by the same system of functions

$$e_q(0,t) = \sum_{m=0}^{\infty} T_{mq}^{(1)}\, l_m\left(\frac{t}{T_0}\right), \quad h_q(0,t) = \sum_{m=0}^{\infty} T_{mq}^{(2)}\, l_m\left(\frac{t}{T_0}\right),$$

$$T_{mq}^{(1)} = \frac{1}{T_0}\int_0^{\infty} e_q(0,t) l_m\left(\frac{t}{T_0}\right)dt, \quad T_{mq}^{(2)} = \frac{1}{T_0}\int_0^{\infty} h_q(0,t) l_m\left(\frac{t}{T_0}\right)dt, \quad q=0,1,2,\ldots$$

Here matrices $\left\|T_{mq}^{(1)}\right\|$ and $\left\|T_{mq}^{(2)}\right\|$ determine the scattering of the m-th Lagerre impulse into the q-th harmonic of the electric and magnetic fields, respectively.

Let's represent the reflected field at z=0 by the expansions

$$E_x^-(0,t)=\sum_{m=0}^{\infty}c_m l_m(\frac{t}{T_0}), \quad H_y^-(0,t)=-\frac{1}{W_0}\sum_{m=0}^{\infty}c_m l_m(\frac{t}{T_0}),$$

where c_m are the unknown coefficients. From the continuity conditions of the tangential field components at the boundary $z = 0$, it follows that

$$\sum_{m=0}^{\infty}(a_m+c_m)l_m(\frac{t}{T_0})=-\frac{A_0}{T_\sigma}\sum_{m=0}^{\infty}\left\{\sum_{q=0}^{\infty}b_q T_{mq}^{(1)}\right\} l_m(\frac{t}{T_0}),$$

$$\frac{1}{W_0}\sum_{m=0}^{\infty}(a_m-c_m)l_m(\frac{t}{T_0})=-\frac{A_0}{WT_\sigma}\sum_{m=0}^{\infty}\left\{\sum_{q=0}^{\infty}b_q T_{mq}^{(2)}\right\} l_m(\frac{t}{T_0}). \tag{3}$$

Algebraization of the system of functional equations (3) based on the properties of orthonormalizeness and completeness of Lagerre functions lead to the infinite system of simultaneous linear algebraic equations for determining the unknown b_q and c_m .

$$\frac{A_0}{T_\sigma}\sum_{q=0}^{\infty}b_q T_{mq}^{(1)}+c_m=-a_m,$$

$$\frac{A_0 W_0}{T_\sigma W}\sum_{q=0}^{\infty}b_q T_{mq}^{(2)}-c_m=-a_m. \quad m=0,1,2,\ldots \tag{4}$$

It is necessary to make some remarks relative to the obtained system. Firstly, in spite of the statements containing in the papers[5,6] equality to zero one of the coefficients a_m doesn't result in equality to zero of corresponding matrix elements. Secondly, the system (4) is the one of the first kind and possibility to apply a reduction method for its solving requires a careful foundation. By our opinion, a preliminary regularization of the system is mostly expedient here (e. g., by using the procedure of matrix operator semiinversion).

ALTERNATIVE APPROACH TO SOLVING A PROBLEM

Here we'll consider a problem of rational determination of a pulsed field in a conducting medium if the field at $z = +0$ is known. As to the field in the section $z = +0$ it can be determined using well-known methods, e.g., spectral ones.

According to (1)

$$E_x^+(0,t)=-\frac{A_0}{T_\sigma}\sum_{q=0}^{\infty}b_q e_q(0,t). \tag{5}$$

Let's expand the given function $-\frac{T_\sigma}{A_0}E_x^+(0,t)$ by orthonormalized system of the functions $\varphi_n(t)$ which are linear combinations of the functions $e_q(0,t)$,

$$\varphi_n(t) = \sum_{q=0}^{n} \lambda_{nq} e_q(0,t). \quad (6)$$

Such system is determined by using standard procedure of Schmidt orthogonalization. It is expedient to choose a weighting function being a part of scalar product determination and orthogonality interval of the functions $\varphi_n(t)$ issued from a characteristic temporal scale of an incident pulse.

$$-\frac{T_\sigma}{A_0} E_x^{+}(0,t) = \sum_{n=0}^{\infty} c_n \varphi_n(t).$$

Now expresion (5) can be represented approximately as

$$\sum_{n=0}^{N} c_n \varphi_n(t) = -\sum_{q=0}^{Q} b_q e_q(0,t).$$

Hence it appears

$$b_q = \sum_{n=q}^{N} c_n \lambda_{nq},$$

here $Q = N$, i.e. number of coefficients b_q under determination is equal to the number of Fourier coefficients c_n and accuracy of their determination depends on the value N.

After determination of the coefficients b_q the field in an arbitrary section $z > 0$ is calculated by the formulae (1),(2). It is essential that when calculating the field in different medium sections the approach stated allows to avoid multiple calculations of inverse Fourier transform and difficulties related to it. The problem related to the solving of infinite system of linear algebraic equations (4) doesn't arise.

NUMERICAL RESULTS

Fig.1-3 illustrate the results of numerical calculations. Fig.1 presents the basic orthonormalized functions $\varphi_n(t)$ for the case of the orthogonality interval $0 \le t \le 10$ and weighting function identically equal to one. Fig. 2 presents the examples of approximation of

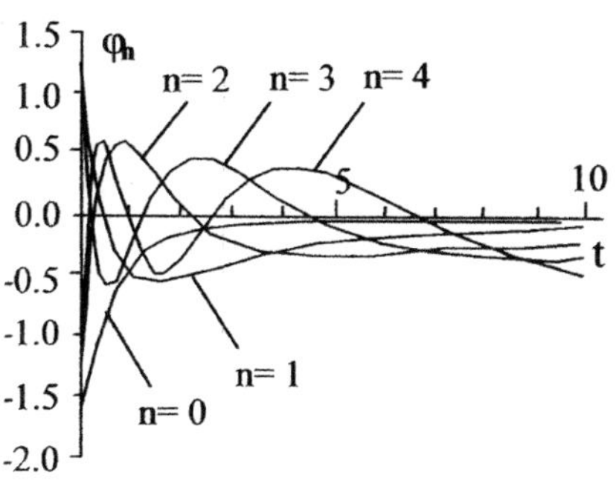

Figure 1. Orthonormalized functions $\varphi_n(t)$.

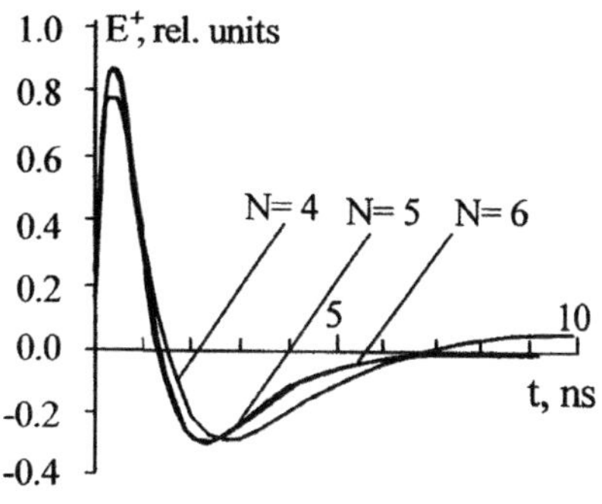

Figure 2. Approximation of the pulse transmitted into medium by orthonormalized functions

an ultrawideband signal transmitted into the medium ($z = 0$) using orthonormalized functions some of which are presented in Fig. 1. Calculations have shown that at $T_0 \sim T_\sigma$, a small number of functions $\varphi_n(t)$ is enough to reproduce the form of the signal that transmitted into the medium.

When calculating a field in a conducting medium, the conclusions of the papers[4-6] have been confirmed about the behavior of harmonic e_q and h_q of electromagnetic field.

Fig.3 illustrates characteristic phenomena at pulse obliquely incident on a interface. It presents the pulse forms for the incident and reflected fields in the plane $z = 0$. The incident pulse was supposed to be in the form

$$E_{in} = E_x(0,t) = exp\{-(t/T_0 - 2)^2\} - exp\{-(t/T_0 - 4)^2\}.$$

The assumed essential differences in the reflected field for a parallel ($E_\parallel$) and perpendicular ($E_\perp$) polarizations are clearly seen.

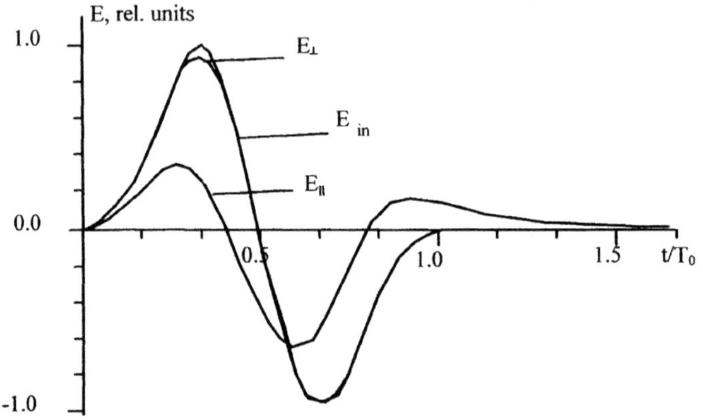

Figure 3. Pulse forms at obliquely incident at the angle 80^0 on the interface $z = 0$: $T_0 = 10^{-9}$ c, $T_\sigma = 2.568 \cdot 10^{-9}$ c, $\varepsilon = 2.9$; conductivity of the medium $\sigma = 0.02$ S/m.

CONCLUSION

A new method of solving the problem concerning a plane electromagnetic pulse scattering at the boundary of a conducting half-space has been discussed. Permittivity and permeability of a half-space dispersive medium are assumed not to depend on frequency. Application of a nonseparable solutions of the Klein-Gordon equation to solving of which at the initial stage reduce the problem of determining a field in a conducting medium has the key meaning in this method. It has been shown that if the field at the medium boundary is determined then it allows to effectively calculate the field in the medium using an orthonormalized function system formed from linearly independent solutions of the Klein-Gordon equation. Scattered field as well as in the known work[6] is presented by the expansions on orthonormalized system of Lagerre functions.

REFERENCES

1. P. Winternitz and I. Frish. Invariant expansions of relativistic amplitudes and subgroups of Lorentz self-group. *Yadernaya phisica.* 1:889 (1965).

2. P. Winternitz, I. Lukach, and Ya. A. Smorodinsy. Quantum numbers in small groups of Poincare group invariant expansions of relativistic amplitudes and subgroups of Lorentz self-group. *Yadernaya phisica.* 7:192 (1968).
3. W., Jr. Miller. "Symmetry and separation of variables," Addison-Wesley Publ. Co., (1977).
4. A.B. Shvartsburg. Nonharmonic electromagnetic waves in plasma. *Doklady Akademii Nauk.* 333:324 (1993).
5. V.A. Danilov and A.B Shvartsburg. Nonharmonic electromagnetic pulses in conducting medium. *Doklady Akademii Nauk.* 341:330 (1995).
6. A.B. Shvartsburg. Video pulses and non periodic waves in dispersive media (exactly solvable models). *Uspekhi phizicheskich nauk.* 168:85 (1998).
7. H. Bateman and A.Erdelyi. "Highest transcendental functions," V.2. Bessel functions, parabolic cylinder functions, orthogonal polinomials. McCraw-Hill Book Company, Inc., New York Toronto London, (1954).

PASSAGE OF SHORT PULSE THROUGHOUT OSCILLATING CIRCUIT WITH DIELECTRIC IN CONDENSER

A.L.Gutman

Voronezh State Forestry Engineering Academy
Department of Physics
Timirryazev St., 8, Voronezh 394613, Russia
E-mail: olc@ns.vglta.ac.ru

INTRODUCTION

This paper contains the continuation of the author's previous research[1,2]. The advance of a short pulse through lumped parameters circuits is interesting in the case when time of the pulse duration and these parameters establishment are commensurable. The circuit's elements with dielectric and magnetic have no time to establish their parameters during the essential part of the pulse continuance. These parameters change their values during the pulse continuance. Therefore the parametric effect acts onto the oscillating circuit. The aim of the work is the pulse form determination in the oscillating circuit where these effects arise due to dielectric in the circuit condenser. We use here a general correlation between a nonstationary field and the polarization in dielectric obtained in previous work [1].

EQUATION FOR NONSTATIONARY FIELD IN OSCILLATING CIRCUIT

To obtain an equation for the field $\boldsymbol{E(t)}$ in a flat condenser we use an equality between a conduction current in the circuit and bias current in the condenser (Fig. 1):

$$\boldsymbol{i} = \frac{1}{4\pi}\frac{\boldsymbol{dD}}{\boldsymbol{dt}}\boldsymbol{S}, \tag{1}$$

where $$\boldsymbol{D} = \boldsymbol{E(t)} + 4\pi \boldsymbol{P(t)}. \tag{2}$$

Therefore a voltage $\boldsymbol{U(t)}$ of a source equals:

Ultra-Wideband, Short-Pulse Electromagnetics 4
Edited by Heyman *et al.*, Kluwer Academic / Plenum Publishers, New York, 1999.

$$U(t) = \frac{1}{4\pi}\frac{dD}{dt}SR + El + \frac{1}{4\pi}\frac{d^2D}{dt^2}LS, \tag{3}$$

where D is an electrical bias, S is an area of the condenser's plates, l is a distance between the condenser's plates.

After division by l this equation gets the following form

$$u = \tau\left\{\frac{dD}{dt} + \theta\frac{d^2D}{dt^2}\right\} + E, \tag{4}$$

where
$$u = \frac{U}{l}, \tag{5}$$

$$\tau = \frac{SR}{4\pi l}, \qquad \theta = \frac{L}{R}. \tag{6}$$

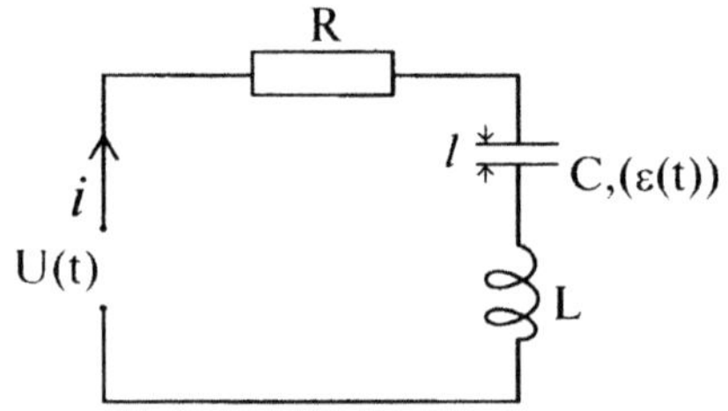

Figure1.

From equation (4) and (2) we obtain the equation:

$$u = \tau\left\{\frac{dE}{dt} + 4\pi\frac{dP}{dt} + \theta\frac{d^2E}{dt^2} + 4\pi\theta\frac{d^2P}{dt^2}\right\} + E. \tag{7}$$

The general correlation for the polarization in dielectric medium is [1]:

$$\begin{aligned} P(t) &= \frac{\omega_p^2}{4\pi\Omega}\int_0^t E(z)e^{-\alpha(z-t)}\begin{cases}\sin\Omega(t-z)\\ \mathrm{sh}|\Omega|(t-z)\end{cases}dz + \\ &+ e^{-\alpha t}P(0)\begin{cases}\cos\Omega t\\ \mathrm{ch}|\Omega|t\end{cases} + \frac{\dot{P}(0) + 2\alpha P(0)}{\Omega}\begin{cases}\sin\Omega t, \ \Omega^2 > 0\\ \mathrm{sh}|\Omega|t, \ \Omega^2 < 0,\end{cases}\end{aligned} \tag{8}$$

where $\omega_p = \frac{4\pi Ne^2}{m}$; m, e are electron mass and charge, N is electron concentration, $\Omega^2 = \omega_0^2 - \alpha^2 - \omega_p^2/3$, ω_0, α are a resonance frequency and an absorption coefficient of the elementary oscillators in dielectric. The substitution equation (8) in (7) gives us the desired integro-differential equation.

SOLUTION OF THE INTEGRO-DIFFERENTIAL EQUATION (7), (8)

Using the Laplace transformation method we obtain the following algebraic equation for the Laplace field transform in the condenser $\overline{E}(\xi)$:

$$\begin{aligned}\overline{U}(\xi) &= \overline{E}(\xi)\left\{1 + \tau\xi(1+\theta\xi)\left[1 + \frac{\omega_p^2}{(\xi+\alpha)^2 + \Omega^2}\right]\right\} + \\ &+ \frac{4\pi\left[(\xi+2\alpha)P(0) + \dot{P}(0)\right](1+\theta\xi)\xi\tau}{(\xi+\alpha)^2+\Omega^2} - \\ &- \tau\left[E(0) + 4\pi P(0)\right](1+\theta\xi) - \tau\theta\left[\dot{E}(0) + 4\pi\dot{P}(0)\right].\end{aligned} \tag{9}$$

And therefore

$$\overline{E}(\xi) = \frac{V(\xi)}{W(\xi)}, \tag{10}$$

where

$$\begin{aligned}V(\xi) &= \overline{u}(\xi)\left[(\xi+\alpha)^2 + \Omega^2\right] - 4\pi\tau\xi\left[(\xi+2\alpha)P(0) + \dot{P}(0)\right] + \\ &+ \left\{\tau(1+\theta\xi)\left[E(0) + 4\pi P(0)\right] + \tau\theta\left[\dot{E}(0) + 4\pi\dot{P}(0)\right]\right\}\left[(\xi+\alpha)^2 + \Omega\right],\end{aligned} \tag{11}$$

$$\begin{aligned}W(\xi) &= \tau\theta\xi^4 + \tau\xi^3(1+2\alpha\theta) + \xi^2\left[1 + \tau\left(2\alpha + \theta\left(\alpha^2 + \Omega^2 + \omega_p^2\right)\right)\right] + \\ &+ \xi\left(2\alpha + \tau\left(\alpha^2 + \Omega^2 + \omega_p^2\right)\right) + \Omega^2 + \alpha^2.\end{aligned} \tag{12}$$

With the zero initial conditions

$$P(0) = \dot{P}(0) = E(0) = \dot{E}(0) = 0$$

we have

$$V(\xi) = \overline{U}(\xi)\left[(\xi+\alpha)^2 + \Omega\right]. \tag{13}$$

The general correlation for polarization transform has the form:

$$\overline{P}(\xi) = \frac{\omega_p^2\overline{E}(\xi)}{4\pi\left[\Omega^2 + (\xi+\alpha)^2\right]} + \frac{(\xi+2\alpha)P(0) + \dot{P}(0)}{\Omega^2 + (\xi+\alpha)^2}, \tag{14}$$

and with zero initial conditions

$$\overline{P}(\xi) = \frac{\omega_p{}^2 \overline{E}(\xi)}{4\pi\left[\Omega^2 + (\xi + \alpha)^2\right]} = \frac{\omega_p{}^2}{4\pi} \frac{\overline{u}(\xi)}{W(\xi)}. \tag{15}$$

We choose for $u(t)$ the function of a single step (Heviside function):

$$u(t) = \begin{matrix} 0, & t < 0, \\ 1, & t \geq 0. \end{matrix} \tag{16}$$

Then

$$\overline{u}(\xi) = \frac{1}{\xi} \tag{17}$$

and

$$\overline{E}(\xi) = \frac{(\xi + \alpha)^2 + \Omega^2}{\xi W(\xi)}, \qquad \overline{P}(\xi) = \frac{\omega_p^2}{4\pi} \frac{1}{\xi W(\xi)}. \tag{18}$$

Now according to decomposition theorem

$$E(t) = \frac{\alpha^2 + \Omega^2}{W(0)} + \sum_{i=1}^{4} \frac{(\xi_i + \alpha)^2 + \Omega^2}{\xi_i W'(\xi_i)} e^{\xi_i t}, \qquad \frac{\alpha^2 + \Omega^2}{W(0)} = 1. \tag{19}$$

$$P(t) = \frac{\omega_p^2}{4\pi(\alpha^2 + \Omega^2)} + \frac{\omega_p^2}{4\pi} \sum_{i=1}^{4} \frac{e^{\xi_i t}}{\xi_i W'(\xi_i)}, \tag{20}$$

where ξ_i is a root of equation

$$W(\xi) = 0 \tag{21}$$

and

$$W'(\xi_i) = 4\tau\theta\xi_i^3 + 3\tau\xi_i^2(1 + 2\alpha\theta) + 2\xi[1 + \tau(2\alpha + \\ + \theta(\alpha^2 + \Omega^2 + \omega_p^2))] + 2\alpha + (\alpha^2 + \Omega^2 + \omega_p^2). \tag{22}$$

Thus the problem is reduced to the calculation of roots of equation (21). This is an algebraic equation of the fourth power. It may be solved exactly through radicals and in the general case.

RESULTS OF NUMERICAL CALCULATIONS

The results of concrete calculations the condenser's field by equation (23)

$$E\left(\tfrac{t}{\tau}\right) = 1 + \sum_{i=1}^{4} T_i e^{\xi_i \frac{t}{\tau}}, \qquad T_i = \frac{(\xi_i + \alpha)^2 + \Omega^2}{\xi_i W'(\xi_i)} \tag{23}$$

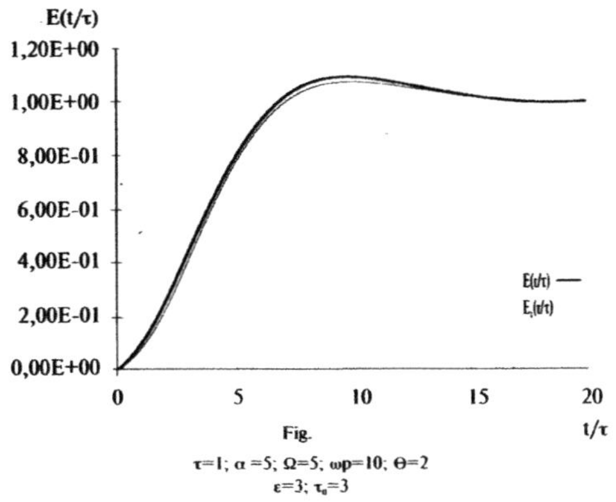

Figure 2.

are given in Fig. 2,3,4,5 and 6 for different values τ, α, Ω, ω_p and θ. These curves are compared on the graphs with the curves $E_1\left(t/\tau\right)$ obtained without the estimate of the establishment process for the established dielectric coefficient. These curves differ slightly, practically coincide in Fig. 2. In this case the absorption coefficient of the elementary oscillators is big ($\alpha = 5$). Therefore the establishment is determined here by macroscopic parameters of the oscillating circuit: by stationary capacity, inductance and resistance. The curves $E\left(t/\tau\right)$ and $E_1\left(t/\tau\right)$ differ essentially in another cases: the oscillating factor has time for the realization of some oscillations during the time of establishment. In the first case (Fig. 2) this time ($1/\alpha$) equals two tenth of time constant

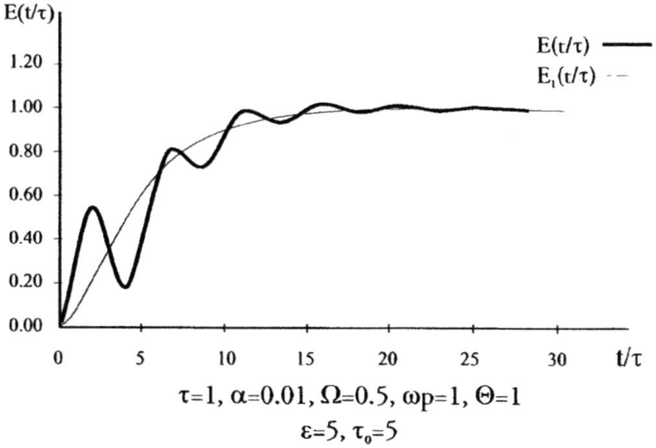

Figure 3.

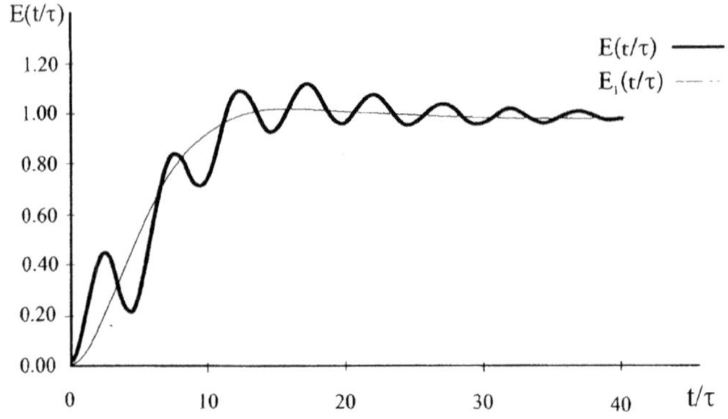

$\tau=1,\ \alpha=0.01,\ \Omega=0.5,\ \omega p=1,\ \Theta=2$
$\varepsilon=5,\ \tau_0=5$

Figure 4.

(τ) of the "empty" condenser, in other cases (Fig. 3,4,5 and 6) these ratios are twenty and more. The collective microscopic oscillations are realized in these cases. But the macroscopic oscillations of the circuit are not detected in the fourth case (Fig. 3), they are slightly detected in the cases Fig. 4 and Fig. 5. The second frequency - the frequency of the macroscopic circuit oscillations is detected in the case of Fig. 6.

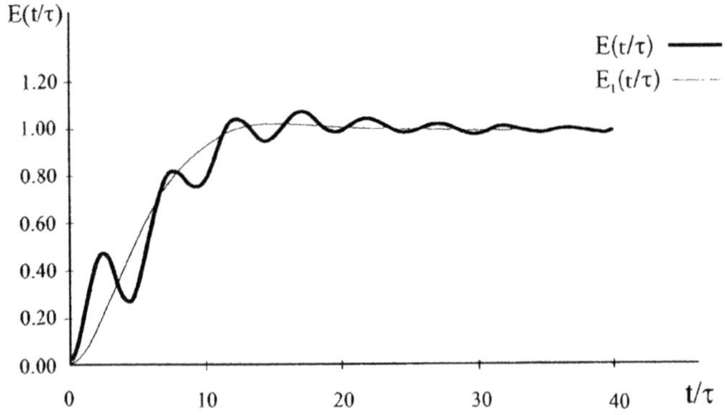

$\tau=1,\ \alpha=0.05,\ \Omega=0.5,\ \omega p=1,\ \Theta=2$
$\varepsilon=4.96,\ \tau_0=4.96$

Figure 5.

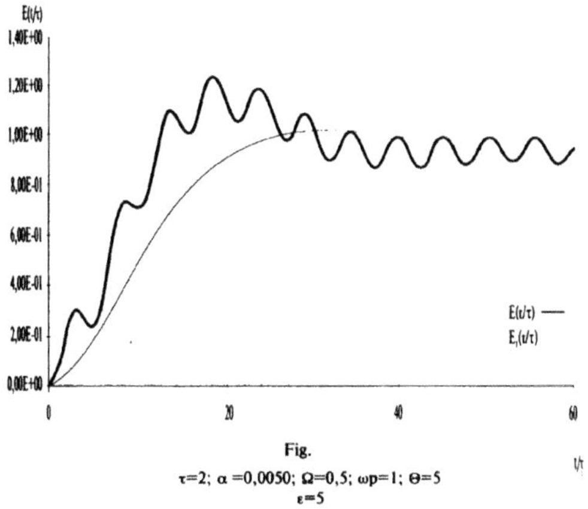

Figure 6.

CONCLUSIONS

The exact solution of the establishment process in the oscillating circuit with dielectric is obtained. Its closed form gives us opportunities to consider different cases of an interaction between the collective microscopic movement of the elementary oscillators and the macroscopic field in the oscillating circuit. This interaction may be slight and then the macroscopic field oscillations are determined with the established parameters of the circuit. But it may be essential and then we observe two different simultanity movements. There may be two oscillations with different frequencies. All these modes of interaction are determined with correlations between microscopic parameters of the dielectric and macroscopic parameters of the oscillating circuit. A possibility of these modes must be taken into consideration and may be used for the pulse shaping.

REFERENCES

1. A. Gutman. Transient Dielectric Coefficient and Conductance in Dielectric Media in Nonstationary Field. Ultra Wideband, Short-Pulse Electromagnetics 3. Edited by Baum et al.., Plemum Press, New York, 1997. pp. 205-209.
2. A. Gutman. Passage of Short Pulse Throughout Condenser with Dielectric. PIERS-97, Cambridge, MA, USA. p. 89.

UWB MEASUREMENTS OF CANONICAL TARGETS AND RCS DETERMINATION

Y. Chevalier, Y. Imbs, B. Beillard, J. Andrieu, M. Jouvet, B. Jecko (1)
M. Le Goff, E. Legros (2)

(1) Institut de Recherche en Communications Optiques et Microondes
IRCOM - 7 rue Jules Vallès - 19100 Brive la gaillarde (France)
E-mail : imbs@brive.unilim.fr
(2) CELAR (DGA) GEOS/SDM - BP 7419 - 35174 Bruz Cedex (France)

INTRODUCTION

An experimentation is described for measurements of impulse scattering responses from different canonical targets. It is intended to validate the transient measurement concept of low frequency (200 MHz - 1 GHz) Radar Cross Section (RCS) and this for large dimensions targets. The RCS determination is performed via a discrete Fourier transform and a calibration technique. This transient method allows to insulate the interesting signal with a time window, so that, the interference signals are eliminated. This is the advantage of the transient method compared to the frequential one. This study was supported by the CELAR (French Technical Centre for Armament Electronics).

MEASUREMENT FACILITY

The basic measurement configuration uses a 36 m x 30 m horizontal ground plane (figure 1). This CELAR test facility was used in the past for electromagnetic compatibility measurements. Two identical wire-antennas are used and their characteristics are determined to radiate an uniform and homogeneous electric field in the test zone. The bistatic angle is about 5° and allows a quasi-monostatic measurement configuration. The target is located at 9 meters of the antennas and is around 40 centimetres above the ground plane level. The distance between the antenna and the target is chosen to obtain a sufficient clear measurement time window. The transmitting antenna is driven by a pulse generator and the receiving antenna is connected to a digital sampling oscilloscope (Tektronix TDS 820 - 6 GHz). The antennas are build with only two wires and they have been designed to radiate and to receive short pulses.

Ultra-Wideband, Short-Pulse Electromagnetics 4
Edited by Heyman *et al.*, Kluwer Academic / Plenum Publishers, New York, 1999.

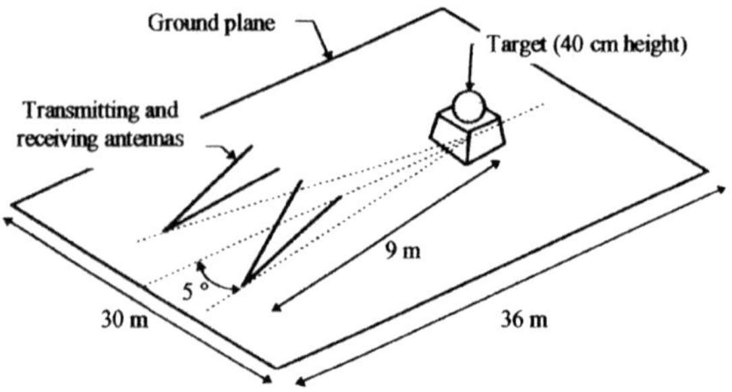

Figure 1. Configuration of measurement facility.

PULSE GENERATOR

The pulse generator used for the experiment was an AVTECH generator which has a pulse output voltage of 77 V, a 10 - 90 % risetime of 400 ps and a 50 % pulse duration (full-width at half-maximum) of 500 ps. This gaussian pulse has a frequency range from DC to about 1.2 GHz (figures 2 and 3).

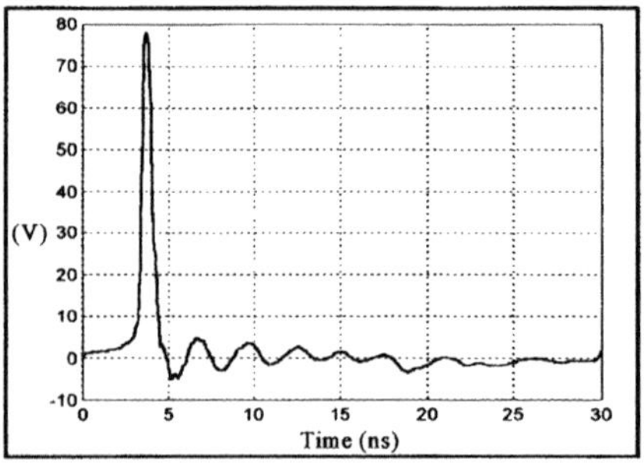

Figure 2. Generator pulse output.

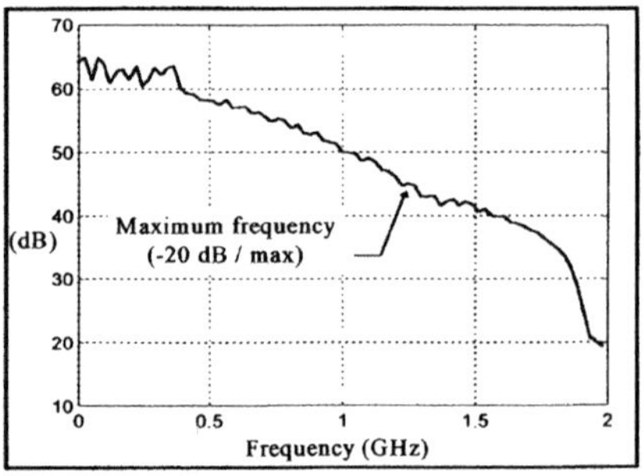

Figure 3. Spectral waveform.

THE WIRE ANTENNAS

The emitting and receiving antennas are identical. Only two wires compose the antenna in order to simplify manufacturing (figure 4). The dimensions have been determined considering the electromagnetic performances (field pattern, radiated pulse). The wire structure is build from a supple electric wire of 4 millimetres diameter.

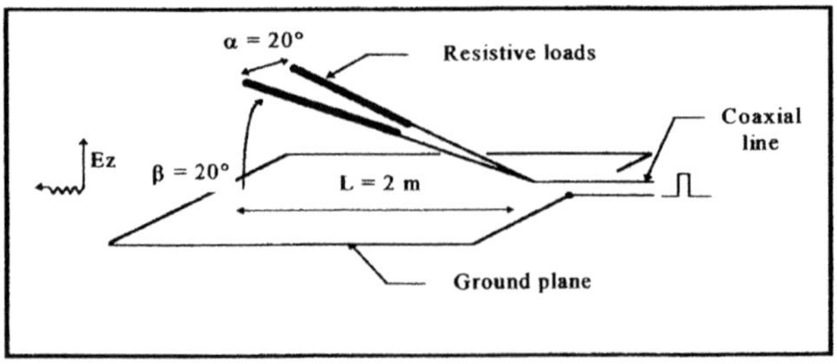

Figure 4. The wire antenna.

Ten resistive loads are uniformly distributed on the upper half-length of each arms (figure 4) to eliminate progressively the reflected currents from the end of the antenna. This technique uses the principle of the Wu-King non reflecting dipole[1]. The benefit of this method is shown in figure 5 where the received impulses for pair of perfectly conducting and resistively loaded antennas are presented. In both cases, the receiving antenna is pointed toward the transmitting antenna at about 15 metres. The difference between the peak levels of the first pulse is the result of the resistive loads and of the antenna wire diameter change. The resistively loaded antenna has a 4 mm diameter while the perfectly conducting antenna has a 1 cm one.

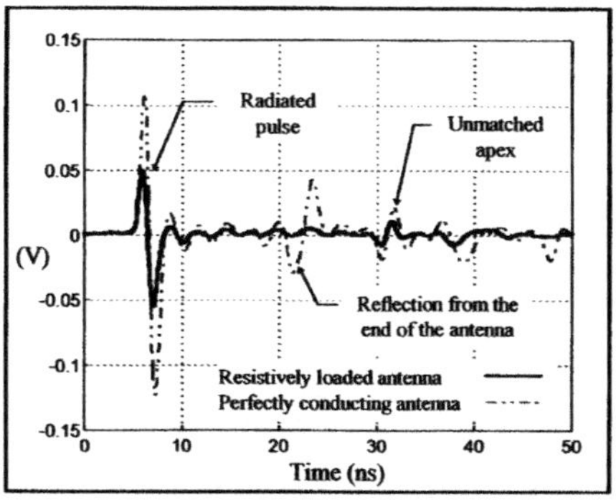

Figure 5. Pulse radiated by a loaded antenna and a perfectly conducting antenna.

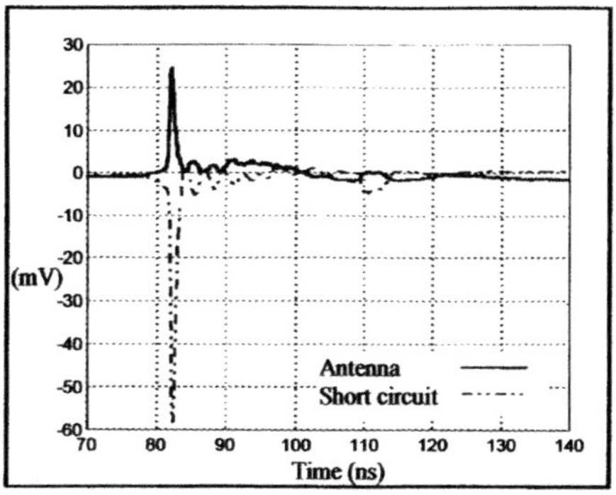

Figure 6. Reflected voltage in the feeding transmission line.

Figure 6 shows the reflected voltage in the feeding transmission line for the antenna and for a short circuit. The antenna feed point impedance is about 110 Ohms. So, the reflected voltage between the generator and the transmitting antenna is visible figure 5 at the time t = 30 ns. A 3 dB attenuator and a long cable are used respectively to minimise this reflection and to put it back from the time measurement window. Thus, the radiate pulse has a frequency range from below 100 MHz to about 850 MHz.

EXPERIMENTAL RESULTS

The measured canonical targets are :

- a 71 cm x 71 cm flat plate,
- a 355 mm metallic sphere,
- a perfectly conducting cylinder (diameter : 25 cm, length : 62 cm).

Flat plates are among the easiest targets to measure. They have large radar cross sections at most frequencies[2]. Aside from being excellent calibration targets, spheres display a creeping surface wave which can be observed using an UWB incident impulse. The horizontal cylinder allows to observe the travelling waves phenomenon.

Impulse scattering responses are measured at different heights around 40 cm. Thus an average is used to reduce the ground-target reflection even if this one is not really significant for all targets. Figure 7 shows a FDTD simulation for the perfectly conducting cylinder above a perfectly conducting ground plane. Excitation is a Gaussian voltage pulse. The ground target height varies from 40 cm to 70 cm. The ground target reflection appears just after the end of the target impulse scattering response. Figure 8 compares the free space FDTD simulation with the signal obtained after the average of the ground FDTD simulations. The difference is very small and it can be reduced if the average is realised with more important measurement number. Thus, ten measurements average is realised for this experimentation.

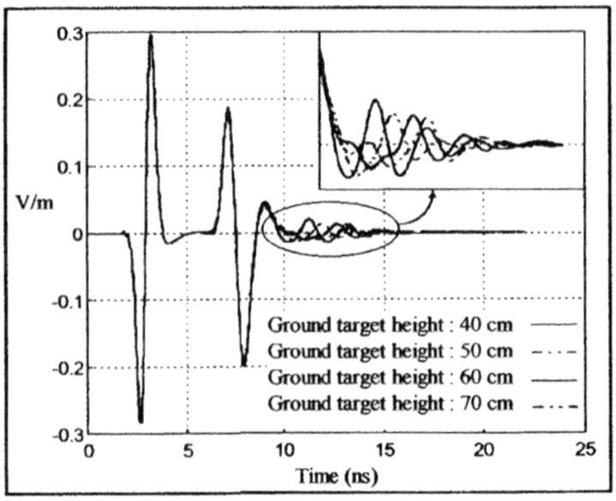

Figure 7. FDTD simulation for the cylinder above a ground plane at different heights.

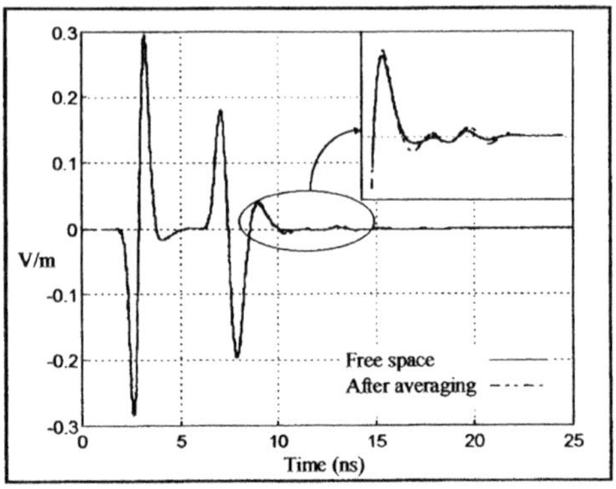

Figure 8. FDTD simulation for the cylinder in free space and after averaging.

Figures 9 shows the signal measured by the receiving antenna for the plate and the sphere. This signal is composed of the direct coupling between antennas and of the time response from the target. So, the signal with the empty chamber is subtracted in all measurements, this eliminates the direct coupling from the transmitting antenna to the adjacent receiving antenna. It also eliminates any returns due to down range scatterers that are inherent to the facility.

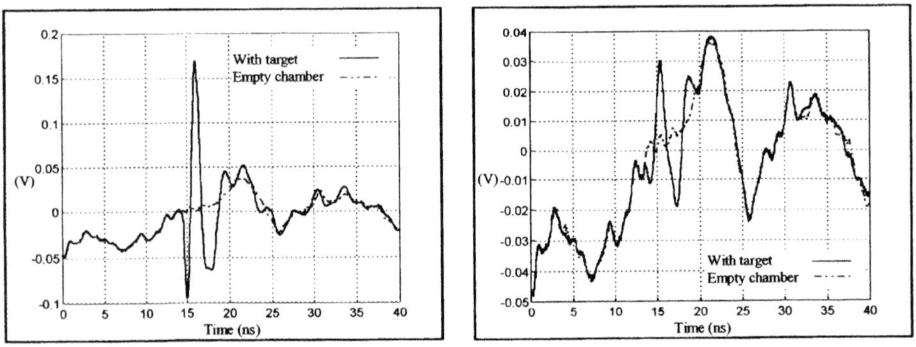

(a) Flat plate (b) Sphere

Figure 9. Pulses measured at the receiving antenna with a target installed and removed

The target signals with background subtraction are shown figure 10. The impulse scattering response of the plate presents a large specular return and a late time ringing. The time domain return from the sphere presents a creeping wave appearing at $(2+\pi).a/c$ seconds after the specular return where a is the sphere's radius and c is the speed of light.

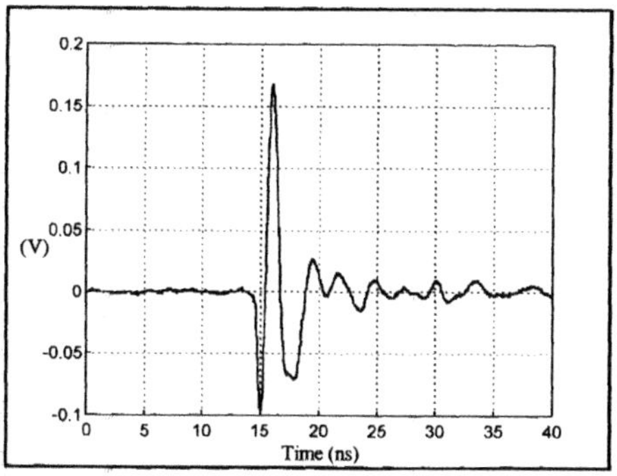

(a) Flat plate

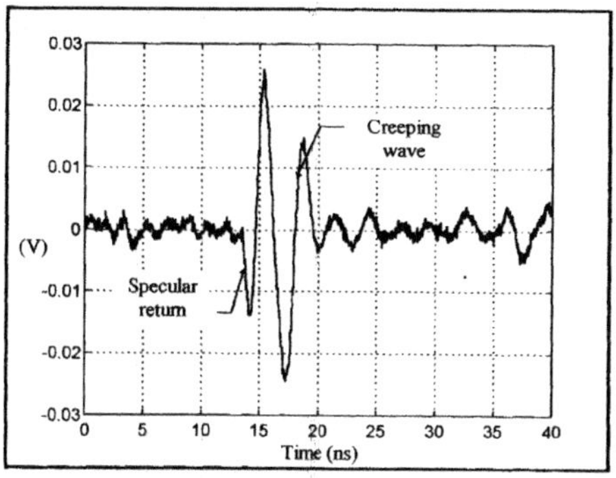

(b) Sphere

Figure 10. Pulses after subtraction

The RCS is determined from the knowledge of the theoretical RCS of a calibration target like a sphere or a plate from the following expression :

$$RCS(f) = \frac{\left|TF(E_{Target\ measurement}(t))\right|^2}{\left|TF(E_{Calibration\ target\ measurement}(t))\right|^2} . RCS_{Theoretical}(f) \qquad (1)$$

$E_{target\ measurement}$ represents the impulse response of each target after subtraction of the empty chamber signal. $E_{calibration\ target\ measurement}$ represents in the same conditions, the impulse response of the calibration target whose RCS is perfectly known either analytically or by means of a computed method like FDTD. This technique allows the determination of the target RCS without to know initially the incident signal.

Then, the RCS of canonical targets are compared to theoretical predictions and differences between measured and expected data is less than 2 dB.m² for the maximum gap in the frequency range (figures 11 and 12). Beyond 800 MHz, the RCS accuracy becomes lower because the pulse radiated by the antenna has not sufficient power for these spectral components.

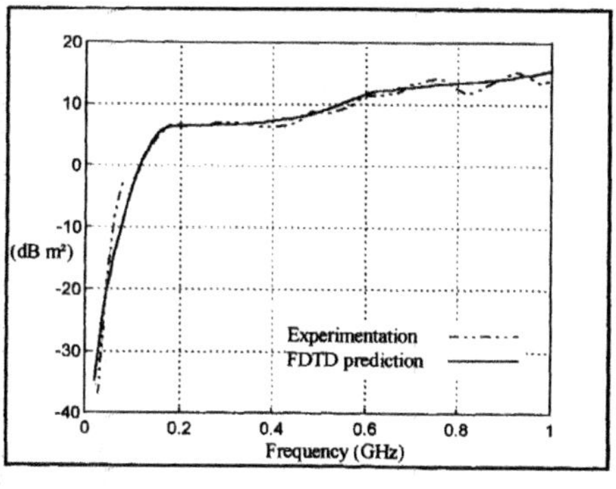

Figure 11. RCS of the plate

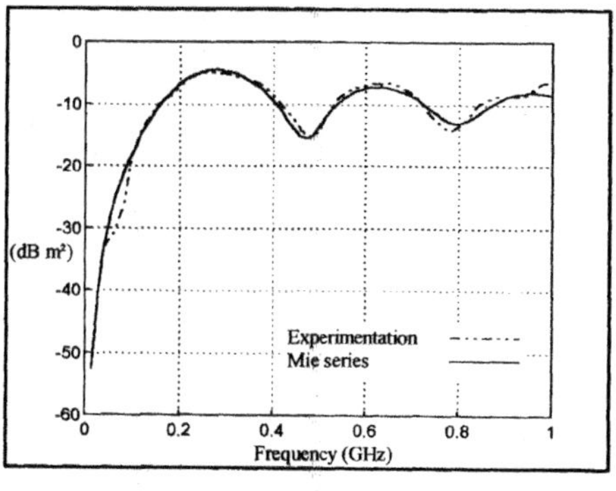

Figure 12. RCS of the sphere

Figure 13 shows the time domain response of the horizontal cylinder. There is a diffraction from the top of the cylinder and a large return from the travelling wave when it reachs the end of the cylinder. The correspondent RCS is shown figure 14.

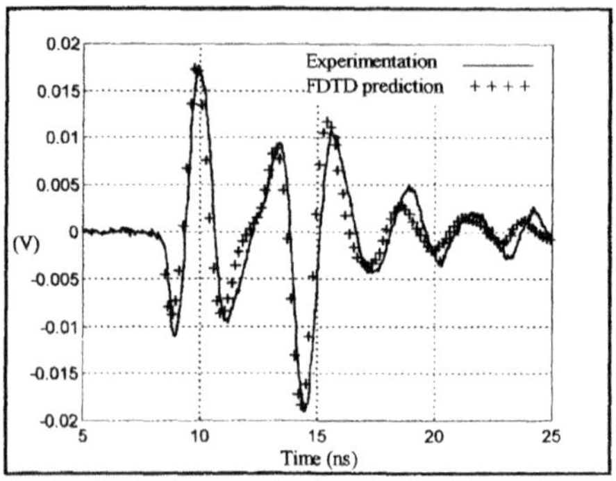

Figure 13. Time domain return from the horizontal cylinder

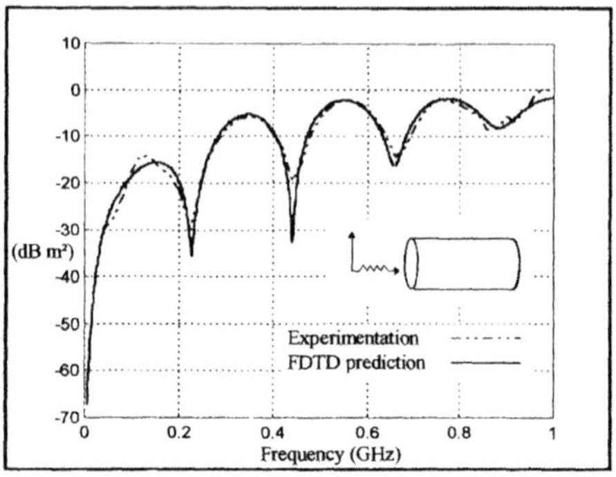

Figure 14. RCS of the cylinder

CONCLUSION

This experimental system allows to measure the impulse scattering responses from a variety of simple targets. The correspondent RCS are determined in the frequency range from 100 MHz to around 800 MHz. The accuracy is about less than 2 dB.m² for the maximum gap in the frequency range. The ground-target interaction is minimised with an average of some measurements made at different heights around 40 centimetres. That first experimentation led at the CELAR and the IRCOM allows to realise a new measurement set-up in a anechoic chamber with better accuracy[3].

REFERENCES :

1. J.G. Maloney, G.S. Smith, « A study of transient radiation from the Wu-King resistive monopole - FDTD analysis and experimental measurements ». IEEE Transactions on Antennas and Propagation - Vol. 41 - N°5 - May 1993.
2. RG. Madonna, P.J. Scheno, G.H. Vilardi, C. Horn, J. Scannepieco, « Scattering, resonance, creeping wave, travelling wave and all that :UWB measurements of various targets ». Ultra-Wideband, Short-Pulse Electromagnetics - Vol. 1 -H.L. BERTONI, L. CARIN and L.B. FELSEN - Ed. Plenum Press, New York, (1993).
3. M. Le Goff, P. Pouliguen, B. Juhel, Y. Chevalier, Y. Imbs, B. Beillard, J. Andrieu, M. Jouvet, B. Jecko, « Ultra wide band short pulse sensor for target electromagnetic backscattering characterisation ». ». Ultra-Wideband, Short-Pulse Electromagnetics - Vol. 4 - In Press.

DUAL POLARIZED, UWB RADAR MEASUREMENTS OF THE SEA AT 9 GHZ

P. Hansen, K. Scheff and E. Mokole

Naval Research Laboratory
Radar Division
4555 Overlook Avenue, S. W.
Washington, DC 20375, USA

INTRODUCTION

The range resolution provided by ultrawideband (UWB) radar measurements contributes to an understanding of the temporal and spatial character of radar backscatter from the natural environment. This paper presents a description of an ultrawideband instrumentation radar, the Microwave Microscope (MWM), along with data examples of backscatter from the sea and from fixed test objects measured at 9 GHz center frequency and 5 GHz instantaneous bandwidth.

The MWM system employs a video impulse excited traveling wave tube transmitter along with dual linear polarized antennas. A coherent receive function is implemented with multiple sampling head, digital sampling oscilloscope hardware. Variable measurement modes include pulse-to-pulse changes in both linear polarization and transmit pulse length. In the ultrawideband mode, the system is capable of transmitting and receiving 150 picosecond wide pulses with a corresponding radar range resolution which is finer than 2 centimeters. With changes in pulse length, the system is also able to contrast real time scattering responses between ultrawideband and conventional, wideband radar waveforms.

Differences in radar clutter responses for ultrawideband and conventional waveforms can be a result of both the resolution of individual features and the constructive (or destructive) interference between scatter from unresolved features. The diagram shown in Figure 1 illustrates the potential differences in responses for perfect ultrawideband and wideband waveforms which are respectively two and ten wavelengths in duration. This diagram emphasizes two characteristics: first, that the ultrawideband waveform can naturally resolve closely spaced scatterers; and second, that the longer waveform is capable of constructively combining the scatter from a greater number of resonantly spaced (one half wavelength) features.

An actual example of dual polarized, ultrawideband (2 cm resolution) sea backscatter data is shown in Figure 2. This data was taken with a range profiling mode where range points spaced every 1.5 cm were sampled within 25 microseconds for both linear polarizations and each 1 meter range profile was completed within 1.8 milliseconds. Ten consecutive range profiles, measured 0.3 seconds apart, are shown. The radar was

Ultra-Wideband, Short-Pulse Electromagnetics 4
Edited by Heyman *et al.*, Kluwer Academic / Plenum Publishers, New York, 1999.

illuminating an upwind rough sea condition at a grazing angle of 4 degrees. Note the resolved sea surface scattering features and the differences in polarization response. Also note that the returns from the group of individually resolved features in this example would tend to interfere (both constructively and destructively) with each other for longer duration radar pulses.

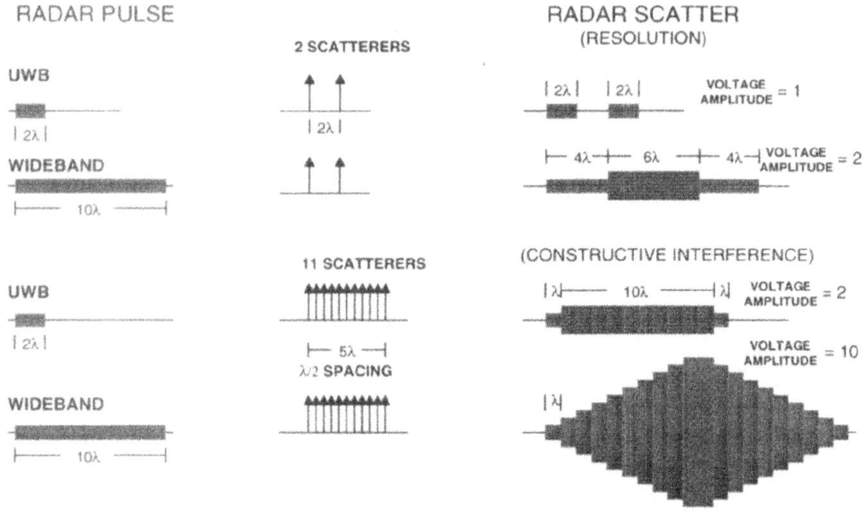

Figure 1. Idealized differences in clutter responses for UWB and wideband waveforms.

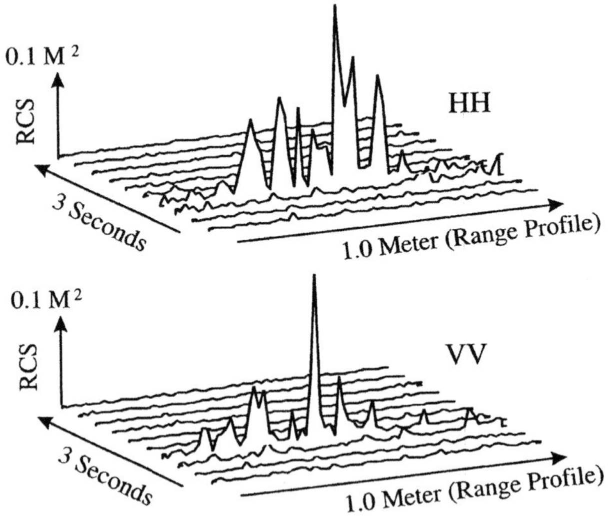

Figure 2. Example of sea backscatter data for UWB waveform (2 cm resolution).

EXPERIMENTAL SYSTEM

A simplified block diagram of the MWM instrumentation radar is shown in Figure 3. The field deployable, dual polarized system utilizes a video impulse excited traveling wave tube (TWT) to produce transmit pulses with 2 kW peak power and 150 picoseconds duration (3 dB power envelope). This ultrawideband transmit waveform provides a nominal 5 GHz instantaneous bandwidth centered at 9 GHz (Early usage of this type of pulse generation technique was cited by Griffin, 1975). A wideband pulse mode of 5 nanoseconds duration is created by filtering the video impulse excitation with a section of waveguide and then using the resulting rf burst as a longer duration input to the TWT. With video impulse excitation, the TWT output phase and amplitude remain constant on every transmit pulse. An oscilloscope recording of the ultrawideband transmit rf voltage waveform and the corresponding detected power envelope is shown in Figure 4. A comparison of the power envelopes for the 150 ps duration and the 5 ns duration transmit pulses is shown in Figure 5. Two separate ultrawideband (2 to 18 GHz response), dual polarized horn antennas are employed for transmit and receive functions in order to provide receiver isolation. These quad-ridged antennas produce a nominal 20 degree beamwidth at the 9 GHz center frequency. Polarization ports are selected on a pulse-to-pulse basis with low loss diode switches.

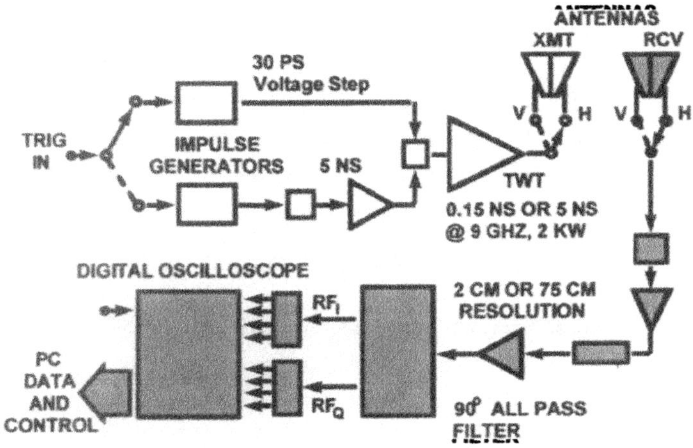

Figure 3. Block diagram of the MWM instrumentation radar.

On receive, the system employs a limiter and low noise amplifier followed by an all pass, 90 degree rf phase shifter. The video driving pulse also serves as the master timing source for the receiver and detector sampling functions. Received signals are detected in coherent baseband quadrature form with a unique direct sampling detector based on a multiple sampling head, eight bit digital sampling oscilloscope. The sampling heads simultaneously sample both the rf in-phase and quadrature phase shifted signal paths. Since the transmit and receive systems are time synchronized, the two simultaneously detected signals represent the baseband I and Q components at any given point in range. Signal pairs sampled at different points in range are also phase related to each other as long as the range separation is kept consistent with direct sampling intervals as defined by

Waters, 1982. Power envelopes are generated computationally by forming the sum of the squares of the digitally recorded quadrature samples. This detection technique supports an overall instantaneous receive bandwidth in excess of 5 GHz. Since only one point in range is sampled per pair of sampling heads per transmit pulse, rapid sampling of single or multiple points is achieved by using relatively high repetition rates. The system can operate at repetition rates as high as 100 KHz. The pulse width modes and polarization states (VV, HH, VH, HV) are also switchable on a pulse-to-pulse basis. Range resolution of the received data is determined by the transmit pulse width and, when using a range scanning mode, by the range sample interval. For the 0.15 ns and 5.0 ns transmit pulses, basic resolutions are approximately 2 cm and 75 cm, respectively.

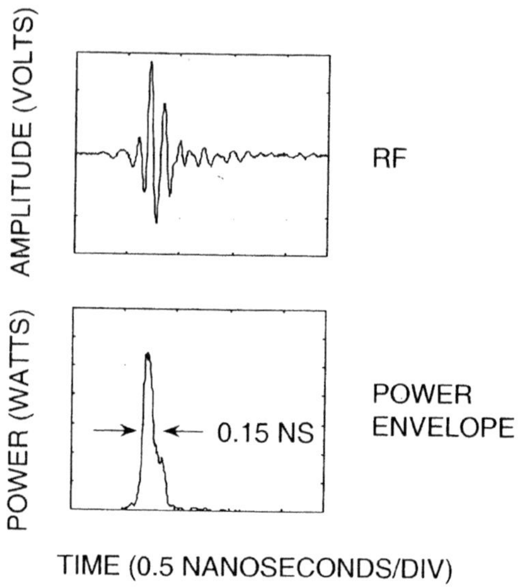

Figure 4. Transmit rf voltage and power envelope of a 0.15 ns pulse.

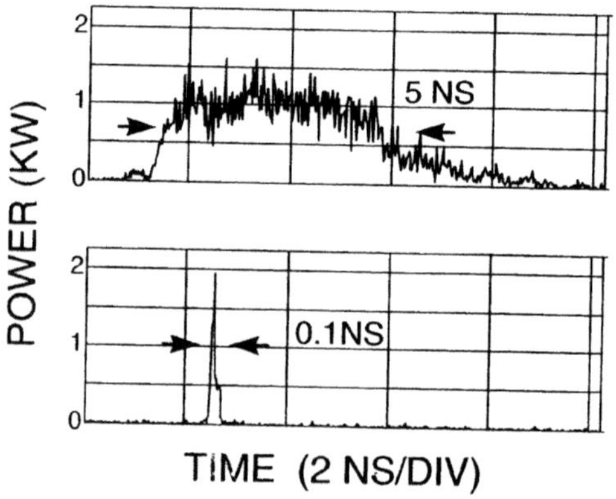

Figure 5. Power envelopes for 0.15 and 5.0 ns transmit pulses.

EXPERIMENTAL MEASUREMENTS

Field measurements were taken from an island site in the Bahamas. Initial system calibration was performed by measuring the first (specular) reflection from a 20 cm diameter aluminum sphere. Figure 6 shows a picture of the tripod mounted sphere situated at a range of 40 meters from the radar. Figure 7 shows measured data using the ultrawideband (0.15 ns) transmit pulse for both linear polarization states (VV and HH). The data segment displays a range profile spanning 15 centimeters in range and illustrates the system gain match for the two polarization states. Note that the single point specular reflection from the sphere effectively produces a return signal which traces out the power envelope of the transmit pulse. The combination of the UWB pulse and the narrow range window is sufficient to gate out both multipath and secondary returns from the sphere. The sphere was positioned at a relatively quiet, ground shadowed region.

Figure 6. Calibration sphere mounted on tripod.

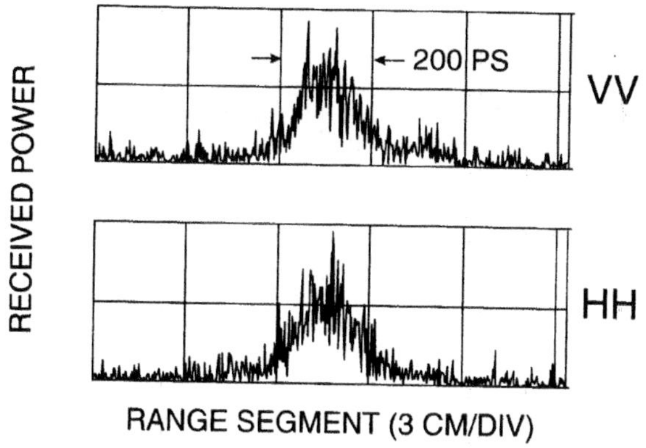

Figure 7. Range profile of the UWB specular reflection from a sphere.

In contrast to the sphere, Figure 8 shows a picture of a polarization preferential target consisting of a group of seventeen equally spaced, vertical, 60 cm length, steel wires. This group of wires was oriented to present a 3.7 cm relative wire-to-wire spacing in range to the radar. Figure 9 shows measured data using the UWB transmit pulse for both linear polarization states (VV and HH). The data segment displays a range profile spanning approximately 75 cm in range. Note that the VV return resolves the individual wires while the HH return is relatively insensitive to the thin vertically oriented targets. This group of wires was also placed at a relatively quiet, ground shadowed region in range with the supporting base physically located just inside the shadowed region in order to supress scatter from the base.

Figure 8. Group of equally spaced vertical wires.

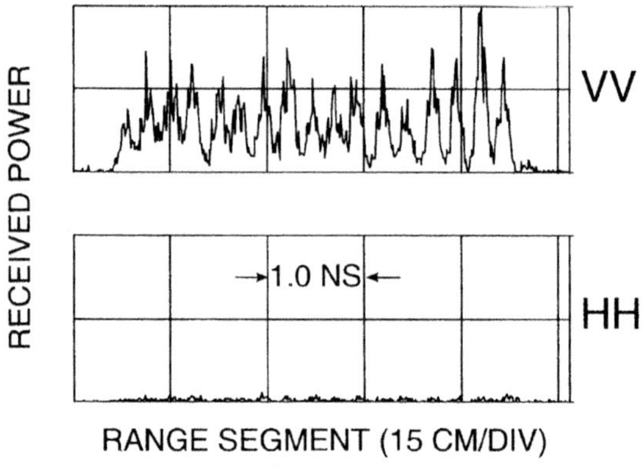

Figure 9. Range profile of the UWB reflection from the group of wires.

As a direct ultrawideband to wideband comparison, Figure 10 shows the VV response of the group of wires for both the 0.15 ns and the 5 ns transmit pulses. In this case, the relative range spacing of the wires is set at approximately 5.6 cm and the range profile spans approximately 300 cm. Note the relative responses for the two waveforms, with the ultrawideband pulse resolving the individual wires in range while the longer wideband pulse exhibits both constructive and destructive interference between overlapping reflections from the multiple wires. (The reflection from a single wire, by itself, would exhibit the same peak amplitude for both waveforms, given that the peak transmit amplitudes of the waveforms are approximately the same.)

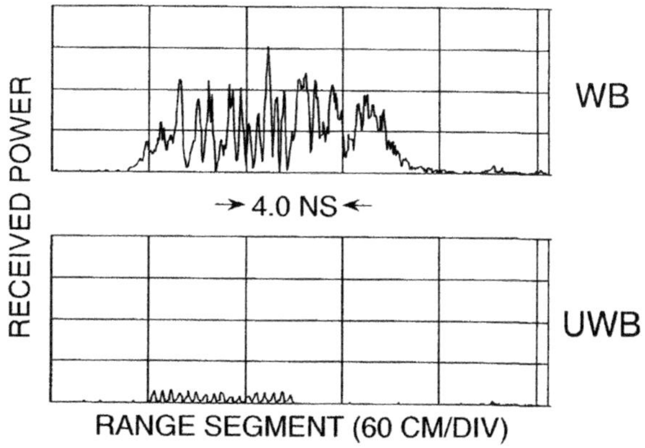

Figure 10. Range profile comparison for UWB and wideband reflections from wires.

Sea clutter measurements were taken with the radar looking upwind into a disturbed sea. The ocean depth in this direction rapidly increased from 35 to several hundred feet. Wind speed was 18 to 20 knots and the nominal radar grazing angle with the surface was 4 degrees. Range cell dimensions were 2 cm by 12 m (cell area of 0.24 square meters) for the 0.15 ns pulse and 75 cm by 12 m (cell area of 9 square meters) for the 5 ns pulse. Figure 11 shows a picture of the radar transmit and receive antenna assembly looking out over the ocean and Figure 12 shows a picture of the typical sea surface as viewed from the radar.

Figure 11. Transmit and receive antennas and boresight video camera.

Figure 13 shows an example of sea clutter data taken by the MWM system while using the 0.15 nanosecond transmit pulse and a windowed, range scanning mode. This data was produced by recording return signals from one range point per transmit pulse over 512 contiguous, 1.5 cm spaced increments for a total range profile which is 7.68 m in extent. The window was situated at a starting range of approximately 50 m from the radar. Each range point was sampled for both linear transmit and receive polarization states (VV and HH) within a time period of 25 microseconds before moving to the next range point and each profile was completed within 12.5 milliseconds. Ten consecutive range profiles, performed at intervals of 0.3 seconds and spanning a total time period of approximately 3 seconds, are shown. Figure axes are range in meters, time in seconds and calibrated, received radar cross section (RCS) in square meters. Calibration was based on measurements of the 20 cm diameter aluminum sphere at the site.

Figure 12. Sea surface with sustained 18 to 20 knot winds.

The data in Figure 13 represents a typical time history of the radar back scatter from a large ocean feature as it travels through the 7.68 m range interval. For this UWB measurement, the near simultaneous, dual polarized return signals show differences in the level and location of scattering points versus polarization, with peak returns for HH system conditions dominating over peak returns for VV system conditions. Relatively quiet periods, located in range before and just after the moving large feature, are also evident for both polarization states. In comparison, Figure 14 displays an example of data taken by the system while utilizing a 5 ns transmit pulse. (Note that the scale of the RCS axis has been increased by a factor of five to accommodate larger peak received amplitudes.) This second data set displays a typical time history of the radar back scatter from another large ocean feature. For this wideband case, the dual polarized return signals still show differences in the level and location of scatterers versus system polarization with peak returns for HH dominating over peak returns for VV. Relatively quiet periods before and after the large feature are also still present. There is, however, a noticeable change in the polarization ratio for lower level signals produced in the region just preceding the large feature, with the average signal level for VV being larger than the signal level for HH.

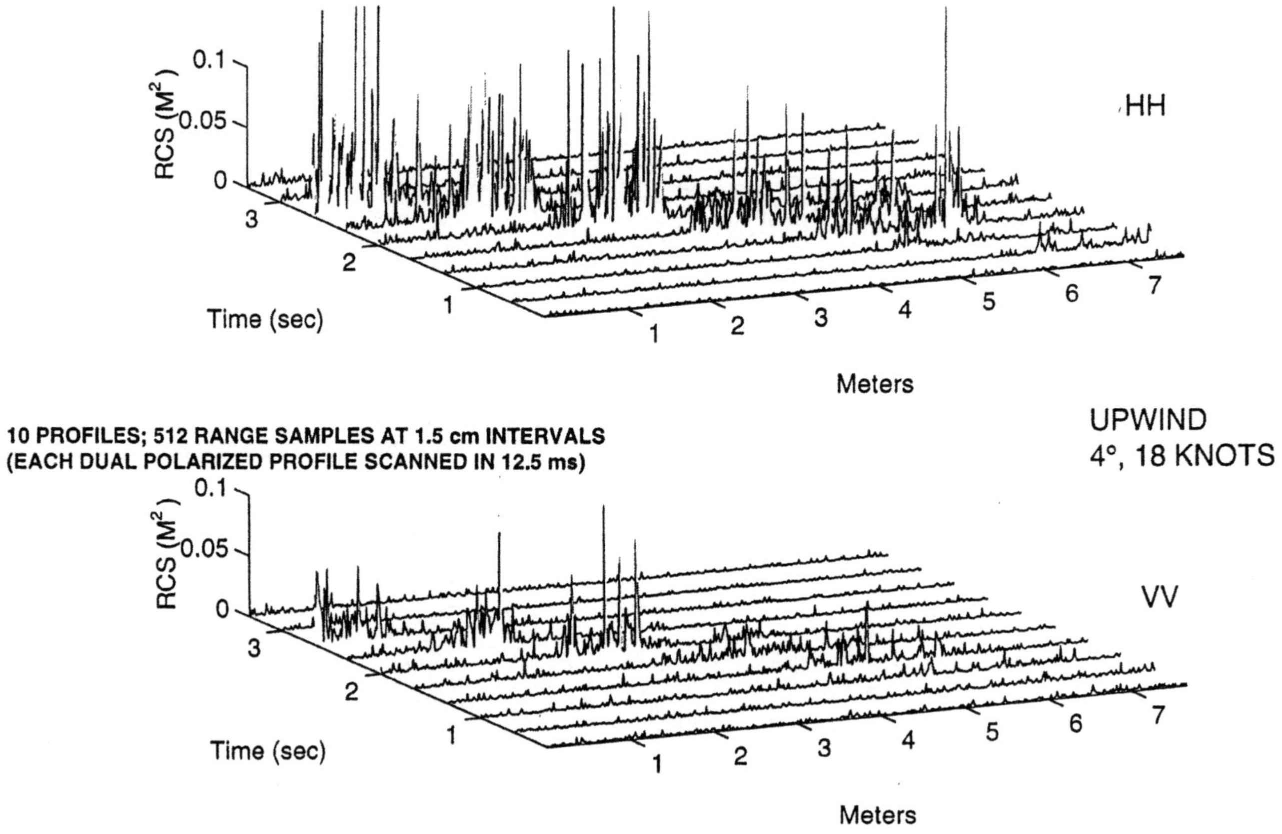

Figure 13. Ultrawideband sea clutter data (0.15 ns pulse)

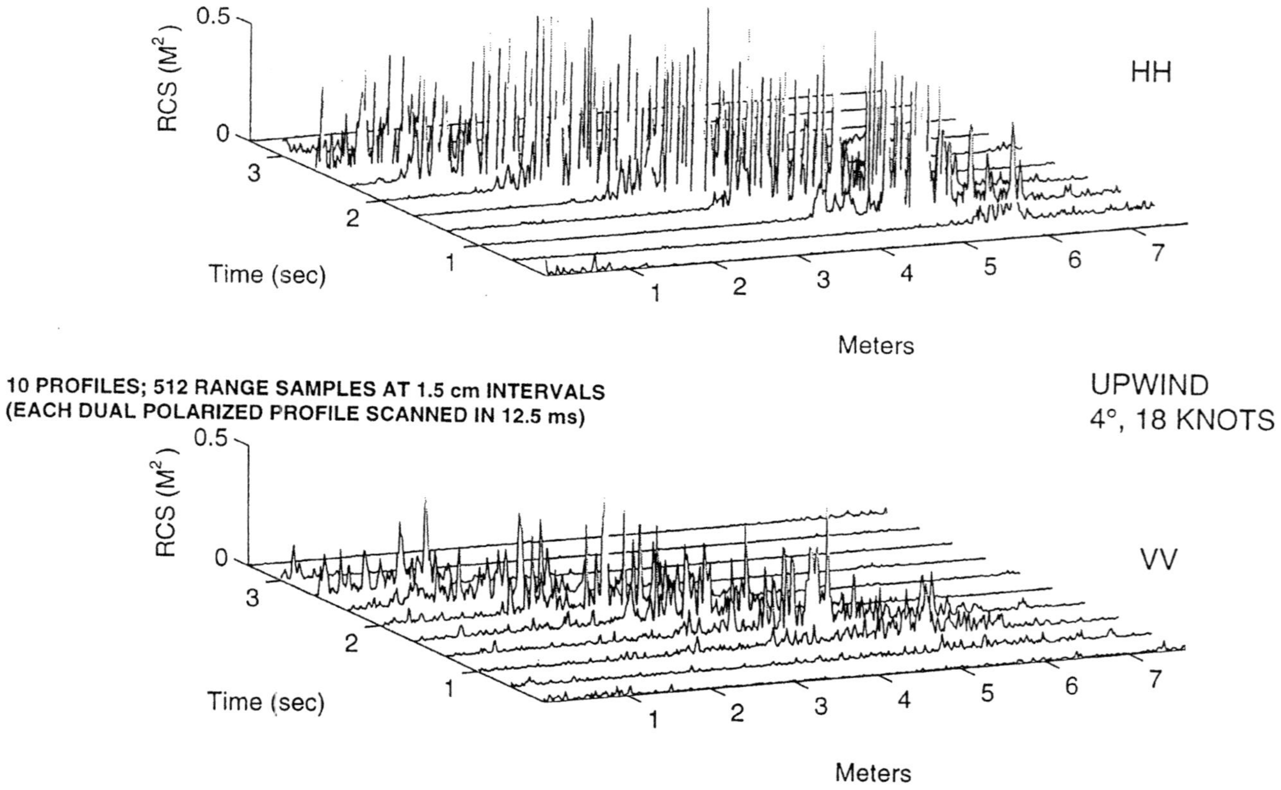

Figure 14. Wideband sea clutter data (5 ns pulse).

Figure 15 shows a single 3 m range profile (formed within 5 milliseconds) taken from the central portion of the UWB data set. Axes are received cross section in square meters and range in meters. This profile highlights the observed differences between VV and HH for the large features. Resolved, high amplitude scattering points are evident.

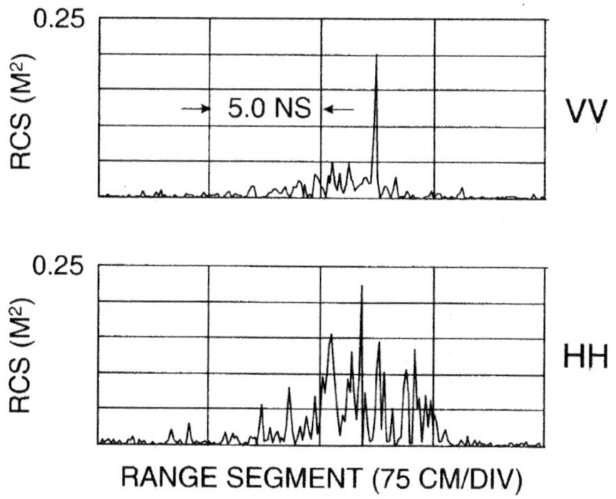

Figure 15. Single, dual polarized range profile of UWB sea clutter.

In the same manner, Figure 16 shows a single 3 m range profile taken from the central portion of the wideband data set. Note that the return signal is showing constructive and destructive interference as evidenced by the breakup of the 5 ns pulse envelopes.

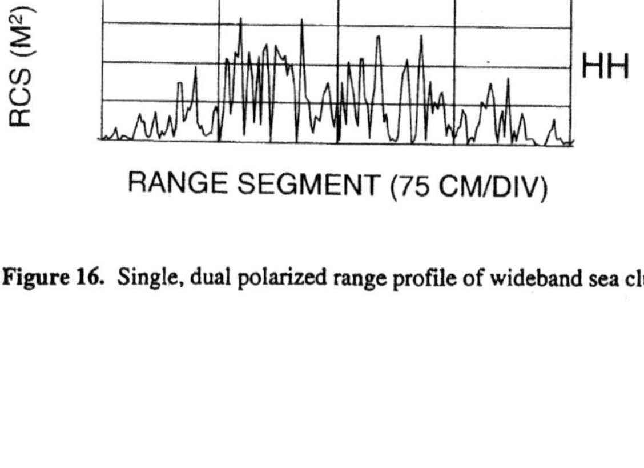

Figure 16. Single, dual polarized range profile of wideband sea clutter.

Comparisons with multiple waveforms can provide indications of interference. Figure 17 shows an example of a single 3 m range profile of a large ocean feature measured simultaneously with both the UWB and the wideband waveforms for VV polarization. Note that this dual waveform sea clutter example is very comparable in general appearance to the profile of the group of wires (previously shown in Figure 10).

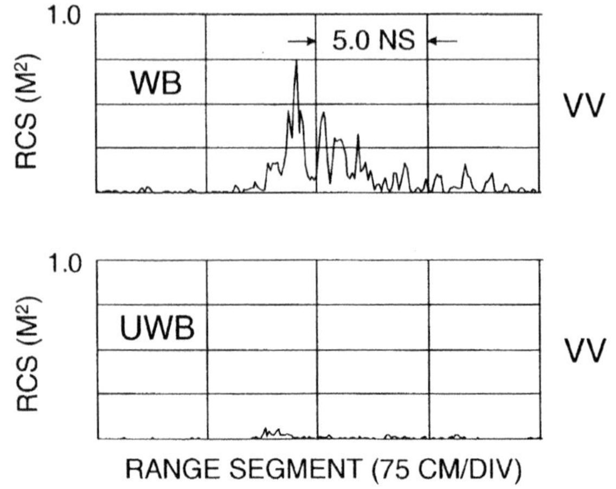

Figure 17. Range profile comparison for UWB and wideband reflections from an ocean feature.

The data of Figure 17 has been plotted on a magnified cross section scale in Figure 18 in order to show the range resolved scattering points revealed by the UWB waveform. As with the spaced group of wires, these individual scattering points are contributing to the combined, higher amplitude return produced by the longer, wideband waveform.

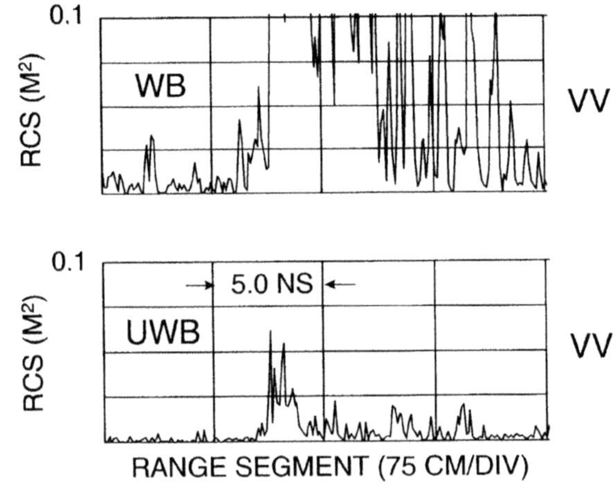

Figure 18. Range profile comparison of Figure 17 on magnified scale.

Figure 19 compares dual polarized amplitude distributions for 30 seconds of data taken with each transmit waveform (ultrawideband, 0.15 ns pulse and wideband, 5 ns pulse). The data is shown as Weibull plots of the cumulative amplitude distribution function for each polarization state. The horizontal axes are measured cross section (5 dB per division with increasing amplitude from left to right) which is related to a 0.5 square meter calibration point.

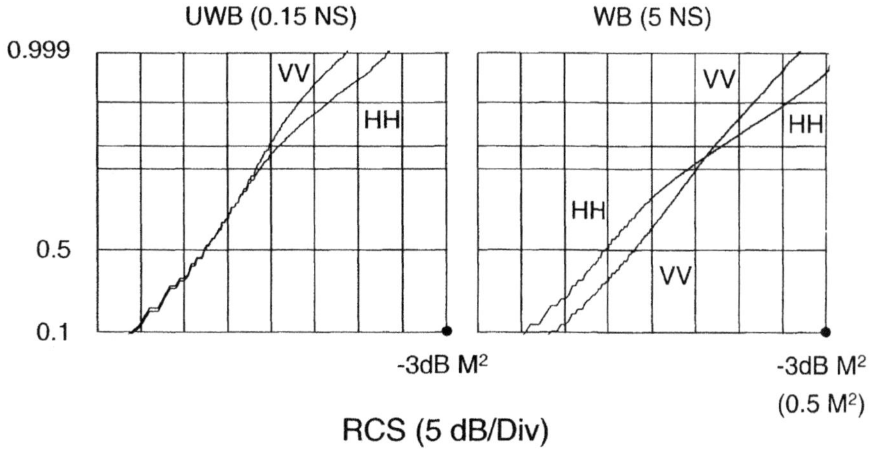

Figure 19. Dual polarized cumulative amplitude distributions.

These distributions highlight the observed differences in the amplitudes of the scatter from the largest features as a function of polarization, the differences in overall level of return for the two waveforms and the definitive increase in the VV to HH polarization ratio of the lower level scatter which is present when using the longer (wideband) waveform.

SEA CLUTTER OBSERVATIONS

A prominent characteristic that appears in the upwind sea clutter measurements is the dominance and the strength of the HH polarization returns from the larger ocean features. Peak cross sections greater than 0.5 square meters were sometimes measured, even for the extremely small footprint (0.24 square meters) of the ultrawideband pulse. These peak cross sections occur in spite of the fact that the short, 0.15 ns pulse limits the effects of constructive interference and also suppresses multipath enhancement past the first interference lobe. The range dimensions of some of the larger scatterers are surprisingly narrow. Sporadic differences between the range locations of the large amplitude scattering points for HH and VV polarizations are also apparent. These differences point to polarization selective scattering features. Similar behavior has been noted in observations by other researchers (Ward, 1990) using substantially lower resolutions. Profile comparisons between the ultrawideband and the wideband waveforms vividly demonstrate that interference (constructive and destructive) between range distributed scattering points changes the character of the radar clutter return. For the longer, wideband waveform, the increase in the VV scattering (from the regions in between the larger ocean features) would seem to indicate that there are many very small, relatively closely spaced, surface features contributing preferentially to the return signal. This phenomena would support a trend for the average level of VV scattering to increase, with increasing radar pulse width, at a faster rate than the average level of HH scattering, since the area physically covered by the numerous small features is substantially greater than that occupied by the large features.

CONCLUSIONS

An ultrawideband, dual polarized, dual waveform, instrumentation radar has been described and experimental measurements of fixed test objects and of a disturbed sea surface have been presented. The two cm range resolution of the system separates the contributions of individual scattering features on the ocean surface and is providing an increased understanding of the composition and behavior of radar sea clutter at low grazing angles. The experimental results have emphasized fundamental differences in the character of the clutter as a function of both polarization and radar resolution.

REFERENCES

Griffin, C. R., et al., 1975, A pico-second pulse radar system for high resolution range measurements of radar targets, *Proceedings of the 1975 IEEE International Radar Conference*, pp. 469-475.

Ward, K. D., Baker, C. J., and Watts, S., 1990, Maritime surveillance radar - part 1: radar scattering from the ocean surface, *IEE Proceedings*, 137, part F, no. 2, pp. 51-62.

Waters, W. M., and Jarrett, B. R., 1982, Bandpass signal sampling and coherent detection, *IEEE Transactions*, AES-23, pp. 731-736.

DIRECT CONSTRUCTION OF A ξ-PULSE FROM NATURAL FREQUENCIES AND EVALUATION OF THE LATE-TIME RESIDUALS

Carl E. Baum
Air Force Research Laboratory
ARFL/DEHP
3550 Aberdeen Ave., SE
Kirtland AFB, NM 87117-5776

1. INTRODUCTION

An important type of target identification utilizes the natural frequencies of the target, these being poles in the complex-frequency s-plane as expressed in the singularity expansion method (SEM).[11] Since the location of these natural frequencies s_α in the s-plane is aspect independent (independent of the properties of the incident field such as direction of incidence and polarization), this simplifies the identification problem to the comparison of the target pole patterns stored in some target library. There is also information in the pole residues (e.g., coupling vectors), but that is another matter.

In order to implement the pole pattern as a target discrimination scheme one needs some scheme to find the s_α in experimental data (scattered fields), either explicitly or implicitly. Some of the early schemes used what is referred to as the Prony method in which one fits a sum of dumped sinusoids to a waveform. This had various limitations which have been overcome to some degree by what is referred to as a matrix pencil method.[12] These and related techniques are explicit in that they determine a set of s_α from one or more waveforms which are then compared to the target library.

In contrast, implicit methods do not generate a set of s_α from the data, but utilize predetermined sets of s_α for various targets which are then manipulated with the data in some way which transforms it to a form in which the discrimination can be more readily performed. Those used to date have used temporal functions with two-sided Laplace transforms with zeros corresponding to the s_α (poles) of preselected targets. Convolving these with the target impulse response removes (annihilates) the late-time response for the properly chosen target. These go by various names such as K-pulse[9] and E-pulse.[10] In the present paper we go further into this latter type of target identification.

2. FORM OF THE RESPONSE FOR ELECTROMAGNETIC SINGULARITY IDENTIFICATION (EMSI)

Consider an incident plane wave of the form

$$\tilde{\vec{E}}^{(inc)}(\vec{r},s) = E_0\,\tilde{f}(s)\,\vec{1}_p\,e^{-\gamma\vec{1}_i\cdot\vec{r}} \quad , \quad \vec{E}^{(inc)}(\vec{r},t) = E_0\,\vec{1}_p\,f\left(t-\frac{\vec{1}_i\cdot\vec{r}}{c}\right)$$

$\vec{1}_i \equiv$ direction of incidence , $\vec{1}_p \equiv$ polarization

$$\gamma = \frac{s}{c} \equiv \text{propagation constant},\ c \equiv [\mu\varepsilon]^{-1/2} \tag{2.1}$$

$\sim\ \equiv$ two-sided Laplace transform (over time t)

$s \equiv \Omega + j\omega \equiv$ complex frequency or Laplace-transform variable

Ultra-Wideband, Short-Pulse Electromagnetics 4
Edited by Heyman *et al.*, Kluwer Academic / Plenum Publishers, New York, 1999.

where the medium of interest is considered lossless and dispersionless (e.g., free space). The far scattered field is given by

$$\vec{E}_f(\vec{r},s) = \frac{e^{-\gamma r}}{4\pi r}\overleftrightarrow{\Lambda}(\vec{1}_o,\vec{1}_i;s)\cdot\vec{E}^{(inc)}(\vec{0},s)$$

$$\vec{E}_f(\vec{r},t) = \frac{1}{4\pi r}\overleftrightarrow{\Lambda}(\vec{1}_o,\vec{1}_i;s)\circ\vec{E}^{(inc)}\left(\vec{0},t-\frac{\vec{1}_o\cdot\vec{r}}{c}\right) \tag{2.2}$$

$$\overleftrightarrow{\Lambda}(\vec{1}_o,\vec{1}_i;s) \equiv \text{scattering dyadic},\ \overleftrightarrow{\Lambda}(\vec{1}_o,\vec{1}_i;s)\bullet\vec{1}_i = \vec{0} = \vec{1}_o\bullet\overleftrightarrow{\Lambda}(\vec{1}_o,\vec{1}_i;s)$$

$$\vec{1}_o \equiv \text{direction to observer},\ \circ \equiv \text{convolution with respect to time}$$

The coordinate origin ($\vec{r} = \vec{0}$) is chosen for convenience near or even within the target (such as the center of the minimum circumscribing sphere[3]). We also often have the case from reciprocity

$$\overleftrightarrow{\Lambda}(\vec{1}_o,\vec{1}_i;s) = \overleftrightarrow{\Lambda}^T(-\vec{1}_i,\vec{1}_o;s)$$

$$\overleftrightarrow{\Lambda}_b(\vec{1}_i,s) \equiv \overleftrightarrow{\Lambda}^T(-\vec{1}_i,\vec{1}_i;s) \equiv \overleftrightarrow{\Lambda}_b^T(\vec{1}_i,s)\ \text{(symmetric in backscattering)} \tag{2.3}$$

The scattering dyadic[3] can be written in SEM form (class 1)

$$\overleftrightarrow{\Lambda}(\vec{1}_o,\vec{1}_i,s) = \sum_\alpha \frac{e^{-(s-s_\alpha)t_i}}{s-s_\alpha}\vec{c}_\alpha(-\vec{1}_o)\,\vec{c}_\alpha(\vec{1}_i) + \text{possible entire function}$$

$$\overleftrightarrow{\Lambda}_b(\vec{1}_i,s) = \sum_\alpha \frac{e^{-(s-s_\alpha)t_i}}{s-s_\alpha}\vec{c}_\alpha(\vec{1}_i)\,\vec{c}_\alpha(\vec{1}_i) + \text{entire function} \tag{2.4}$$

$$\overleftrightarrow{\Lambda}_b(\vec{1}_i,t) = \sum_\alpha u(t-t_i)\,e^{s_\alpha t}\,\vec{c}_\alpha(\vec{1}_i)\,\vec{c}_\alpha(\vec{1}_i) + \text{entire function}$$

$$t_i \equiv \text{initial time or turn-on time}$$

For the backscattering dyadic (in contradistinction to the currents induced on the scatterer) an entire function (no singularities in the finite s-plane) is definitely required. In time domain this is a time-limited function related to transit times on the body. Note that the analysis[3] is for the case of finite-size perfectly conducting scatterers (also allowing certain types of sheet-impedance loading).

Defining

$$L_0 \equiv \text{maximum linear dimension of target}\ ,\ t_0 \equiv \frac{L_0}{c} \tag{2.5}$$

it is shown[3] that in a worst-case sense for backscattering there is a time window of length t_0, beginning with the first return from the target, for which the pole series does not converge, and hence cannot be used as a valid representation in this time window. One can also define a discrimination time t_d as a minimum time for a target identification based on the time after the first signal return that information from everywhere on the target can reach the observer. For a "transparent" target, defined such that such information can travel in straight lines (allowing some attenuation) at the speed of light, c, we have for backscattering

$$t_d = 2\,t_0 = 2\frac{L_0}{c} \tag{2.6}$$

So even for times greater than t_0 there is some ambiguity in the discrimination process in the sense that different shapes (impedances, etc.) can be ascribed to the target, implying a different set of s_α and associated modes, but giving the same temporal response up to a time t_d. For non-transparent targets such as perfectly conducting targets, the time required to observe the far side of the target in backscattering is increased so that (2.6) is rewritten as

$$t_d \geq 2t_0 \tag{2.7}$$

i.e., as a bound. Given a particular target one can of course calculate this minimum discrimination time maximized over all directions of incidence. Furthermore t_d might be indexed as t_{dm} , i.e., a separate value for each mth target type.

Another important case for EMSI with wavelengths of the order of target dimensions concerns targets buried in soil. This introduces some new difficulties in the identification problem for both perfectly conducting[4] and dielectric[7] targets. For not-too-conducting soil one can operate in the "high-frequency window" and utilize some approximate scaling relationships to allow for the dependence of the s_α on the external-medium parameters. This complicates the calculation of the t_d and even introduces branch cuts for which one should allow in the discrimination scheme.

While the present discussion has been in the context of the electromagnetic fields, the concepts can also be applied in the context of acoustic and elastodynamic fields.[8] For acoustic singularity identification (ASI), particularly if the response is dominated by p-waves in the external medium (e.g., water), similar considerations concerning discrimination time t_d apply.

3. FORM OF THE RESPONSE FOR MAGNETIC SINGULARITY IDENTIFICATION (MSI)

For highly, but not perfectly, conducting targets there is a set of natural frequencies s_α which are based on diffusion of fields in the target, and scale as proportional to $(\mu \sigma d^2)^{-1}$ where d is some characteristic linear dimension of the target and σ is the target conductivity (high as in the case of typical metals).[5,6] The external medium is assumed to have permeability μ_0 and conductivity typical of soil or water, and as such does not significantly affect the target response in the low-frequency range.

The incident field for this purpose would generally not be a plane wave as in (2.1), but the near magnetic field of one or more loops. Skin depths in the external medium for the low frequencies of interest are assumed large compared to all distances of interest. It is then the induced magnetic dipole moment $\vec{m}(t)$ that one uses for the waveform to be fed into the target-discrimination process. This is related to the magnetic-polarizability dyadic of the target via

$$\tilde{\vec{m}}(s) = \tilde{\overleftrightarrow{M}}(s) \cdot \tilde{\vec{H}}^{(inc)}(s) \tag{3.1}$$

where the incident magnetic field is evaluated at some convenient location in the vicinity of the target. The scattered magnetic field is

$$\tilde{\vec{H}}^{(sc)}(\vec{r},s) = \frac{1}{4\pi r^3}\left[3\,\vec{1}_r\,\vec{1}_r - \overleftrightarrow{1}\right] \cdot \tilde{\overleftrightarrow{M}}(s) \cdot \tilde{\vec{H}}^{(inc)}(s) \tag{3.2}$$

$$\overleftrightarrow{1} \equiv \vec{1}_x\,\vec{1}_x + \vec{1}_y\,\vec{1}_y + \vec{1}_z\,\vec{1}_z \equiv \text{identity dyadic} \;,\quad \vec{1}_r = \frac{\vec{r}}{r} \;,\quad r = |\vec{r}\,|$$

where $\vec{r}$ is the observer position measured with respect to the target.

The general form for the magnetic polarizability dyadic[5,6] is

$$\tilde{\overleftrightarrow{M}}(s) = \tilde{\overleftrightarrow{M}}(\infty) + \sum_\alpha M_\alpha \vec{M}_\alpha \vec{M}_\alpha \left[s - s_\alpha\right]^{-1} \;,\quad \overleftrightarrow{M}(t) = \tilde{\overleftrightarrow{M}}(\infty)\,\delta(t) + \sum_\alpha M \vec{M}_\alpha \vec{M}_\alpha\, e^{s_\alpha t} u(t)$$

$$s_\alpha < 0 \;,\quad \text{i.e., real and negative}$$

$$\vec{M}_\alpha \cdot \vec{M}_\alpha = 1 \;,\quad \vec{M}_\alpha = \text{real vector} \;,\quad M_\alpha = \text{real scalar} \tag{3.3}$$

$$\tilde{\overleftrightarrow{M}}_\alpha(\infty) = \sum_{\upsilon=1}^{3} M_\upsilon^{(\infty)} \vec{M}_\upsilon^{(\infty)} \vec{M}_\upsilon^{(\infty)} \;,\quad \vec{M}_{\upsilon_1}^{(\infty)} \cdot \vec{M}_{\upsilon_2}^{(\infty)} = 1_{\upsilon_1,\upsilon_2} \;\text{(orthonormal eigenvectors)}$$

$$M_\upsilon^{(\infty)} \equiv \text{real eigenvalues (non positive, not necessarily distinct)}$$

$$\vec{M}_\upsilon^{(\infty)} = \text{real eigenvectors (three)}$$

This is the form for the delta-function response. One can also use the step response which takes the form

$$\frac{1}{s}\overleftrightarrow{M}(s) = \frac{1}{s}\overleftrightarrow{M}(0) + \sum_{\alpha}\frac{M_\alpha}{s_\alpha}\vec{M}_\alpha\vec{M}_\alpha[s - s_\alpha]^{-1}$$

$$\int_{-\infty}^{t}\overleftrightarrow{M}(t')\,dt' = \overleftrightarrow{M}(0)\,u(t) + \sum_{\alpha}\frac{M_\alpha}{s_\alpha}\vec{M}_\alpha\vec{M}_\alpha\,e^{s_\alpha t}\,u(t)$$

$$\overleftrightarrow{M}(0) = \sum_{\upsilon=1}^{3}M_\upsilon^{(0)}\vec{M}_\upsilon^{(0)}\vec{M}_\upsilon^{(0)} \quad , \quad \vec{M}_{\upsilon_1}^{(0)}\cdot\vec{M}_{\upsilon_2}^{(0)} = 1_{\upsilon_1,\upsilon_2} \text{ (orthonormal eigenvectors)} \tag{3.4}$$

$M_\upsilon^{(0)}$ ≡ real eigenvalues (non negative, not necessarily distinct)

$\vec{M}_\upsilon^{(0)}$ = real eigenvectors (three)

The DC (zero frequency) polarizability applies only to permeable scatterers characterized by $\overleftrightarrow{\mu}(\vec{r})$ different from μ_0 1 (free space, or more generally, the external medium).

Returning to the subject of discrimination time t_d discussed in the previous section, in the present context (MSI) the problem is considerably simplified. On the present time scale (s_α^{-1} associated with diffusion in the target) transit times across the target (as well as to and from the target) are neglected. So we can write

$$t_d = 0_+ \tag{3.5}$$

i.e., just a little larger than zero. In (3.3) the entire function contribution is $\delta(t)$, but in practice will have some small non-zero width consistent with (3.5). One can also use the step response in (3.4) for the discrimination which has no entire-function contribution, but has an additional (and useful) pole at $s = 0$.

4. A SIMPLE WAY OF CONSTRUCTING A ξ-PULSE

Consider a sampling pulse $f_s(t)$ which is to be convolved with the data. This can be a filter, say to remove high-frequency noise, or whatever. For convenience this can be normalized to a unit area unless one wishes this pulse to have zero area. This can be used as a basis function, such as a subsectional basis function on an interval from 0 to T_0 as

$$f_s(t) = \begin{cases} T_0^{-1} & \text{for } 0 < t \le T_0 \\ 0 & \text{otherwise} \end{cases} \quad , \quad \int_{-\infty}^{\infty} f_s(t)dt = 1 \tag{4.1}$$

Here T_0 would be a sampling interval and such a basis function is often used with K/E pulses.

Consider an elementary pulse (an annihilation pulse)

$$a_\ell(t) = d(t) - e^{s_\ell T_0}\,d(t - T_0) \quad , \quad \tilde{a}_\ell(s) = 1 - e^{[s_\ell - s]T_0} \quad , \quad \tilde{a}_\ell(s_\ell) = 0 \tag{4.2}$$

where $T_0 > 0$ is any convenient time (including possibly a sampling interval). Convolving this with an elementary target response $e^{s_\ell t}u(t)$ for any single pole gives zero after time T_0, i.e.

$$a_\ell(t) \circ \left[e^{s_\ell t}u(t)\right] = e^{s_\ell t}\left[u(t) - u(t - T_0)\right] \tag{4.3}$$

This is fine for a single pole with s_ℓ real (as is the case for all poles in MSI). In a more general case (e.g., EMSI and ASI) s_ℓ can be complex. One can generalize one's thinking to allow complex temporal waveforms, or recognize that such cases come in complex-conjugate pairs. In such a case we can associate

$$s_{\ell'} = s_\ell^* \tag{4.4}$$

and then form

$$\tilde{a}_\ell(s)\,\tilde{a}_{\ell'}(s) = \left[1 - e^{[s_\ell - s]T_0}\right]\left[1 - e^{[s_{\ell'} - s]T_0}\right] \tag{4.5}$$

which can be used to annihilate real valued waveforms of the form $\left[a\,e^{s_\ell t} + a^*\,e^{s_{\ell'} t}\right]u(t)$ after a time 2 T_0.

Using this procedure one can construct a pulse with zeros at

$$s = s_\ell \text{ for } \ell = 1, 2, \ldots, L \tag{4.6}$$

for L even or odd, including both conjugate pairs and individual poles on the negative Re[s] axis of the s-plane. Defining for convenience

$$S_\ell \equiv s_\ell T_0 \quad , \quad S \equiv s T_0 \quad , \quad X_\ell \equiv e^{S_\ell} \quad , \quad X \equiv e^S \tag{4.7}$$

we have

$$\begin{aligned} \tilde{A}(s) &= \prod_{\ell=1}^{L} \tilde{a}_\ell(s) = \prod_{\ell=1}^{L}\left[1 - X_\ell X^{-1}\right] = \sum_{n=1}^{L} A_n X^{-n} \\ A(t) &= \mathop{O}_{\ell=1}^{L} a_\ell(t) = \mathop{O}_{\ell=1}^{L}\left[\delta(t) - X_\ell \delta(t - T_0)\right] = \sum_{n=0}^{L} A_n \delta(t - \ell T_0) \end{aligned} \tag{4.8}$$

Properly including conjugate pole pairs, all the A_n are real.

These coefficients can be explicitly calculated as

$$A_0 = 1 \quad , \quad A_1 = \sum_{\ell=1}^{L} X_\ell$$

$$A_2 = \sum_{\substack{\text{all pairs} \\ \ell^{(1)} = \ell^{(2)}}} X_{\ell^{(1)}} X_{\ell^{(2)}} \quad \text{i.e., sum over all pairs with distinct indices}$$

$$A_n = (-1)^n \sum_{\substack{\text{all distinct} \\ \text{ntuplets}}} X_{\ell^{(1)}} X_{\ell^{(2)}} \ldots X_{\ell^{(n)}} \quad \text{i.e., sum over all ntuplets with } n \text{ distinct indices} \tag{4.9}$$

$$A_L = (-1)^L \prod_{\ell=1}^{L} X_\ell$$

The products are the combination of L objects taken n at a time without repetition.[13] All such products are summed for each n. So we have all the required coefficients to construct A(t) for a given set of natural frequencies without inverting a matrix. Furthermore, all of the A_n are real since $A(t)$ is real, even though the X_ℓ can in general be complex, since they are included in conjugate pairs. Note that if we set all the $X_\ell = 1$ we obtain the binomial coefficients for the A_n.

Now recall our sampling pulse $f_s(t)$ to form

$$\xi(t) = f_s(t) \circ A(t) = \sum_{n=0}^{L} A_n f_s(t - nT_0) \quad , \quad \tilde{\xi}_s(s) = f_s(t) \circ A(t) = \sum_{n=0}^{L} A_n f_s(t - nT_0) \tag{4.10}$$

where we have chosen the symbol ξ (pronounced ksee) for this late-time-annihilating waveform (not being sure of whether to use K or E). Note in this form that the A_n are independent of the choice of $f_s(t)$ which can be chosen for various useful properties.

Relating this form to previous considerations,[11] enforce the L zeros as

$$\tilde{A}(s_\ell) = \sum_{n=0}^{L} A_n X_\ell^{-n} = 0 \text{ for } \ell = 1, 2, \ldots, L \tag{4.11}$$

giving the matrix $(L \times (L+1))$ equation

$$\begin{pmatrix} X_1^0 & X_1^{-1} & \cdots & X_1^{-L} \\ X_2^0 & X_2^{-1} & \cdots & X_2^{-L} \\ \vdots & \vdots & \cdots & \vdots \\ X_L^0 & X_L^{-1} & \cdots & X_L^{-L} \end{pmatrix} \cdot \begin{pmatrix} A_0 \\ A_1 \\ \vdots \\ A_L \end{pmatrix} = \begin{pmatrix} 0 \\ 0 \\ \vdots \\ 0 \end{pmatrix} \tag{4.12}$$

which, by moving the first column (all elements as well as A_0 being 1), gives

$$\begin{pmatrix} X_1^{-1} & \cdots & \cdots & X_1^{-L} \\ \vdots & & & \vdots \\ \vdots & & & \vdots \\ X_L^{-1} & \cdots & \cdots & X_L^{-1} \end{pmatrix} \cdot \begin{pmatrix} A_1 \\ \vdots \\ \vdots \\ A_L \end{pmatrix} = \begin{pmatrix} 1 \\ \vdots \\ \vdots \\ 1 \end{pmatrix} \tag{4.13}$$

so that the matrix is now square ($L \times L$).

Solving for the A_ℓ requires that this last matrix have an inverse. This means that we must have L independent rows and L independent columns. This matrix (of the Vandermonde type) then must have distinct X_ℓ. For real S_ℓ (as for MSI in Section 3) this requires that all the S_ℓ be distinct. Turning this around, under what conditions do distinct S_ℓ give non-distinct X_ℓ, i.e.

$$e^{S_{\ell_1}} = e^{S_{\ell_2}} \quad , \quad \mathrm{Re}\left[S_{\ell_1}\right] = \mathrm{Re}\left[S_{\ell_2}\right] \quad \text{(same damping)} \tag{4.14}$$

$$\mathrm{Im}\left[S_{\ell_1}\right] = \mathrm{Im}\left[S_{\ell_2}\right] + 2m\pi \quad (m = \text{any integer})$$

Noting that a complex pole pair has

$$\Omega_{\ell_1} + j\omega_{\ell_1} = \Omega_{\ell_2} - j\omega_{\ell_2} \tag{4.15}$$

and applying (4.15) to the case gives

$$\omega_{\ell_1} T_0 = \omega_{\ell_2} T_0 + 2m\pi \quad , \quad \omega_{\ell_1} T_0 = -\omega_{\ell_2} T_0 = m\pi \quad , \quad T_0 = \frac{m\pi}{\omega_{\ell_1}} \tag{4.16}$$

The smallest T_0 implies $m = 1$ and maximum ω_ℓ. This relates to the "natural" E-pulse if we have L poles $= 2N$ poles or N pole pairs (neglecting poles in the negative real s-axis), and the duration of $A(t)$ is just (for $L + 1$ delta functions)

$$LT_0 = 2NT_0 = \frac{2N\pi}{\max\limits_{\ell} \omega_\ell} \tag{4.17}$$

This ties our present development to previous considerations.[11] Note, however, that in (4.11) if we have, say, equal X_{ℓ_1} and X_{ℓ_2}, then there are only L (not $L + 1$) equations when one goes to the matrix in (4.12). So the A_n are still uniquely determined as in (4.9) whether or not all the X_ℓ are distinct. The "natural" E-pulse occurs when an $a_\ell(t)$ as in (4.2) annihilates more than one pole, indicating that there are two values of ℓ giving the same $a_\ell(t)$, one thereby being redundant.

5. APPLICATION OF ξ-PULSE

The ξ-pulse has a time-duration

$$t_\xi = LT_0 + t_s$$

$$t_s \equiv \text{time duration of } f_s(t) \quad (\text{e.g., } T_0 \text{ for subsectional basis function}) \tag{5.1}$$

When applied to a target impulse-response waveform containing the target natural frequencies s_α as its zeros, the late-time response becomes zero (except for noise). The time at which this occurs depends on the target discrimination time and the duration of the ξ-pulse. The ξ-pulse has a duration proportional to L, the number of natural frequencies used in its construction. While targets generally have an infinite number of natural frequencies, only a finite number, the important natural frequencies are used in practice. So it is only an approximation to regard the late-time response to be zero after the application of the ξ-pulse due to this truncation. Of course, it is necessary to have such a truncation of the ξ-pulse so that it does not have an infinite duration (for fixed $T_0 > 0$) and thereby remove the concept of a late-time region with zero response.

The ξ-pulse need not only have zeros at target natural frequencies; others can be added for convenience. A transient radar will transmit some incident waveform which is not a perfect delta function (finite height, non-zero width, zero time integral (antennas not radiating DC), etc.), or the waveform might be quite different from the delta function. The target impulse response can be recovered by deconvolution techniques, but of course only approximately so. An alternate technique would be to construct a ξ-pulse which annihilates the late-time portions of the transmit and receive impulse responses, and combine this with the ξ-pulses for the individual targets. The ξ-pulse can also incorporate some noise filtering by choice of $f_s(t)$, to remove high frequencies beyond some frequency of interest. Inclusion of a zero at $s = 0$ in the ξ-pulse also serves as a low-frequency filter.

Sections 2 and 3 have considered the target discrimination time t_d. Times (after initial backscatter) need to be greater than this to establish the particular set of s_α characteristic of a particular target. For EMSI this is bounded below by $2t_0$ where t_0 is the transit time for the target maximum linear dimension. For MSI this is a negligible time compared to the diffusion time through the target. For some set of N targets we have N sets of s_α and N corresponding ξ-pulses which we can designate $\xi_n(t)$. Corresponding to this we have durations t_{ξ_n} as in (5.1).

Then we can define

$$T_n \equiv t_{\xi_n} + t_d \tag{5.2}$$

giving a time interval (after first signal return with worst case aspect) after which the $\xi_n(t)$ will annihilate the late-time of the target impulse response. We can also find the longest such time as

$$T_{\max} = \max_{1 \le n \le N} T_n \tag{5.3}$$

Let $f_m(t)$ be a target scattered waveform which has been processed to remove the radar transfer functions (to some extent), and which presumably contains noise. Then form for this mth target ($m = 1, 2, ..., M$)

$$F_{n,m}(t) \equiv \xi_n(t) \circ f_m(t) \quad , \quad \tilde{F}_{n,m}(s) = \tilde{\xi}_n(s)\tilde{f}_m(s) \tag{5.4}$$

As illustrated conceptually in fig. 5.1 we have some waveform which continues on beyond T_m. When operated on by $\xi_n(t)$ the late-time behavior ($t > T_m$) is dramatically different depending on whether or not $n = m$.

One way that one may wish to quantify the judgment of which target is present is by forming

$$U_{n,m}^{(0)} = \left[\frac{\int_{T_n}^{t_n^{(\max)}} F_{n,m}^2(t)dt}{\int_0^{t_n^{(\max)}} F_{n,m}^2(t)dt}\right]^{\frac{1}{2}} = \frac{\left\| F_{n,m}(t)\left[u(t-T_n) - u\left(t-t_n^{(\max)}\right)\right]\right\|_{2,t}}{\left\| F_{n,m}(t)\left[1 - u\left(t-t_n^{(\max)}\right)\right]\right\|_{2,t}} \tag{5.5}$$

which is a kind of 2-norm evaluation of the residual (energy related). Since $f_m(t)$ is assumed to begin at $t = 0$ for present purposes, negative times are excluded from the integrals. We also need to truncate for times greater than $t_n^{(\max)}$ to avoid an infinite value due to integrating the squared noise out to ∞. An optimal choice for $t_n^{(\max)}$ concerns when the residual from $f_m(t)$ (after filtering by $\xi_n(t)$) becomes less than the remaining noise (as filtered). The choice of T_n is as previously discussed. Then we find

$$\begin{aligned} U_{\min,m}^{(0)} &\equiv \min_n U_{n,m}^{(0)} \\ n_{\min} &\equiv \text{value of } n \text{ achieving this minimum} \equiv \text{value assigned to } m \end{aligned} \tag{5.6}$$

If there are other m approaching this minimum, i.e., other $U_{n,m}^{(0)}/U_{\min,m}^{(0)}$ not too much greater than one, then there is some uncertainty in our choice of m for the correct target. At this point one may wish to introduce probabilistic estimates.

An alternate definition of $U_{n,m}$ might be

$$U_{n,m}^{(1)} = \left\| F_{n,m}(t)\, e^{-\frac{t}{t_w}} u(t-T_n) \right\|_{2,t} \left\| F_{n,m}(t)\, e^{-\frac{t}{t_w}} u(t) \right\|_{2,t}^{-1} \tag{5.7}$$

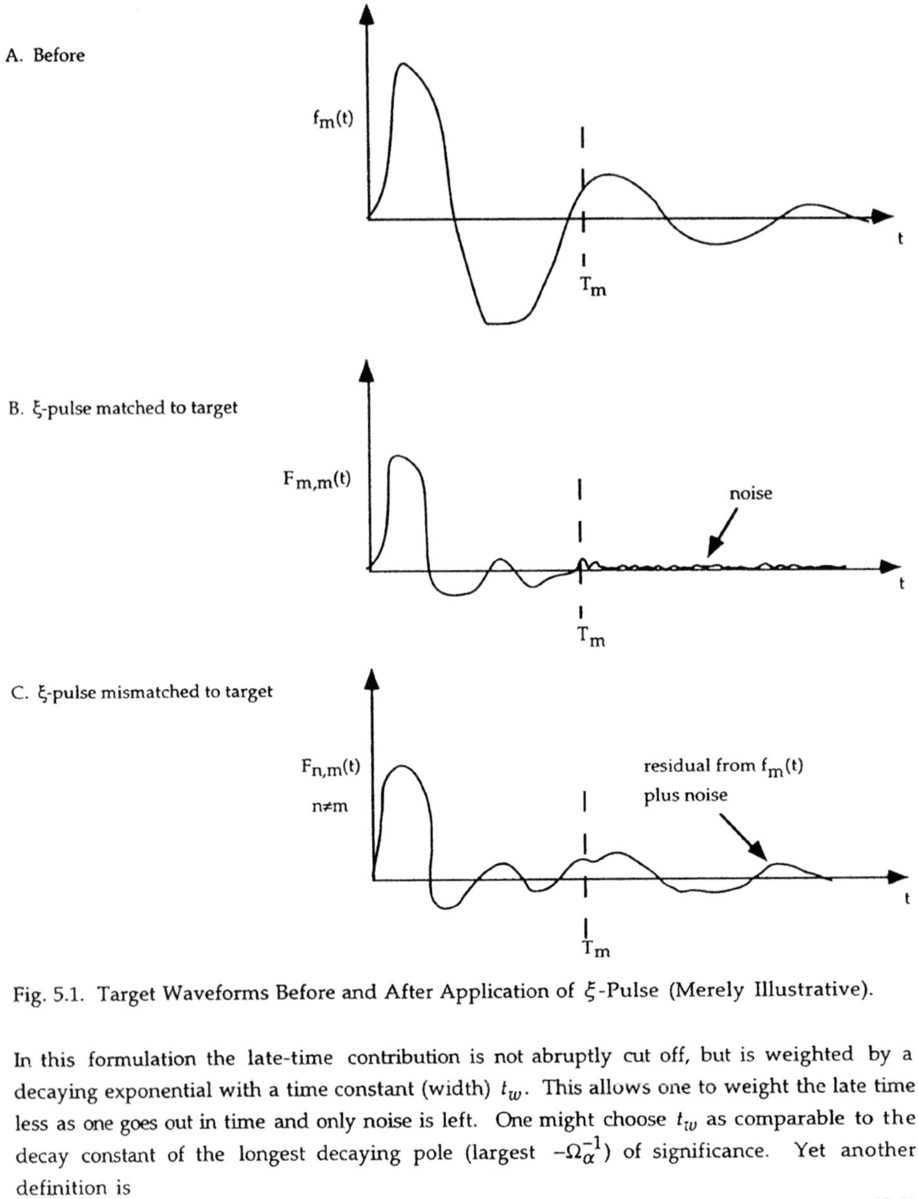

Fig. 5.1. Target Waveforms Before and After Application of ξ-Pulse (Merely Illustrative).

In this formulation the late-time contribution is not abruptly cut off, but is weighted by a decaying exponential with a time constant (width) t_w. This allows one to weight the late time less as one goes out in time and only noise is left. One might choose t_w as comparable to the decay constant of the longest decaying pole (largest $-\Omega_\alpha^{-1}$) of significance. Yet another definition is

$$U_{n,m}^{(2)} = \left\| F_{n,m}(t)\, e^{-\frac{t-T_n}{t_w}}\, u(t-T_n) \right\|_{2,t} \left\| F_{n,m}(t)\, e^{-\frac{t}{t_w}}\, u(t) \right\|_{2,t}^{-1} \tag{5.8}$$

This gives a stronger weight to the late-time regime used for the target discrimination.

6. USE OF WINDOW LAPLACE/FOURIER TRANSFORM TO EVALUATE LATE-TIME RESIDUALS

Instead of considering the "total energy" remaining in the late-time interval as in (5.5), one can look more in detail in this time interval by various transforms. The window-Laplace and wavelet transforms use a triwave-transform kernel[14] of the form

$$\frac{1}{t_w}\, g\!\left(\frac{t-t_s}{t_w}\right) e^{-st} \ ,\quad t_w > 0 \ ,\quad t_s \equiv \text{time shift (real)} \tag{6.1}$$

$s \equiv$ complex frequency , $t_w \equiv$ time dilation (width of window or wavelet)

An interesting form to choose for this for present purposes is

$$g(\zeta) = e^{-\zeta} u(\zeta) \ ,\quad \zeta \ \text{real} \ ,\quad \int_{-\infty}^{\infty} g(\zeta)\, d\zeta = 1 \tag{6.2}$$

With this kind of normalized window function this is appropriate to think of as a window Laplace transform.

Apply this to the filtered waveforms as

$$\tilde{F}_{n,m}(s,t_s,t_w) \equiv \int_{-\infty}^{\infty} F_{n,m}(t)\frac{1}{t_w} e^{-\frac{t-t_s}{t_w}} u(t-t_s)\, e^{-st} dt \tag{6.3}$$

As one increases ts from zero, we first have the contribution from the large response for all targets. As t_s passes T_m the contribution to the integral in (6.3) should drop for $n = m$ by comparison to other $n \neq m$. Setting

$$s \equiv j\omega \tag{6.4}$$

we can have the phase-space diagrams as illustrated in Fig. 6.1 where we plot in a rough illustrative sense the magnitude of the transform versus both ω and t_s (for some selected t_w). For $t_s > T_m$ and the correct choice $n = m$ we have noise, whereas for $n \neq m$ we have poles from the target and zeros from the filter appearing with the noise. This may be a more sensitive discriminant than a simple 2-norm of the late-time, but one would still like to reduce this information to a few numbers (ultimately from a computer) that distill this phase-space information, say by looking at peaks (and perhaps valleys) in the late time.

There is still the question of normalization for better comparison of the $F_{n,m}(t)$. One choice might be the 2-norm denominator in (5.5). For present purposes one might define a normalized transform related to (6.3) as

$$\tilde{F}_{n,m}^{(n)}(s,t_s,t_w) = \left\| F_{n,m}(t)\frac{1}{t_w} e^{-\frac{t}{t_w}} u(t) \right\|_1^{-1} \tilde{F}_{n,m}(s,t_s,t_w)$$

$$\left\| F_{n,m}(t)\frac{1}{t_w} e^{-\frac{t}{t_w}} u(t) \right\|_{1,t} \equiv \int_{-\infty}^{\infty} \left| F_{n,m}(t) \right| \frac{1}{t_w} e^{-\frac{t}{t_w}} u(t)\, dt \tag{6.5}$$

The 1-norm is used to make the result dimensionless, but other norms could also be used. A norm is desirable to avoid a possible zero denominator. Note the use of the same exponential decay in the denominator (avoiding a discrete late-time cut off) as in the numerator.

The window width t_w is a parameter available to us to choose for optimal results. It should be comparable to or somewhat larger than the decay time of the dominant target resonances (those of significant amplitude which last for the longest times). However, t_w should not be too large so as to reduce the influence of the late-time noise (when the target information is negligible)..

At this point let us note that this phase-space view of the late time may be more appropriate in the case of EMSI for which we have large $j\omega_\alpha$ parts of s_α which can carry over in late time. For MSI with only Ω_α parts of s_α, resolution along the $j\omega$ axis may not be too useful. Note that if we set $s = 0$ in (6.3) and (6.5) then the normalized transform in (6.5) begins to look like (5.8) except that different norms are used the present numerator is a linear operation on the filtered waveform. With $s = 0$, this leaves two parameters t_w and t_s which one might also use. This can be viewed as a wavelet transform with t_w^{-1} (frequency like) for the vertical axis and t_s for the horizontal axis to give the phase space.[14]

A. $\left|\tilde{F}_{m,m}(j\omega, t_s, t_w)\right|$ for correct target

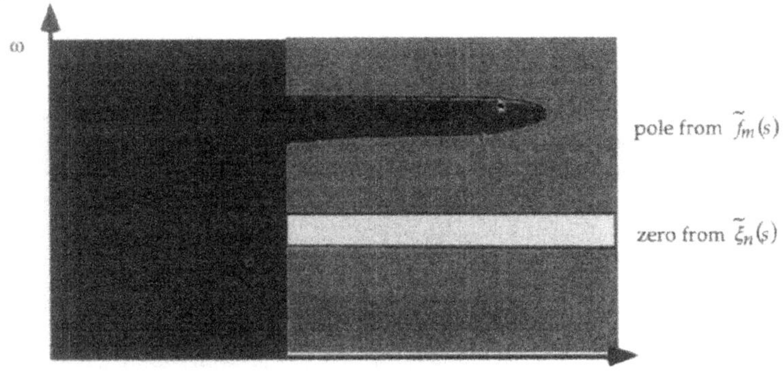

B. $\left|\tilde{F}_{n,m}(j\omega, t_s, t_w)\right|$ for incorrect target

Fig. 6.1. Phase Space for Target Discrimination (Merely Illustrative).

Additional properties of the transformed filtered waveform can be found by manipulating (6.3) as

$$\begin{aligned}\tilde{F}_{n,m}(s,t_s,t_w) &= \frac{e^{-st_s}}{t_w}\int_{-\infty}^{\infty} F_{n,m}(t)\, u(t-t_s)\, e^{-\left[s+t_w^{-1}\right]\left[t-t_s\right]}dt \\ &= \frac{e^{-st_s}}{t_w}\int_{-\infty}^{\infty} F_{n,m}(t+t_s)\, u(t)\, e^{-\left[s+t_w^{-1}\right]t}dt\end{aligned} \tag{6.6}$$

This effectively replaces s by $s+t_w^{-1}$ in the transform. With $s = j\omega$ this is like looking off in the right half s-plane a distance t_w^{-1}. As the sampling pulse width t_w is made larger and larger the Fourier transform ($s = j\omega$) of the late time (after t_s) is pushed closer and closer to the $j\omega$ axis. The factor $e^{-j\omega t_s}$ in front is a phase shift with magnitude one.

At this point we can note a connection between the window-Laplace/Fourier transform and the time-domain form. Considering the energy-related 2-norm the Parseval theorem gives

$$\left\|\tilde{F}_{n,m}(j\omega,t_s,t_w)\right\|_{2,\omega} = \left[\int_{-\infty}^{\infty}\left|\tilde{F}_{n,m}(j\omega,t_s,t_w)\right|^2 d\omega\right]^{1/2} = \left[2\pi\int_{-\infty}^{\infty}\left[F_{n,m}(t)\frac{1}{t_w}e^{-\frac{t-t_s}{t_w}}u(t-t_s)\right]^2 dt\right]^{1/2}$$

$$= [2\pi]^{1/2}\left\|F_{n,m}(t)\frac{1}{t_w}e^{-\frac{t-t_s}{t_w}}u(t-t_s)\right\|_{2,t} \tag{6.7}$$

So integrating the magnitude squared of the transform over all ω gives the integral of the square of the late-time ($t_s = T_m$) weighted filtered waveform as in (5.8). In this form 2-norms are appropriate, including the normalization.

7. GENERALIZATION TO DYADIC WAVEFORMS

Depending on the design of the radar, one can have more than one waveform from a given target in backscatter. One can, of course, use multiple copies of the same waveform (stacking, same polarization). Here, however, one may have two polarizations (h,v radar coordinates) in transmission combined with two in reception to establish the 2 x 2 scattering dyadic (only components transverse to $\vec{1}_i$) for EMSI as in (2.2). Since the natural frequencies s_α are aspect independent (i.e., are the same for all directions of incidence and polarizations), then the four waveforms inherent in the scattering dyadic operator all have the same s_α set. So form

$$\overleftrightarrow{F}_{n,m}(t) \equiv \xi_n(t) \circ \overleftrightarrow{\Lambda}_{bm}(\vec{1}_i,t) \quad , \quad \tilde{\overleftrightarrow{F}}_{n,m}(s) = \tilde{\xi}_n(s)\,\tilde{\overleftrightarrow{\Lambda}}_{bm}(\vec{1}_i,s) \tag{7.1}$$

with subscript m for the scattering dyadic operator (impulse response) of the mth target (subscript m) analogous to (5.4). Then the norms used in (5.5), (5.7), and (5.8) can be readily generalized to dyadic functions.[1, 2] Furthermore, the transforms used in Section 6, being scalar linear operations which commute with dyadics, can also be directly applied to (7.1). In phase space (as in fig. 6.1) the magnitude of the transform is readily replaced by the 2-norm in matrix sense. While (7.1) is defined in terms of backscattering, it is readily generalized to multistatic scattering.

For MSI as in (3.3) or (3.4) we deal with SEM representation of a 3 x 3 magnetic-polarizability dyadic (in general symmetric). Depending on the type of experimental data available one may have some or all of the components of this dyadic available for analysis. So form (to the extent components are available)

$$\overleftrightarrow{F}_{n,m}(t) = \xi_n(t)\circ\overleftrightarrow{M}(t) \text{ or } \overleftrightarrow{F}_{n,m}(t) = \xi_n(t)\circ\int_{-\infty}^{t}\overleftrightarrow{M}(t')\,dt' \tag{7.2}$$

similar to (7.1). The late-time evaluation including transforms then proceeds as discussed previously.

8.0 CONCLUDING REMARKS

There is still a lot to be considered. Noise is an obvious limitation. To suppress the effects of noise one can filter out frequencies outside the range of interest. By enhancing the transmit waveforms in the frequency range of interest one can increase the signal-to-noise ratio for the frequencies. The receive waveforms then need to be deconvolved to produce an approximate delta-function response (or some other convenient form such as step-function response).

Another subject of interest is a possible adaptive ξ-pulse. In the case of metal targets (approximated as perfectly conducting) buried in soil there are scaling relationships for the natural frequencies.[4] If one knows the ε and σ of the soil (say from measurement using the same radar) then one can alter the A_n coefficients by the scaling relationship. Another approach would consider the approximate affine scaling of the s_α (in a simple soil) which preserves the pattern of the s_α in the s-plane, and iteratively adjust ε and σ to achieve the late-time annihilation. For dielectric targets in the soil the problem is yet more complicated, but some asymptotic formulas for the s_α movement in the s-plane may be helpful.[7] This problem of the movement of the s_α in the s-plane due to the variability of the external-medium constitutive parameters is a significant one, perhaps indicating that a quite different approach to the target identification (say explicit instead of implicit) may be useful.

REFERENCES

1. C.E. Baum, Norms of Vectors of Time-Domain Signals Passing Through Filters and Norm Limiters at Subshields, Interaction Note 469, January 1988, and Proc. EMC Symposium, Zurich, March 1991, pp. 589-594.
2. C.E. Baum, The Theory of Electromagnetic Interference Control, Interaction Note 478, December 1989, and pp. 87-101 in J. Bach Andersen (ed.) *Modern Radio Science 1990*, Oxford University Press, 1990.
3. C.E. Baum, Representation of Surface Current Density and Far Scattering in EEM and SEM with Entire Functions, Interaction Note 486, February 1992.
4. C.E. Baum, The SEM Representation of Scattering from Perfectly Conducting Targets in Simple Lossy Media, Interaction Note 492, April 1993.
5. C.E. Baum, Low Frequency Near-Field Magnetic Scattering from Highly, but not Perfectly Conducting Bodies, Interaction Note 499, November 1993.
6. C. E. Baum, The Magnetic Polarizability Dyadic and Point Symmetry, Interaction Note 502, May 1994.
7. C. E. Baum, Concerning the Identification of Buried Dielectric Targets, Interaction Note 504, July 1994.
8. C. E. Baum, The SEM Representation of Acoustic and Elastodynamic Scattering, Interaction Note 512, June 1995.
9. E. M. Kennaugh, The K-Pulse Concept, IEEE Trans. Antennas and Propagation, 1981, pp. 327-331.
10. E. Rothwell, D. P. Nyquist, K.-M. Chen, and B. Drachman, Radar Target Discrimination Using the Extinction-Pulse Technique, IEEE Trans. Antennas and Propagation, 1985, pp. 929-937.
11. C. E. Baum, E. J. Rothwell, K.-M. Chen, and D. P. Nyquist, The Singularity Expansion Method and Its Application to Target Identification, Proc. IEEE, 1991, pp. 1481-1492.
12. T. K. Sarkar and O. Pereira, Using the Matrix Pencil Method to Estimate the Parameters of a Sum of Complex Exponentials, IEEE Antennas and Propagation Mag., Vol. 37, No. 1, February 1995, pp. 48-55.
13. C. A. Korn and T. M. Korn, *Mathematical Handbook for Scientists and Engineers, 2nd Ed.*, McGraw Hill, 1968.
14. C. E. Baum, Symmetry and Transforms of Waveforms and Waveform Spectra in Target Identification, ch. 7, pp. 309-343, in C. E. Baum and H. N. Kritikos (eds.), *Electromagnetic Symmetry*, Taylor & Francis, 1995.

A MIXED TIME-FREQUENCY-SCALE ANALYSIS OF THE HYBRID WAVEFRONT-RESONANCE REPRESENTATION

Amos Arev, Ehud Heyman and Ben-Zion Steinberg

Faculty of Engineering, Tel-Aviv University,
Tel Aviv 69978, Israel

Abstract

Several time-frequency-scale processing schemes for the analysis of short pulse scattering fields are explored and calibrated in connection with the problem of reflections from multilayered dispersive media. The time-frequency signature consists of a (generally nonuniform) grid of wavefront arrivals and resonances. It is shown that the time-frequency resolution is bound by the unit-cell area, hence the signature of a particular wave process can be detected only if the processing windows are scale-matched to corresponding unit cells. The short-time Fourier transform (STFT) has therefore an inherent limitation because it has a built in scale, while the wavelet transform with a Morlet wavelet provides a framework that adapts to all scales with no a-priori information while retaining the frequency signature of the resonances. Localized scrutiny of a specific scale may then be achieved using the STFT, while super resolution may be achieved using a model based analysis.

INTRODUCTION

The problem of extracting and classifying the scattering signature of an object or a medium is receiving an increased attention in connection with various applications involving target identification. The various approaches to this problem rely on the

Ultra-Wideband, Short-Pulse Electromagnetics 4
Edited by Heyman *et al.*, Kluwer Academic / Plenum Publishers, New York, 1999.

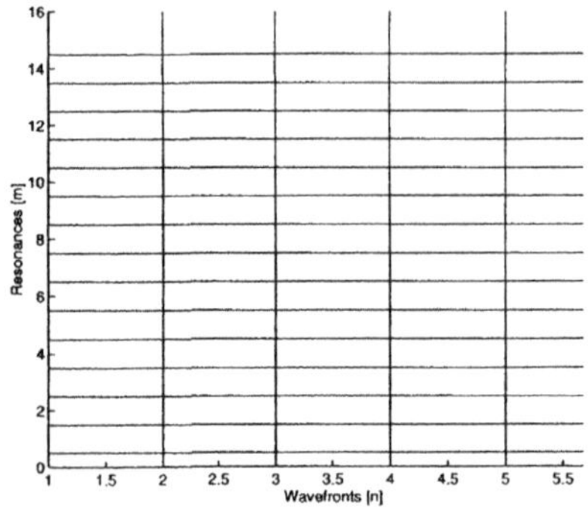

(a) a non dispersive layer

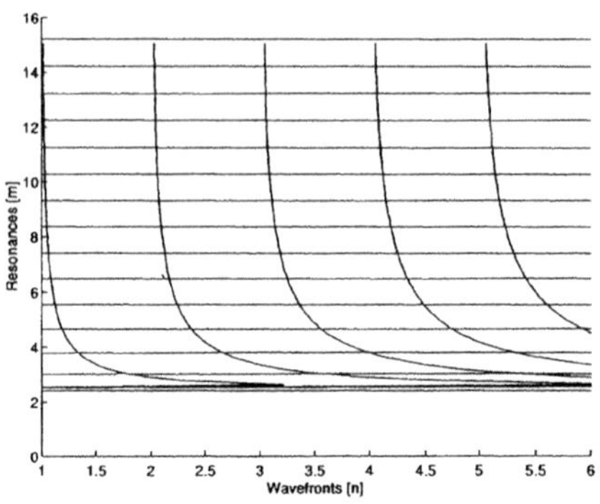

(b) a dispersive layer

Figure 1: Phase space skeleton for a non-dispersive layer (a) and a dispersive layer (b) . The horizontal and vertical lines depict the resonant frequencies ω_m and the the wavefront arrivals T_n, respectively. Note in (b) that near the cutoff frequency v_g tends to zero and the resonant frequencies grow closer, while at higher frequencies the skeleton is similar to that in (a). In both cases, the unit cell area is $\Delta\omega\Delta T = 2\pi$.

existence of physical observables in the data that can be detected and identified. There are basically two types of scattering observables, namely *global* and *local*: The local ones are conveyed by the traveling wave fields that sense the local features of the medium along their propagation paths, while the global ones have the form of late time resonances that represent the collective effect of the multiple interactions of the traveling waves with the medium. The local observables are therefore localized in the early time signal while the global ones are localized in the frequency spectrum [1]. An effective classification scheme may therefore utilize both types of discriminants by identifying the scattering signature in a mixed time-frequency domain. From a theoretical perspective, the time-frequency format provides a framework for describing the wavefront-resonance interplay, e.g., the process by which the local wavefront events at early time are gradually transformed into the partial resonances corresponding to substructures of the composite scatterer, and eventually to the global resonances of this composite [1].

Time-frequency analysis of target signatures has been explored recently via several phase space processing techniques, including the short time Fourier transform (STFT), the multi resolution analysis (MRA) and the adaptive model-based representations (MBR) [2, 3]. Extracting the signature via these techniques depends on the choice of the processing window: Narrow windows may resolve the wavefront arrivals, while resolving the resonance signature usually requires wider windows. As will be shown here, choosing the "optimal" window that extracts both features, if possible, depends on the scale of the wave interactions. The presence of multiple interacting wave species (e.g., in multilayered medium) complicates this choice because of the scale diversity, while dispersion introduces a different type of difficulty because

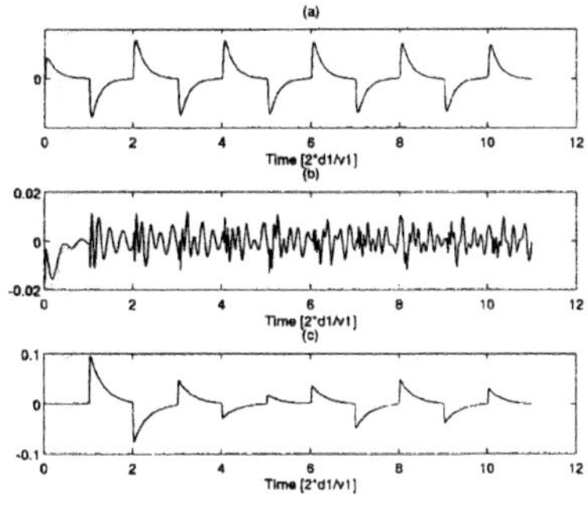

Figure 2. The reflected signals. (a) and (b): nondispersive and dispersive layers; (c): double layer with $\Delta T_2 = 5\Delta T_1$.

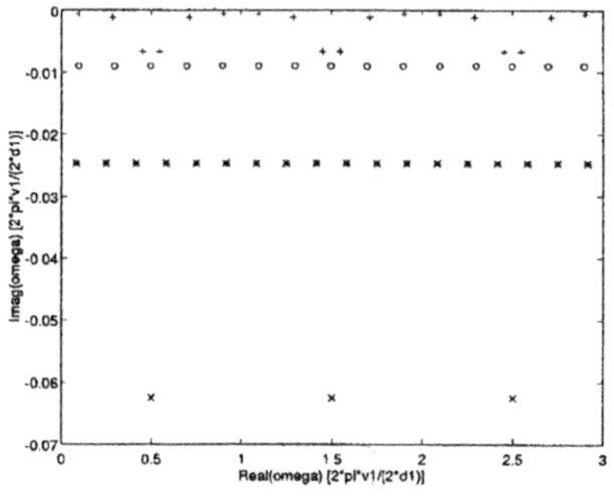

Figure 3. Complex resonances of a double layer medium: + Global (true) resonances; $\times$, $\circ$ and $*$: Partial resonances of layers 1, 2 and of a path combining layers 1 and 2.

of the frequency dependence of the characteristic scales. This suggests a possible advantage of the MRA since it does not require an a priori knowledge of the problem scale. The discussion above calls for proper calibration of the processing tools in order to identify the phase space signature of *both* the wavefronts *and* the resonances. This problem is addressed here in the context of the simplest, yet most fundamental problem of transient reflections from layered medium [4].

WAVEFRONT AND RESONANCE OBSERVABLES IN LAYERED MEDIA

To analyze the tradeoff between the time domain discriminants, i.e., the wavefront arrivals and the damped resonances, we consider the problem of a pulsed plane wave $f(t - z/v_0)$ impinging from $z < 0$ on a multilayered medium at $z > 0$ composed of N uniform layers of width d_j and wavespeeds $v_j(\omega)$, where $j = 1, \ldots N$ and ω is the radian frequency. The reflected signal can readily be constructed in the frequency domain by conventional techniques and then transformed into the time domain. There are two types of expressions: One expresses the field explicitly as a series of total internal reflections, giving rise to the series of wavefront arrivals in the time domain, while the other is a collective expression that replaces the wavefronts by the resonances.

1. A single layer

As a simple example consider a pulsed plane wave $f(t - z/v_0)$ impinging from $z < 0$ on a single uniform layer of width d_1 and wavespeed v_1 which may, in general, be a function of ω. The wavefront representation of the reflected field at $z = 0$ is given, in the frequency domain, by

$$\hat{u}_r(z,\omega)\big|_{z=0} = \underbrace{\hat{f}(\omega)\hat{\Gamma}_{01}}_{\text{Direct reflection}} + \underbrace{\hat{f}(\omega)\hat{\Gamma}_{12}\hat{T}_{01}\hat{T}_{10}\, e^{2ik_1 d_1} \sum_{n=0}^{\infty} \left[\hat{\Gamma}_{10}\hat{\Gamma}_{12} e^{2ik_1 d_1}\right]^n}_{\text{Multiple internal reflections}} \qquad (1)$$

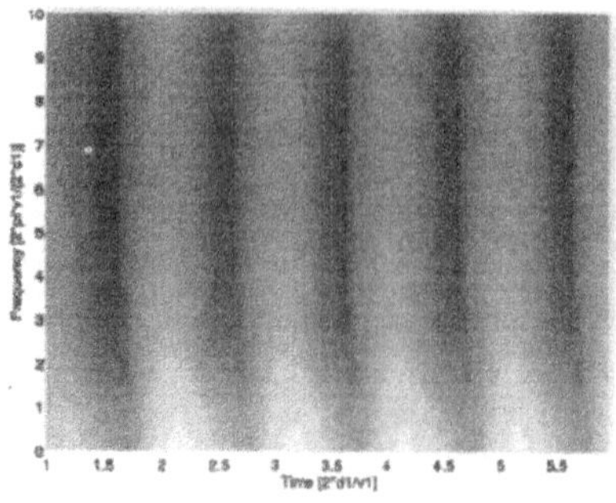

(a) A narrow window: $0.8\Delta T$

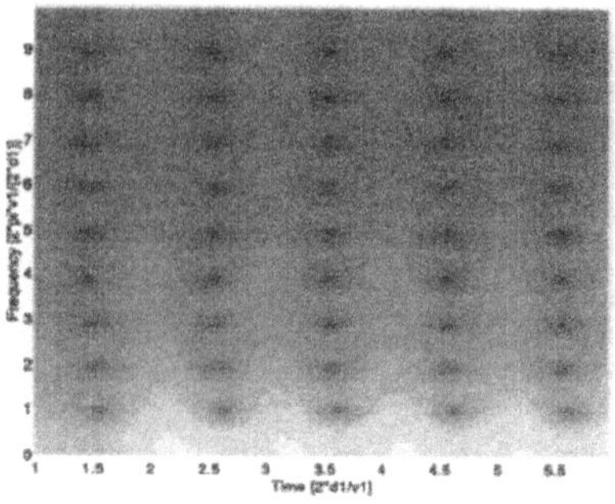

(b) A matched window: ΔT

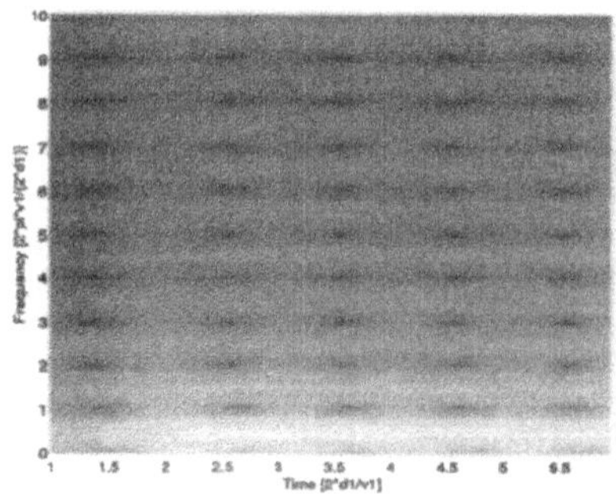

(c) A wide window: $1.2\Delta T$

Figure 4: STFT analysis of a single non-dispersive slab with narrow, matched and wide windows. The matched window in (b) extracts both the resonances and the wavefronts while the narrow and wide windows in (a) and (c), emphasize the wavefronts *or* the resonances, respectively.

where here and henceforth we use the over-hat to denote frequency domain constituents, $\hat{f}(\omega)$ is the source spectrum and $k_1(\omega) = \omega/v_1(\omega)$. Here $\hat{\Gamma}_{j\ell}$ and $\hat{T}_{j\ell}$ are the reflection and transmission coefficients at the interface between layer j and ℓ: If u and its normal derivative are continuous at the interfaces then $\hat{\Gamma}_{j\ell} = \frac{v_\ell - v_j}{v_\ell + v_j}$ and $\hat{T}_{j\ell} = \hat{\Gamma}_{j\ell} + 1$.

The collective expression is obtained by summing the series in (1) in closed form, obtaining

$$\hat{u}_r(0,\omega) = \underbrace{\hat{f}(\omega)\hat{\Gamma}_{01}}_{\text{Direct reflection}} + \hat{f}(\omega)\hat{\Gamma}_{12}\hat{T}_{01}\hat{T}_{10}\underbrace{\frac{e^{2ik_1d_1}}{1-\hat{\Gamma}_{10}\hat{\Gamma}_{12}e^{2ik_1d_1}}}_{\text{Collective term}} \tag{2}$$

(a) Non-dispersive case

From (1) one finds that the nth wavefront has the form $A_n f(t - T_n)$, where A_n represents the effect of all reflection and transmission coefficients that occur along the

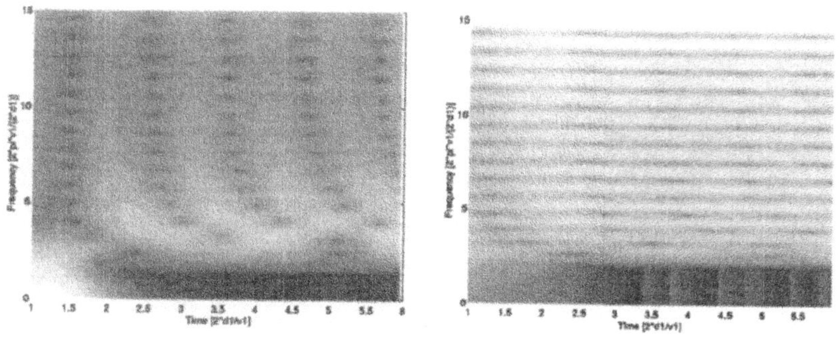

(a) STFT with a high-frequency matched window

(b) STFT with a near cutoff matched window

Figure 5: STFT analysis of a single dispersive slab, using a windows matched to the high-frequency regime (a) or to the near cutoff regimes (b).

wave path and $T_n = 2nd_1/v_1$. The collective term in (2) has complex resonances at $\omega_m = (v_1/d_1)[m\pi + \frac{i}{2}\ln\Gamma_{01}\Gamma_{12}]$, $m = 1, 2, \cdots$. From these relations one finds that the wavefronts and resonance observables form a rectangular grid in the time-frequency domain, with a unit cell area (see fig. 1(a))

$$\Delta T \Delta\omega = 2\pi. \tag{3}$$

(b) Dispersive case

We start with a brief analysis of the space-time propagation trajectory for a pulse in dispersive medium. The propagating pulse is described by the inverse Fourier transform

$$u(z,t) = \frac{1}{2\pi}\int_{-\infty}^{\infty} d\omega \hat{f}(\omega) e^{ik(\omega)z - i\omega t} \tag{4}$$

where $k(\omega) = \omega/v(\omega)$. Using a stationary point approximation, one obtains the *contributing frequency* $\omega_s(t)$ at a given time $\frac{\partial\psi}{\partial\omega} = t - z\frac{\partial k(\omega)}{\partial\omega}\Big|_{\omega_s} = 0$. The space-time-frequency trajectory is therefore described by the relation

$$\frac{z}{t} = \frac{\partial\omega}{\partial k} \equiv v_g(\omega_s), \quad \text{at } \omega_s(t) \tag{5}$$

The final result for the propagating field

$$u(z,t) \sim \frac{1}{\sqrt{2\pi}}\mathrm{Re}\left\{\hat{f}(\omega)\frac{e^{-i[\omega t - k(\omega)z] + i\frac{\pi}{4}}}{\sqrt{z\partial_\omega^2 k(\omega)}}\right\}_{\omega_s(t)} \tag{6}$$

Using these results for a uniform dispersive layer, we find that the arrival times are given by the time-frequency trajectories (see the "vertical" lines in Fig. 1(b))

$$T_n(\omega_s) = (n+1)2d_1/v_g[(\omega_s(t)] = 2(n+1)d_1 k_1'[\omega_s(t)] \tag{7}$$

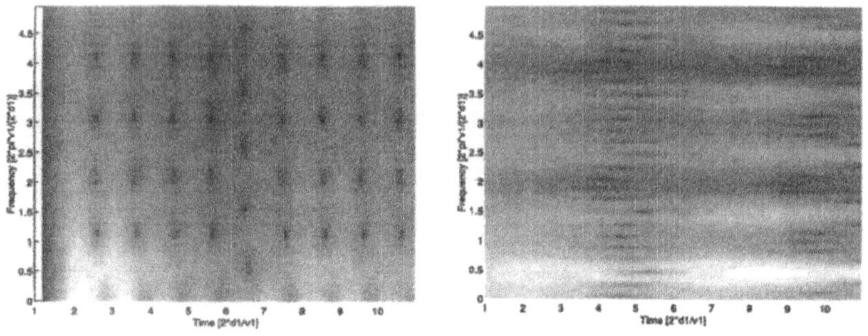

(a) STFT with a window matched to ΔT_1 (b) STFT with a window matched to ΔT_2

Figure 6: STFT analysis of a double layers medium with $\Delta T_2 = 5\Delta T_1$. In (a) and (b) the processing window is matched to the travel time ΔT_1 and ΔT_2 of the first and second layers, respectively.

and from (2), the resonance frequencies are calculated via

$$2k_1(\omega_m)d_1 = 2\pi m \quad \Rightarrow \quad \Delta\omega_m = \pi/k_1'(\omega_m)d_1. \tag{8}$$

In view of (7) and (8) one concludes that the area of the unit cells remains 2π as in (3) even in the non-dispersive case where the phase space skeleton is not rectangular (see Fig. 1(b)). As an example, we consider in Fig. 1(b) the waveguide-type dispersion $k_1(\omega) = (1/v_1^\infty)\sqrt{\omega^2 - \omega_c^2}$ where ω_c is the cutoff frequency and v_1^∞ is the wavespeed at $\omega \to \infty$. In the numerical example we choose ω_c to be $\frac{1}{2}(\omega_3 + \omega_2)$ where $\omega_{2,3}$ are the resonances of the corresponding non-dispersive layer (with $\omega_c \to 0$). Using the expressions for these resonances (see discussions preceding (3)) we obtain an expression for the medium parameters in this case: $\omega_c = \frac{5}{2}\pi v_1^\infty/d_1$. Note that the non-rectangular grid in Fig. 1(b) satisfies the phase space constraint in (3). The time-domain data for the non-dispersive and the dispersive layers is shown in Fig. 2.

2. A double layers medium

The multi-layer is characterized by many time-scales corresponding to different multiply reflected ray paths. Furthermore, in this case the set of resonance observables consists of both *global* (or true) and *partial* resonances. The first ones are the true resonances of the global structure, representing the collective effect of *all* wave interactions, while the others correspond to a partial set of the interactions. For example, at early times, before the arrival of the wavefronts from the deep layers, one may expect to observe resonances that represent multiple interactions in the first layer (or layers). Clearly, the partial resonances are not the *true* ones. In order to explore the various alternatives for dealing with this multiscale problem we consider the example of a double layers medium. The medium is chosen so that the travel time in the front layer is 5 times shorter than that of the deeper layer, i.e., $\Delta T_2 = 5\Delta T_1$. The two layers were strongly coupled with $\Gamma_{1,2} = 0.75$. The reflected signal is shown in Fig. 2(c) while the *global* resonances in the

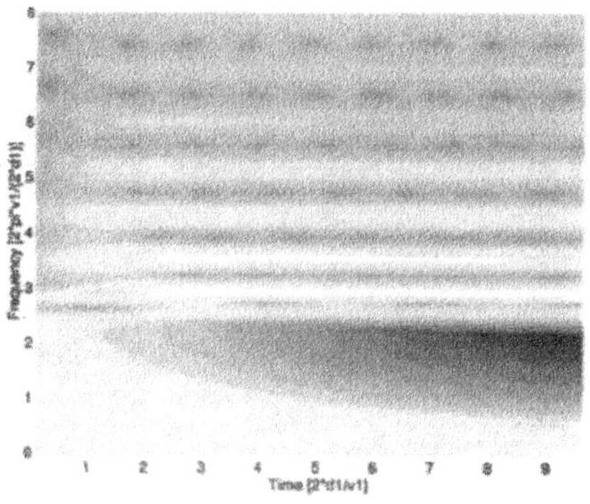

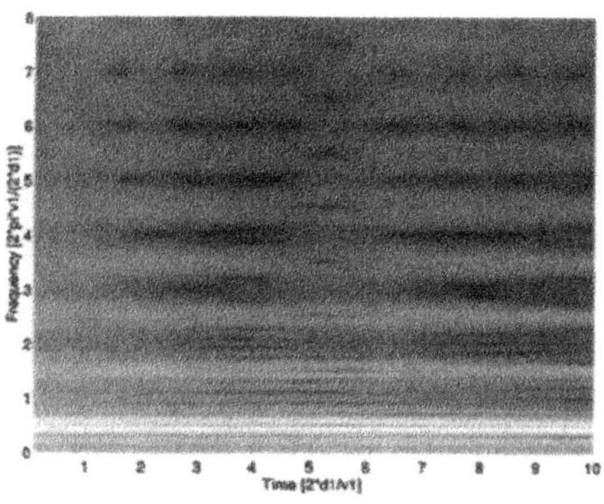

(a) CWT of a single dispersive slab (b) CWT of a double layers medium

Figure 7: Continuous wavelet transform (CWT) analysis of: (a) a single dispersive slab and (b) a double layer medium, using a Morlet wavelet with $N = 8$. The vertical axes express the scale in terms of the equivalent frequency ω_σ of (12).

fourth quadrant of the complex ω plane are shown in Fig. 3 (the resonances in the third quadrant are symmetrical). Fig. 3 also depicts the *partial* resonances corresponding to the front layer (layer 1), the back layer (layer 2), and those corresponding to a ray path that involves both layers but without cross coupling at the interface between these layers. Note that the global resonances have higher Q than the partial ones. Note that the periodicity of the global resonances is determined essentially by the $*$ partial resonances, and that there is also a distinct effect of the $\times$ and $\circ$ interactions.

TIME-FREQUENCY PROCESSING: DISCUSSION AND RESULTS

The time-frequency processing aims at discerning the following observables in the data. First, the wavefronts, resonances or both should be discriminated. Second, a selected group of resonances should be isolated in order to focus on specific characteristics of the reflecting target, and last for a composite resonator (e.g. a double slab), the early time effects (e.g. partial resonances associated with the closer layer) and the global effects (e.g. turn-on time of the global resonances) should be discriminated and identified. In order to address these requirements we shall consider time-frequency (phase space) processing tools.

The Short Time Fourier Transform (STFT) is defined as

$$(T^{STFT} f)(\omega, \bar{t}) \stackrel{\text{def}}{=} \int_{-\infty}^{\infty} f(t)\, W^*_{\omega,\bar{t}}(t)\, dt\,, \qquad W_{\omega,\bar{t}}(t) = e^{-i\omega t} W(t - \bar{t}) \tag{9}$$

This operation is identified as the Fourier transform of $f(t)$ *windowed* about $\bar{t}$ with a window W of a *fixed* width. Due to the 2π phase space resolution limit implied by the processing window, the STFT may discern the wavefront and resonance simultaneously *only if the window is matched to the data*, i.e., if the width of W equals ΔT (while its spectral width equals $\Delta\omega$). This fundamental limit on the STFT is demonstrated

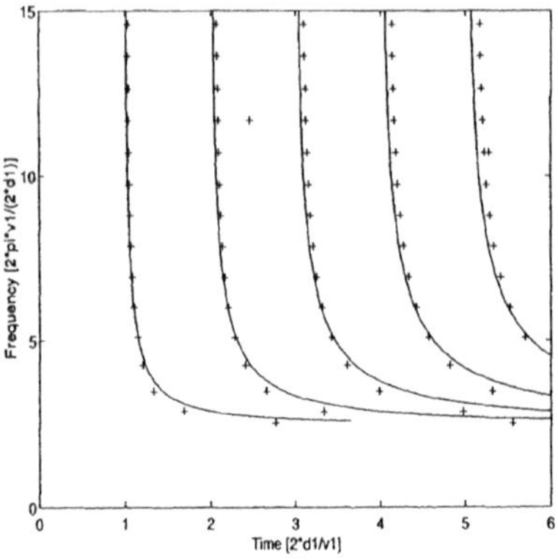

Figure 8: Pencil of Functions analysis of the single dispersive slab model. Note the accumulating error in the time delays at late times which is a result of the fact that each arrival time is calculated using the difference in respect to the one before.

in Figs. 4(a,b,c), depicting, respectively, the STFT analysis of the reflected pulse from a single non dispersive layer, using windows of $0.8\,\Delta T$, ΔT and $1.2\,\Delta T$ widths. Only in the matched window case (Fig. 4(b)), both the wavefronts and the resonances are discerned, while the narrow and wide windows provide either the temporal or the frequency observables, respectively.

The existence of medium dispersion breaks the rectangular structure since each frequency component now travels at a different wavespeed $v(\omega)$ (see Fig. 1(b)). Yet the areas of the unit cells are still 2π. In Figs. 5(a) and 5(b), the processing windows have been chosen to match the unit cells at the high frequencies and at frequencies near cutoff, respectively. One readily discerns in Fig. 5(a) poor spectral resolution at the near-cutoff zone, while the results in Fig. 5(b) provide sharp spectral resolution near the cutoff but poor resolution for the high-frequency wavefront.

In multilayered configurations where several resonance processes may take place, one should tune the window width to match the wave process of interest. Figs. 6(a,b) depict the STFT distributions for a two layer medium considered in Fig. 2 and 3, where the travel time in the first layer is 5 times shorter than in the second one. The processing window width in Fig. 6(a) has been matched to the travel time of the first layer, thereby revealing the early time wavefronts as well as the existence of the *partial* resonances corresponding to the first layer. This window however fails to resolve the much finer structure of the resonances corresponding to the farther layer. In Fig. 6(b), on the other hand, the window width has been matched to the travel time of the farther layer, thereby revealing the related partial resonance structure but not the fine temporal details corresponding to the first layer.

The examples above reveal the difficulties of the STFT in detecting wavefront-resonance processes with non-uniform grids, since the window width can at most be matched to a particular unit cell. In order to accommodate such problems we explored the MRA with a continuous wavelet transform (CWT), using a time-frequency-scale window. The CWT is defined as

$$(T^{CWT} f)(\sigma, \bar{t}) \stackrel{\text{def}}{=} \int_{-\infty}^{\infty} f(t)\, \Psi_{\sigma,\bar{t}}(t)\, dt\,, \qquad \Psi_{\sigma,\bar{t}}(t) = \frac{1}{\sqrt{\sigma}}\psi\left(\frac{t-\bar{t}}{\sigma}\right) \tag{10}$$

where $\psi(t)$ is a properly chosen wavelet function. The width of kernel of the CWT, $\Psi_{\sigma,\bar{t}}(t)$ adapts to all scales while the STFT kernel $W(t-\bar{t})e^{i\omega t}$ forces an a-priori assumption on the scale. The CWT enables a more robust analysis of the data with less a-priori knowledge. The wavelet may be chosen so that it also senses the frequency characteristics of the data. By choosing a wavelet of the form

$$\psi_N(t) = (2\pi)^{-1/2} e^{\frac{-t^2}{2}} e^{-i\pi N t} \tag{11}$$

The wavelet is tuned to detect frequencies

$$\omega_\sigma = \pi N/\sigma \tag{12}$$

Fig. 7(a) and 7(b) depict the CWT of the single dispersive layer and double layers examples considered in Figs. 5 and 6, respectively, using a Morlet window with $N = 8$. An improved resolution is observed in the cut-off zone while still maintaining the resolution away from that zone. Comparing the results in Fig. 7(a) to those in Figs. 5 we note the ability of the CWT to resolve the near-cutoff resonances where $\Delta\omega$ grows smaller. Beneath the cut-off, the slow decaying detection is a spurious effect of the wide window resolving the first arrival of the wavefront reflected off the face of the slab. Note that the width of the wavelet cannot be *exactly* matched to the unit cells, since the latter grows to infinity at the cutoff, whereas the former grows to infinity as ω_σ tends to zero.

Comparing the results in Fig. 7(b) to those in Figs. 6 we observe that at large scales (small ω_σ) the CWT detects the low order global resonances, while at the higher frequencies it detects the early-time wavefront arrivals and the *partial resonances*. Thus, the CWT cannot reveal the entire phase-space skeleton, but it does extract the scales of all relevant wave processes. The STFT, on the other hand, may reveal the entire phase-space skeleton as in Fig. 1, but only at a the specified scale (which has to be determined first).

Finally, pinpointing of the above results via MBA is demonstrated in Fig. 8, which depicts the phase space signature obtained using a matrix pencil of functions (POF) analysis of the dispersive layer example, using an adaptive set of decaying resonances.

SUMMARY

Scattered fields from composite targets may be described by *wavefronts* (early time) or *resonances* (late time). The wavefront-resonance descriptors can be jointly explored in phase space where their skeleton has the form of generally nonuniform *grid*. Different wave species or dispersion generate unit-cells with different shapes, residing in different regions of the phase space (e.g. early vs. late time or high vs. low frequency). The unit-cells' area is compatible with the resolution limit of the processing tools; one can resolve wavefronts and resonances simultaneously only with matched windows. A properly matched short time Fourier transform (STFT) can be used to explore a specific region in the phase space. The continuous wavelet transform (CWT), on the other hand, can extract the scales of all relevant wave processes, but it cannot reveal the entire phase-space skeleton. A combined process, involving the CWT and the STFT or the CWT with a model based technique like the pencil of functions (POF) is required.

ACKNOWLEDGEMENTS

This work is supported in part by the US-Israel Binational Science Foundation, Jerusalem, Israel, under Grant No. 95-00399, and by the U.S. Air Force Office of Scientific Research under Grant No. 96-1-0039.

References

[1] E. Heyman and L.B. Felsen, "A wavefront interpretation of the singularity expansion method," *IEEE Trans. Antennas Propagat.*, **AP-33**, 706–718, 1985.

[2] L. Carin, L.B. Felsen, D.R. Kralj, H.S. Oh, W.C. Lee, and U. Pillai, "Wave-oriented signal processing of dispersive time domain scattering data," *IEEE Trans. Antennas Propagat.*, **AP-45**, 599–600, 1997.

[3] L.C. Trintinalia and H. Ling, "Joint time frequency ISAR using adaptive processing," *IEEE Trans. Antennas Propagat.*, **AP-45**, 221–227, 1997.

[4] K.A. Nabulsi and D.G. Dudley, "Hybrid formulations for scattering from the dielectric slab," *IEEE Trans. Antennas Propagat.*, **AP-40**, 959–965, 1992.

JOINT PHYSICAL OPTICS AND FDTD ANALYSIS OF MICROWAVE SCATTERING

B. Zhang[1], S. Primak,[1] J. LoVetri,[1] and S. Kashyap[2]

[1]Department of Electrical Engineering
The University of Western Ontario,
London, Ontario, Canada N6A 5B9
joe@gauss.engga.uwo.ca

[2]Department of National Defence
Defence Research Establishment Ottawa,
3701 Carling Ave., Ottawa, Ontario, Canada K1A 0K2

INTRODUCTION

The use of ultra-wideband radar for identification of targets has received considerable attention during the past two decades. In [1] a single target profile (*i.e.* the impulse response of a target taken at one aspect angle) was used to identify targets using the classical M-ary correlation receiver technique which was adapted to serve this purpose. However, it has been shown in [2] that even an optimal receiver produces poor results if an unknown signal is corrupted by noise. The performance of such a receiver also decays very fast with the number of possible targets [2]. As a consequence, multiple measurements must be performed. An alternative possibility is to build a synthetic aperture radar (SAR) image of the unknown target and to attempt to identify its major scattering centres, which are related to the geometry of the target [3-4]. Most of the models which have been used to build the SAR image of a target ignore the dispersive character of the scattering centres and any natural resonance of the target itself [5]. As a result, the image which is obtained is difficult to interpret and classify. Here we suggested a simple technique allowing one to obtain more tractable images by combining real measured data (or in our case, simulated data using FDTD) with those obtained from the Physical Optics (PO) approximation.

THE SIGNAL MODELS

The simplest radar model for a target is to model it as an ideal point scatterer of strength A located at position (x_0, y_0). Excited by a monochromatic electromagnetic plane-wave of frequency ω an ideal scatterer produces the normalized scattered field

Ultra-Wideband, Short-Pulse Electromagnetics 4
Edited by Heyman *et al.*, Kluwer Academic / Plenum Publishers, New York, 1999.

$$S(\omega, \varphi_i) = \frac{E^{sc}}{E^{inc}} e^{-2\boldsymbol{k} \cdot \boldsymbol{r}} = A\exp\left(-2j\frac{\omega}{c}x_0\cos\varphi_i - 2j\frac{\omega}{c}y_0\sin\varphi_i\right) \tag{1}$$

where $\boldsymbol{r}$ is the vector position of an *imaginary* centre of the target and $\boldsymbol{k}$ is the wave vector. Taking the 2-D inverse Fourier transform of (1) with respect to ω and observation angle φ_i (in the polar coordinate system), the following equation, giving rise to SAR imaging, can be obtained:

$$s(x, y) = A\delta(x - x_0)\delta(y - y_0). \tag{2}$$

In other words, ideal scattering centres correspond to ideal points in the SAR image [6]. Thus the reflectivity of the object in the $\omega - \varphi_i$ plane is simply the 2-D inverse Fourier transform of the scatterer geometry.

The later statement remains valid for a perfectly conducting body under the high frequency Physical Optics (PO) approximation [7]. For a perfectly conducting body illuminated by a plane monochromatic wave of wave number $\boldsymbol{k}$ and scatterers that are large compared to the wavelength of the excitation, the PO approximation states that the normalized monostatic response is given as [7]

$$\rho(\boldsymbol{k}) = \frac{S(\boldsymbol{k})}{I(\boldsymbol{k})} = \frac{j}{\sqrt{\pi}} \int_{\boldsymbol{k} \cdot \boldsymbol{n} < 0} e^{-2j\boldsymbol{k} \cdot \boldsymbol{x}}\, \boldsymbol{k} \cdot d\boldsymbol{s}\,. \tag{3}$$

Here $I(\boldsymbol{k})$ is the incident plane wave, $S(\boldsymbol{k})$ is the scattered response, and $\boldsymbol{n}$ represents the outward normal to the surface of the scatterer and thus the condition $\boldsymbol{k} \cdot \boldsymbol{n} < 0$ represents the illuminated part of the body under consideration. If both $\boldsymbol{k}$ and $-\boldsymbol{k}$ are taken into account, the object is completely illuminated[1] and thus [8]

$$S(\omega, \varphi_i) = \rho(\boldsymbol{k}) + \rho^*(-\boldsymbol{k}) = \frac{2k^2}{\sqrt{\pi}} \int_V e^{-2j\boldsymbol{k} \cdot \boldsymbol{x}} dv\,. \tag{4}$$

In general, scattering from a real target cannot be approximated well by either an ideal point scatterer approximation, as in equation (1), or by the PO approximation, as given by equation (4). Two other physical phenomena, inherently present and not described by the PO approximation, are the frequency dependence of the scattering parameters and their dependence on the polarization of the incident wave. An advanced insight into these properties can be obtained in the context of the Geometrical Theory of Diffraction (GTD), valid under the assumption that the wavelength of the incident excitation is small relative to the targets extent [8]. It represents the normalized scattered field as originating from a set of M discrete scattering centres located at points (x_m, y_m), $1 \le m \le M$, *i.e.*

$$\rho(\omega, \varphi_i) = \sum_{m=1}^{M} S_m(\omega, \varphi_i)\exp\left[-2j\frac{\omega}{c_0}(x_m\cos\varphi_i + y_m\sin\varphi_i)\right], \tag{5}$$

where c_0 is the wave propagation velocity. The amplitude $S_m(\omega, \varphi_i)$ is a frequency and angle dependent coefficient, determined by the geometry, composition, and orientation of the corresponding scatterer. For the frequency dependence, GTD predicts [8] that the scattering amplitude is of the form $(j\omega)^{\alpha_m}$ where α_m is an integer multiple of $1/2$. Table 1 shows the relation between different values of the parameter α_m and the geometry of the scatterer [3]. The angle dependence of the magnitude of the scattered field is a much more complicated question. Most man-made targets, containing sharp edges can be characterized by a *sinc*-like aspect dependence of the scattering coefficient. In practice, especially, for the

1. The assumption that the target is a convex body is used at this step. However, a number of publications suggest that the results, obtained on the basis of equation (4) work well even if a target is not convex [9].

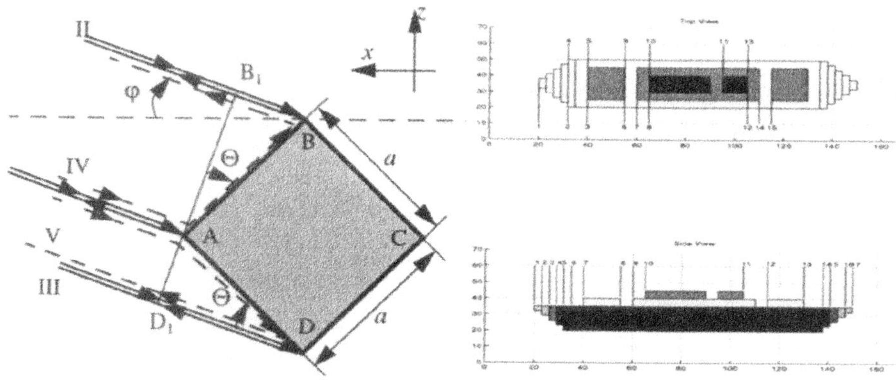

Figure 1. Geometry of the targets: a) rectangular plate, b) scaled ship model

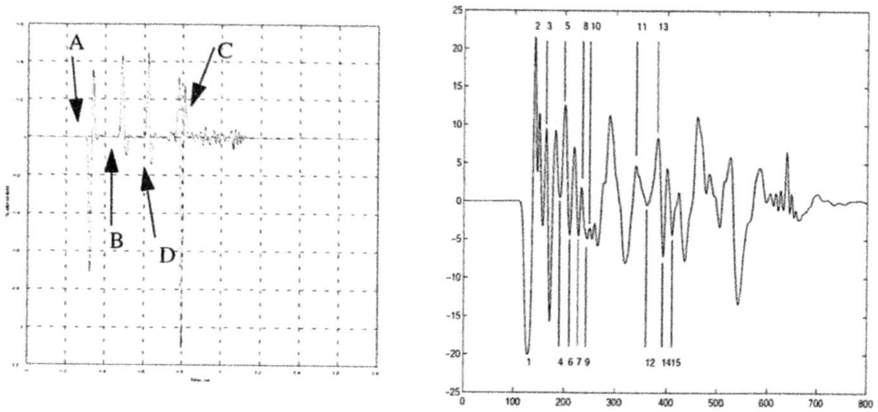

Figure 2. Scattering from a) rectangular plate, 30° aspect angle and b) simplified ship model, 30° aspect angle. Both calculated using FDTD.

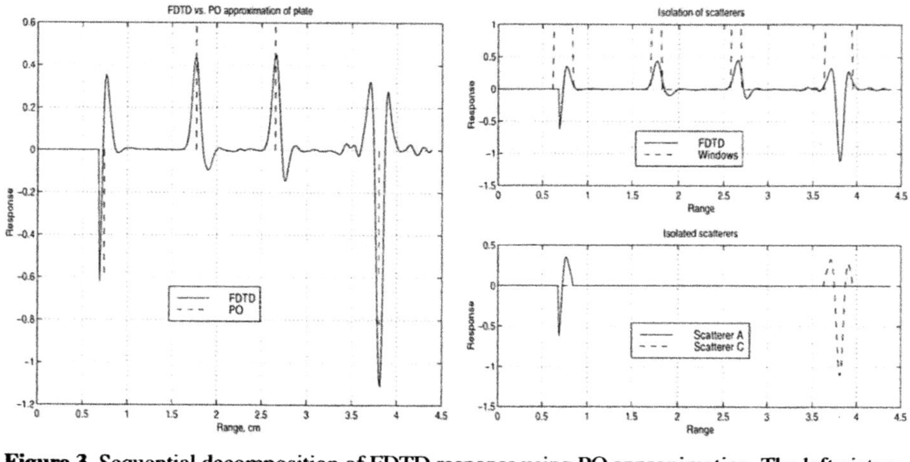

Figure 3. Sequential decomposition of FDTD response using PO approximation. The left picture shows "exact" FDTD response and locations and magnitudes of scattering centres estimated from the PO approximation. The top right figure shows the original signal with rectangular windows centred at the estimated scattering centres. The bottom right figure shows two isolated scattering centres, corresponding to $\alpha_1 = 2$ (edge A) and to $\alpha_4 = 4$ (edge C);

narrow-band of angles SAR, this dependence can be accurately approximated by an exponential, *i.e.* [3]

$$S(\omega, \varphi_i) = A_m(j\omega)^{\alpha_m} e^{\beta\varphi_i}. \tag{6}$$

The image of a target built on the basis of GTD is usually *different* from those obtained using the PO approximation. A single physical scatterer (for example plate corner) corresponds to three separate scatterers on the SAR image.

Table 1. Relation between parameter α and scatterer geometry

Value of α	**Canonical scatterers**
1	Flat plate at broadside; dihedral
1/2	Singly curved surface reflection
0	point scatterer; doubly curved surface reflection, straight edge specular
-1/2	curved edge diffraction
-1	corner diffraction

More problems can be encountered for relatively low frequencies (when the wavelength of the incident wave is larger then a target extent). The Singularity Expansion Method (SEM) [10] predicts that scattered waveforms can be represented as a sum of damped sinusoids in the time domain which results in "non-real" quasi-multiple scattering centres in the SAR image, which are not related to any real discontinuity in the target shape. This significantly complicates object identification.

JOINT PHYSICAL OPTICS - REAL DATA PROCESSING

Equation (4) can be simplified for the case of standard scatterers such as a rectangular plate. Consider a perfectly conducting plate of size $a \times b \times c$, then the scattering response derived from (4) can be simplified to

$$S(\omega, \varphi) = -\frac{16c}{\sqrt{\pi}} \frac{1}{\cos\varphi\sin\varphi}[e^{j(T_a+T_b)\omega} + e^{-j(T_a+T_b)\omega} - e^{j(T_a-T_b)\omega} - e^{-j(T_a-T_b)\omega}] \tag{7}$$

where c_0 is the speed of the wave and we have used the following notations

$$T_a = 2\frac{a\cos\varphi}{c_0},\ T_b = 2\frac{b\sin\varphi}{c_0}. \tag{8}$$

Similar results can be obtained for a circular cylinder of radius R and length L

$$S(\omega) = \frac{2l\omega}{c_0\sqrt{\pi}} R J_1\left(\frac{2R\omega}{c_0}\right) \tag{9}$$

and a sphere of radius R

$$S(\omega) = -2\sqrt{\pi}R\cos\left(\frac{2R}{c_0}\omega\right) + \frac{c_0\sqrt{\pi}}{\omega}\sin\left(\frac{2R}{c_0}\omega\right). \tag{10}$$

It can be seen from (7-10) that this approximation properly predicts the locations of the scattering centres (for example at the edges of a plate and the centre of the sphere) and gives a good estimate for their magnitudes. At the same time it ignores the frequency dependence of the scatterer's response and the low-frequency portion of the response. It can be seen from equation (3) that the polarization of the scattered field yielded by the physical optics

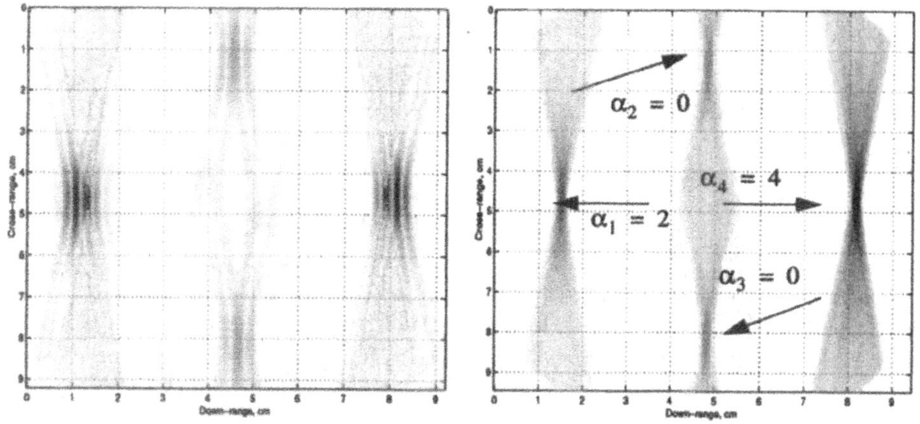

Figure 4. SAR image of the rectangular plate: a) obtained from the original FDTD data, b) obtained from PO approximation with GTD attributes attached.

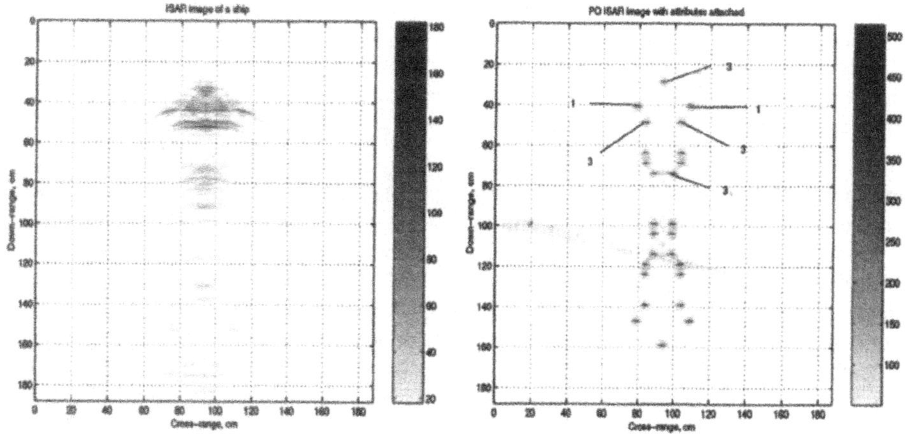

Figure 5. SAR image of a scaled model of a ship: a) obtained from the original FDTD data, b) obtained from PO approximation with GTD attributes attached.

approximation is co-linear with the incident field polarization. Thus, no polarization information can be obtained from the physical optics approximation.

Yet another important property of the PO approximation is the fact that if a body occupying volume V is subdivided into two, occupying volumes V_1 and V_2, such that the intersection of the two volumes is zero, *i.e.* $mes(V_1 \cap V_2) = 0$, according to (4), the normalized frequency response is just the sum of the frequency responses from each part V_1 and V_2 considered separately[1]. This gives us an opportunity to calculate the PO approximation for a complex object by dividing it into a set of simplified objects (for example in a set of rectangular, spheres or cylinders).

USING PO TO LOCATE SCATTERING CENTRES

Description of PO Based Technique

All the properties mentioned above, usually considered as a disadvantage of PO, can be used to obtain more tractable SAR images. As the first step, the main scattering centres are properly predicted by the PO approximation and extracted by any of the spectral estimation methods (Prony's, ESPRIT, Matrix Pencil, *etc.* [11]). The simple character of PO modelled scatterers (since PO does not model dispersive scatterers and resonances, see (7-10) for example) improves the performance of these spectral estimation methods which are known to be very sensitive to the number of poles to be extracted and their frequency dependence. In fact, a scatterer with $\alpha_m = 2$ and $\alpha_m = 3$ (points A and C in Fig. 2) can be considered as two or three closely located scatterers. At the same time there are two more simple scatterers (points B and D on the same figure). This fact leads to poor conditioned matrices used in estimation methods [11] which yields inaccurate results. This is not the case in PO approximation; thus the results are more numerically stable. The information extracted from PO data gives the position T_i and strength A_i of the i-th scatterer. A narrow time-domain window $w_{\alpha_m}(t, T_i)$ can be used to isolate a single scatterer.

$$r_w(t) = w_{\alpha_m}(t, T_i) r(t) \tag{11}$$

Here the window $w_{\alpha_m}(t, T_i)$ is assumed to be centred at $t = T_i$ and be $2\alpha_m$ times wider than the width of the excitation to obtain a more realistic response (FDTD, measurements, *etc.*). If α_m is not known *a priori*, the width of the window is iteratively increased until no changes in the estimated parameters are observed. Main steps of this technique are shown in Fig. 3. Using more realistic data allows to improve estimation of scatterer location and magnitude. SAR images obtained by joint PO and realistic data processing are closer to "optical" picture of a target than those obtained from raw realistic data (FDTD, measurements, *etc.*). However, they contain more information than those obtained just from PO.

The information extracted can also be used in identification of unknown targets. Classical approach, suggested in [10], requires estimation of scatterers locations directly from raw data with no *a priori* knowledge of their location nor of their dispersive nature. Such algorithms are very sensitive to noise, especially if the number of scatterers is not chosen properly. In addition, the dispersive nature of many man-made scatterers also causes difficulties in using such algorithms. Assuming that an unknown target is one of those included in a database one can try to estimate the corresponding scattering locations and the parameters α_m for each of them and compare the results to those stored in the database. It is

---

1. Once again a convexity of the target has to be assumed. However, if a target does not contain "strongly" non-convex parts, such as cavities, then convexity is not a very important assumption.

worthy to note that this can be done for multiple aspect measurements (since they all belong to the same target) and thus can be used to improve the performance of discrimination schemes. A similar situation exists if a SAR image is built from raw data. In many cases it can be overshadowed by resonances. If some assumptions are made about the target, a scatterer's location and strength can be estimated from the raw data, which will result in much cleaner SAR image. Let us emphasize again that if the right target is assumed, these estimates will be very consistent and will result in highly localized image. On the contrary if a wrong guess has been made the image which will be obtained will be very poor.

Examples: PEC Plate and Coarse Model of a Ship

The procedure suggested above was applied to the scattering response of a rectangular plate at the aspect angle $\varphi_i = 30^o$ (see Fig. 2). It can be seen from this figure that the scattering from the corners B and D look like simple reflections while the scattering from the corner A looks like the derivative of the incident pulse and the scattering from the corner C looks like the second derivative of the incident pulse. Application of Prony's method shows a different number of scattering centres which can be extracted from the windowed data. The corresponding results are shown in Table 2.

A number of SAR images, obtained using the FDTD code [12] and joint physical optics real data preprocessing are shown in Fig. 4-5. As was expected, this kind of preprocessing emphasizes the presence of the point scatterers at the corners of the plate and suppresses the oscillating nature of the scatterers. This can be clearly seen by comparing the corresponding pictures. At the same time, the ISAR image with preprocessing differs from those, obtained using physical optics approximation, involving real, unequal magnitude of the different scattering centres (see Table 2). This can be used in the identification problem, especially if some information about the properties of each scatterer is incorporated. For example, Moses [3] suggests to label each spot, corresponding to strong scatterers in accordance to Table 1. This is also shown in the Fig. 4-5 and gives more information about the object under investigation. Let us also note that the information contained in the low frequency (oscillatory) response can be ignored as the result of such preprocessing and "non-real" scattering centres can be avoided.

Table 2. Scattering centres information for a rectangular plate

Corner	Number of centres	Location, ns FDTD code	Predicted (PO) Location, ns	Normalized Amplitudes FDTD code
A	2	0.32, 0.34	0.33	0.65, -0.29
B	1	0.49	0.5	-0.32
C	3	0.78, 0.80, 0.82	0.81	-0.28, 1, -0.24
D	1	0.61	0.63	-0.34

CONCLUSIONS

A simple preprocessing algorithm of SAR data has been suggested. As the first step the information about the location of important non-dispersive scatterers predicted by the Physical Optics approximation is extracted and the corresponding SAR image is built. This image in many cases is similar to the "optical" image of the target and can be useful in target identification and training of human operators. As the next step more information can be extracted from more accurate approximations of the real target response, such as the GTD approximation, FDTD or other numerical simulation. This additional information describes a shape of a specific scatterer predicted by PO. Extraction of the magnitude of the scatterers and their dispersive properties by conventional spectral estimation techniques can be simplified if time localization is used. The resonant behaviour of the target response can also be ignored which results in more clean and recognizable SAR images.

REFERENCES

1. H.J. Li and S.H. Yang, "Using Range Profiles as Feature Vectors to Identify Aerospace Objects," *IEEE Tran. Antennas and Propagation*, vol. 41, No.3, pp 582-588, March, 1993, pp. 261-268.
2. S. Primak, J. LoVetri, S. Kashyap, "A Comparative Study of UWB Radar Discrimination Schemes in a Noisy Background", presented on *1997 IEEE-AP International Symposium*, Montreal, Canada, July, 1997.
3. L. C. Porter, and R.Moses, "Attributed Scattering Centers for SAR ATR", *IEEE Transactions on Image Processing*, Vol. 6, No. 1, January, 1997, pp. 79-91.
4. C. Jakowatz, D. Wahl, P. Eichel, D. Ghiglia, and P. Thomson, *Spotlight-Mode Synthetic Aperture Radar: A Signal Processing Approach*, Kluwer, Boston, 1996.
5. S. Primak, J. LoVetri, Z. Damjanschitz, and S. Kashyap, "The E-pulse Technique for Dispersive Scatterers," in *Ultra-wideband, Short Pulse Electromagnetics* III, 1996, pp 327-334.
6. D.C. Munson, and R.L. Visentin, "A Signal processing View of Strip-Mapping Synthetic Aperture Radar", *IEEE Transaction on Acoustics, Speech and Signal Processing*, Vol. 37, No. 12, December, 1989, pp. 2131-2147.
7. N. Bojarski, "A Survey of the Physical Optics Inverse Scattering Identity", *IEEE Transactions on Antennas and Propagation*, Vol. AP-30, No. 5, September, 1982, pp. 980-989.
8. C. Balanis, *Advanced Engineering Electromagnetics*, John Wiley and Sons, New York, 1989.
9. E. Rothwell, K.M. Chen, D.P. Nyquist and J.E. Ross, "Time-Domain Imaging of Airborne Targets Using Ultra-Wideband or Short-Pulse Radar," *IEEE Transactions on Antennas and Propagation*, Vol. 43, No. 3, March 1995, pp.327-329.
10. C. Baum, E. Rothwell, K.-M. Chen, and D. Nyquist, "The Singularity Expansion Method and Its Applications to Target Identification", *Proc. IEEE*, Vol. 79, No. 10, October, 1991.
11. L. L. Sharf, *Statistical Signal Processing: Detection, Estimation, and Time Series Analysis*, Addison-Wesley, New York, 1991.
12. D. Mardare, J. LoVetri and R. Siushansian, *3-D EMFDTD User's Manual*, Developed at the Department of Electrical Engineering, Faculty of Engineering Science, The University of Western Ontario, 1996.

TOWARDS A THEORY OF SUBMARINE MAST DETECTION

S. Primak,[1] J. LoVetri,[1] and S. Kashyap[2]

[1]Department of Electrical Engineering
The University of Western Ontario,
London, Ontario, Canada N6A 5B9
[2]Department of National Defence
Defence Research Establishment Ottawa,
3701 Carling Ave., Ottawa, Ontario, Canada K1A 0K2

INTRODUCTION

Complex radar targets are often modelled as a number of individual scattering elements randomly distributed throughout the spatial region containing the targets. A common assumption in this case requires independent scatterers with a uniformly distributed phase [1]. Asymptotically, this leads to a Gaussian distribution of the scattered signal or a Rayleigh distributed Radar Cross Section (RCS). However, a class of problems exists where the target must be modelled as a random scatterer and the number of scatterers is not large enough to safely apply the Large Numbers Theorem. One such problem is the detection of submarine masts in a clutter. Here we consider an alternative approach to describe the statistics of the RCS based on deterministic target response modelling and a statistical description of the target movement.

CONVENTIONAL MODEL OF AN EXTENDED TARGET

A common model for complex or extended radar targets is to consider them as a collection of randomly distributed scattering elements [1]. Each element making up the extended target is assumed to be a point isotropic scatterer (*i.e.* a scattering centre). The general statistics of the resulting radar cross-section is very difficult to derive. A physical interpretation of the scattering in the very high frequency region suggests that a deterministic target consists of a relatively small number of scattering centres (bright points) [2, 3]. In one of the models considered in [1], a radar resolution cell is assumed to contain a collection of n elemental points randomly distributed throughout the resolution cell with each scatterer position distributed independently of the position of the other scatterers. This distribution was assumed to be uniform on the surface of the scattering object. Each backscattered electric field component E_j from the j-th scatterer has a constant amplitude A_j proportional to the size or reflection strength of the j-th scatterer and a random phase ϕ_j

Ultra-Wideband, Short-Pulse Electromagnetics 4
Edited by Heyman *et al.*, Kluwer Academic / Plenum Publishers, New York, 1999.

uniformly distributed over the interval $[0, 2\pi]$. The superposition of the radar responses from each of the n elementary scatterers gives rise to the total backscatter electric field from the resolution cell as

$$M_n = \sum_{j=1}^{n} E_j = \sum_{j=1}^{n} A_j e^{j\phi_j}. \tag{1}$$

The overall intensity, S_n, measured by a radar (RCS) is proportional to the square of the magnitude of M_n, *i.e.* it is given by

$$S_n = \left| \sum_{j=1}^{n} A_j e^{j\phi_j} \right|^2. \tag{2}$$

Exact expressions for the Probability Density Function (PDF) of the RCS of one and two point scatterers were obtained in [1] as

$$p_1(s_1) = \delta(s_1 - A_1^2), \tag{3}$$

$$p_2(s_2) = \begin{cases} \dfrac{1}{\sqrt{((\sqrt{s_2} + A_2)^2 - A_1^2)(A_1^2 - (\sqrt{s_2} - A_2)^2)}} & \text{if } (\sqrt{s_2} - A_2)^2 < A_1^2 < (\sqrt{s_2} + A_2)^2 \\ 0 & \text{elsewhere} \end{cases} \tag{4}$$

Even more complicated results can be obtained in the case of three scatterers [1]. If the number of scatterers exceeds four, the exact solution is impossible to obtain, nevertheless, a very good approximation can be obtained through the Gram-Chalier expansion [1].

Unfortunately, assumptions of an independent scatterer with constant intensity and uniformly distributed phase has no real *physical* background, in contrast to communication applications, where such a representation was found to be realistic [4]. We will consider in detail why the aforementioned schemes cannot be applied to our specific problem, that is mast detection, as was formulated in [5, 6].

REALISTIC MODEL OF EXTENDED TARGETS

One of the possible applications of ultrawide band radar is the detection of a submarine by the detection of its masts (periscopes, communications masts, *etc.*), usually located above the sea surface, as shown in Fig. 1. The RCS of this target is a *random* function, because it depends on the sea state, which induces a rocking motion and uncertainty in the submarine location (depth). In addition, the returned signal may be embedded in a strong sea clutter. Due to these factors, detection of a submarine is a problem of detecting a random signal in clutter. The general solution is too complicated to handle analytically. However, some simplification can be made in order to reduce the complexity of the problem to a manageable size. The following are the main issues: a) determining a statistical model of RCS; b) a statistical model of the clutter; and c) finding detection algorithms. We will concentrate our attention on the first problem only. Also, we confine ourselves to the following simplified model of a submarine: we will consider only one mast and assume that it can be modelled as a cylindrical rod. In addition, only a rocking type motion, induced by the sea will be considered.

It was found, that the most suitable analytical technique for describing the high-frequency scattering from a man-made target is the so called Geometrical Theory of

Diffraction (GTD) [7]. In the frame of this theory, a target can be represented as a set of point scatterers, with frequency response $F(\omega, \varphi)$ given as

$$F(\omega, \varphi) = A(j\omega)^{\alpha} e^{|\beta\varphi|}. \tag{5}$$

Here ω is the operating frequency of the radar, φ is the aspect angle of the scatterer with respect to the incident wave, and α is a half-integer parameter (*i.e.* $\alpha = 0, \pm 0.5, \pm 1, \pm 1.5, \ldots$), representing different types of scattering mechanisms (see [3, 7] for details). Equation (5) already shows that elementary scatterers cannot be considered as isotropic. On the other hand, GTD predicts that scattering centres must appear where a conducting body has a singularity like a sharp edge. This contradicts the assumption that scatterers are uniformly distributed on the surface of a body. Finally, the assumption that all partial sources are independent is also contradictory. In fact, the roughest model of a submarine, shown in Fig. 2, requires that the relation between the amplitude and the phase of both the real and the imaginary (*i.e.* mirrored) scattering centres are deterministically related, *i.e.* just the opposite to the standard assumption.

The previous considerations show that a different approach should be taken in order to describe the statistical properties of the RCS. Such a model should involve a deterministic analysis of the target, producing a representation of a target's RCS as a response from a set of point scatterers. This problem, at least for very high frequencies, can be solved using either the Physical Optics (PO) approximation [7] or GTD, mentioned above. Yet another option is the numerical simulation using any of a variety of methods, such as FDTD, Finite Elements, *etc.* In the second step, randomness has to be introduced through a statistical description of the position of the target under consideration. A number of models for the sea surface can be of great help. For the sake of simplicity we will use the simplest model which results in the assumption that the deflection angle γ, defined as shown in Fig. 2 is considered as a uniformly distributed random variable in the interval $[\gamma_{min}, \gamma_{max}] \in [-\pi/2, \pi/2]$. Further refinements may be made in future work.

The difference between our models and conventional ones can be shown immediately on the simplest case of a single scatterer. In this case, its RCS can be expressed according to (5) as

$$S(\omega, \varphi) = |F(\omega, \varphi)|^2 = |A(j\omega)^{\alpha} e^{|\beta\varphi|}|^2 = A^2\omega^{2\alpha}\exp(2|\beta\varphi|). \tag{6}$$

Now if the aspect angle φ is a random variable (due to the rocking motion of the boat), and is uniformly distributed on the interval $[-\Delta\varphi, \Delta\varphi]$, *i.e.*

$$p_{\varphi}(\varphi) = \begin{cases} \dfrac{1}{2\Delta\varphi} & \varphi \in [-\Delta\varphi, \Delta\varphi] \\ 0 & \varphi \notin [-\Delta\varphi, \Delta\varphi] \end{cases} \tag{7}$$

then the distribution of RCS is given by

$$p_{RCS}(S) = \begin{cases} \dfrac{1}{2S\exp(\beta\Delta\varphi)A^2\omega^{2\alpha}} & S \in [\exp(\beta\Delta\varphi)A^2\omega^{2\alpha}, A^2\omega^{2\alpha}] \\ 0 & S \notin [\exp(\beta\Delta\varphi)A^2\omega^{2\alpha}, A^2\omega^{2\alpha}] \end{cases} \tag{8}$$

which is quite different from that given by (3).

Let us now turn our attention to the case of two *independent* scatterers, each of them represented by the model given by (5),

$$F(\omega, \varphi) = A_1(j\omega)^{\alpha_1} e^{|\beta_1\varphi_1|} + A_2(j\omega)^{\alpha_2} e^{|\beta_2\varphi_2|} \tag{9}$$

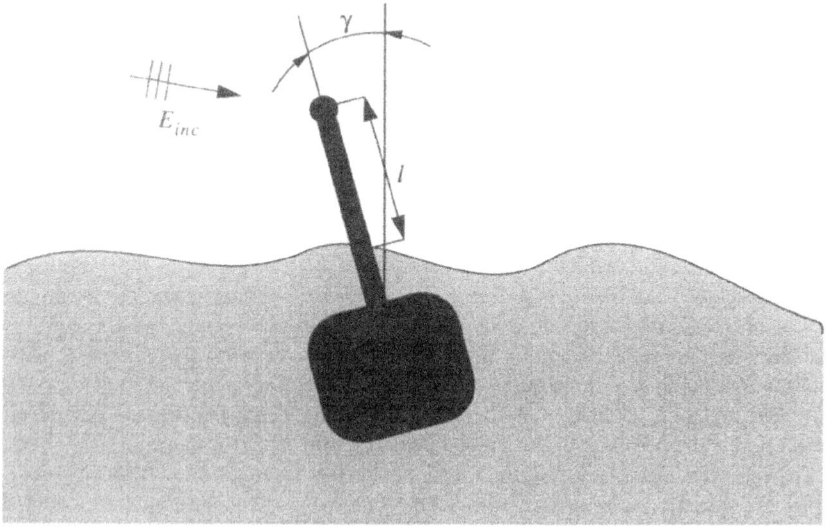

Figure 1. Typical set-up for submarine detection.

and thus

$$\begin{aligned} S(\omega, \varphi) &= \left|A_1(j\omega)^{\alpha_1} e^{|\beta_1\varphi_1|} + A_2(j\omega)^{\alpha_2} e^{|\beta_2\varphi_2|}\right|^2 = \\ & A_1^2\omega^{2\alpha_1}\exp(2|\beta_1\varphi_1|) + A_2^2\omega^{2\alpha_2}\exp(2|\beta_2\varphi_2|) + \\ & 2A_1A_2\omega^{\alpha_1+\alpha_2}\exp(|\beta_1\varphi_1| + |\beta_2\varphi_2|)Re\{j^{\alpha_1+\alpha_2}\} \end{aligned} \tag{10}$$

The complete analytical investigation of this case is rather impossible, however numerical simulations for a particular case can be easily implemented.

Finally we will consider a very special case of the geometry: when two scatterers are identical, but not independent. This setup has a very close relation to the problem of mast RCS statistics. In fact, the strongest contribution to the backscattered fields are from the edge of a mast and from its mirror image as shown in Fig. 2. This also means that the two scatterers are dependent and

$$\begin{aligned} S_2(\omega, \varphi) &= \left|A_1(j\omega)^{\alpha_1} e^{|\beta_1\varphi|} + \rho(\omega, P)A_1(j\omega)^{\alpha_1} e^{|\beta_1\varphi|}\right|^2 = \\ & A_1^2\omega^{2\alpha_1}|1 + \rho(\omega, P)|^2 e^{|\beta_1\varphi|} = A^2 e^{|\beta_1\varphi|} \end{aligned} \tag{11}$$

Here the reflection coefficient $\rho(\omega, P)$ is included to describe the conducting sea surface and the polarization P of the incident wave. The last equation indicates that, for this special case, the distribution of the RCS should follow the shape of the distribution derived for a single scatterer, however, its deterministic dependence on frequency is, of course, different.

Further refinement can be achieved by taking into account that the rocking motion of a boat can be well approximated by the following differential equation excited by a White Gaussian Noise (WGN) $\xi(t)$ [8]:

$$\ddot{\varphi} + \varepsilon(1 + f(\varphi))\dot{\varphi} + \omega_0^2\varphi = K\xi(t) \tag{12}$$

Here, $-\pi/2 < \varphi < \pi/2$ is the angular deviation of a boat's position from the vertical position, K is the intensity of the noise, ω_0 is the average frequency of fluctuation of the sea

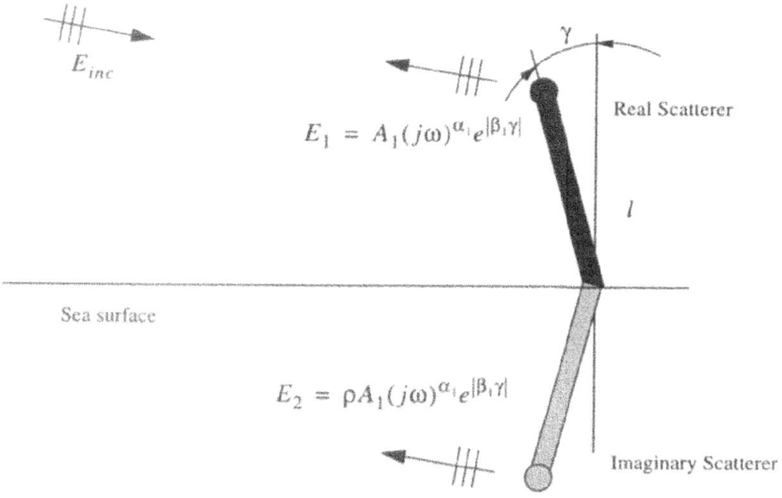

Figure 2. Rough model of a submarine mast.

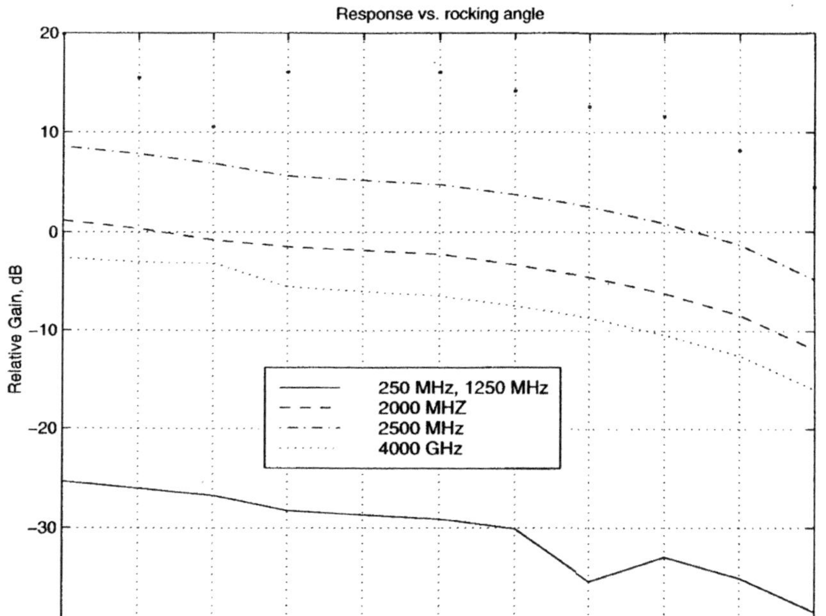

Figure 3. Dependence of the scattering response on the rocking angle. Frequency is a parameter.

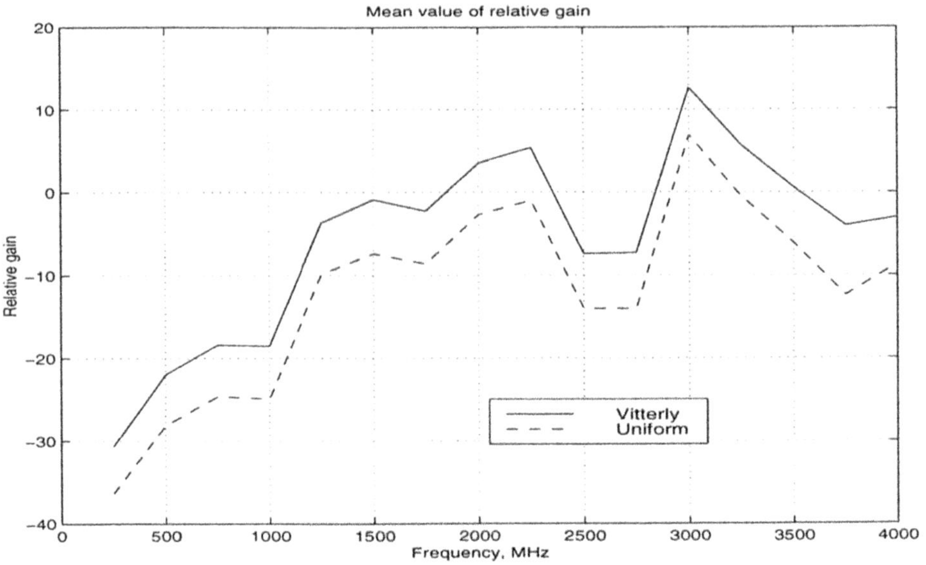

Figure 4. Mean value of the response for a uniformly distributed phase on the interval $[-10°, 10°]$ and for the Vitterbi distribution given by (13).

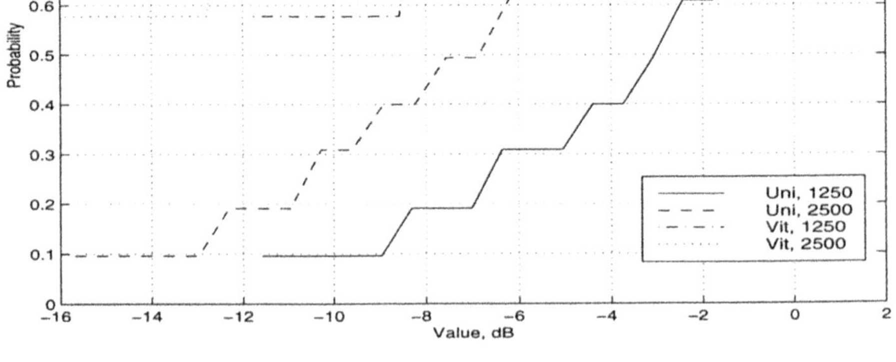

Figure 5. Cumulative distributions for two frequencies and uniform and Vitterby distributions.

surface, ε is a small parameter, and $f(x)$ is a non-linear function. Any solution of equation (12) is an almost periodic function of frequency ω_0 with a distribution of instantaneous values depending on the form of the non-linearity of $f(x)$. One of the possible distributicns for $\varphi(t)$, predicted by (12), is the so called Vitterbi distribution [9], given as

$$p_{\varphi}(\varphi) = \frac{\exp(D\cos\varphi)}{2\pi I_0(D)}, \; -\frac{\pi}{2} < \varphi < \frac{\pi}{2} \tag{13}$$

where the parameter D is proportional to ε. For a very small ε this distribution is almost Gaussian, and converges to a uniform one when ε approaches unity. The model implied by (13) has a more physical meaning than just a uniform distribution: the probability of large deviations from the vertical position is small, *i.e.* the angle distribution cannot be a uniform one. Combining (13) and (11) one can obtain the distribution of the RCS as

$$p_{RCS}(S) = \begin{cases} \dfrac{\exp\left(D\cos\ln\dfrac{S}{A^2}\right)}{4\pi I_0(D) S \exp\left(\dfrac{\beta\pi}{2}\right) A^2\omega^{2\alpha}} & S \in \left[\exp\left(\dfrac{\beta\pi}{2}\right)A^2\omega^{2\alpha},\, A^2\omega^{2\alpha}\right] \\ 0 & S \notin \left[\exp\left(\dfrac{\beta\pi}{2}\right)A^2\omega^{2\alpha},\, A^2\omega^{2\alpha}\right] \end{cases} \tag{14}$$

The models described here are very simple ones, but being analytical models they allow us to properly approximate the behaviour of the RCS of seaborne objects. More accurate approximations can only be performed numerically as described in the following section.

PROCESSING OF REAL DATA

Realistic data are much more complex than that represented by a simple or double scatterer. However similar methods can be applied to describe a target's behaviour. Two different models can be considered: frequency domain and time domain. The first can be applied to all kinds of narrowband radar, including stepped frequency and chirp waveform radars. The latter has attracted more attention recently due to the intensive development of ultra-wideband radar radiating very short pulses instead of synthesizing the time domain responses from a number of single frequency measurements.

Both methods can be implemented in the very same way. Let $R(\omega, \gamma)$ and $r(t, \gamma)$ be the frequency and time domain responses of the target under test as a function of the "rocking" angle γ (see Fig. 1). This angle itself has a certain distribution $p_{\gamma}(\varphi)$, for example given by (7) or (13). Then for every single frequency ω_j and frequency response of the target $R(\omega, \gamma)$ the corresponding histograms for phase and magnitude can be estimated as

$$p_{FM}(\rho, \omega_j) = Prob\{|R(\omega_j)| < \rho\} \tag{15}$$

$$p_{FPh}(\Phi, \omega_j) = Prob\{\angle(R(\omega_j) < \Phi|)\}. \tag{16}$$

The mean values of the magnitude and phase, as well as confidence intervals for every frequency can then be calculated from the empirical distributions (15-16) and used in detection algorithms. For the case of time domain data, the only difference is that time samples must be processed instead of frequency samples. Corresponding results are shown in Fig.'s 3-5.

CONCLUSIONS AND FUTURE WORK

This paper addresses the first steps of the complex problem of detection of fluctuating targets such as submarine masts; we have found that the models are very different from those suggested in the literature. The next step in our research will be the development of detection algorithms based on these analytical and empirical models.

Another opportunity to improve the detection algorithm is to incorporate model based approximations of targets. It was stated that the time response at any angle should contain the dominant scattering from "bright points" suggested in the previous section--a property totally ignored by most algorithms. One possible and powerful tool will be time-frequency analysis. Instead of looking at the distribution of the signal itself it is possible to look at the distribution of different time-frequency components. The data processing will be very similar to that suggested for frequency domain data. A possible advantage may be the better localization of energy in the time-frequency domain than in the frequency domain alone.

REFERENCES

1. J. Daba, and M. Bell, "Statistics of Scattering Cross-section of a Small Number of Random Scatterers", *IEEE Transaction on Antennas and Propagation*, Vol. 43, No. 8, August, 1995, pp. 773-783.

2. D. Moffat, J, Young, A. Ksienski, H.-C. Lin, and C. Rhoads, "Transient Response Characteristics in Identification and Imaging", *IEEE Transaction on Antennas and Propagation*, Vol. 29, No. 2, March, 1981, pp. 192-204.

3. L. C. Porter, and R.Moses, "Attributed Scattering Centers for SAR ATR", *IEEE Transactions on Image Processing*, Vol. 6, No. 1, January, 1997, pp. 79-91.

4. D. Middleton, *An introduction to statistical communication theory*, McGraw-Hill, New York, 1960.

5. S. Kashyap, A. Louie, and A. Drosopoulos, "RCS and Radar Detection of Submarine Masts", *IEEE-URSI AP Conference*, 1997, p. 675.

6. I. Zabel, and P. Phu, "UHF Periscope Detection and Discrimination", *IEEE-URSI AP Conference*, 1997, p. 674.

7. C. Balanis, *Advanced Engineering Electromagnetics*, John Wiley and Sons, New York, 1989.

8. T. Ozaki, "Statistical Identification of Nonlinear Random Vibration Systems," *Journal of Applied Mathematics*, Vol. 56, No. 3, March, 1985, pp. 186-191.

9. W. Lindsey, *Synchronization systems in Communications and Control*, Englewood Cliffs, N.J.: Prentice-Hall, 1972.

10. S. Primak, *Modelling, Synthesis and Numerical Simulation of Non-Gaussian Random Processes in Application to Communications*, Ph.D. Thesis, BGU, Beer-Sheva, Israel, 1996.

RADIO FREQUENCY INTERFERENCES SUPPRESSION FOR NOISY ULTRA WIDE BAND SAR MEASUREMENTS

Bruno Juhel[1], Georges Vezzosi[1], Marc Le Goff[2]

[1] LTSI (Laboratoire de Traitement de Signal et de l'Image)
Université de Rennes 1 Campus de Beaulieu
35042 Rennes cédex FRANCE
Email : Bruno.Juhel@univ-rennes1.fr
[2] CELAR (DGA) Section GEOS/SDM 35170 Bruz FRANCE

INTRODUCTION

A radar image is representative of the electromagnetic reflectivity of terrain surfaces and targets. In widespread use throughout both civil and military observation systems, narrow band SAR (Synthetic Aperture Radar) imagery based on Chirp signals is limited (mainly in terms of resolution). Thanks to the broad instantaneous spectrum of UWB signals (very short pulses of about a nanosecond without carrier), the UWB radar may gain in resolution and penetrate the natural screens as the low frequency radars. Such systems combine the penetration enhancement associated with UHF frequencies with the resolution of wide bandwidth. Today, interest in UWB radar is growing. Measurements became possible due to fast response equipment which could generate pulses of nanosecond duration. It is accompagnied by the appearance of new applications, such as ground penetrating radar, construction control, pipe and cable detection. We investigate the possibilities of Ultra Wide Band radar imaging in order to scan the advantages over conventionnal radar. One of the potential advantage is the accurate target discrimination because of the short pulse duration, but one problem that we encounter is the radio frequency interferences (RFI) present in the UWB return.

Analyses from the results of the processing that emerged from the initial UWB trials, carried out at CELAR (French Technical Center for Armament Electronics) using a prototype radar, show that electromagnetic noise level is very high because the required bandwidth share some RF bands occupied by other communication systems (0,1 GHz - 1 GHz). The

Ultra-Wideband, Short-Pulse Electromagnetics 4
Edited by Heyman *et al.*, Kluwer Academic / Plenum Publishers, New York, 1999.

advance of telecommunications technology creates an increasingly hostile environment for radio frequency measurements. The RFI appear in the UWB signal. They can obscure the target response and make detection difficult. If we want a good Signal to Noise Ratio (SNR) on SAR image, RFI must be extracted and removed. An algorithm to remove RFI, based on Least Mean Square method, is then tested.

PRELIMINARY UWB SAR EXPERIMENTATION

UWB SAR imagery processing

The geometric configuration of a SAR mounted on a vehicle travelling at speed v is shown in figure 1. The short pulse, with its wide instantaneous spectrum, is transmitted using UWB antenna and then backscattered by the elementary target reflectivity *sigma0*.

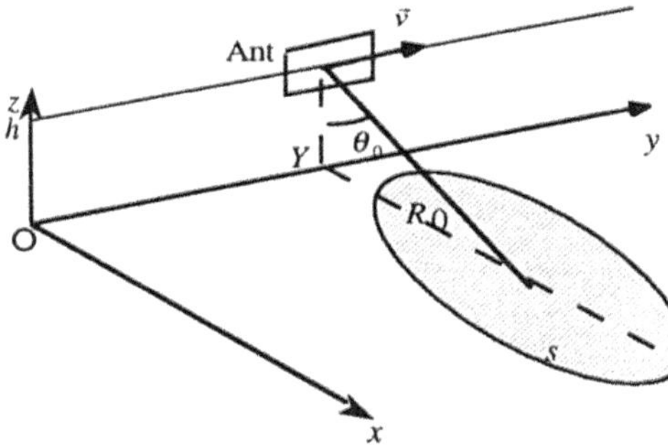

Figure 1. Geometric configuration of a SAR

The algorithm developed[1] for data processing is based on focusing by time domain back projection. The method involves the coherent summation of sampled pulse responses of targets and terrain measured over the angular sector corresponding to radar displacement. Unlike the narrow band radar, UWB responses termed as pulse result directly from the measurement. The coherent summation can be summarised by the following formula :

$$I(x_i, y_i) = \sum_{k=1}^{N} R_x\left(d(x_i, y_i), p_k\right). \qquad (1)$$

Image computation is carried out pixel by pixel. For each pixel $I(x_i, y_i)$, the algorithm identifies, for each pulse response R_x, and using the real distance $d(x_i,y_i)$ between the antenna and the pixel under consideration, the corresponding range bin. As this point is most unlikely to correspond exactly with a measured range bin, its value is calculated by linear interpolation between the values of the two nearest bins. A value for each pixel is obtained by summing these values obtained from all the pulse responses which made up the measurement and which correspond to the N various aperture positions of the radar p_k (hyperpola of range migration). The following options can be added for a more accurate focusing : range and azimuth compression, terrain cartography, motion compensation.

UWB SAR experimentation results

A trial programme took place[2] at CELAR in december 1996 with a prototype UWB radar installed on the external STRADI platform. The principle of the measurements was to emit a UWB pulse, for a given position, and to digitize the signal reflected from the scene directly in the time domain with a sequential equivalent-time sampling oscilloscope. This kind of oscilloscope needs as many pulse transmissions as sampling points to obtain the returned signal. The operation was then restarted every 60 cm on the trail, and for various scenes. Only largest targets as trihedral could be detected in the scenes. Other areas were measured, but they were made of clutter and electromagnetic noise.

As a results, RFI appears in the received signal. Thus, interference extraction and removal is needed. But, the sequential equivalent-time sampling leads to average and modify interference characteristics because of the big number of measurements. That's why single shot sampling will be used for the next experimentations and for the following simulated signals.

Interferences are now studied and RFI suppression are developped to increase the probability of detecting while keeping the output power requirement low.

ELECTROMAGNETIC INTERFERENCES

The frequency range of the UWB signal is also used extensively by many radio frequency services : TV, radio, cellular phones, etc, which are relatively narrow-banded, as can be seen on figure 2 where a spectrum of typical RFI is shown in the band from 0 Hz to 1 GHz. The electromagnetic noise was measured at CELAR through an UWB antenna with a spectrum analysor. The envelop shows that peak levels on analysor reach - 11 dBm. High levels are located in FM and TV bands around 100 MHz and from 400 to 700 MHz. Two other bands are detected : 160-200 MHz and 930- 960 MHz.

Next, we present an algorithm developped by Miller, McCorkle and Potter[3] for the removal of narrow band interferences from Ultra Wide Band signal.

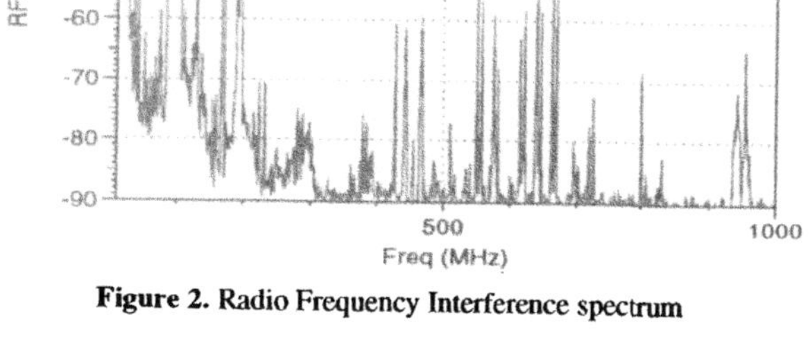

Figure 2. Radio Frequency Interference spectrum

RFI SUPPRESSION ALGORITHM

In this section, the UWB received signal is assumed[3] to be

$$x(k) = s(k) + n(k) + r(k), \quad k = 0,...,N-1, \tag{2}$$

with $s(k)$ the UWB signal, $n(k)$ a white noise, $r(k)$ RFI signals, and N the number of points of the measured signal. If M values of interference frequencies w_i are assumed to be known and theirs levels $\{a_i, b_i\}$ unknown, $r(k)$ is given by

$$r(k) = \sum_{i=1}^{M} a_i \sin(w_i k) + \sum_{i=1}^{M} b_i \cos(w_i k), \quad k = 0,..., N-1. \tag{3}$$

We want to estimate the signal $r(k)$ in a way to minimize the quadratic error criterion

$$\sum_{k=0}^{N-1} |x(k) - r(k)|^2 . \tag{4}$$

In this objective, it gives $\underline{r} = A\underline{z}$ with A a matrix Nx$2M$ containing terms $\sin(w_i k)$ and $\cos(w_i k)$ and z a column of $2M$ lenght containing coefficients a_i and b_i. Least mean squares problem is $A\underline{z} = \underline{x}$, with :

$$A = \begin{bmatrix} \sin(w_1 k_1) & \cos(w_1 k_1) & . \; . \; . & \sin(w_M k_1) & \cos(w_M k_1) \\ . & . & . \; . \; . & . & . \\ . & . & . \; . \; . & . & . \\ . & . & . \; . \; . & . & . \\ \sin(w_1 k_N) & \cos(w_1 k_N) & . \; . \; . & \sin(w_M k_N) & \cos(w_M k_N) \end{bmatrix}, \tag{5}$$

$$\underline{z} = \begin{bmatrix} a_1 \\ b_1 \\ . \\ . \\ a_M \\ b_M \end{bmatrix}, \qquad \underline{x} = \begin{bmatrix} x(k_1) \\ x(k_2) \\ . \\ . \\ x(k_N) \end{bmatrix}. \tag{6}$$

The solution consits of finding $\hat{\underline{z}}$ to minimize

$$J(\underline{z}) = \|\underline{x} - A\underline{z}\|^2 = (\underline{x} - A\underline{z})^t(\underline{x} - A\underline{z}) = \underline{x}^t\underline{x} - 2\underline{x}^t A\underline{z} + \underline{z}^t A^t A\underline{z}. \tag{7}$$

The derivative expression $\frac{\partial J(\underline{z})}{\partial \underline{z}} = -2\underline{x}^t A + 2\underline{z}^t(A^t A) = 0$ leads to $\hat{\underline{z}} = (A^T A)^{-1} A^T \underline{x}$ and the Least Mean Square solution gives

$$\hat{\underline{r}} = A(A^T A)^{-1} A^T \underline{x}. \tag{8}$$

Then, $\hat{\underline{r}}$ is substracted to the signal x. LMS method has only a solution if A^tA is a non-singular matrix. We also must verify $\dim(\underline{z}) < \dim(\underline{x})$ that's to say more equations than unknowns.

RESULTS AND PERFORMANCES

First, ten RFI frequencies corresponding to figure 2 with a random phase and the same level were added to an UWB indoor ISAR measurement of a target. Figure 3 presents the simulated signal with RFI and the signal obtained after RFI removal. This algorithm allows to find the initial UWB signal without distorsion.

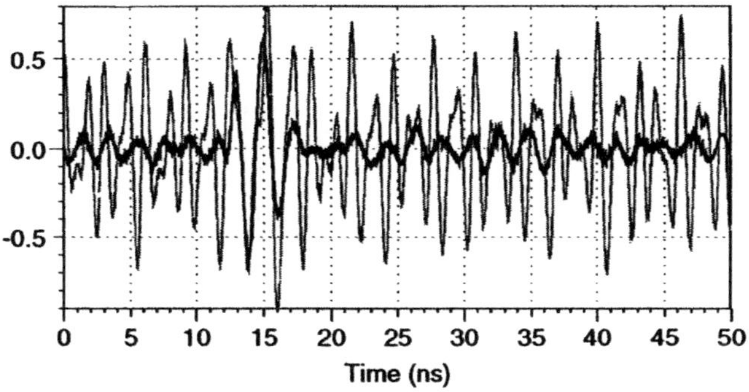

Figure 3. Signals before and after RFI removal

If the RFI signals have not a stationnarity characteristic, the estimation of the frequencies can be biased. Then, some frequencies can not be removed. For example, considering an UWB signal (-18 dB SNR) with 15 RFI frequencies, the figure 4 shows the SNR after RFI suppression when *Nber* frequencies are removed. The example proves the importance to have a good estimation of all the RFI frequencies before removal algorithm processing. The Signal to Noise Ratio of the noisy UWB signal was defined as :

$$SNR = 10\log\left(\frac{\sum_{k=0}^{N-1} s^2(k)}{\sum_{k=0}^{N-1} r^2(k)}\right), \tag{9}$$

and the SNR after RFI suppression was given by :

$$SNR = 10\log\left(\frac{\sum_{k=0}^{N-1} s^2(k)}{\sum_{k=0}^{N-1}\left(x(k)-\hat{r}(k)-s(k)\right)^2}\right). \tag{10}$$

We seek maximum reduction of interferences energy, minimum loss and minimum distorsion. Losses were defined as :

$$LOSS = 10\log\left(\frac{\sum_{k=0}^{N-1} s^2(k)}{\sum_{k=0}^{N-1}\left(x(k)-\hat{r}(k)\right)^2}\right). \tag{11}$$

Losses for the maximum of the signal are represented on figure 4. We can conclude that the losses increase with the number of removed frequencies. It may be a limitation of this processing.

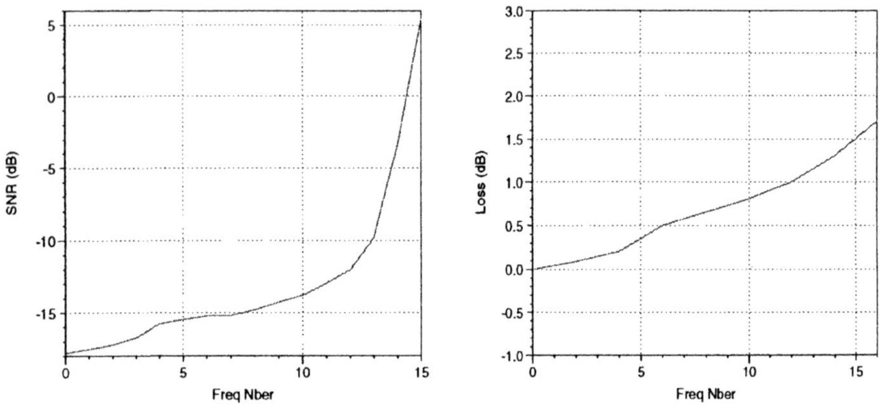

Figure 4. SNR and loss after RFI removal

Then, the figures 5 and 6 represent simulated images of a target (small missile) before and after RFI suppression and show the capability to detect the target after RFI removal.

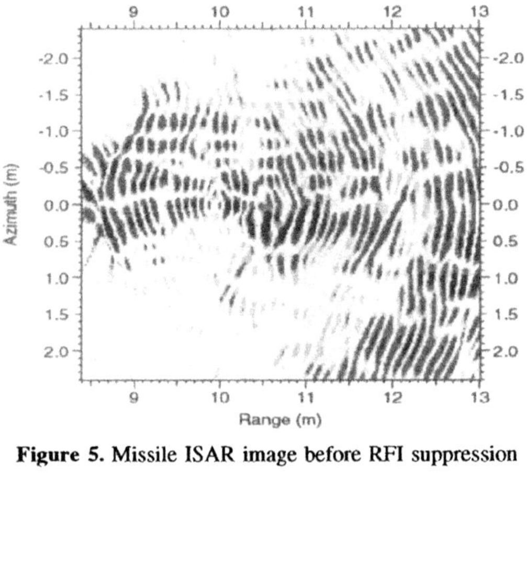

Figure 5. Missile ISAR image before RFI suppression

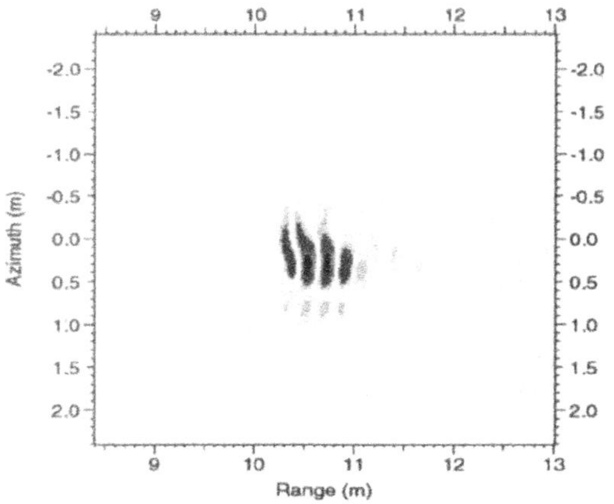

Figure 6. Missile ISAR image after RFI suppression

CONCLUSION

In this document, back projection algorithm dedicated to Ultra Wide Band signals is described. The time domain aspect is all the more important that the transient nature of UWB signal supplies a vast amount of information on the electromagnetic scattering mechanisms. But, during UWB measurements, radio frequency interferences can appear and obscure target detection. We have described how RFI can be an important noise source in UWB system and how they can be suppressed. The Least Mean Square method for extracting RFI give good results, which are quantified with simulated data. Next, we would like to improve our sinusoidal model of the RFI to include a more complete description of the RFI signal and to apply the LMS algorithm to measured data.

REFERENCES

1. B. Juhel, E. Legros, M. Le Goff "Algorithms and experimental results in UWB SAR imaging" Oct. 1997, *IEE Int. Conf. Radar 97*, Edinburgh
2. M. Le Goff, B. Juhel Rapport d'essais CELAR n°GEOS/97.233/SDM/25262/DR "Mesures SAR Ultra Large Bande sur STRADI" Sept. 1997
3. T. Miller, J. McCorkle, L. Potter "Near-least squares radio frequency interference suppression" *SPIE Proceedings Algorithm for SAR imagery* Vol. 2487

THE PROBLEMS OF SMALL BASE ULTRAWIDEBAND RADAR

Vladimir I. Koshelev, Sergey E. Shipilov, and Vladimir P. Yakubov

High Current Electronics Institute RAS
4, Akademichesky Ave., 634055 Tomsk, Russia

INTRODUCTION

Recently, thank to essential progress in the field of generation and radiation of short ultrawideband (UWB) electromagnetic pulses[1,2,3], an interest to the analysis of ultrawideband radar potential possibility including radar object (RO) recognition has become stronger. Often, solving of the problem of object recognition is directly connected with the possibility to obtain the data concerning its form. Analysis of the published papers has shown that theoretically and practically one manages to reconstruct the RO form only at an essentially large space (angle) base of its observation. In this case, possibility to obtain many aspect angles is realized and the problem is reduced to usage of tomographic methods with deconvolution of Radon conversion. In case of a small-base radar system an application of well-known methods of RO form reconstruction becomes problematic.

Possibility to synthesize an aperture in the process of RO tracking during some time allows to realize, as a rule, a small angle base. In such cases solving of the problem of RO form reconstruction excites a special interest. To realize the aperture synthesizing, an information about the object coordinates is necessary. Possibility to obtain them in case of small-base radar system is discussed in this paper. In case of realization of a small angle base of the RO surveillance, the approaches concerning its form evaluation are suggested.

DETERMINATION OF COORDINATES AND OBJECT VELOCITY

RO position is determined by three coordinates. In order to find them it is suggested to use a system consisting of an ultrawideband signal transmitter and fore spatially separated receivers. Fig.1 presents the scheme of such system in which a transmitter and one of the receivers are combined. Measurements of the pulse delay from the object relative to some definite moment of time are made by means of the given system. The delays measured by means of the receivers being away each other can be used to determine the RO coordinates $\vec{R}$. Delay measurement in each pulse has relative value. It is convenient to take as a count beginning a pulse delay in one of the receivers considered as a main one which is placed at

Ultra-Wideband, Short-Pulse Electromagnetics 4
Edited by Heyman *et al.*, Kluwer Academic / Plenum Publishers, New York, 1999.

the beginning of coordinates in Fig. 1. Calculation of a relative delay τ_i is reduced to determining of the position of convolution maximum of main and given receiver signals.

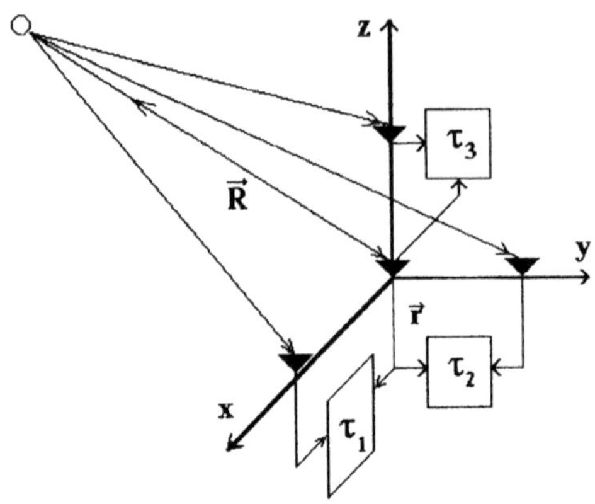

Figure 1. Scheme of radar system for a coordinate determining.

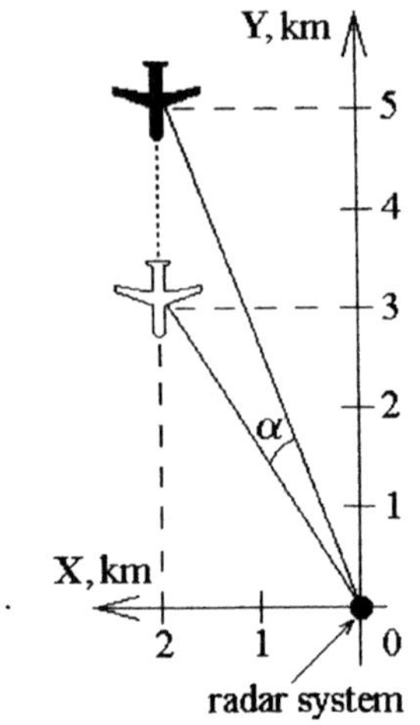

Figure 2. Aperture synthesis.

In case when the receivers have the coordinates (a,0,0), (0,a,0), (0,0,a) relative to the main one, the expression for the components of the vector $\vec{R}$ looks like

$$r_i = \frac{2c\tau_i R - a^2 - (c\tau_i)^2}{2a}.$$

Here, R is the distance to the object under sounding determined through the convolution of signal from the transmitter and receiver; c is the velocity of light. The error of coordinate determination Δr will depend on the error in determing the delay $\Delta\tau$ as $\Delta r = \Delta\tau c R / a$. Application of regularizing factor based on the similarity of temporal forms of registrated pulses of scattered radiation in the given scheme essentially increases the accuracy of RO coordinate determining under noise conditions.

When sounding the object by pulses with the pulse rate f, to determine its velocity at each moment of time, one can use the data of coordinate calculations at the current and previous moments of time:

$$V(t) = \left(\vec{R}(t) - \vec{R}(t - 1/f)\right)f.$$

The obtained value will be an averaged velocity for the time $1/f$ and the error will look like $\Delta V = 2\Delta r f$.

RECOGNITION AT APERTURE SYNTHESIZING

Using a suggested approach for determining the coordinates and velocity of RO, it is possible to realize aperture synthesizing. Angle base α (Fig.2) is realized one possibility to reconstruct the RO form depends on its value. Geometrical object form is the most informative parameter in the problem of recognition and alongside with the data concerning the velocity it increases the possibility of correct attributing the object under sounding to one or another class.

At the UWB sounding the problem of the RO form reconstruction can be solved by tomographic methods. It should be noted that at the limited angle base such methods are not applied directly and their essential modification is required. We have developed an algorithm of the RO form reconstruction based on the well-known Lewis-Boyarsky relation. Given relation has been modified for a case of a small angle sounding base and small number of angle aspects[4]. Numerical simulation has shown a possibility to reconstruct the unshaded object part at the angle surveillance $\alpha=45°$ and noise level ε up to 10% and at $\alpha=10°$ with the noise level less than 1%. The used sounding pulse has a fourfold overlapping in frequencies. The accuracy of the form reconstruction will increase with the frequency band broadening in a sounding signal.

If aperture synthesizing realizes the less angle base and tomographic methods becomes not capable, object form approximation by a set of genetic functions[5] is possible. The selection of a necessary genetic function is realized through a search of a convolution maximum of a current realization of a signal reflected from RO with a set of genetic functions calculated beforehand. At a high spatial resolution of a sounding pulse the given approach allows to evaluate the dimensions of the object and its characteristic feature. As the number of different genetic functions participating in approximation increases, more exact evaluation of the RO form is possible. The simplest genetic function is a mirror signature from the scatterer in the form of a rectangular plate.

To test the given approach, a numerical simulation has been carried out in which the RO form approximation was made by a set of 6 scatterers of this kind. A three-dimensional 10-m long plane model was used as a scattering RO. In the course of aperture synthesizing during 10 seconds with a 2 Hz pulse rate at the RO motion at a constant height Z=1000 m and velocity V_y=200 m/sec (Fig.2) 21 observation aspect angles were realized. Determining of coordinates of each scatterer was made by the previously describe scheme. Obtaining and processing of reflected signal in each of the angle aspects were carried out using a radar system consisting of four receivers and one transmitter of UWB pulses (Fig.1). The direct problem of calculation of the reflected signals was solved using a special algorithm. The scattering body surface was divided into a great number of rectangular areas, the dimensions of which are small as compared to the minimum wavelength in the sounding signal spectrum and to the curvature radius of body surface in the given point. The problem was solved in the approximation of a one-dimensional scattering taking into account object self-shading. Comparison of the calculation results by the given algorithm with the results obtained by means of packaged programs of a scattering calculation at a wire structure[6] has shown a good agreement. The time spent for signature calculation by the suggested algorithm turn out to be two orders of magnitude less as compared to program for wires.

In the course of numerical simulation, the influence of the duration of the sounding pulses τ_p, the distance between the receivers in a radar system a and the noise level ε on the accuracy of the RO approximation has been studied. Figs. 3, 4, 5 and 6 present the numerical simulation results with the bipolar pulse duration τ_p=8 ns and τ_p=2 ns with the distance between the receivers 10 m and 50 m and noise level ε=0%, 5% and 10%. The dimensions of the points in figures are proportional to the ones of the scatterers. In each variant a dispersion of finding the coordinates σ_r (m) and velocity σ_v (m/sec) by all the realization obtained in the process of aperture synthesizing was evaluated.

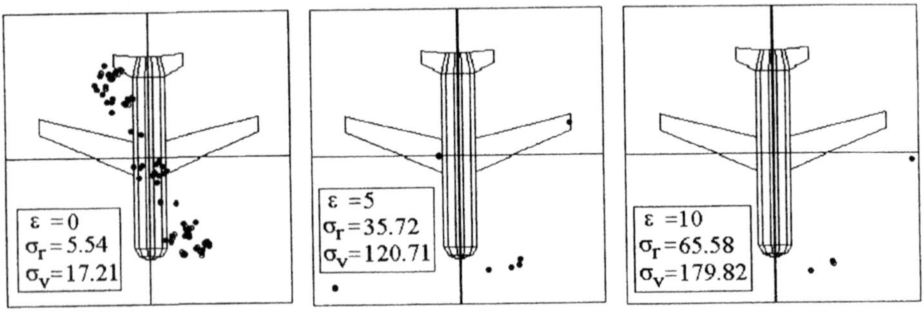

Figure 3. Approximation of the RO form in the process of aperture synthesizing with τ_p =8 ns and a=10 m.

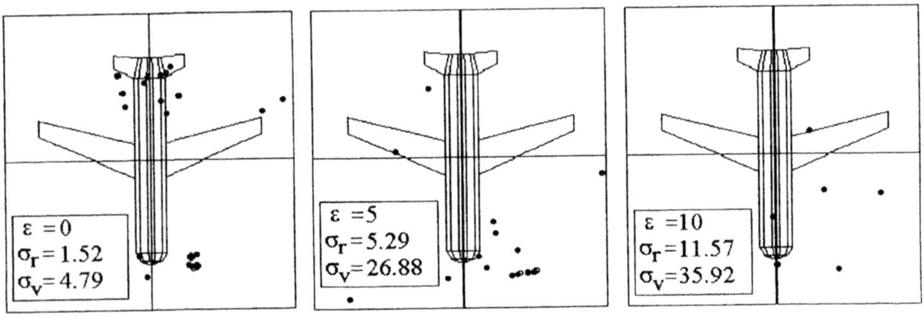

Figure 4. Approximation of the RO form in the process of aperture synthesizing with τ_p =8 ns and a=50 m.

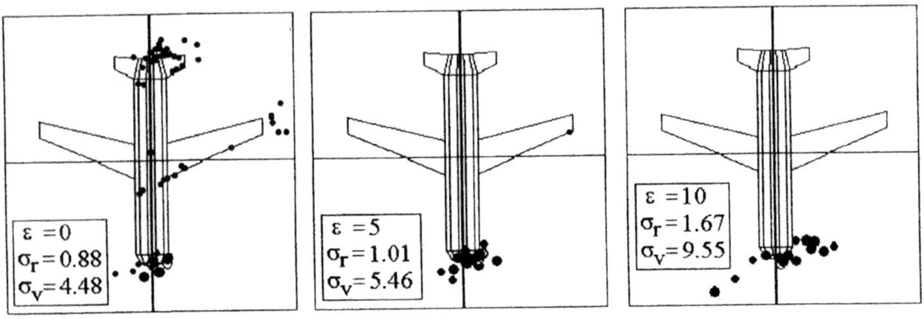

Figure 5. Approximation of the RO form in the process of aperture synthesizing with τ_p =2 ns and a=10 m.

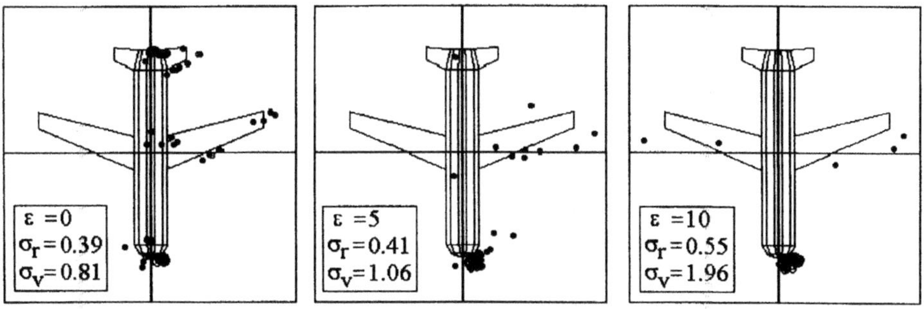

Figure 6. Approximation of the RO form in the process of aperture synthesizing with τ_p =2 ns and a=50 m.

CONCLUSION

Simulation results have shown a perspectiveness of application of approach for the RO form evaluation with usage of genetic functions when aperture synthesizing realizes a small angle base of surveillance. The accuracy of the object form evaluations increase with the decrease of a sounding pulse duration and increase of the distance between the receivers. Application of a suggested scheme of coordinate determination for this aim allowed to obtain the object form evaluations when the receiving system base is 10 m. Influence of measuring noises on the accuracy of approximation and determining of coordinates makes necessary to increase the relation signal/noise, e.g. owing to the increase of the radiated signal power. Besides, the accuracy of the RO form evaluation will increase and processing velocity will decrease when adding new genetic functions. A search of optimum composition of genetic functions is a significant task.

REFERENCES

1. J.A. Oicler, J.R. Grant, and M.H. Herman, Realizing the Potential of Photoconductive Switching for HPM Applications, SPIE Proc. 2557:225 (1995).
2. V.G Shpak, S.A Shunailov, M.R. Ulmaskulov, and M.I. Yalandin, Generation of High-Power Broadband Electromagnetic Pulses with PRF of 100 PPS, Proc. of 10-th IEEE Inter. Pulsed Power Conf. Albuquerque, New-Mexico. 1:666 (1995).
3. V.I. Koshelev, Y.I. Buyanov, B.M. Kovalchuk, Y.A. Andreev, V.P. Belichenko, A.M. Efremov, V.V. Plisko, K.N. Sukhushin, V.A Vizir, and V.B. Zorin, High-Power Ultrawideband Electromagnetic Pulse Radiation, SPIE Proc. 3158:209 (1997).
4. V.I. Koshelev, S.E. Shipilov, and V.P. Yakubov, Object Form Reconstruction at Small Angle Base oı Ultrawideband Radar, Proc. of Conf. on Direct and Inverse Problems of Electromagnetic and Acoustic Wave Theory, Lviv: Pidstryhach Institute of Applied Problems of Mechanics and Mathematicsof the Ukrainan National Academy of Sciences. 47 (1997).
5. V.E.Ginsburg, "Forming and Processing of Image in Real Time", Radio i svjaz, Moscow (1986).
6. V.M. Lugovtsov, Spline-Collocation Method for Numerical Analysis of Nonstationary Electromagnetic Radiation and Scattering of Thin-Wired Structures, *Radiotekh. Electron.* 38:1559 (1993).

TIME-FREQUENCY ANALYSIS AND NOISE SUPPRESSION WITH SHIFT-INVARIANT WAVELET PACKETS

Israel Cohen, Shalom Raz and David Malah

Department of Electrical Engineering
Technion — Israel Institute of Technology
Technion City, Haifa 32000, Israel

INTRODUCTION

Time-frequency representations map one-dimensional signals into two-dimensional images that indicate their energy content in the joint time-frequency plane [1, 2]. These representations combine time-domain and frequency-domain analyses to yield a more revealing picture of the temporal localization of spectral components. They have proven indispensable in a wide range of applications, including the study of non-stationary phenomena in high power microwave tubes, such as mode build-up and competition and pulse shortening [3].

A popular time-frequency analysis tool is the spectrogram, a squared magnitude of the short-time Fourier transform. Its major limitation is the inherent trade-off between time and frequency resolution for a particular window function. Since good time resolution requires a narrow window while good frequency resolution requires a wide window, high resolution simultaneously in both directions is unattainable. This limitation promoted the development of bilinear time-frequency representations that attempt to match the window function to the analyzed signal [1]. Unfortunately, the bilinear nature of the latter representations results in a high noise sensitivity and presence of interference terms, which restrict their practical application.

In this paper, we present a wavelet-based method for constructing an efficient time-frequency representation, which is characterized by high time-frequency resolution, noise immunity and reduced interference terms. This method also provides a robust nonlinear technique for estimating a discrete signal from its noisy measurement.

SHIFT-INVARIANT WAVELET PACKET DECOMPOSITION

Overcomplete libraries of waveforms that span redundantly the signal space encourage adaptive signal representations. Instead of representing a prescribed signal in a fixed basis, it is often useful to choose a suitable basis that facilitates a desired application,

Ultra-Wideband, Short-Pulse Electromagnetics 4
Edited by Heyman *et al.*, Kluwer Academic / Plenum Publishers, New York, 1999.

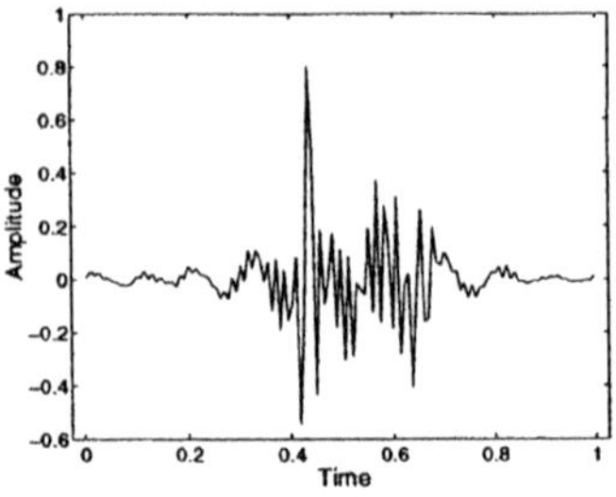

Figure 1: Test signal $g(t)$ consisting of a short pulse, a tone and a nonlinear chirp.

such as compression, identification, classification or noise removal (denoising). The library of wavelet packets is a huge library of bases that consists of translations and dilations of wavelet packets [4, 5]. The basis functions are localized in the time-frequency plane, and organized in a binary tree structure where efficient search algorithms for the best basis can be implemented.

The *shift invariant wavelet packet decomposition* (SIWPD) [6] is an adaptive representation in an *extended* library of wavelet packet bases. The extended library includes an additional degree of freedom that adjusts the time-localization of the basis functions. This degree of freedom is incorporated into the best-basis search algorithm by adaptively selecting between even and odd down-sampling. Specifically, following the low-pass and high-pass filtering, when expanding a parent-node, either all the odd samples or all the even samples are retained, according to the choice which minimizes the cost function.

Let $\{\psi_n(t) : n \in \mathbb{Z}_+\}$ be a wavelet packet family [4] generated by

$$\psi_{2n}(t) = \sqrt{2}\sum_{k\in\mathbb{Z}} h_k\psi_n(2t-k) \tag{1}$$

$$\psi_{2n+1}(t) = \sqrt{2}\sum_{k\in\mathbb{Z}} g_k\psi_n(2t-k) \tag{2}$$

where $g_k = (-1)^k h_{1-k}$, and $\psi_0(t) \equiv \varphi(t)$ is an orthonormal scaling function, satisfying

$$\langle\varphi(t-p), \varphi(t-q)\rangle = \delta_{p,q}, \quad p, q \in \mathbb{Z}\,. \tag{3}$$

The extended library of wavelet packets is defined as the collection of all the orthonormal bases which are subsets of

$$\left\{B_{\ell,n,m} \;:\; 0 \le \ell \le L,\; 0 \le n, m < 2^{L-\ell}\right\}, \tag{4}$$

where L denotes the finest resolution level, and

$$B_{\ell,n,m} \equiv \left\{2^{\ell/2}\psi_n\left[2^\ell(t-2^{-L}m)-k\right] \;:\; 0 \le k < 2^\ell\right\}. \tag{5}$$

This library is larger than the standard wavelet packet library by a square power, but can be still structured into a tree configuration which supports fast search algorithms [6]. The additional parameter m provides the crucial degrees of freedom required for the time-adjustment of the basis functions. When an analyzed signal is translated in time by $\tau = q \cdot 2^{-L}$ ($q \in \mathbb{Z}$), a new best-basis is selected whose elements are also translated by τ compared to the former best-basis. Thus the expansion coefficients remain, and the time-frequency representation is shifted in time by the same period.

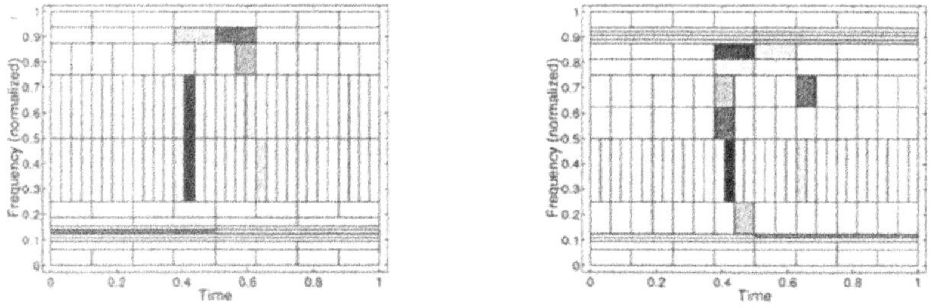

(a) (b)

Figure 2: Effect of a temporal shift on the time-frequency representation using the WPD with 8-tap Daubechies wavelet filters: (a) $g(t)$ in its best basis, Entropy= 2.69. (b) $g(t-2^{-6})$ in its best basis, Entropy= 2.72.

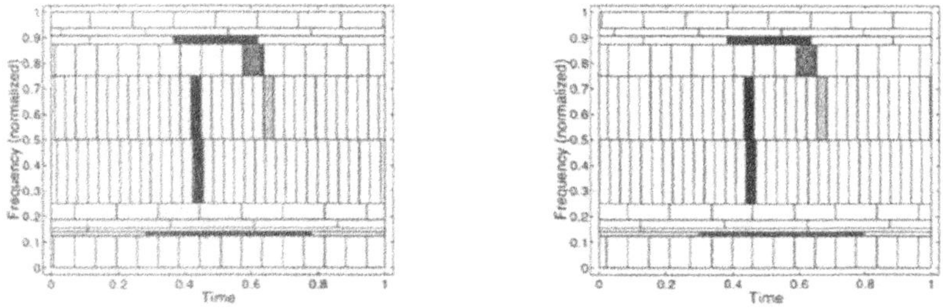

(a) (b)

Figure 3: Time-frequency representation using the SIWPD with 8-tap Daubechies wavelet filters: (a) $g(t)$ in its best basis, Entropy= 1.72. (b) $g(t-2^{-6})$ in its best basis, Entropy= 1.72. Compared with the WPD (Fig. 2), beneficial properties are shift-invariance and lower information cost.

Compared with the ordinary wavelet packet decomposition (WPD) [4], the SIWPD is determined to be advantageous in the following respects [6]: 1) Shift-invariance; 2) Lower information cost; 3) Improved time-frequency resolution; 4) More stable information cost across a prescribed data set; 5) Controlled computational complexity (at the expense of the information cost down to $O(Nlog_2N)$). To illustrate the shift-invariant properties of the SIWPD and its enhanced time-frequency representation compared to the standard WPD, we refer to the expansion of the signal $g(t)$, depicted in Fig. 1. This signal, containing $2^7 = 128$ samples, comprises a short pulse, a tone and a component with nonlinear frequency modulation. For definiteness, we choose D_8 to serve as the scaling function (D_8 corresponds to 8-tap Daubechies least asymmetric wavelet filters [7, page 198]) and the Shannon entropy as the cost function, defined by [4] $\mathcal{M}(\{x_i\}) = -\sum_{i:x_i\neq 0} x_i^2 \log x_i^2$. Figs. 2 and 3 display the best-basis expansions under the WPD and the SIWPD algorithms, respectively, for the signals $g(t)$ and $g(t-2^{-6})$. The sensitivity of WPD to temporal shifts is obvious, while the best-basis SIWPD representation is indeed shift-invariant and characterized by a lower entropy and improved time-frequency resolution.

MODIFIED WIGNER DISTRIBUTION

The Wigner distribution (WD) possesses a number of desirable mathematical properties relevant time-frequency analysis. However, the presence of interference terms renders the WD of multicomponent signals extremely difficult to interpret [1]. Several methods, developed to reduce noise and cross-components at the expense of reduced signal concentration, employ some kind of smoothing kernel or windowing [2]. The choice of the kernel dramatically affects the appearance and quality of the resulting time-frequency representation. Hence adaptive representations often exhibit performance far surpassing that of fixed-kernel representations. However, they are either computationally expensive or have a very limited adaptation capability.

Recently, we have introduced a *modified Wigner distribution* (MWD) [8], which obtains high resolution and concentration in time-frequency, and is superior in eliminating interference terms. We showed that using extended libraries of orthonormal bases, interference terms can be reduced by adaptively thresholding the cross WD of pairs of basis functions.

Let $g = \sum_\lambda c_\lambda \varphi_\lambda$ be the SIWPD of the signal g. Then its MWD is defined by

$$T_g(t,\omega) = \sum_{\lambda\in\Lambda} |c_\lambda|^2 W_{\varphi_\lambda}(t,\omega) + 2 \sum_{\{\lambda,\lambda'\}\in\Gamma} \mathrm{Re}\{c_\lambda c_{\lambda'}^* W_{\varphi_\lambda,\varphi_{\lambda'}}(t,\omega)\} \tag{6}$$

where W_{φ_λ} is the auto WD of φ_λ and $W_{\varphi_\lambda,\varphi_{\lambda'}}$ is the cross WD of φ_λ:

$$W_{\varphi_\lambda}(t,\omega) = \int \varphi_\lambda(t+\tau/2)\varphi_\lambda^*(t-\tau/2)e^{-j\omega\tau}d\tau \tag{7}$$

$$W_{\varphi_\lambda,\varphi_{\lambda'}}(t,\omega) = \int \varphi_\lambda(t+\tau/2)\varphi_{\lambda'}^*(t-\tau/2)e^{-j\omega\tau}d\tau \tag{8}$$

The summations in (6) are limited to basis-functions whose coefficients are large enough, and to pairs which are "close" in time-frequency plane. Let ϵ and D denote respectively thresholds of relative amplitude and time-frequency distance. Then the sets Λ and Γ are given by

$$\Lambda = \{\lambda \mid |c_\lambda| \geq \epsilon M\}, \quad M \equiv \max_\lambda\{|c_\lambda|\} \tag{9}$$

$$\Gamma = \{\{\lambda,\lambda'\} \mid 0 < d(\varphi_\lambda,\varphi_{\lambda'}) \leq D,\ |c_\lambda c_{\lambda'}| \geq \epsilon^2 M^2\}, \tag{10}$$

The distance d between a pair of basis-functions is measured by their degree of adjacency:

$$d(\varphi_\lambda,\varphi_{\lambda'}) = \left[\frac{(\bar{t}_\lambda - \bar{t}_{\lambda'})^2}{\Delta t_\lambda \Delta t_{\lambda'}} + \frac{(\bar{\omega}_\lambda - \bar{\omega}_{\lambda'})^2}{\Delta\omega_\lambda \Delta\omega_{\lambda'}}\right]^{1/2} \tag{11}$$

where $(\bar{t}_\lambda, \bar{\omega}_\lambda)$ is the time-frequency position of the basis-function φ_λ, and Δt_λ and $\Delta\omega_\lambda$ are the corresponding time and frequency uncertainties. By adjusting the distance threshold D and amplitude threshold ϵ, one can effectively balance the cross-term interference, the useful properties of the distribution (time/frequency marginals, energy conservation, instantaneous frequency, *etc.*), and the computational complexity [9].

Here, the basis-functions are of the form

$$\psi_{\ell,n,m,k}(t) = 2^{\ell/2}\psi_n\left[2^\ell(t - 2^{-L}m) - k\right] \tag{12}$$

where ℓ is the resolution-level index ($0 \leq \ell \leq L$), n is the frequency index ($0 \leq n < 2^{L-\ell}$), m is the shift index ($0 \leq m < 2^{L-\ell}$) and k is the position index ($0 \leq k < 2^\ell$). Each basis-function is associated with a rectangular tile in the time-frequency plane which is positioned about

$$\bar{t} = 2^{-\ell}k + 2^{-L}m + (2^{L-\ell} - 1)C_h + (C_h - C_g)R(n), \tag{13}$$

$$\bar{\omega} = 2^{\ell-L}[GC^{-1}(n) + 0.5], \tag{14}$$

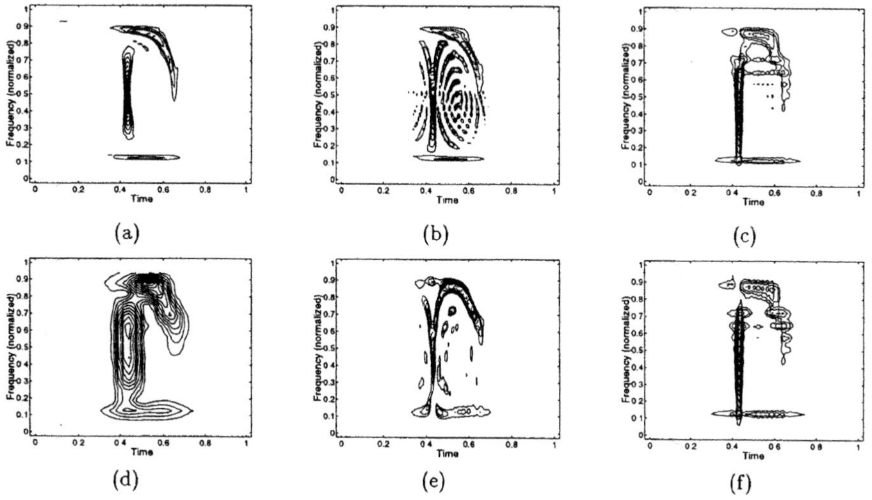

Figure 4: Contour plots for the signal $g(t)$: (a) Modified Wigner distribution; (b) Wigner distribution; (c) Choi-Williams distribution; (d) Spectrogram; (e) Cone-kernel distribution; (f) Reduced interference distribution. The modified Wigner distribution eliminates the interference terms while retaining high energy concentration of signal components.

where C_h and C_g are respectively the centers of energy of the low-pass and high-pass quadrature filters h and g [5], defined by

$$C_h = \frac{1}{\|h\|^2} \sum_{k \in \mathbb{Z}} k|h_k|^2, \quad C_g = \frac{1}{\|g\|^2} \sum_{k \in \mathbb{Z}} k|g_k|^2, \tag{15}$$

$R(n)$ is an integer obtained by bit reversal of n in a $L - \ell$ bits binary representation, and GC^{-1} is the inverse Gray code permutation. The width and height of the tile are given by

$$\Delta t = 2^{-\ell}, \quad \Delta\omega = 2^{\ell - L}. \tag{16}$$

Fig. 4(a) shows the MWD for the signal $g(t)$, attained by utilizing expression (6) combined with the thresholds $D = 1.5$ and $\epsilon = 0.1$. Figs. 4(b)–(f) describe respectively the WD, the Choi-Williams distribution, the spectrogram, the cone-kernel distribution and the reduced interference distribution [2]. Clearly, the MWD achieves high resolution and concentration in time-frequency, and is superior in eliminating interference terms associated with the WD.

TRANSLATION-INVARIANT DENOISING

The use of wavelet bases for estimation of signals embedded in noise has been the object of considerable recent research. While traditional methods often remove noise by low-pass filtering, thus blurring the sharp features in the signal, wavelet-based methods show good performance for a wide diversity of signals, including those with jumps, spikes and other nonsmooth features [10, 11, 12]. The transform-based thresholding method consists of three steps: transformation of the noisy data into a time-scale domain, soft or hard thresholding to the resulting coefficients, and transformation back into the original space. This scheme necessitates determination of the "best" basis and threshold value, leading to the best signal estimate. It is useful to employ the library of wavelet-packet bases as a collection of competing models, and select the best model according to the *Minimum Description Length* (MDL) criterion [11, 13]. However, denoising based on the conventional WPD may exhibit visual artifacts, attributable to the lack of shift-invariance [12].

One approach to attaining shift-invariance is to average the translation dependence: applying a range of shifts to the noisy data, denoising the shifted versions with the wavelet transform, then unshifting and averaging the denoised data [12]. This procedure, termed *Cycle-Spinning*, generally yields better visual performance on smooth parts of the signal. However, transitory features may be significantly attenuated [14].

In this section, we present a translation-invariant signal estimator, which is based on the SIWPD and the MDL criterion. A collection of signal models is generated using the *extended* library of orthonormal wavelet-packet bases, and an additive cost function, approximately representing the MDL principle, is derived. We show that the minimum description length of the noisy observed data is achieved by utilizing the SIWPD and thresholding the resulting coefficients. This estimator is efficiently combined with the MWD, yielding robust time-frequency representations that are characterized by high resolution and suppressed interference-terms.

Let $y(t) = f(t) + z(t)$ represent the noisy observed data, where $f(t)$ is the unknown signal to be estimated, and $z(t)$ is a white Gaussian noise with zero mean and a known power spectral density σ^2. Denote by $\mathcal{B}$ the extended library of wavelet packet bases. The description length of y represented on a basis $B \in \mathcal{B}$ is given by [9]

$$\mathcal{L}(By) = \sum_{(\ell,n,m)\in E} \mathcal{L}(B_{\ell,n,m}y) \tag{17}$$

where

$$\mathcal{L}(B_{\ell,n,m}y) = 3 + \frac{1}{2\sigma^2 \ln 2} \sum_{k=1}^{2^\ell N} \min\left\{C^2_{\ell,n,m,k}(y)\,,\ 3\sigma^2 \ln N\right\} \tag{18}$$

is the codelength associated with a terminal node (ℓ, n, m), and

$$B_{\ell,n,m}y = \left\{C_{\ell,n,m,k}(y) = \langle y\,,\ \psi_{\ell,n,m,k}\rangle\ :\ 1 \le k \le 2^\ell N\right\} \tag{19}$$

are the expansion coefficients of the observed data. Since the codelength in Eq. (17) constitutes an additive cost function, the SIWPD gives the optimal basis according to the MDL principle. The optimal estimate of $f(t)$ is obtained by expanding the observed data $y(t)$ on the optimal basis $A = \left\{\hat{\phi}_k\right\}_{1\le k\le N}$ and hard-thresholding the coefficients by $\tau \equiv \sigma\sqrt{3\ln N}$. Specifically,

$$\hat{f}(t) = \sum_{k=1}^{N} \eta_\tau(y_k)\hat{\phi}_k(t) \tag{20}$$

where $y_k = \left\langle y, \hat{\phi}_k\right\rangle$, and $\eta_\tau(c) \equiv c\mathbf{1}_{\{|c|>\tau\}}$ is the *hard-threshold* function. The time-frequency distribution estimate of y is obtained by computing the MWD for the signal estimate:

$$\hat{T}_y(t,\omega) = T_{\hat{f}}(t,\omega) = \sum_{k\in\Lambda} |y_k|^2 W_{\hat{\phi}_k}(t,\omega) + 2 \sum_{\{k,k'\}\in\Gamma} \mathrm{Re}\{y_k y^*_{k'} W_{\hat{\phi}_k,\hat{\phi}_{k'}}(t,\omega)\} \tag{21}$$

where

$$\Lambda = \left\{k\ :\ |y_k| > \sigma\sqrt{3\ln N},\ 1 \le k \le N\right\}, \tag{22}$$

$$\Gamma = \left\{\{k,k'\}\ :\ k,k' \in \Lambda,\ 0 < d(\hat{\phi}_k, \hat{\phi}_{k'}) \le D\right\}. \tag{23}$$

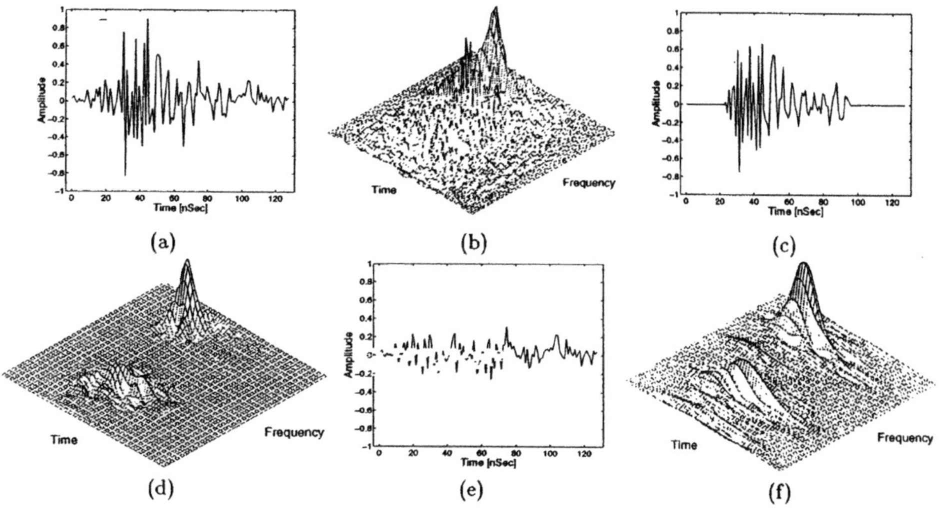

Figure 5: Electromagnetic pulse in a relativistic magnetron (heterodyne detection; local oscillator= 2.6GHz): (a) Noisy measurement $y(t)$. (b) Wigner distribution for $y(t)$. (c) The signal estimate $\hat{f}(t)$ by the MDL principle. (d) The estimate of the modified Wigner distribution for $y(t)$. (e) Residual between $y(t)$ and $\hat{f}(t)$. (f) Smoothed pseudo Wigner distribution for $y(t)$.

Fig. 5(a) shows a noisy measurement of an electromagnetic pulse ($\approx$ 100 nanoseconds long) generated by high power ($\approx$ 100 MegaWatts) relativistic magnetron. The measurement involves heterodyning at 2.6GHz, filtering at 500kHz and sampling at 1GHz [3]. The Wigner distribution, depicted in Fig. 5(b), is clearly ineffective as a time-frequency analysis tool, for its high noise sensitivity. Yet, the estimates of the signal and the MWD, as shown in Figs. 5(c) and (d), are potentially valuable when analyzing the measurements and studying the non-stationary phenomena, such as mode build-up and competition and pulse shortening [15], which are common in such high power microwave tubes. The residual between the noisy measurement and the signal estimate is depicted in Fig. 5(e). To ascertain that this residual is actually the noise component, we compare the estimate of the MWD with the smoothed pseudo Wigner distribution of the noisy measurement (Fig. 5(f)). Since these two distributions are similar, in view of the fact that smoothing in the Wigner domain reduces the noise at the expense of smearing the signal components, it is reasonable to assume that the signal estimate contains all the signal components and the residual is mostly noise. A detailed derivation and implementation of the proposed estimator, examples illustrating its performance, and a discussion of the relation to other work is given in [9].

CONCLUSION

Cross terms associated with bilinear distributions are not necessarily interpretable as interference terms. Any signal can be broken up in an infinite number of ways, each of which generates different cross terms. Therefore, it is important to choose an appropriate decomposition that separates the parts which are well delineated in the time-frequency plane. We have presented a modified Wigner distribution, where undesirable interference-terms can be eliminated while still retaining high energy concentration.

A prescribed signal is expanded into a redundant library of orthonormal wavelet-packet bases, from which the best decomposition is selected, and subsequently transformed into the Wigner domain. The discrimination between beneficial cross terms, which primarily enhance the useful properties of the time-frequency representation, and undesirable interference terms is determined according to the degree of adjacency and relative amplitudes of the interacting basis functions; Only adjacent pairs whose coefficients are large enough are related to the same component of the signal. The balance between interference terms, concentration and computational complexity is achieved by adjusting the distance and amplitude thresholds.

A translation-invariant denoising method, which uses the SIWPD and the MDL criterion has been described . The MDL principle is applied for approximating the description length of the noisy observed data and for choosing the optimal wavelet-packet basis. The proposed signal estimator, combined with the modified Wigner distribution, generates robust time-frequency representations.

REFERENCES

[1] L. Cohen, *Time-Frequency Analysis*, Prentice-Hall Inc., 1995.

[2] F. Hlawatsch and G. F. Boudreaux-Bartels, "Linear and quadratic time-frequency signal representations", *IEEE SP Magazine*, Apr. 1992, pp. 21–67.

[3] I. Schnitzer, A. Rosenberg, C. Leibovitch, M. Botton, I. Cohen and J. Leopold, "Evolution of spectral power density in grounded cathode relativistic magnetron", *Proc. SPIE, Intense Microwave Pulses IV*, Vol. 2843, Aug. 1996.

[4] R. R. Coifman and M. V. Wickerhauser, "Entropy-based algorithms for best basis selection", *IEEE Trans. Inform. Theory*, Vol. 38, No. 2, Mar. 1992, pp. 713–718.

[5] M. V. Wickerhauser, *Adapted Wavelet Analysis from Theory to Software*, AK Peters, Ltd, Wellesley, Massachusetts, 1994.

[6] I. Cohen, S. Raz and D. Malah, "Orthonormal shift-invariant wavelet packet decomposition and representation", *Signal Processing*, Vol. 57, No. 3, Mar. 1997, pp. 251–270.

[7] I. Daubechies, *Ten Lectures on Wavelets*, CBMS-NSF Regional Conference Series in Applied Mathematics, SIAM Press, Philadelphia, Pennsylvania, 1992

[8] I. Cohen, S. Raz and D. Malah, "Eliminating interference terms in the Wigner distribution using extended libraries of bases", *Proc. Int. Conf. on Acoustics, Speech and Signal Processing, ICASSP-97*, Munich, Germany, 20–24 Apr. 1997, pp. 2133–2136.

[9] I. Cohen, *Shift-Invariant Adaptive Wavelet Decompositions and Applications*, D.Sc. Dissertation, Technion - Israel Institute of Technology, Haifa, Israel, May 1998.

[10] D. L. Donoho and I. M. Johnstone, "Ideal spatial adaptation via wavelet shrinkage", *Biometrica*, Vol. 81, 1994, pp. 425–455.

[11] N. Saito, "Simultaneous noise suppression and signal compression using a library of orthonormal bases and the minimum description length criterion", in: E. Foufoula and P. Kumar, eds., *Wavelets in Geophysics*, Academic Press, 1994, pp. 299–324.

[12] R. R. Coifman and D. L. Donoho, "Translation-invariant de-noising", in: A. Antoniadis and G. Oppenheim, ed., *Wavelet and Statistics*, Lecture Notes in Statistics, Springer-Verlag, 1995, pp. 125–150.

[13] H. Krim, and J.-C. Pesquet, "On the statistics of best bases criteria", in: A. Antoniadis and G. Oppenheim, ed., *Wavelet and Statistics*, Lecture Notes in Statistics, Springer-Verlag, 1995, pp. 193–207.

[14] N. A. Whitmal, J. C. Rutledge and J. Cohen, "Wavelet-based noise reduction", *Proc. Int. Conf. on Acoustics, Speech and Signal Processing, ICASSP-95*, Detroit, Michigan, 8–12 May 1995, pp. 3003–3006.

[15] J. Benford and J. Swegle, *High Power Microwaves*, Artech House, Norwood, 1992.

SUMMARY OF THE DARPA BACKGROUND CLUTTER DATA COLLECTION EXPERIMENT

Vivian George and Thomas W. Altshuler

Walcoff and Associates
Arlington, VA USA
Institute for Defense Analyses
Alexandria, VA USA

ABSTRACT

Most technologies in use or proposed for use to detect landmines and unexploded ordnance (UXO) suffer from unacceptably high false-alarm rates, even at modest probabilities of detection. High false-alarm rates are a consequence of the inability to discriminate real UXO and landmines from man-made and naturally occurring clutter. Experimental data on expected signatures of UXO and landmines have been collected by various sensor developers and research agencies. However, an equivalent characterization of clutter returns has been lacking. As a result, background signatures have not been considered in sensor or algorithm design.

The goal of the two-phase, DARPA-sponsored Background Clutter Data Collection Experiment is to begin to fill this void. The first phase, completed during the fall of 1996, consisted of high aerial density site surveys using the following sensor types: magnetometer, infrared, electromagnetic induction, and ground-penetrating radar. The second phase, which is currently underway, consists of analysis of the collected sensor data to include excavation of clutter objects. This paper describes the clutter data collection program and provides a cursory summary of the analysis done to date.

INTRODUCTION

The current state of the art in detection technology is generally incapable of distinguishing responses that are a result of clutter as opposed to those that are caused by ordnance items; the clutter usually results in a false alarm[1, 2, 3]. Expected signals from mines and UXO are at least moderately well understood, and theoretical models can, in many cases, reliably predict target signatures. In contrast, clutter signatures

Ultra-Wideband, Short-Pulse Electromagnetics 4
Edited by Heyman *et al.*, Kluwer Academic / Plenum Publishers, New York, 1999.

have not been well studied and, as a result, many false alarm mitigation efforts are conducted in relative ignorance. In this paper clutter is defined to be any object, feature, or event that induces a sensor response that is similar to the response caused by the presence of a landmine or UXO item.

The objective of the DARPA-sponsored Background Clutter Data Collection Experiment is to advance the state of the art in UXO and landmine detection. The data collected will aid in understanding the effects of clutter on system performance and will also be useful in evaluating detection systems because such evaluations require knowledge of both clutter and target statistics [2, 3].

THE DARPA BACKGROUNDS PROGRAM

The backgrounds program consists of two phases. During Phase I (Sept. 1996–Nov. 1996), contractors surveyed the backgrounds sites to collect high-resolution data with various sensor systems. Each site includes a 100 m × 100 m clutter area as well as 1,300 m^2 of UXO and landmine areas, termed target Side Bars. During Phase II, currently in progress, the sensor data collected in Phase I is being analyzed. This phase of the program is unique in the sense that the Defense Advanced Research Projects Agency (DARPA) will make this data set widely available to sensor developers, universities, and other private companies that have an interest in developing improved detection/clutter rejection methods or algorithms. DARPA intends to encourage these development activities by providing all of the data collected—and a subset of the corresponding ground truth—to anyone who requests it. In return, DARPA will impose limited requirements on the requester to provide feedback on the type of work being performed and any improvements to sensors or algorithms that result from the data collected. The remainder of the ground truth data will be held by the government and used to test and evaluate any new or improved detection/clutter rejection algorithms.

Background experiment design

The world-wide UXO and landmine problem is not limited to specific geologies, climates, or amounts of anthropic contamination, all of which are expected to contribute to the clutter problem. Selecting specific sites required a trade between the amount of variation to represent and the amount of area to be characterized during the experiment. Four sites were chosen at two U.S. locations (Fort Carson, Colorado, and Fort A.P. Hill, Virginia), representing three soil moisture combinations: wet sand, dry sand, and dry clay. One of the four sites, located at Fort A.P. Hill, contained a substantial amount of suspected anthropic debris, but none of the sites were expected to contain UXO.

Sensor Selection

Four sensor types are typically used for detecting landmines and UXO: ground-penetrating radar (GPR), electromagnetic induction (EMI), magnetometer, and infrared (IR). Because it was not feasible to bring every variation of each of these sensors to the sites, an attempt was made to select systems that reflected variations of operation. In addition, for all systems an attempt was made to include both a fielded system and a prototype, or the next advance in system design. For example, EMI sensors can be designed to record responses in either the time domain or the frequency domain. The EMI systems selected represent both time- and frequency-domain systems as well as prototype and mature systems.

The contractors who participated in the data collection program and a brief description of the selected sensor systems follows:

Coleman Research Corporation. Coleman provided vehicle-towed time domain electromagnetic (TDEM) and GPR sensor arrays. The vehicle-towed GPR system consists of two transmit and three receive antennas that are dragged to maintain contact with the ground. Data was collected with two antenna configurations: co-polarized and cross-polarized. The system is described as a 2 MHz stepped-frequency radar utilizing frequencies from 100 MHz to 610 MHz, with a 150 μs dwell time per frequency. In addition to the array of GPR antennas, a separate hand-held GPR unit was attached to the vehicle. This GPR was configured with a 10 MHz frequency step, operating from 1,000 MHz to 4,000 MHz, with a 300 μs dwell time per frequency. The hand-held antenna was mounted to maintain a 6-cm vertical standoff. The TDEM array consisted of three 1-m Geonics, Ltd. EM61 metal detectors with a pulse rate of 15 Hz. The lower coil (transmit) maintained a 25-cm vertical standoff and the upper coil (receive) was mounted 57 cm above the ground.

Geo-Centers, Inc. Geo-Centers provided vehicle-mounted GPR and IR sensors. The forward looking IR sensor consists of a 3–5 mm Amberview camera that provides a 14 bit, 256 by 256 pixel, video image every 2 seconds. The GPR system is mounted 20 in. above the ground. The 27-in. wide multi-antenna (four transmit, four receive) ground-penetrating radar array, termed a Focused Array Radar (FAR), focuses and sweeps energy into the ground to detect buried objects. This GPR operates in a frequency range of 700–1,300 MHz.

Geometrics. Geometrics provided a hand-towed cart with an arrangement of six G–858 Cesium vapor magnetometers configured to provide both high-density magnetometer and vertical magnetic gradiometer measurements. In addition, a separate cart with a 1-m EM61 was used to survey a portion of the Fort Carson sites.

Geophex, Ltd. Geophex provided a hand-held GEM–3 and a two-element G–858 total field magnetometer system configured as a vertical magnetic gradiometer. The GEM–3 is a frequency-domain electromagnetic sensor operated with a 10–Hz sampling rate and two transmit frequencies: 4,050 Hz and 12,270 Hz.

Lawrence Livermore National Laboratories. LLNL developed a Calibrated Autonomous Radar Testbed (CART) to collect synthetic aperture radar data in the field. This laboratory system is comprised of a vector network analyzer attached to a dual-ridged conical waveguide horn antenna and a control computer. It has the capability of sweeping the frequency band from 45 MHz to 18 GHz and capturing data at up to 801 discrete frequency steps, at better than 40 dB above the environmental noise floor.

Parsons Engineering Science, Inc. Parsons fielded two TDEM systems: (1) A custom-built 0.5 m Geonics EM61 employing a bipolar, rectangular current with a 50-percent duty cycle and a repetition rate of 75 Hz. (2) A prototype Geonics EM61-3D. The latter system includes three nested orthogonal circular receive coils (1.0 m dia.) mounted in the center of the standard 1-m transmit coil. The EM61–3D measures the decay of the secondary magnetic field at 20 geometrically spaced time gates in each of the three orthogonal directions.

Science Applications International Corporation. SAIC provided a TDEM array and an IR camera. The TDEM sensor is an array of eight partially overlapping Schiebel induction coils that is dragged on the ground. The IR camera is a down-looking 3–5 mm Amberview camera.

Site Characterization

A comprehensive site characterization was performed by the Army Corps of Engineers Waterways Experiment Station at both Fort Carson, Colorado and Fort A.P. Hill, Virginia, before the burial of registration targets, calibration sources, UXO, and mine targets. Because it was not known which environmental parameters would play a critical role in later data analyses an attempt was made to provide an all-inclusive site characterization to the depth of interest (1.0 m). The site characterization included the following:

- Soil characterizations.
- Soil dielectric properties at different frequencies and moisture contents.
- EM31 electromagnetic induction survey.
- Proton precession magnetometer survey.
- Electrical resistivity soundings.
- In-situ determinations of the soil dielectric properties.
- Ground-penetrating radar surveys at 50, 100, 200, and 900 MHz.
- Details of geology.

In addition, a number of environmental measurements were collected continually during the contractor site surveys:

- Meteorological data.
- Soil moisture data.
- Mine and surrounding soil temperature data for selected targets.

Site design—center square and target side bars

Each site is identical in design, measuring 125 m by 100 m, and consists of three areas (see figure 1):

- The Center Square, where the background clutter data was collected. The Center Square contains five groups of registration targets described later. This area measures 100 m × 100 m.

- The Side Bars, which contain real targets and target surrogates. The BLUE Side Bar measures 100 m × 5 m, the YELLOW Side Bar measures 100 m × 3 m, and the ORANGE Side Bar measures 100 m × 5 m. Portions of the BLUE and YELLOW Side Bars are set aside for system calibration.
- The Calibration Area, includes portions of the BLUE and YELLOW Side Bars and the RED lane. The total Calibration Area (RED, YELLOW, and BLUE portions) is 30 m × 15 m.

To eliminate contamination of the clutter area with the signatures of emplaced targets other than the registration targets, the Center Square and Target Side Bars were created as separate areas. Furthermore, buffer regions separated the target Side Bars from the Center Square to eliminate the potential of target signatures bleeding into the Center Square and contaminating the clutter data. (The data analysis is greatly simplified when the target data is physically separated from the clutter data.)

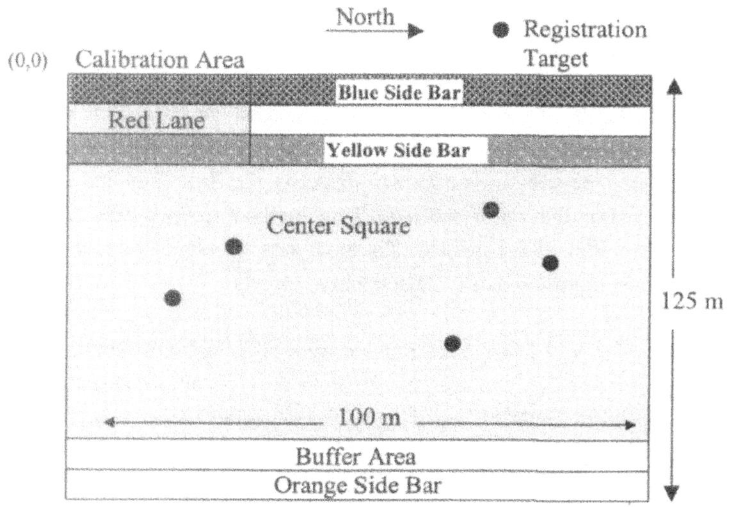

Figure 1. The background site layout.

Details About Target Selection and Burial

The side bars are color coded based on the types of targets they contain.

- The RED Lane (5 m by 30 m) contains a copy of the registration targets, targets for calibration, and system-stressing targets.
- The YELLOW Side Bar contains mines, infrared calibration sources, and fabricated targets.
- The BLUE Side Bar contains UXO and a demagnetized sphere. UXO targets are provided in varying orientations to illustrate the effect that orientation has on the resulting magnetic moment.
- The ORANGE Side Bar contains UXO buried deeper, in general, than the targets in the BLUE Side Bar.

Registration Targets. The clutter data were collected from the 100-m × 100-m section of the site, termed the Center Square. Five identical groups of registration targets were buried in surveyed locations in the Center Square. These targets were designed with the intent that each of the sensors will easily observe at least one of the registration targets at each location. The registration targets permit the sensor data sets to be referenced to a known coordinate system and mutually registered.

At each registration location, the same three targets were buried or placed on the surface. The three registration targets used for this test were the following:

Registration Sphere—4 7/8-in. iron sphere buried at a depth of 5 cm. This is the registration target for both the magnetometers and EMI systems.

Registration Large Aluminum (Al) Plate—an 8-in. square, 1-in. thick, aluminum plate buried at a depth of 5 cm. This target is intended to ensure that both the high- and low-frequency GPR systems will register a measurable signal when the GPR sensor passes over the target.

Registration Small Aluminum Plate—a 4-in. square, 1/2-in. thick, aluminum plate placed on the surface. This registration target is for IR sensors.

Stressing Targets. A "stressing" target was emplaced for each of the sensors. The sensor operators were asked to adjust system settings, if any, until the stressing target was detected. Once the stressing target was detected, the Side Bars and Center Square were surveyed using the same settings. This method defined the signal strength threshold of the clutter data collected. The goal was to avoid collecting clutter data with the system set to preferentially detect "easy" targets.

Mine Targets. The YELLOW Side Bar contains both real and fabricated mine-like targets. The purpose of the fabricated targets is to provide simple target geometries of known composition to facilitate modeling of the expected sensor response. All of the fabricated mine-like targets were placed in the calibration portion of the YELLOW Side Bar. Two fabricated surface targets were also provided: two 8-in. square aluminum plates, one painted white and the other painted black. Targets RED–3, RED–2, YEL–3, YEL–4, YEL–5, YEL–6 and YEL–9 were monitored with thermistors. For all of the targets that were monitored, three thermistors were emplaced, one on the surface of the target one in the surrounding soil and one in the nearby undisturbed soil. Only the larger metallic and completely nonmetallic calibration targets were instrumented to prevent alteration of the target signature.

The YELLOW side bar also contains real mines, with no explosive content but instead a filler with dielectric properties similar to explosives.

UXO Targets. The BLUE and ORANGE Side Bars were designed to provide UXO target signatures at various depths and orientations. Four different types of UXO, buried at different depths and orientations, provided targets with a broad distribution of peak magnetic signatures. The targets were chosen to represent a sample of UXO that might be found in the first 1 m of soil.

The depths of the ordnance items were chosen to provide a cross section of signatures. To permit analysis of signatures that are typically at or slightly above the clutter level, the signature size is more heavily populated in the 10 to 40 γ range. By weighting the population in this signature regime, algorithm development can focus on identification of UXO signatures that are hard to discriminate from the background.

Data Analysis

Data analysis efforts completed to date include generation of single pixel distributions for the Parson's 0.5 meter EM61, the GEM-3, and the Geometrics magnetometer array for the both target and clutter pixels in order to determine the efficacy of thresholding as a discrimination algorithm. The distributions almost completely overlap which indicates that simple thresholding techniques are inadequate. In light of this finding more sophisticated algorithm development is being pursued by Dr. Tom Altshuler (Institute for Defense Analyses), Dr. Leslie Collins (Duke University), and Dr. Paul Gader (university of Missouri-Rolla). Initial results show a decrease in the number of false alarms at equal probabilities of detection when compared to a simple thresholding algorithm. Much work remains to be done. The purpose of publishing this paper at the Euroem conference is to expose more potential algorithm developers to the data.

Digging

To obtain more information regarding the sources of clutter signatures, 101 anomalies were excavated at FP20 resulting in the discovery of 203 objects. All of the anomalies were selected based on either a magnetometer or metal detector anomaly. The GPR data analysis has not progressed far enough to warrant the selection of anomalies for further investigation. Of the 203 objects recovered 14 were inert landmines. Of those, 12 were identical M12 landmines and 7 of the objects were mine field warning signs. The objects found at the site, in order of prevalence, include: metal banding (64%), metal scrap, car parts, communications wire, training rounds, metal plates, 2 fence posts, 1 soda can, 1 piece of foil, 1 55 gallon drum, and 1 fuel tank. The digging data also provided known target locations to apply towards the problem of increasing location accuracy. That work is yet to be published.

CONCLUSIONS AND DATA AVAILABILITY

Phase I of the experiment was successful. We collected approximately 35 CD-ROMs of data (about 15 GB) that elucidate clutter statistics as compared to target statistics in three different geologies with two levels of anthropic clutter. All of the contractors surveyed the required areas of the sites, with many of the contractors surveying much more than the required area. Some of the contractors have data gaps in the final data product but the gaps are insignificant. So overall the report is glowing, but the experiment was not perfect.

The single biggest problem with the data set is the location data. All of the contractors were required to store a synchronized time stamp with their data for future post-processing that would permit unambiguous application of the laser tracker location information to the sensor system data. The various contractors used different methods to provide location data, and some of the contractors did not store location data at all, choosing instead to store a synchronized time stamp only and thus rely on the laser tracker for the location information. One contractor arrived at the site with no capability to store the time stamp or location data.

The laser-tracker system used to track all of the sensors during the experiment did not store the time stamp correctly. The magnitude of the problem was not discovered until after the all of the contractors had completed the site surveys.

Currently, algorithms are under development to correct the laser-tracker error. Still, to accelerate data analyses, the backgrounds data is being released with the contractor location information.[1] A second release with increased location accuracy will be created, if possible, at a later date.

The laser-tracker location data is included in the data release and can be used as a logical mask to determine those locations that were missed. In other words, the laser-tracker data provides location data that tells us precisely where the system was on the site, just not exactly when. The clutter data provides a unique opportunity to explore clutter mitigation algorithms in order to improve single sensor performance. The clutter data with the current location inaccuracies cannot be used to explore sensor fusion across contractor data sets (fusion using sensors on the same platform is possible).

ACKNOWLEDGMENTS

The need to collect background clutter data was originally identified by Dr. Regina Dugan, who at the time worked for IDA. The continued fervent support by Dr. Dugan, currently at DARPA, made collection of the data not only possible but paramount.

REFERENCES

1. Altshuler, Thomas W., et. al. 1997. "Mine and UXO Detection: Measures of Performance and Their Implication in Real-World Scenarios." SPIE Vol. 3079, April 1997.

2. Altshuler, Thomas W., A.M. Andrews, R. E. Dugan, V. George, M. P. Mulqueen, and D. A. Sparrow. October 1995. "Demonstrator Performance at the Unexploded Ordnance Advanced Technology Demonstration at Jefferson Proving Ground (Phase I) and Implications for UXO Clearance." IDA Paper P-3114.

3. Mulqueen, Micheal, V. George, A.M. Andrews, D.A. Sparrow, and R. E. Dugan. April 1995. "Performance Assessment at the Jefferson Proving Ground Demonstration of Systems for the Detection and Identification of Buried Unexploded Ordnance." SPIE Vol. 2496, p. 649, 0-8194-1852-8.

[1] Data is available at the cost of reproduction. Contact Vivian George, Walcoff and Associates, vgeorge@walcoff.com.

NUMERICAL AND MEASUREMENT BASED STUDY OF THE RESONANCE FREQUENCY TECHNIQUE TO DETECT ANTI-PERSONNEL MINES

F. Olyslager[1], J. Fortuny[2], A. Franchois[2], G. Nesti[2], and A.J. Sieber[2]

[1] Department of Information Technology, University of Ghent
Sint-Pietersnieuwstraat 41, B-9000 Gent, Belgium
[2] Space Applications Institute, Joint Research Centre
Via E. Fermi 1, 21020 Ispra (VA), Italy

Introduction

An anti-personnel mine is often mainly a dielectric object which shows specific resonance frequencies. It has been sugested [1] that these resonance frequencies could be used to detect such an object. There are however a number of problems which could impede the usefulness of this technique. The resonance frequencies will change when the surrounding medium changes. Losses in the soil will result in a damping of the resonances and a strong attenuation of the transmitted and reflected waves, making the detection of the resonance frequencies more difficult. Often there are also other objects in the neighbourhood of the mine, such as branches and stones. These objects could have similar resonant behaviour than the mine for which one is looking.

In the present contribution we investigate the effect of each of these complications. For this purpose we use a Finite Difference Time Domain (FDTD) simulation technique which gives us a large degree of freedom in the geometries that can be studied. To validate the technique simulations of existing, plastic and metallic, mines are performed and compared with measurements in well defined circumstances obtained in the European Microwave Signature Laboratory (EMSL) in Ispra, Italy.

The different effects above mentioned are investigated with the FDTD technique applied on a typical dielectric mine. The change and detectability of resonances is investigated for different parameters: depth of the mine, dielectric properties of the soil, conductivity of the soil, roughness of the soil surface and the influence of other objects. Also the attenuation of waves penetrating through the soil is investigated. The simulations reported in this paper give us some insight in the validity of the resonance frequency technique. The main objective in this work is to establish the practical limitations and feasibility of the technique.

Ultra-Wideband, Short-Pulse Electromagnetics 4
Edited by Heyman *et al.*, Kluwer Academic / Plenum Publishers, New York, 1999.

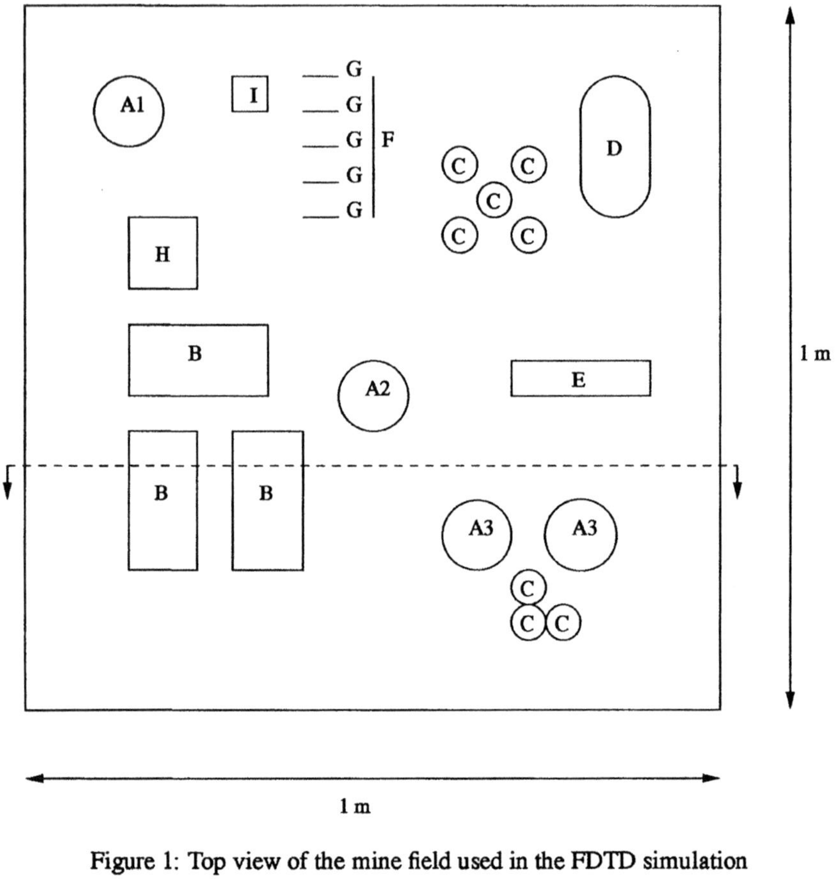

Figure 1: Top view of the mine field used in the FDTD simulation

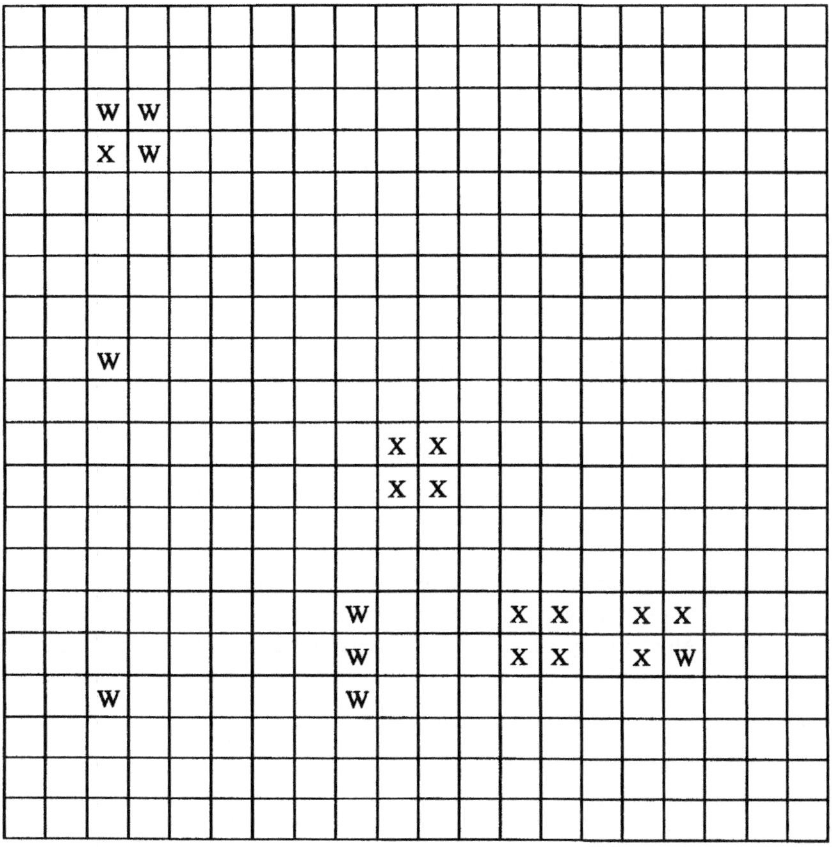

Figure 2: Results of the classification algorithm based on the detection of the resonance frequencies in the FDTD simulation: the W and X marks indicate the identification of two and three resonance frequencies in the backscattered pulse, respectively

FDTD Simulations

As numerical simulation tool we used the Finite Difference Time Domain (FDTD) technique which allows the simulation of a realistic environment. An example of such an environment is shown in Fig. 1. We have a piece of terrain of 1 m$\times$1 m in which a number of objects are buried. The soil is modelled as a lossy medium with relative permittivity $\epsilon_r = 5$ and conductivity $\sigma = 0.28$ S/m. This could correspond to wet sand with about 10% water content. The objects labeled as A1, A2 and A3 are mines modeled as dielectric cylinders with diameter 10 cm, thickness 3 cm and $\epsilon_r = 3$. Mine A1 is buried 6 cm deep, mine A2 is buried 4 cm deep and the mines A3 are buried 2 cm deep. The objects B, C, D and E are stones with different shapes, different permittivities and buried at different depths. The objects F and G are perfectly conducting wires, H is wire loop and I is a small metal plate. Below the dashed line the soil is covered with a 2 cm thick conducting layer with $\epsilon_r = 1.2$ and conductivity $\sigma = 0.005$ S/m. This could model a vegetation layer. The FDTD mesh size was 5 mm and a modified Mur 2nd order absorbing boundary condition was used. For the present problem PML's are not necessary.

As antennas we used two crossed dipole antennas in our simulations which are both parallel to the surface. One dipole is used as transmission antenna and the other as receiving antenna, i.e. a simulation of a near field cross-polarisation measurement is performed. By recording the cross-polarisation the strong reflection of the surface is eliminated provided that the surface is flat. In the simulations we placed the antennas 3 cm above the surface and we did simulations at different positions in a grid with steps of 5 cm in both directions. To accelerate the simulations we only included for each simulation a region of 30 cm$\times$30 cm around the antenna of the total environment. A wideband Gaussian excitation pulse was used with center frequency of 6 GHz and a bandwidth of about 9 GHz.

By division with the input pulse in the frequency domain we derived the impulse response from the Gaussian pulse response. By time gating and the Prony technique the natural complex resonance frequencies were derived. We compared the real part of the three most dominant complex resonance frequencies obtained from a reference simulation above the object A2. If the three resonance frequencies were close to the reference values an X was plotted in Fig. 2. Alternatively, when only two resonance frequencies were in agreement a W was plotted. We only used the real part of the resonance frequencies since it was not possible to get accurate and stable values for the imaginary part using the Prony technique. This has been confirmed for metal objects in [2].

It is clear that the result in Fig. 2 is very encouraging. The resonance frequencies seem to be independent of the depth of the object. This was also confirmed with more specific simulations. For dielectric objects the resonance frequencies depend less on the soil characteristics than for metal objects because an important part of the resonant field is located inside the object. The higher the contrast of the soil with the object the better it can be seen. When the contrast is higher the ringing time will be longer, i.e. the Q-factor of the resonances will be higher. This means that on the one hand it should be easier to find plastic mines in a lossy soil than in a dry soil due to the higher contrast. On the other hand the resonant signal coming from the mine is more attenuated when penetrating through the lossy soil making it more difficult to detect.

We also did simulations with a rough surface. The rough surface generates cross-polarisation making it more difficult to detect the signal coming from the buried object. If the dimensions of the ondulations are small enough such that the cross-polarisation generated by the surface is at sufficiently high frequencies then simulations showed that it is still possible to detect the resonance behaviour of the object. In the simulations we considered ondulations with a size of about 1 cm which corresponded to cross-polarisation frequencies above 8 GHz.

For such ondulations it was still possible to detect the resonance frequencies.

Although the above results look very promising we have to make some remarks regarding the practical applicability. These are simulations in ideal conditions. The dynamic range of the FDTD simulations is very high, something not attainable in reality. Perhaps the most restricting factor comes from the antennas. In the FDTD simulation the isolation between the two antennas is perfect, in reality this will not be the case. It is very difficult to manufacture antennas with a high isolation grade, with sufficient bandwidth and with sufficient dynamic range. Note that with respect to metal objects the reflected signal of a dielectric buried object is very small.

Radar Measurements

In order to validate the numerical simulations with the FDTD technique a number of radar measurements with targets both in free space and buried in an homogeneous soil have been carried out. All measurements have been performed in anechoic chamber using a fully polarimetric stepped frequency radar which can be operated both in monostatic and bistatic modes. This system has a number of unique features. The target under test is mounted on a rotating platform inside an anechoic chamber. The same platform can also be displaced along the cross-range direction in order to synthesize a linear aperture. The looking angle of the antennas may vary within the angular range from -150 to 150 deg keeping a constant range to the target of about 10 m. Fig. 3 shows an inside view of the anechoic chamber.

Two types of tests have been performed in order to assess the quality of the measurement data acquired in the EMSL. The first test was intended to estimate the sensitivity of the measurement system. To this end, two consecutive measurements of the empty room were undertaken in order to see the differences between them. The second test consisted in a comparison between measurements and the theoretical predictions for two different objects: a metallic sphere and a finite metallic cylinder. Such a comparison provided us with an estimate for the accuracy of the measurement system.

The sensitivity of the measurement system (EMSL) has been assessed by looking at the coherent difference between two consecutive empty room measurements spaced in time about one hour. Two series of measurements covering the low and high frequency bands (i.e., 0.5-2.0 GHz and 2.0-9.5 GHz) have been performed. Initial measurements had established that the low band was the most critical area. An upper frequency of 9.5 GHz adequately covers the range of measurement frequencies envisaged and allowed some reduction in calibration time compared to the 12 GHz limit of the selected antennas.

The accuracy the the measurement system has been assessed by comparing measurements and theoretical results for two canonical targets (a metallic sphere and a finite metallic cylinder). Having a reference, one can easily obtain the errors associated to the measurements. We usually apply a time gating to the measurement data in order to filter out spurious reflections from the target support. The measurement errors in the frequency domain have been calculated using always the same time window span. To go from the time domain to frequency domain and vice versa we use the Chirp-Z algorithm [3].

The calculated mean rms error within the frequency ranges 0.5-2.0 GHz and 2.0-9.0 GHz are 0.62 dB and 0.34 dB, respectively. These error bounds show that the acquired data are well calibrated. As expected, the performance of the system is slightly higher in the high frequency band. Note that, due to the time domain windowing, data at the ends of the frequency window are discarded because of the errors introduced in the calibration process. Fig. 4 shows the computed (via a MoM technique) and measured backscattered fields of a finite metallic cylinder of length 10 cm and diameter 3 cm. The agreement between measurements

Figure 3: Inside view of the anechoic chamber used in the experimental validation

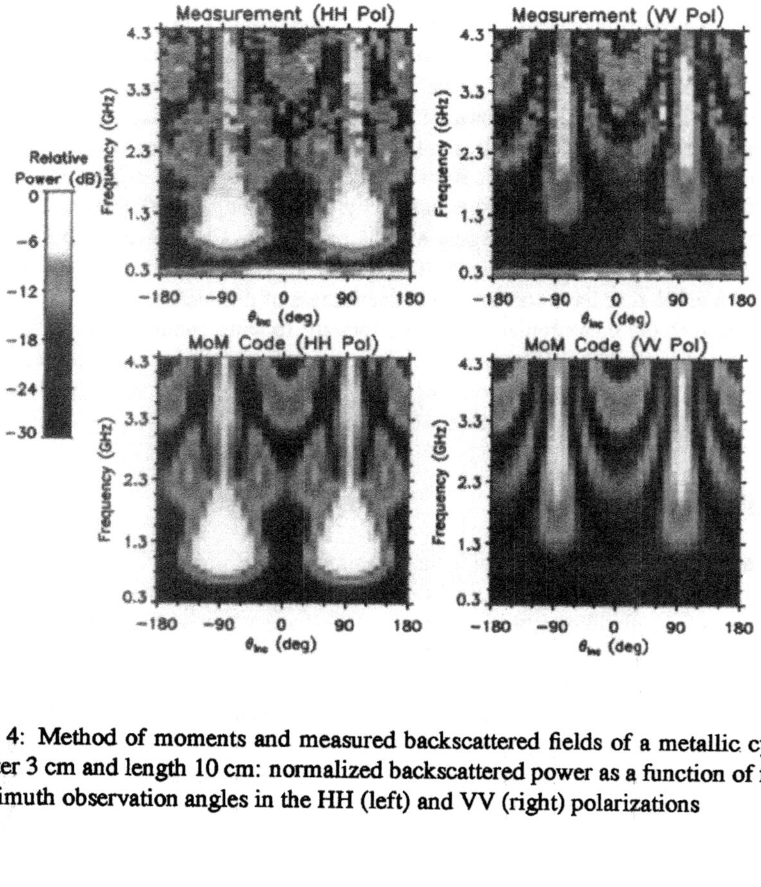

Figure 4: Method of moments and measured backscattered fields of a metallic cylinder of diameter 3 cm and length 10 cm: normalized backscattered power as a function of frequency and azimuth observation angles in the HH (left) and VV (right) polarizations

and theory within the frequency range from 0.5 to 2.0 GHz is excellent. Note that this type of measurement needs an extremely careful empty room subtraction as well as a very precise positioning of the object inside the anechoic chamber.

Measurements have been split into two groups. First, the mine-like objects have been measured in free-space. Then, the same objects have been buried in different types of soil at varying depths. Two frequency bands have been covered: the low band 0.3-4.3 GHz, which has been covered using two dual-polarized logperiodic antennas, and the high band 1.5-9.5 GHz (using dual-polarized horn antennas). For the measurements of the buried objects, a cylindrical container of diameter 2 m and depth 0.5 m has been used. At first, the container has been filled up with dry sand which dielectric constant was about $\epsilon_r = 2.55$.

The mine-like objects buried in dry sand using a cylindrical container used in the measurements vary in shape (cylinders, discs, cubes, ...) and material (metal, teflon, rohacell, ...). These measurements present a number of difficulties. With a forward looking geometry, the return from the surface is still present even if the interface is perfectly flat mainly because the area of the surface illuminated by the antenna is by far larger than that of the buried object. Thus the isolation of the response of the object (including the interaction with the surface) becomes extremely difficult unless the contrast between the object and the surrounding medium is high. Note that the returns from the surface and the buried object will ovelarp in the time domain. This problem may be tackled by synthesizing a finer antenna beam using SAR techniques. The coherent averaging associated to the synthetic aperture filters out the surface and therefore increases the signal to clutter ratio. Measurement results with a metal cylinder and a teflon disc have confirmed the uselfulness of this technique.

In addition to the measurements with the forward looking geometry, some first experiments using a down-looking system which measures only the cross-polar backscatter have been performed. With these test we will try to validate experimentally the FDTD simulations. Data from these measurements are being processed at present.

Concluding Remarks

In all modesty we would like to conclude that the simulations show that it is usefull to explore this technique further towards a practical design. Currently some first experiments seem to support some of the findings from simulations. We have identified a number of practical problems with the laboratory measurements that need to be tackled. We need to find an optimum polarization state that enhances the presence of buried man-made objects (symmetric). A combination of both imaging and extraction of natural frequencies information needs to be defined. We think that laboratory measurements under controlled conditions is a key issue in the establishment of the practical limitations of the proposed techniques.

References

[1] M. C. Worthy and C. E. Baum, "A Library of the Natural Frequency Responces for Cylindrical Shaped Buried Plastic Mines", Phillips Laboratory's Interaction Note 530, July 1997.

[2] C. C. Chen, "Buried Unexploded Ordnance Identification with Complex Natural Resonances,"IEEE Trans. on Antennas and Propagation, Vol. 45, No. 11, pp. 1645-1654, 1997.

[3] Alan V. Oppenheim and Ronald W. Schafer, "Discrete-Time Signal Processing", Prentice-Hall, New Jersey, 1989.

FEASIBILITY STUDY INTO THE IDENTIFICATION OF LANDMINES USING UWB RADAR: AN ANALYSIS USING SYNTHESIZED DATA

J. LoVetri[1,2], S. Primak[1], B.J.A.M. van Leersum[2], and A.P.M. Zwamborn[2]

[1]Department of Electrical Engineering
The University of Western Ontario,
London, Ontario, Canada N6A 5B9

[2]TNO Physics and Electronics Laboratory
P.O. Box 96864, 2509 JG Den Haag, The Netherlands

INTRODUCTION

The goal of this research is to detect and identify buried land mines (LM). The technique being investigated is the use of ultra-wideband (100 MHz to 3 GHz) ground penetrating radar (UWB/GPR). The first step is to gain an understanding of the scattered signal received from the use of such a system. A major point of interest is: which electromagnetic scattering features and signal processing techniques can be used in the detection/identification process? Possible candidates of scattering features are late-time resonances and diffusion poles, whereas signal processing techniques are synthetic aperture radar (SAR) imaging [1], correlation and/or E-pulse receivers [2]. It is important to have good detection and identification algorithms, which work well in the ideal situation, before one can deal with the practical situation, in which there will be uncertainty in various physical parameters such as: the local electrical characteristics of the ground, the local heterogeneity of the ground (rocks, layers, tree and plant roots), surface roughness of the ground and covering foliage, position (depth and orientation) of the LM, and relative position of the radar with respect to the ground and the LM. Thus, in this work, we investigate the post-processing of impulse type radar signals, where the signals are synthesized, for relatively ideal situations, using the finite-difference time-domain (FDTD) technique. Much has been published about this technique so the details will be left for the references [3].

Both the ground and the LM's are modelled as lossy dielectrics. Synthetic impulse responses are obtained using a multistatic configuration (*i.e.* one transmitter point above the ground, many receiver points just above the ground). First we construct SAR images from the multistatic impulse responses. The effect of the air-ground interface is partially removed by running the FDTD code without the LM (system response), as well as with the LM (target response), and then deconvolving the system response from the target response. This produces an approximation for the impulse response for the LM itself.

As a possible identification technique, the complex natural resonances of the radar responses are tabulated for various angles. Prony's method is used to extract the late-time poles from the radar signals.

FDTD SIMULATIONS OF THE GPR PROBLEM

The geometry of the problem which was analyzed using the FDTD technique is shown in Figure 1. A cubical mesh of size 100X100X100 having a cell size of 1 cm was used. A transient dipole (**S1** or **S2** in the figure), modelled by an ideal *y*-directed current source which is 1 cm in length, is located above a lossy ground at grid location (50, 40, 50) for **S1** and location (65, 40, 50) for **S2**, as shown. Only one source was used during any run of the FDTD code. A lossless dielectric cube is placed in the ground and the electric field is sampled at test points **T1-T9** which are 1 cm above the ground. The electrical parameters of the cube and the ground are as shown in Figure 1.

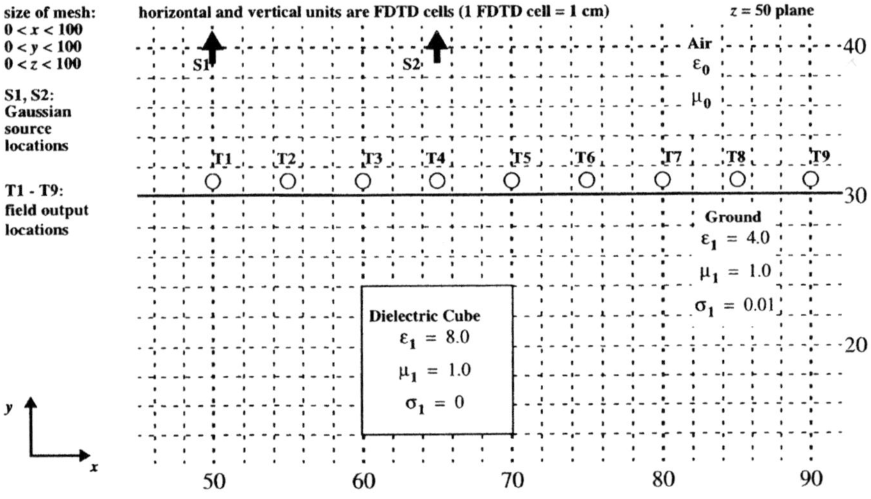

Figure 1. FDTD geometry for buried dielectric cube inside a lossy ground. The horizontal and vertical units are FDTD cells

The time domain variation of the small dipole, **S1** or **S2**, is given by the derivative of a Gaussian function, specifically:

$$g(t) = \begin{cases} 0, & t<0 \\ -2((t-t_0)/b^2)\exp(-(t-t_0)^2/b^2), & t\geq 0 \end{cases} \quad (1)$$

with $b = 5\times 10^{-10}$ s, and $t_0 = 2\times 10^{-9}$ s. A scaled version of this waveform and its Fourier transform are shown in Figure 2. Example plots of the electric fields calculated at some of the observation points are shown in Figure 3.

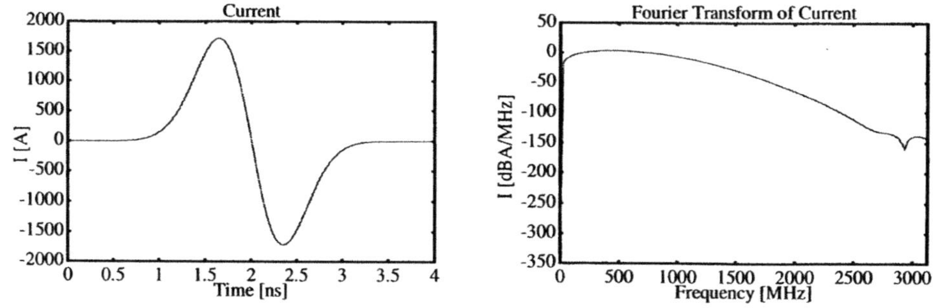

Figure 2. Current waveform and its Fourier transform.

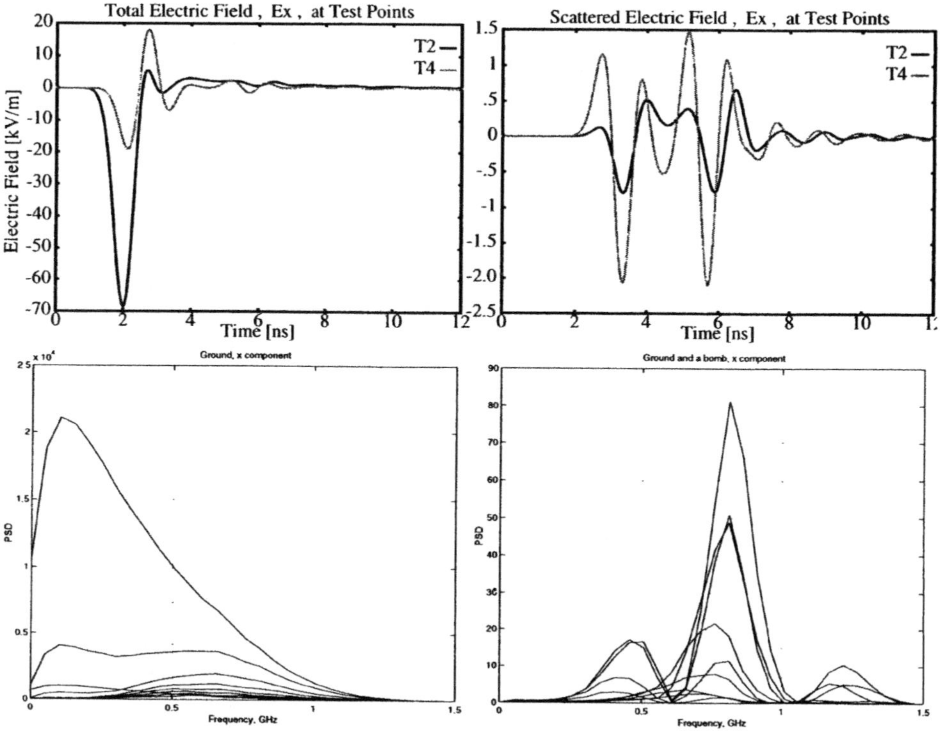

Figure 3. Top: example fields at the observation points: T2 and T4. Bottom: Fourier transform of E_x fields at all observation points. Left: Total fields; Right: Scattered Fields.

ISAR IMAGING

As the first step we have built a set of classical bistatic ISAR imaging [4], assuming that time-domain signal, received at each observation point corresponds to a specific angle of arrival of reflected waves. The following reconstruction algorithms can be found in [4]

$$f_m(\rho, \theta) = \int_{\phi_{min}}^{\phi_{max}} R_m[\rho\cos(\theta - \phi)]d\phi \quad (2)$$

for monostatic measurements, and

$$f_b(\rho, \theta) = \int_{\beta_{min}}^{\beta_{max}} R_b\left[\rho \cos\left(\theta - \frac{\phi}{2}\right)\right] d\phi \tag{3}$$

for bistatic measurements. Here $R_m(t)$ is the monostatic target range profile in the time-domain, taken at aspect angle ϕ, ρ and θ are polar coordinates in the image plane. Observations (slices) are available for aspect angles between ϕ_{min} and ϕ_{max}. $R_b(t)$ is the bistatic target range profile which is measured in the bistatic angular range of β_{min} to β_{max}. As can be seen from equations (2) and (3) the procedure for image reconstruction using bistatic measurements is exactly the same as for monostatic imaging except that instead of rotating the image by the angle of observation and then backprojecting, one uses one half of this angle. For the case considered herein, the values of these bistatic angles are unknown since the phase centre of the effective scatterer is unknown. A set of bistatic SAR images were constructed, each assuming a different phase centre of the effective scatterer. The geometry which was used to calculate the bistatic angle is shown in Figure 4. Some of the results are shown in Figure 5.

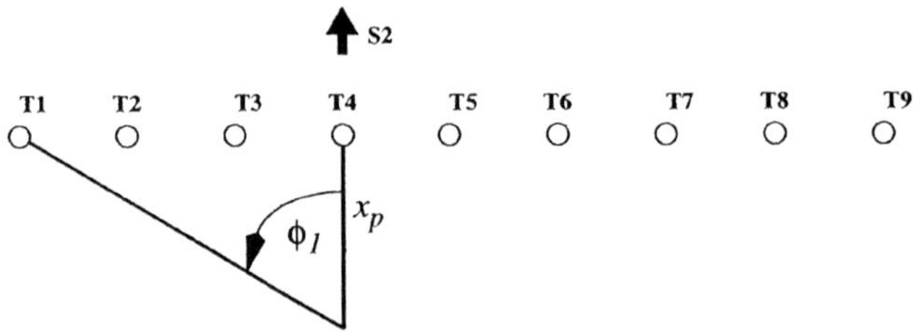

Figure 4. Geometry used to estimate bistatic angles, ϕ_1 to ϕ_9, when source was at **S2**. The distance to the assumed phase centre, x_p, was varied.

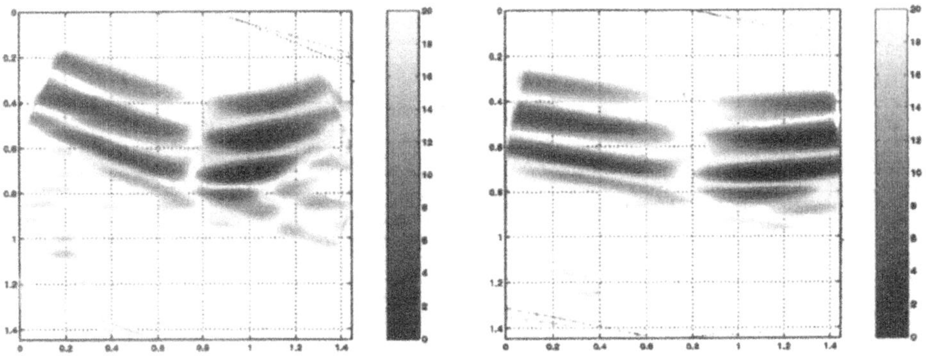

Figure 5. ISAR image corresponding to x_p = 0.75 m, left, and x_p = 1.75 m, right.

PRE-PROCESSING: NEAR-FIELD FOCUSING OF RADAR SIGNALS

The results obtained in the previous section can hardly be considered as satisfactory. This is not surprising since the assumptions inherent in ISAR reconstruction using equations (2-3) are far from being satisfied. First of all, at a single observation point all directions contribute to the scattered field. In this section, an attempt is made to preprocess the scattering data obtained from the FDTD code to get better compliance with the requirements inherent in the ISAR imaging algorithm summarized by equations (2-3).

The observation points in the previously described problem can be considered as the elements of an antenna array as shown in the Figure 6. Time domain responses $r_k(t)$, $k = 1, \ldots, 9$, measured at each observation point are converted into the frequency domain, and are denoted $R_k(\omega)$. A set of weighting coefficients $w_k^{\phi}(\omega)$ for near-field focusing is developed for every frequency ω and direction of the main lobe ϕ. A combined frequency domain response is calculated as

$$R^{\phi}(\omega) = \sum_{k=1}^{9} w_k^{\phi}(\omega) R_k(\omega) \tag{4}$$

and then converted back to a time domain response using the inverse Fourier transform. A schematic view of the complete procedure is depicted in Figure 7.

The weights, or coefficients $w_k^{\phi}(\omega)$, are developed in such way, that the corresponding directivity of the corresponding antenna array has a maximum at aspect angle ϕ as shown in Figure 6. This can be achieved (see [5]) by setting the magnitude of each coefficient equal to unity

$$\left|w_k^{\phi}(\omega)\right| = 1 \tag{5}$$

and setting the phase using a quadratic dependence on the position n of the element in the array, that is

$$\angle w_k^{\phi}(\omega) = -2\pi f_X' nd + \frac{k}{2r'}(nd)^2 . \tag{6}$$

Here, d is the distance between two sequential elements of the array, $k = \omega/v$ is the wave vector, v is velocity of propagation, f_X' is a spatial frequency related to the aspect angle ϕ and frequency ω as

$$f_X' = \frac{\omega \cos\phi}{2\pi v} . \tag{7}$$

Since this is a near-field design, a distance r' must be specified which gives the distance from the centre of the array at which the focusing is desired. Various values of r' were used, from 10 cm to 1000 m, in order to see the effect in the SAR images.

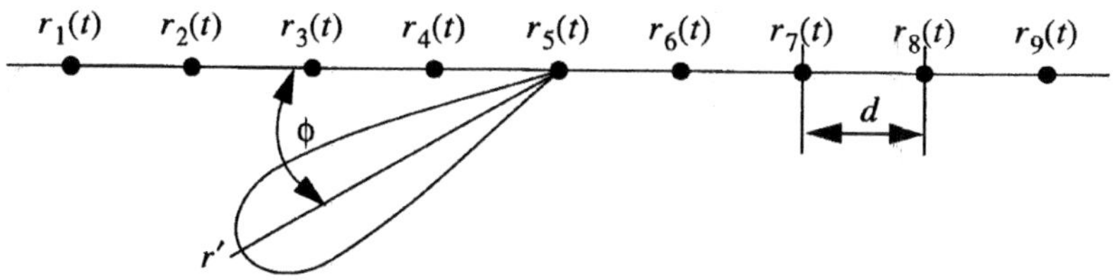

Figure 6. Test points as an antenna array which can be focused to a point in the near-field.

The returned signal, observed at each point, has frequency content limited by the frequency spectrum of the excitation. This reduces the number of frequency bands where the weighting coefficients must be calculated. A typical synthesized time-domain waveform for $\phi = 90°$ is shown in Figure 8. Signals were synthesized for 25 angles, $\phi = 30° \ldots 150°$ in steps of $5°$ and the corresponding SAR images using these signals are shown in Figure 9 for $r' = 10$ cm, and $r' = 10$ m.

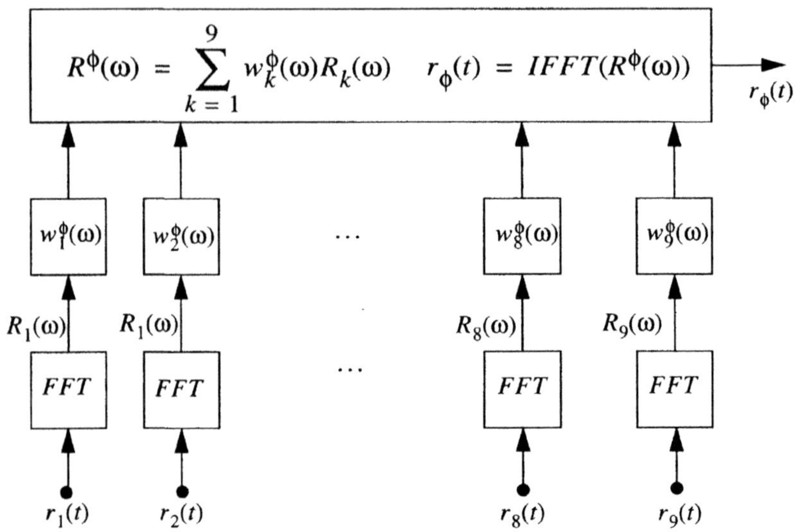

Figure 7. Weighted sum preprocessing of the data.

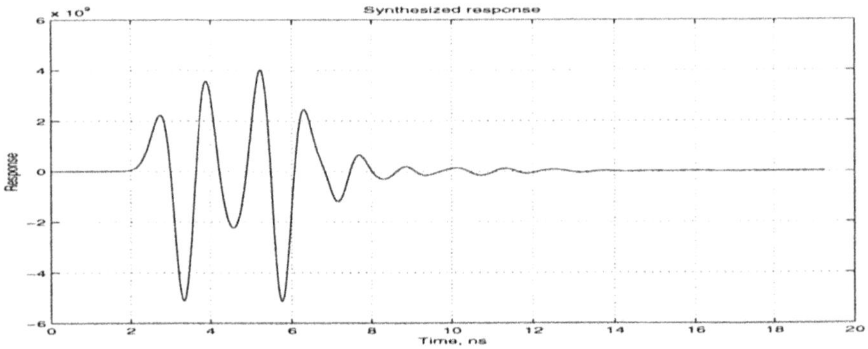

Figure 8. Synthesized signal using near-field focusing algorithm, $\phi = 90°$.

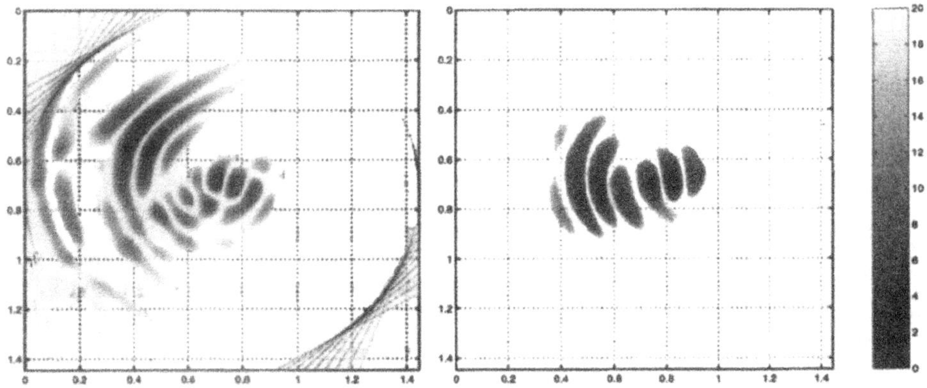

Figure 9. ISAR image obtained using beam-forming algorithm. Left $r' = 10$ cm, Right $r' = 10$ m.

IDENTIFICATION USING LATE-TIME POLES

In frequency ranges where the scatterer produces a considerable natural response (that is, when the wavelength of the incident field is in the order of the size of the target) the late-time scattering response can be approximated as a finite sum of damped sinusoids. The damping time constant and the frequency correspond to complex poles which may be used to identify the scatterer. This is the basis of the so called singularity expansion method (SEM) [6]. According to this model, in the late-time, the time-domain scattered waveform can be represented as

$$r(t) = \sum_{k=1}^{N} A_k \exp(\sigma_k t)\cos(2\pi f_k t + \varphi_k),\ t > T_L. \tag{8}$$

Here, $s_k = \sigma_k + j2\pi f_k$ represent characteristic poles of the target, A_k is the intensity of the corresponding pole, and T_L describes the beginning of late-time at which this model is appropriate. A well known technique for determining the complex poles and residues is Prony's method. The method used is this paper follows that of van Blaricum *et al.* [7, 8].

Before we apply Prony's method to the FDTD data, an attempt is made at removing the system response from the data. The system response is defined as that due to the ground and the source, including the time-domain waveform of the source excitation. A picture of this is shown in Fig. 10. The measured target response can be related to the measured system response and the impulse responses by convolution, that is $s(t)*h(t) = c(t)$. Therefore, in order to obtain the impulse response, $h(t)$, of the target we must deconvolve the system response, $s(t)$, from the measured target response, $\delta(t)$. The procedure we follow for getting the impulse response is the same as that given by Rothwell *et al.* [9].

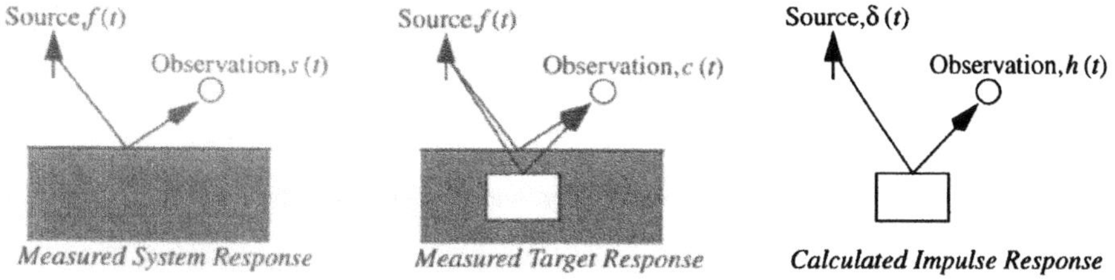

Figure 10. Definition of measured system, measured target, and calculated impulse responses.

Once the impulse response is computed, Prony's method is used to obtain the poles and residues of the late-time portion. The start of the late-time, T_L, is determined in an iterative fashion by varying T_L, computing the damped sinusoid approximation for that T_L using Prony's method, and determining the relative Euclidean norm of the error between the real impulse response and the approximation. When the relative Euclidean norm falls below a specified criterion, we accept the value of T_L as suitable. Example plots of the synthesized system response, the synthesized target response, the calculated impulse response with the approximate signal reconstructed using Prony's method, and the complex poles which were obtained are shown in Figure 11.

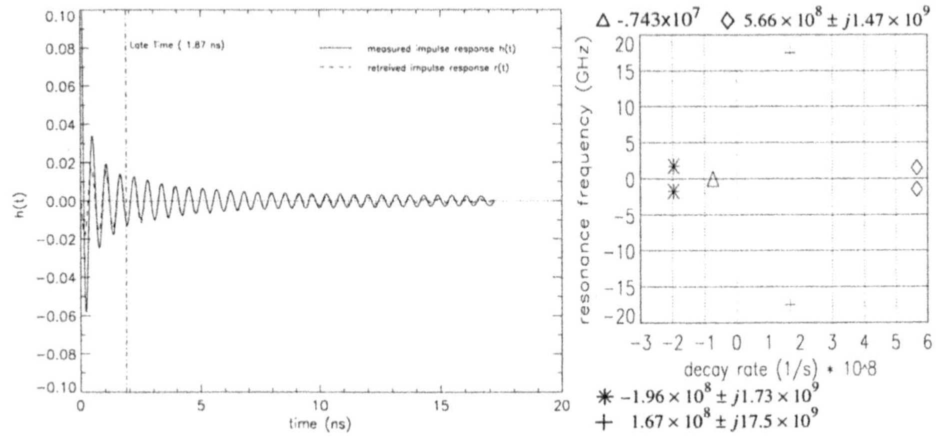

Figure 11. Example plots of the calculated impulse response with the approximate signal reconstructed using Prony's method, and the complex poles for the impulse response.

CONCLUSIONS

Synthetic data of the GPR problem was created using the FDTD technique. Signals for the scattering from various configurations of buried dielectric objects were calculated and SAR images were created from a bistatic set-up consisting of a single wide-band dipole positioned 10 cm above the ground and 9 receiver points just above the ground. A pre-processing algorithm which focuses the scattered signal which was obtained at all 9 observation points was used with the focusing angle and focusing depth chosen at various values to create a set of new signals. The new signals were used to create SAR images of the buried target. This procedure seems feasible for the detection problem since only one wide-band shot was needed, with 9 receiver points. The impulse response at a single receiver point was calculated using a stable deconvolution scheme and Prony's method was applied in order to extract the complex poles. Preliminary investigations are positive that this technique will prove valuable in the identification problem.

REFERENCES

1. E.M. Johansson, J.E. Mast, "Three-dimensional ground penetrating radar imaging using synthetic aperture time-domain focusing," *SPIE* Vol. 2275, pp. 205-214, 1994.
2. L.C. Chan, L. Peters, Jr., D.L. Moffat, "Improved performance of a subsurface radar target identification system through antenna design," *IEEE Trans. AP*, vol. 29, no. 2, pp. 307-311, March 1981.
3. K.S. Kunz and R.J. Luebbers, *The Finite Difference Time Domain Method for Electromagnetics*, CRC Press, 1993.
4. H.-J. Li, and F.-L. Lin, "A generalized interpolation and prediction in microwave imaging involving frequency and angular diversity", *Journal of Electromagnetic Waves and Applications*, Vol. 4, No. 5, 1990, pp. 415-430.
5. L. Ziomek, *Fundamentals of Acoustic Field Theory and Space-Time Signal Processing*, CRC Press, Ann Arbor, 1995.
6. C. Baum, E. Rothwell, K.-M. Chen, and D. Nyquist, "The singularity expansion method and its applications to target identification", *Proc. IEEE*, Vol. 79, No. 10, October, 1991.
7. M.L. van Blaricum, and R. Mittra, " A technique for extracting the poles and residues of a system directly from its transient response," *IEEE Trans. AP*, vol. 23, no. 6, pp. 777-781, 1975.
8. M.L. van Blaricum, and R. Mittra, " Problems and solutions associated with Prony's method for processing transient data," *IEEE Trans. AP*, vol. 26, no. 1, pp. 174-182, 1978.
9. E.J. Rothwell, and W. Sun, "Time domain deconvolution of transient radar data," *IEEE Trans. AP*, vol. 38, no. 4, pp. 470-475, 1990.

ULTRA-WIDEBAND SYNTHETIC APERTURE RADAR FOR MINE FIELD DETECTION

Lawrence Carin, Norbert Geng and Mark McClure

Department of Electrical and Computer Engineering
Duke University
Durham, NC 27708-0291

Jeffrey Sichina and Lam Nguyen

Army Research Laboratory
AMSRL-SE-RU
Adelphi, Maryland 20783

I. INTRODUCTION

Ground-penetrating radar (GPR) constitutes one of the oldest technologies for subsurface sensing [1-9]. Most such systems are placed in direct or near-direct contact with the earth surface [3,4,8,9]. A significant drawback of this approach is the lack of "standoff", a particular problem for the detection of buried ordnance (*e.g.*, mines [10]). Moreover, it is time consuming to use such hand-held systems for large-area interrogation over variable terrain and foliage. The US Army Research Laboratory (ARL) has therefore undertaken the development of a synthetic aperture radar (SAR) system, with which significant standoff can be achieved [11]. This system provides surveillance of large areas, the results from which can dictate the smaller regions over which conventional hand-held systems (*e.g.*, electromagnetic induction [10], conventional GPR [1-9], and magnetometers [10]) should be deployed. Therefore, while ideally we would like to use the SAR system to detect and distinguish *each* mine, we are particularly interested in mine *field* detection, over very large areas. Consequently, we need not detect each mine, but rather mine *clusters*. This simplified problem statement still constitutes a significant technological challenge, particularly in highly cluttered (natural and anthropic) environments.

To enhance discrimination between targets and clutter, the excitation should be as wideband as possible. However, it is also essential to operate at frequencies for which adequate soil (and foliage) penetration can be effected, particularly for deeply buried targets or regions in which there has been significant foliage growth (for humanitarian

Ultra-Wideband, Short-Pulse Electromagnetics 4
Edited by Heyman *et al.*, Kluwer Academic / Plenum Publishers, New York, 1999.

demining, one often must deal with mine fields that have been in the ground for decades [10]). The ARL ultra-wideband (UWB) SAR therefore transmits and receives waveforms with usable bandwidth from 50-1200 MHz. Moreover, this system is fully polarimetric, providing access to HH, HV, VH, and VV imagery.

While radar-based subsurface sensing is an old technology, until recently there has been very little modeling done to characterize expected system performance as a function of mine type and environment. In practice there can be wide variability in the mines deployed and in environmental (background) conditions. Therefore, system performance has historically been anecdotal, with excellent performance under certain conditions and unsatisfactory performance elsewhere. Recently, however, we have developed rigorous electromagnetic scattering models [12-16] for conducting and plastic mines buried in lossy, layered soil. Our models are based on a full-wave formulation of Maxwell's equations, with solution via the method of moments (MoM). The principal challenge to making such an analysis tractable is computation of the halfspace or layerered-medium Green's function [17], which we have performed efficiently via the method of complex images [18]. Moreover, we have restricted ourselves to targets that can be simulated as a body of revolution (BOR) [11-16], simplifying computational complexity significantly. While not all mines can be modeled as a BOR, the vast majority of land mines (both conducting and plastic) are accurately so modeled [10]. The details of these models are described elsewhere, while here we concentrate on their application to measured SAR imagery.

In addition to using wave modeling to elucidate phenomenology and guide system design, the target models play an important role in detection and identification algorithms. In particular, optimal detectors require an accurate model for the signature of the target in question [19]. For the case of buried mines, until recently one was required to use *measured* signatures for such applications. However, empirical target models for buried mines are of limited value due to the strong dependence of the target signature on the mine type, target depth, and soil properties. We have therefore utilized our simulated target models as integral parts of a detector, with results presented here for data measured by the BoomSAR at Yuma Proving Ground (Yuma, AZ, USA). We also address wavelet-based compression techniques which reduce algorithm complexity.

II. MEASUREMENT SYSTEM AND THEORETICAL MODEL

The Army Research Laboratory (ARL) has been pursuing the use of UWB SARs for a number of years, to help understand foliage penetrating and ground penetrating radar and to develop and qualify suitable wide-band models for predicting performance of proposed systems. These radars provide 1 GHz of instantaneous bandwidth and the full polarization matrix. The current implementation, hereafter referred to as the BoomSAR, is a mobile extension of the original design, allowing data collection over a wide range of clutter and target-in-clutter scenarios. The majority of the hardware is mounted to the basket of a telescoping boom lift capable of moving at approximately 1 km/hr while the basket is elevated to 45 m (see Fig. 1). For typical collection geometries, down-look angles to the target vary from 45° to approximately 10° (from grazing), depending on the range to the target and the height of the boom.

As in conventional SAR, the BoomSAR is moved along a straight path, and for each position the scattered response is measured. In its simplest embodiment, the bipolar image amplitude $I_{m,n}$ at pixel position (m,n) is given by

$$I_{m,n} = \sum_{k=1}^{K} s_k(t = T_{k,m,n}) \tag{1}$$

where $s_k(t)$ represents the time-domain scattered field measured at the kth sensor position and $T_{k,m,n}$ is the round-trip time delay between sensor position k and the physical position in the image represented by pixel (m,n). The focused image has a resolution of 0.15 m in the range dimension, and 0.3 m in the cross-range dimension.

The phenomenology associated with UWB scattering from plastic and conducting land mines has been investigated through development of a full-wave scattering model for a body of revolution (BOR) buried in a lossy, dispersive, layered medium. In addition to studying UWB scattering from such targets [15], the algorithm has been modified to model their natural resonances [16] as well. The principal numerical hurdle involves efficient and accurate computation of the background-medium Green's function. The efficient computation of the layered-medium Green's function is effected via the method of complex images [18], with careful attention for the surface-wave and leaky-wave pole positions encountered in the Sommerfeld integral. The latter issue is of particular concern for the complex frequencies characteristic of the natural target resonances [16].

Figure 1. Photograph of the ARL BoomSAR.

III. COMPARISON OF THEORY AND MEASUREMENT

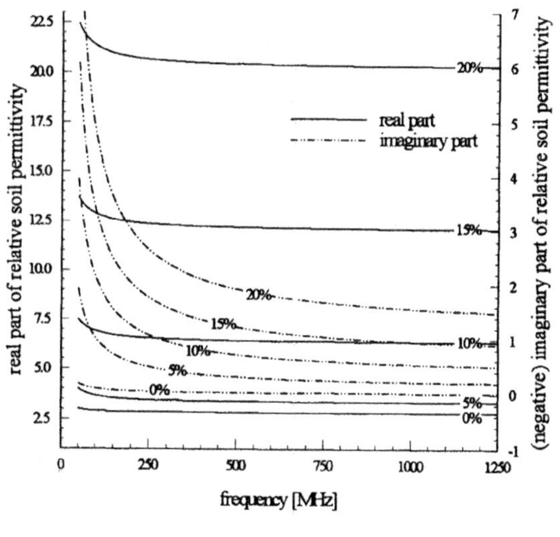

Figure 2. Measured real and imaginary part of the frequency-dependent complex dielectric constant of soil from Yuma Proving Ground, Yuma, AZ (USA).

We consider targets buried in the soil characterized in Fig. 2, measured using soil samples taken from Yuma Proving Ground, Yuma, AZ (USA). The soil characterization was performed with a Hewlett-Packard (HP) coaxial-probe test fixture in conjunction with an HP network analyzer, with measurements performed as a function of water content (by weight). The initial set of scattering examples below are for soil with approximately 5% water, representing relatively low loss (these were the conditions encountered when we performed our SAR measurements).

Equation (1) gives the connection between the measured (time-domain) fields and the subsequent SAR image. For a comparison of theoretical and measured SAR imagery, it is essential to use (in the theory) an incident waveform consistent with that of the actual UWB SAR system. A through measurement was therefore performed, from which an approximation to the incident pulse was determined. It should be noted, however, that in general the antennas have frequency-dependent properties that have a strong angular dependence [20]. Moreover, each of the waveforms $s_k(t)$ measured along the SAR aperture constitutes a different angle between the sensor and the target. Therefore, in actuality, there is *not* a single pulse shape incident on the target, from all aperture positions. Similar issues hold with regard to the polarization purity – an antenna may be vertically polarized at boresight, but the polarization can change markedly with angle. To account for this, only a limited number of aperture positions are used to image a given pixel in the SAR image (such that the properties of the incident field on that pixel are similar for all waveforms used in image formation). In this work the image is segmented into a collection of contiguous tiles (sub-images), with 60° of aperture used for each tile, ±30° with respect to the normal between the SAR aperture and the tile center.

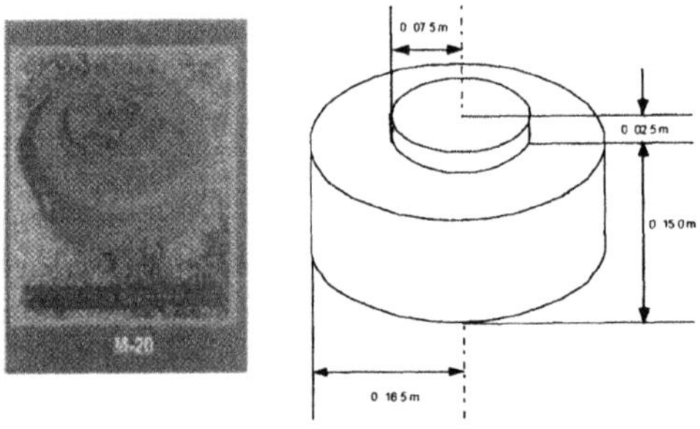

Figure 3. M20 anti-tank mine used in measurements and calculations. The diameter is 32 cm, and the height 15 cm.

We compare the scattering model with measurements by forming theoretical SAR images. In computing $s_k(t)$, we account for the k-dependent angle of incidence to a given pixel, assuming plane-wave excitation, pure polarization properties and a fixed incident-pulse shape. The latter two conditions are dictated by our incomplete knowledge of the antenna properties over UWB frequencies, while the plane-wave-incidence approximation is appropriate for the SAR applications of interest. Finally, the wveforms $s_k(t)$ used to form the theoretical image are weighted as a function of aperture position, to reflect the properties of the antenna pattern. A Gaussian weighting was used. In particular, for aperture position y, we use the weighting $w(y)=\exp[-(y-y_i)^2/2\sigma^2]$, where y_i represents the aperture point at which a normal from the pixel in question intersects the aperture, and σ =21.34 m was found to yield theoretical cross-range performance similar to that found from the measured SAR imagery.

We consider the M20 anti-tank mine, a relatively large conducting target, shown in Fig. 3. The measured and computed (bipolar) SAR imagery are compared in Fig. 4 for the M20 on the surface of the Yuma soil, for VV polarization (vertical incident polarization, vertical received polarization). Similar accuracy was found when the target was buried. From these results we see that the agreement between the measured and computed imagery is good. These results bolster our confidence that scattering from the three-dimensional mine is modeled accurately. Note that the imagery is characterized by significant structure, a result of the UWB bandwidth and the cross-range resolution provided by the SAR aperture.

IV. MINE AND MINE-FIELD DETECTION

If the clutter were white, Gaussian and additive, and the target signature known exactly, the optimal detector is a matched filter [19]. In reality, for the data we have measured, the clutter is generally nonstationary, characterized by discrete scatterings from localized naturally occurring obstacles (bushes, rocks, petrified wood, animal burrows, etc.). Nevertheless, the results of Sec. III indicate that the scattering model is in accord with the measurements, suggesting the development of a matched-filter-type detection scheme. It should be emphasized that this approach makes good use of our model but is suboptimal in that it does not exploit the statistics of the clutter (*i.e.*, it improperly assumes the clutter is white and Gaussian). The strong non-stationary quality of the clutter makes determination of the clutter statistics difficult, and therefore we do not attempt such here.

Even if the aforementioned simplifications with regard to the clutter are utilized, the simple matched-filter detector has limitations. In particular, the target signature for the buried land mine is a function of the burial depth and the background soil properties. Therefore, if $I_t(x,y;\gamma)$ represents the model target image, as a function of the parameter vector γ (symbolically representing the target depth and soil properties), we implement the generalized matched filter as [19]

$$I(x_o, y_o) = \int dx \int dy \int d\gamma M(x,y) I_t(x - x_o, y - y_o; \gamma) p_\gamma(\gamma) \qquad (2)$$

where $M(x,y)$ represents the measured SAR imagery, (x_o, y_o) denotes the shifted position of the *computed* template $I_t(x,y;\gamma)$, and $p_\gamma(\gamma)$ represents the estimated density function on the parameters γ. The test statistic $I(x_o, y_o)$ is compared to a threshold t, and a target is declared at point (x_o, y_o) if $I(x_o, y_o) > t$, while clutter is declared (*i.e.*, no-target) if $I(x_o, y_o) < t$. By varying the threshold t, the receiver operating characteristic (ROC) is generated [19], characterized as the probability of detection versus the probability of false alarm.

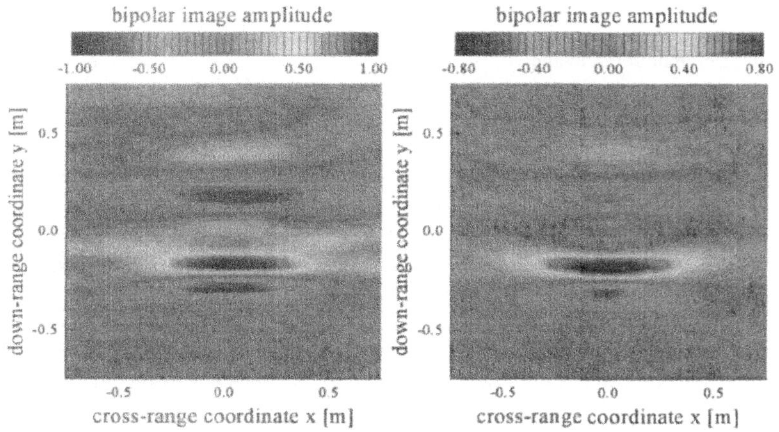

Figure 4. Comparison of measured (left) and computed (right) synthetic aperture radar images for mine in Fig. 3 situated in the soil in Fig. 2. The mine is on the soil surface; similar results were found when the mine was flush-buried just under the surface and buried 15.24 cm.

There are several issues with regard to (2) that require further discussion. First, there are many examples for which the clutter response is much stronger in amplitude than that of the desired target. Consequently, even though the *shape* of $M(x,y)$ is not matched to $I_t(x,y;\gamma)$ in the vicinity of the clutter, the former's strong amplitude can yield a large l (and hence a false alarm). For the clutter considered here, this is a particular problem. Therefore, when computing (2), we normalize $M(x,y)$ such that $\int\int dx\, dy M^2(x,y)=1$, where the integration is performed over the principal support of $I_t(x,y;\gamma)$. A second issue with regard to (2) is the possibility of obtaining multiple "hits" on the same target (or clutter), in which several closely situated *discrete* positions (x_o,y_o) exceed the threshold. Therefore, after running the detector, we distinguish localized pixel clusters that exceed the threshold, and

Figure 5. Synthetic aperture radar imagery for a 0.46 km^2 section of the total 1.9 km^2 area. Locations of detection are marked with a white dot. Left image: Target declared (detected) at all points for which the image amplitude is consistent with that of a buried mine (energy detector). Right image: Results for the detector in (2), with the threshold t set for a probability of detection equal to 0.7 and a false alarm rate of 1/km^2. To make the results clear, in the right image the rectangle locates the region in which mines exist. In the right image detections were only found in this region, while on the left detections (false alarms) were found throughout the image.

declare the total cluster as a detection (correct or otherwise). This cluster therefore defines a "halo", and if the halo intersects the position of a target (known from "ground truth"), then a correct detection is declared, otherwise it is scored as a false alarm. The size of the clusters, and hence the halo, is defined to be the putative size of the buried target.

The parameter density function $p_\gamma(\gamma)$ is used to account for environmental uncertainty. In particular, while the target types that may be encountered are often known *a priori* in demining, the exact soil properties and target depth are generally unknown. However, $p_\gamma(\gamma)$ accounts for the anticipated variability in these parameters. In the examples considered here, we seek, for example, M20 mines buried to a depth of 15.24 cm. For those results $p_\gamma(\gamma)$ is designed such that the depth varied normally between 12.70 to 17.78 cm (to account for real-world variability in the burial depth). Moreover, we considered soil moisture content (Fig. 2) varying normally between 0 to 6%. We also considered uniform distributions for these parameters, and little change was witnessed in the results.

The procedure in (2) exploits the accurate models discussed in Sec. II, while accounting for realistic environmental variability. However, the requisite computations are intensive for a large swath of SAR imagery (two-dimensional spatial integration, or summation, for each realization of γ, with the number of γ required dictated by the density $p_\gamma(\gamma)$). However, there are techniques available to make such two-dimensional spatial integrations tractable. In particular, we perform the two-dimensional integration, for *all* shifts (x_o, y_o) *simultaneously*, through exploitation of the properties of the Fast Fourier transform (FFT). Both the templates $I_t(x,y; \gamma)$ and measured imagery $M(x,y)$ are converted to the two-dimensional wavenumber domain via a two-dimensional FFT in the coordinates x and y. The product of the two spectral-domain functions is performed, followed by the two-dimensional inverse FFT, from which all spatial shifts are acquired simultaneously. This procedure is performed separately for each γ required in (2), as dictated by $p_\gamma(\gamma)$.

While the two-dimensional FFT results in a significant reduction in computational complexity, matters can be simplified further through use of a multi-resolution search strategy. In particular, we initially consider the templates $I_t(x,y; \gamma)$ and image $M(x,y)$ at low resolution, by filtering their spectral content, thereby allowing down-sampling. The test statistic (2) is computed at low resolution, and only those pixels for which a given threshold is exceeded are processed at the next, higher-resolution level (for which greater spectral content is used in both $I_t(x,y; \gamma)$ and $M(x,y)$). At each stage of the multi-resolution hierarchy, the regions in $M(x,y)$ at which mines are anticipated is pruned, and through this pruning process the highest resolution (full-band) computation of (2) is only performed over a relatively limited domain of $M(x,y)$. We have performed this multi-resolution scheme via use of the Daubecheis perfectly reconstructing wavelet [21] (in two dimensions). For the imagery of interest here, we have found four levels useful (full band, 1/4, 1/16 and 1/64 full band). Therefore, through this procedure, (2) is only computed for *all* shifts (x_o, y_o) at 1/64 the resolution, followed by pruning. This process is repeated at 1/16 full band, then at 1/4 full band, and, finally, at full band. Therefore, the full-band, most computationally expensive computations in (2) are performed over a relatively limited portion of $M(x,y)$. In this process one must carefully account for the shift-dependent properties of the wavelet transform [22].

We present here results for detection of M20 mines. There were several other mine types placed in the test zone, which are deemed false alarms if detected by the algorithm; *i.e.*, these additional mines were considered (anthropic) clutter to the detector designed for the M20. Using the detector in Sec. IV, designed for both flush-buried and 15.24 cm deep mines, we plot in Fig. 5 the locations in the image that exceed a prescribed threshold t (with t selected for a probability of detection of 0.7 and a false alarm rate of 1/km^2). These results indicate that the mine field has been located properly. Note, each mine need not be detected to constitute a mine field, and stray detections can be discarded as false alarms. As a comparison of the quality of the detector in (2), relative to a simpler approach, in Fig. 5 we also identify portions of the SAR image that have signature amplitudes consistent with that expected for the mine in question. From Fig. 5 we see that such an energy-based detection criterion (which does not exploit the detailed target signatures, in Fig. 4) results in an unacceptably high false-alarm rate.

V. CONCLUSIONS

In demining applications, one generally must interrogate a large swath of terrain. Moreover, standoff is of obvious benefit when attempting to locate these insidious targets. Therefore, ARL has focused on development of a SAR-based system that affords both of these desired attributes. Moreover, the ARL BoomSAR operates over ultra-wideband frequencies (50-1200 MHz), simultaneously providing soil penetration and resolution. Synthetic-aperture radar results have been presented for data measured at Yuma Proving Ground, Yuma, AZ. These data were used to confirm our MoM-based scattering models as well as predictions with regard to the polarimetric properties of most land mines. In particular, we confirmed that mines that approximate a body of revolution (BOR) excite no cross-polarized backscattered fields.

After substantiating the model accuracy, we used the computed target signatures to effect a detector. This detector accounted for the real-world situation in which the expected mines are generally known, but the exact (spatially varying) soil properties and target depths are often not known exactly. We accurately detected a large percentage of the mines, at a relatively low false alarm rate. Moreover, the false alarms tended to be spread sporadically throughout the SAR image (based on *a posteriori* ground-truth knowledge), while the correct mine detections tended to be clustered, as expected of a mine field. Hence, SAR appears to be an attractive option for mine-*field* detection. Finally, a wavelet-based optimization scheme was investigated to improve detector efficiency. Results from this preliminary multi-resolution study appear attractive, but significantly greater computational savings can be achieved by exploiting a multi-resolution scheme that is shift invariant (unlike the spatial wavelet transform).

REFERENCES

[1] R.W.P. King and C.W. Harrison, Jr., "The transmission of electromagnetic waves and pulses into the earth," *J. Appl. Phys.*, vol. 39, pp. 4444-4452, Aug. 1968.
[2] J.A. Fuller and J.R. Wait, "Electromagnetic pulse transmission in homogeneous dispersive rock," *IEEE Trans. Antennas Propag.*, vol. AP-20, pp. 530-533, July 1972.
[3] D.L. Moffat and R.J. Puskar, "A subsurface electromagnetic pulse radar," *Geophysics*, vol. 41, pp. 506-518, June 1976.
[4] L. Peters, Jr. and J.D. Young, "Applications of subsurface transient radars," in *Time-Domain Measurements in Electromagnetics*, E.K. Miller, Ed., New York: Van Nostrand Reinhold, 1986.
[5] G.S. Smith and W.R. Scott, Jr., "A scale model for studying ground penetrating radars," *IEEE Trans. Geosci. Remote Sensing*, vol. GE-27, pp. 358-363, July 1989.
[6] C. Liu and C. Shen, "Numerical simulation of subsurface radar for detecting buried pipes" *IEEE Trans. Geosci. Remote Sensing*, vol. GE-29, pp. 795-798, Sept. 1991.
[7] N. Osumi and K. Ueno, "Microwave holographic imaging of underground objects," *IEEE Trans. Antennas Propag.*, vol. AP-33, pp. 152 - 159, Feb. 1985.
[8] L. Peters Jr., J.J. Daniels and J.D. Young, "Ground penetrating radar as an environmental sensing tool," *Proc. IEEE*, vol. 82, pp. 1802-1822, Dec. 1994.
[9] J.M. Bourgeois and G.S. Smith, "A fully three-dimensional simulation of ground penetrating radar: FDTD theory compared with experiment," *IEEE Trans. Geosci. and Remote Sensing*, vol. GE-34, pp. 36-28, Jan. 1996.
[10] A.C. Dubey and R.L. Barnard (Eds.), Detection and Remediation Technologies for Mines and Minelike Targets, *SPIE Proc.*, vol. 3079, Orlando 1997
[11] S. Vitebskiy, L. Carin, M.A. Ressler and F.H. Le, "Ultra-wideband, short-pulse ground-penetrating radar: simulation and measurement," *IEEE Trans. Geosci. Remote Sensing*, vol. GE-35, pp. 762-772, May 1997.
[12] S. Vitebskiy and L. Carin, "Moment-method modeling of short-pulse scattering from and the resonances of a wire buried inside a lossy, dispersive half-space," *IEEE Trans. Ant. and Prop.*, vol. 43, pp. 1303-1312, Nov. 1995.
[13] S. Vitebskiy, K. Sturgess and L. Carin, "Short-pulse plane-wave scattering from buried perfectly conducting bodies of revolution," *IEEE Trans. Ant. and Prop.*, vol. AP-44, pp. 143-151, Feb. 1996.
[14] S. Vitebskiy and L. Carin, "Resonances of perfectly conducting wires and bodies of revolution buried in a lossy dispersive half-space," *IEEE Trans. Ant. and Prop.*, vol. AP-44, pp. 1575-1583, Dec. 1996.

[15] N. Geng and L. Carin, "Wideband electromagnetic scattering from a dielectric BOR buried in a layered lossy, dispersive medium," submitted to *IEEE Trans. Ant. and Prop.*
[16] N. Geng, D.R. Jackson and L. Carin, "On the resonances of a dielectric BOR buried in a dispersive layered medium," submitted to *IEEE Trans. Ant. and Prop.*
[17] K. A. Michalski and D. Zheng, "Electromagnetic scattering and radiation by surfaces of arbitrary shape in layered media, Parts I and II," *IEEE Trans. Antennas Prop..*, vol. AP-38, pp. 335 - 352, March 1990.
[18] R. M. Shubair and Y. L. Chow, "A simple and accurate complex image interpretation of vertical antennas present in contiguous dielectric half-spaces," *IEEE Trans. Antennas Prop.,* vol. AP-41, pp. 806 - 812, June 1993.
[19] H.L. Van Trees, Detection, Estimation, and Modulation Theory, Part 1, pp. 261-264, Wiley, New York, NY, 1968.
[20] M.A. Ressler and J.W. McCorkle, "Evolution of the Army Research Laboratory ultra-wideband test bed," in Ultra-Wideband Short-Pulse Electromagnetics 2, L. Carin and L.B. Felsen eds., Plenum Press, NY, pp. 109-123, 1995.
[21] I. Daubechies, Ten Lectures on Wavelets, Society for Industrial & Applied Mathematics, May 1992.
[22] H. Sari-Sarraf and D. Brzakovic, "A shift-invariant discrete wavelet transform," *IEEE Trans. Signal Proc.*, vol. 45, pp. 2621-2626, Oct. 1997.

MODIFICATION OF THE APERTURE SYNTHESIZING METHOD FOR UNDERGROUND SOUNDING DATA PROCESSING*

V.O. Kovalenko, S.A. Masalov

Institute for Radio Physics and Electronics of NAS of Ukraine
12, Ak. Proskury st., Kharkiv, Ukraine, 310085
e-mail: smith@ire.kharkov.ua

Creation of the subsurface radar systems is the actual problem in modern radio physics and thus is investigated by many of authors. The ultrawideband (videopulse) scanning systems usually produce data which are hard to interpret. This leads to necessity in developing of efficient data processing algorithms. A Method of Synthetic Aperture (MSA) seems to be one of the most powerful technique of data processing. One of the first attempt to apply it to the subsurface data is described in [1,2].

ASM is based upon the following two-stage procedure:

1.application of compression filter to the reflected by the medium signals (echo-returns); the filter indicates the presence of the object reflected signal;

2.spatial convolution of the obtained set of the echo-returns.

The first stage stands for increasing of the depth resolution and the second one — for increasing of horizontal resolution.

ASM seems to be one of the most powerful technique for radio-location data processing. It is successfully used in the fields where Fresnel-Kirhgoff scalar diffraction theory satisfactory acts (this includes location in air and non-dissipative media sounding). Although efficiency of its application to the data, obtained over stratified media having non-negligible resolution is rather limited. One of the basis reasons is strong attenuation of the pulse and its shape distortion while propagation in such media. These processes lead to the failure of the compression filter and thus algorithm does not work correctly.

* This work is partly supported by the Science and Technology Center of Ukraine (project #366)

Ultra-Wideband, Short-Pulse Electromagnetics 4
Edited by Heyman *et al.*, Kluwer Academic / Plenum Publishers, New York, 1999.

On the other hand, consideration of approximate a-priory information regarding the medium investigated enables to estimate the pulse shape on the given depth. In this paper we suggest the modification of the SAM that uses construction of the compression filter based on the estimated shape of the pulse. This approach permits to increase vertical resolution in lossy media.

Another problem connected with using of ASM is significant computational time needed during echo-returns spatial convolution on the second stage. Modification of the computational technique that 3 times shorts computational efforts at this stage is suggested.

We conclude with the natural experiment results, which indicate efficiency of the algorithms.

PROBLEM STATEMENT

The results of the ultrawideband subsurface radar system work usually are implemented as the set of the echo -returns $\{v(x_r,t)\}_{r=1}^{M}, t \in [0;\Theta]$ (Θ is the quantity o the time reaming) which describe the underground structure in the rectangular area which is defined by the length of the scanning route and the Θ. They are implemented as the matrix $N \times M$, where N is the quantity of the time samples distinguished in each realization and M is the quantity of the realizations in the route. The classical SAM supposes that connection between the received at the point $(x_r,0)$ and transmitted at the point $(x_r + L,0)$ is established as follows:

$$v(x_r,t) = \frac{1}{2\pi V}\int_{\Gamma}\frac{z(L_T + L_R)}{(L_T L_R)^2}p(x,z)e^{-\alpha(L_T+L_R)}u'(t - \frac{L_T + L_R}{V})dx \quad (1)$$

where V is the velocity in the medium, Γ is the object contour, $p(x,z)$ is the complex reflectance amplitude, x_r is the point at the route, $L_T = \sqrt{(x - x_T)^2 + z^2}$, $L_R = \sqrt{(x - x_R)^2 + z^2}$ are the distances between the antennas and the contour point, α is the attenuation constant. The relation (1) corresponds to the Fresnel-Kirhgoff diffraction theory.

The data procession using the SAM reduced to the applying to the $v(x_r,t)$ inverse-matched filter with the transfer function:

$$H_1(f) = \frac{V_1{}^*(f)}{1 - \eta + \eta|V_1(f)|^2}\cdot\sum_{i=1}^{I} c_i \cos 2\pi f t_i \quad (2)$$

where f is the frequency and $0 < \eta < 1, \{c_i\}_1^I, \{t_i\}_1^I$ are the set of the defined parameters, * means the complex conjunction and $V_1(f)$ is the spectrum of the time derivative of the radiated signal. And then the spatial convolution of the echo-returns set is provided:

$$b(x,z) = \int_{A} w(x_r, \frac{L_T + L_R}{V})\cdot\frac{z\cdot(L_T + L_R)}{(L_T L_R)^2}e^{-\alpha(L_T+L_R)}dx_r \quad (3)$$

here A is the implementation of the route, $w(x,t)$ is the output of the filter defined by (2), $b(x,z)$ is so named imaging function [2] - the result of the data procession.

INVERSE-MATCHING FILTER MODIFICATION

The first problem considered is the scanning pulse shape distortion in conducting media (see fig. 1). The Fresnel-Kirhgoff diffraction model, represented by (1) does not take it into account, and thus filter (2) does not work properly. The shape of the radiated pulse can be estimated using preliminary information about the medium being under the investigation.

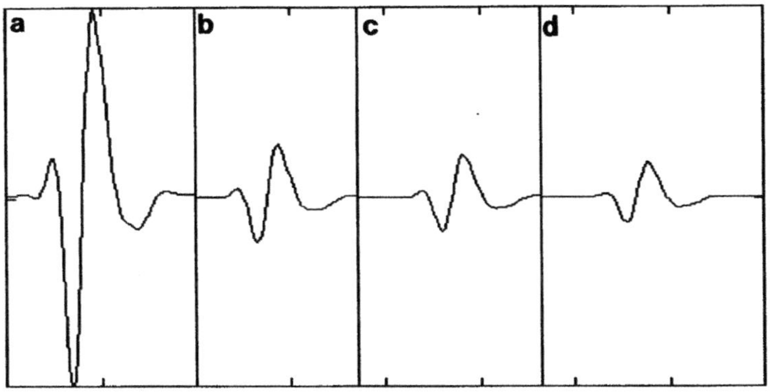

Figure 1. Radiated pulse and its form (out of scale) in the depths of 1,2,5 meters in the loam ($\varepsilon_r = 8$, $\sigma = 1.2 \cdot 10^{-2}$)

One can consider the conducting medium as the filter with transfer that can be obtained as follows [3]:

$$G(d,\omega) = |\Gamma(\omega)| \exp\{-d\alpha(\omega) - i(\varphi_\Gamma(\omega) - d\beta(\omega))\} \tag{4}$$

here is the transfer function at the depth d, $\omega = 2\pi f$, f is the frequency, ε_0 - free space permittivity, ε_1 - related permittivity of the medium, σ_1 - conductivity, c - free space velocity,

$$\Gamma(\omega) = |\Gamma(\omega)| e^{i\varphi_\Gamma(\omega)}, \qquad \Gamma(\omega) = \frac{2\frac{\omega}{c}}{\frac{\omega}{c} + \beta(\omega) + i\alpha(\omega)}, \tag{5}$$

$$\alpha(\omega) = \frac{\omega}{c}\sqrt{\varepsilon_1(\omega) sh(\frac{1}{2} arcsh \frac{\sigma_1(\omega)}{\omega\varepsilon_0\varepsilon_1(\omega)})}, \quad \beta(\omega) = \frac{\omega}{c}\sqrt{\varepsilon_1(\omega) ch(\frac{1}{2} arcsh \frac{\sigma_1(\omega)}{\omega\varepsilon_0\varepsilon_1(\omega)})}$$

The shape of the electromagnetic pulse estimated accordingly to (5) in loam is presented at the fig.1. The filter (2) uses $u'(t)$ as the reference function and thus may to give unpredictable errors in its output. However estimation of the $G(d,\omega)$ demands to use only the values of the ε_1 and σ_1 which both can be estimated beforehand. It is suggested to use following function instead of $u'(t)$ as the reference one in (2) in order to achieve more precisely information from the depth d:

$$V_d(t) = F^{-1}\left[G(d,\omega)F\left[\frac{d}{dt}F^{-1}\left[F[u(t)]G(d,\omega)\right]\right]\right] \tag{6}$$

where $F[\ ]$ means Fourier transform and $u(t)$ is the scanning signal. Figure 2 represents the real echo-profile , and result of its processing using (2) with $V_d(t)$, $d = 2$.

It can be seen from (2) that modified filter more exactly represents data from the range of depths 1.8-2.2 m.

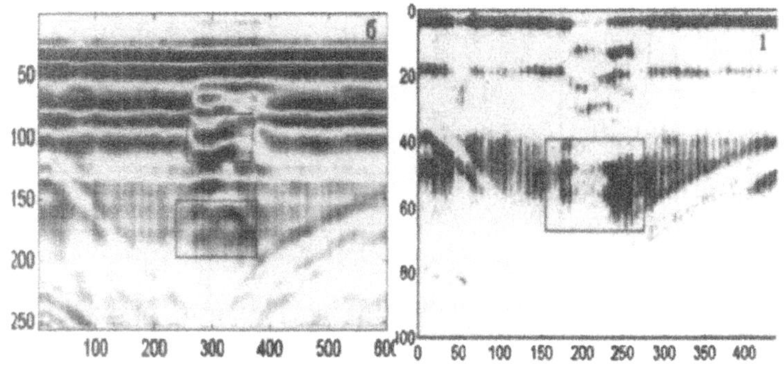

Figure 2. Real echo-profile and result of its filtering by the modified inverse-matched filter

THE CONVOLUTION PROCEDURE MODIFICATION

Replacing in (3) integrals with finite sums one can obtain:

$$B_{xz} = \sum_{r=1}^{N} z \cdot \frac{L_T + L_R}{(L_T L_R)^2} w_{r\tau} e^{\alpha(L_T + L_R)} \tag{7}$$

where B_{xz} is the sampled imaging function, M is the quantity of the realizations on the route, $\tau = \left[\frac{L_T + L_R}{2}\right]$, and $[\ ]$ means the integer part,

$$w_{r\tau} = \begin{cases} w(z,t)\big|_{\substack{z=r \\ t=\tau}} & ,\text{if } t \in [0;\Theta] \\ 0 & ,\text{otherwise} \end{cases} \tag{8}$$

Here w_{rt} is the output of the filter (2). Further

$$L_T = \sqrt{(x-r)^2 dx^2 + z^2 dz^2}$$

$$L_R = \sqrt{(x - r - \lambda)^2 dx^2 + z^2 dz^2} \tag{9}$$

where $dx = \dfrac{L}{M}$, L is the length of the route, $dx = \dfrac{\Theta}{N}$, $\lambda = \dfrac{T - R}{dx}$, if λ is integer (this means following: every point of radiation becomes later the one of reception; so this requirement can be can be satisfied physically) then the following equalities hold:

$$L_R = L_T(x, z, r + \lambda),$$

$$L_T(x+1, z, r) = \sqrt{(x - (r-1))^2 dx^2 + z^2 dz^2} = L_T(x, z, r-1), \tag{10}$$

So one does not need to recount the quantities of L_R on each step and $L_T(x,r,z)$ is counted the $\dfrac{2N}{N+\lambda}$ times less. Since the typical values of the N and λ are about of 256 and 10 respectively, the last modification leads to the significant improvement of the algorithm.

Furthermore, taking into account the facts of $\tau \in [1;N] \Leftrightarrow t \in [0,\Theta]$ and hyperbola-like character of varying of τ (see fig.3) one can replace (7) with:

$$B_{xz} = \sum_{r=x}^{\tau \le N} z \cdot \frac{L_T + L_R}{(L_T L_R)^2} w_{r\tau} e^{\alpha(L_T + L_R)} + \sum_{r'=x}^{\tau \le N} z \cdot \frac{L_T + L_R}{(L_T L_R)^2} w_{r\tau} e^{\alpha(L_T + L_R)}, \tag{11}$$

where $r' = x - r$. In the figure 3 one can see the graph of $\tau(r)$. Points over the horizontal line correspond to the zeros in the (8) and thus would be neglected in calculations. Dotted line in the figure corresponds to summation by the formula (7) and solid one - to the formula (11). The replacement is resulting as follows: one starts the summation from the point corresponding to x instead of 1 and provides the summations in the opposite directions until the horizontal line is reached and then computation stops. It is obvious that this approach allows to shorten the computations.

Modified in such a manner algorithm is working approximately three times rapidly than classical one and produces the same results.

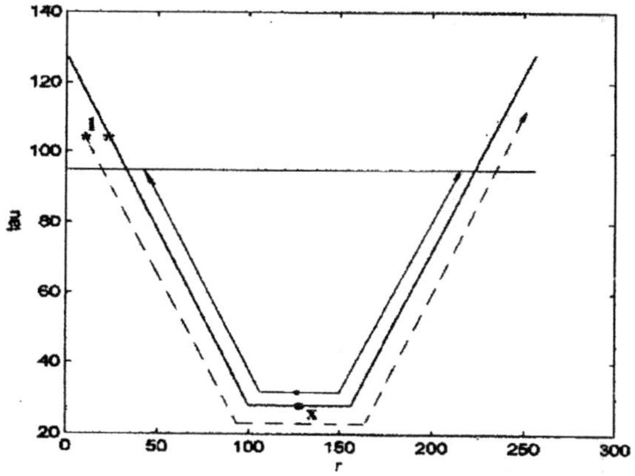

Figure 3. Graph of the function $\tau(r)$ and ways of summation. Dotted line classical scheme, solid line - suggested scheme.

CONCLUSION

Natural and numeric experiment results show that useful algorithms for subsurface data processing can be developed on the basis of the synthetic aperture algorithm. Nevertheless some modifications of this algorithm have to be done in order to achieve accessible results in real media data procession. In this paper the algorithms aimed at the correction of the inverse-matched filter and computation time shortening are presented.

ACKNOWLEDGMENT

Authors are truly grateful to Prof. Henning F. Harmuth (Catholic University of America) for his attention and collaboration.

REFERENCES

1. N.Osumi, K.Ueno, Microwave holographic imaging method with improved resolution, *IEEE Transaction on Antennas and Propagation*, pp.1018-1026, no 10, oct. 1984
2. N.Osumi, K.Ueno, Microwave holographic imaging of underground objects, *IEEE Transaction on Antennas and Propagation*, pp.152-159, no 2, feb. 1985
3. R.King , M.Smith, Antennas in Matter, Moscow, Mir, 1984 (Russian Translation).
4. Henning F. Harmuth, Sequency Theory: Foundations and Applications, Moscow, Mir, 1980 (Russian Translation)

TIME FOCOUSING OF ELECTROMAGNETIC SOUNDING IN CONDUCTIVE MEDIA USING SIMILARITIES BETWEEN WAVE AND DIFFUSION PROPAGATION

Meir Gershenson
Coastal Systems Station, Code R22
Dahlgren Division, Naval Surface Warfare Center
Panama City Florida 32407-7001
GershensonME@ncsc.navy.mil

ABSTRACT

By using similarities between EM sounding in dielectric and conductive media, it is shown that one can transform between solutions in one type of propagation to the other. The method is based on the similarities of the Laplace transform between the diffusive and the nondiffusive cases. In the diffusive case the equation involves the Laplace variable s in the first power while for the nondiffusive cases similar equations occur with s^2. Three alternative implementations are developed and their use is demonstrated. The first implementation is based on substitution s^2 for the Laplace transform variable s using forward and inverse numerical Laplace transform. The second implementation is based on expanding the diffusive time response on an exponential time base and replacing it with its image function in the wave case, namely a sinusoidal function. The third implementation is based on direct transformation in the time domain using exponential time interval sampling. The advantages of the techniques are their simple implementation. Other advantages and limitations of the method and each of the implementations are discussed. Application of common techniques used in processing of seismic, radar for processing and EM sounding in conductive media are discussed.

INTRODUCTION

Time domain reflection seismology and GPR dominates by far any of electromagnetic (EM) sounding in subsurface remote sensing when the media is dominated by conductivity. A major reason for the preference of acoustics or GPR over EM is the ease of interpreting wave reflections. Similar to radar in free space, GPR satisfies the wave equation, which allows the identification of individual reflections. In contrast, EM sounding in conducive media satisfies the diffusion equation, which makes identification of individual reflections difficult. Analogies between wave propagation and diffusion propagation have been recognized in earlier papers, Kunetz G. (1972), Isaev (1981), Filatov (1984). A method for calculating the electrical response of conductive media was presented by Lee et al. (1989). The method is based on similarities between the differential equation governing signal propagation in dielectric media and propagation in a conductive media. It allows the calculation of the time response in a conductive media based on a free space solution. An earlier paper by Levy, et al. (1987) presented a method

Ultra-Wideband, Short-Pulse Electromagnetics 4
Edited by Heyman *et al.*, Kluwer Academic / Plenum Publishers, New York, 1999.

for inverse transformation of magnetotelluric data that resulted in wave-like reflections. Use of such similarities to construct the equivalent of migration is developed by Lee et al. (1987.) In a paper by Lee and Xie (1993), a similar transformation was used to tomographically reconstruct EM cross borehole data. In this paper, three simple transformations are presented. These transformations map diffusive EM waves in conductive media into propagating waves in a nondiffusive media. The methods also allow for the transformation from a nondiffusive into a diffusive medium. Such a transformation allows for easy identification of individual reflections. It also enables the methods developed for non-dispersive wave propagation to be applied to EM sounding in resistive media. This work is a part of a Navy Independent Research Program into Underwater Mine Counter Measures.

THEORY

Lee et al. (1989) worked out the similarity between the wave and the diffusion equations. For the sake of thoroughness and to introduce additional details, some of the derivation will be repeated here. In the time domain, the wave equations for electric and magnetic fields in a linear media have the form

$$\nabla^2 \vec{e}_n - \varepsilon \cdot \mu \cdot \frac{\partial^2}{\partial t^2} \vec{e}_n = \nabla \times \vec{m} + \mu \cdot \frac{\partial}{\partial t} \vec{j} \tag{1}$$

$$\nabla^2 \vec{h}_n - \varepsilon \cdot \mu \cdot \frac{\partial^2}{\partial t^2} \vec{h}_n = \nabla \times \vec{j} + \sigma \cdot \vec{m} \tag{2}$$

For dielectric and conductive media

$$\nabla^2 \vec{e}_d - \sigma \cdot \mu \cdot \frac{\partial}{\partial t} \vec{e}_d = \nabla \times \vec{m} + \mu \cdot \frac{\partial}{\partial t} \vec{j} \tag{3}$$

$$\nabla^2 \vec{h}_d - \sigma \cdot \mu \cdot \frac{\partial}{\partial t} \vec{h}_d = \nabla \times \vec{j} + \sigma \cdot \vec{m} \tag{4}$$

The symbol *e* stands for electric field, *h* for magnetic field and *t* for time. Subscript *n* denotes non-diffusive media, and *d* denotes diffusive media. The terms *m* and *j* on the right hand side represent the magnetic moment and current distribution of the excitation sources. Using the Laplace transform equations (1) through (4) become

$$\nabla^2 \vec{E}_n - \varepsilon \cdot \mu \cdot s^2 \cdot \vec{E}_n = \nabla \times \vec{M} + \mu \cdot s \cdot \vec{J} \tag{5}$$

$$\nabla^2 \vec{H}_n - \varepsilon \cdot \mu \cdot s^2 \cdot \vec{H}_n = \nabla \times \vec{J} + \sigma \cdot \vec{M} \tag{6}$$

$$\nabla^2 \vec{E}_d - \sigma \cdot \mu \cdot s \cdot \vec{E}_d = \nabla \times \vec{M} + \mu \cdot s \cdot \vec{J} \tag{7}$$

$$\nabla^2 \vec{H}_d - \sigma \cdot \mu \cdot s \cdot \vec{H}_d = \nabla \times \vec{J} + \sigma \cdot \vec{M}, \tag{8}$$

where capital letters are used for the transformed function and s is the Laplace transform variable. The left-hand side for the diffusive media can be transformed into the nondiffusive one by the relation

$$\sigma \cdot s \rightarrow \varepsilon \cdot s^2. \tag{9}$$

The right-hand side of equations (7) and (8) describe the nature of the source of excitation (magnetic, electric). Unfortunately, under the transformation described by equation (9), the excitation depends on *s*, or on time. Therefore, the right hand side of equations (3) and (4) {or (7) and (8)} will transform into time dependence which is different from that of the corresponding equations (1) and (2) {or (5) and (6)}. To handle that, the excitation will be limited to the form

$$F(r)\cdot G_i(s), \tag{10}$$

where $F(r)$ is the space distribution of the excitation, r is position vector and $G_i(s)$ is the Laplace transform of the time dependence of the excitation, i stands for diffusive or nondiffusive (d or n). Different excitations will be allowed between the diffusive and the nondiffusive cases. Equations 5 through 8 for the electric and the magnetic fields can be written in the general form

$$\nabla^2\vec{X}_n - s^2\cdot\varepsilon\cdot\mu\cdot\vec{X}_n = F(r)\cdot G_n(s) \tag{11}$$

and

$$\nabla^2\vec{X}_d - s\cdot\sigma\cdot\mu\cdot\vec{X}_d = F(r)\cdot G_d(s). \tag{12}$$

After certain algebraic manipulations and assigning the value of σ to the value of ε in the transformed domain ($\varepsilon = \sigma$) the following occurs

$$X_d(r,s) = X_n(r,s^{\frac{1}{2}})\cdot\frac{G_d(s^{\frac{1}{2}})}{G_n(s)}, \tag{13}$$

or

$$X_n(r,s) = X_d(r,s^2)\cdot\frac{G_n(s^2)}{G_d(s)}. \tag{14}$$

The first inversion technique that allows one to determine $x_n(t)$ from $x_d(t)$ is based on the above relation and is performed by numerically calculating the Laplace transform, substituting s^2 for s, and then performing numerical inversion of the Laplace transform. An alternative method for calculating the Laplace transform of s^2 uses the relation

$$F(s^2) = \int_0^{\infty} f(t)\cdot\exp(-s^2\cdot t)\,dt. \tag{15}$$

It should be kept in mind that inverse Laplace transform is an ill-posed problem (Craig and Thompson 1994). This technique will be called *Direct Laplace Transform Inversion.* It would be advantageous to be able to use the Fourier transform instead of the Laplace Transform by replacing the complex variable $i\omega$ with $-\omega^2$ obtaining

$$F(i\omega) = \int_{-\infty}^{\infty} f(t)\cdot\exp(\omega^2\cdot t)\,dt. \tag{16}$$

But the transform would involve the negative portion of the complex plane where it is not defined.

If the excitation G is limited to the form

$$G_d(s) = s^{n_d} \quad G_n(s) = s^{n_n} \tag{17}$$

which includes the most common time-dependent excitations, i.e. delta, step, and ramp functions, then equation (13) becomes

$$X_d(r,s) = X_n(r,s^{\frac{1}{2}})\cdot s^n \tag{18}$$

$$n = \frac{n_d}{2} - n_n \quad \text{or} \quad m = 2\cdot n + 1. \tag{19}$$

Following the same steps as the direct Laplace transform inversion, but using an analytical rather than numerical derivation (Appendix A) yields

$$x_d(r,t) = \frac{1}{2^{\frac{m}{2}}\sqrt{\pi}\,t^{\frac{m+1}{2}}}\int_0^{\infty} e^{-\frac{q^2}{4t}} \cdot He_m\left(\frac{q}{\sqrt{2t}}\right)\cdot x_n(r,q)dq$$

$$m = 0, 1, 2... \qquad (20)$$

Here $He_m(x)$ are Hermite polynomials. (These are discussed in more detail in appendix A.) Equation (20) enables the transformation between the lossless media and a lossy media. Lee et al. (1989) coined the term *q domain* for the *equivalent Wave field.* While the above relation allows the calculation of $x_d(t)$ from $x_n(t)$, the inverse problem of calculating $x_n(t)$ from $x_d(t)$ requires solution of an integral Fredholm equation of the first kind. This integral equation is an ill-posed integral transform and its solution usually involves inversion of large matrices. An important conclusion of equation (20) is that the transformation given by equation (9) is linear.

Equation (20) can be transformed to a more useful form by introducing the exponential time series
$t_i = t_o \exp(2d\, i)$, will lead to (Gershenson 1997):

$$x_d(t)\cdot t^{\frac{m+1}{2}-\alpha}\cdot\frac{2^{\frac{m}{2}}\sqrt{\pi}}{\delta q_o} = \left\{He_m\left[\frac{q_o}{\sqrt{2t}}\right]t^{-\alpha}\cdot\exp\left[-\frac{q_o^2}{4t}\right]\right\} * \left\{x_n\left[q_o\sqrt{\frac{t}{t_o}}\right]\left(\frac{t}{t_o}\right)^{-\alpha+\frac{1}{2}}\right\} \qquad (22)$$

m is determined by equation (19). Equation (22) is equivalent to equation (20) but has advantage that efficient convolution and deconvolution routines are available. Equation (22) defines the second inversion method that will be called *Telluric Deconvolution.* The uniqueness of the deconvolution depends wether the Fourier transform of

$$A = He_m\left[\frac{q_o}{\sqrt{2t}}\right]t^{-\alpha}\cdot\exp\left[-\frac{q_o^2}{4t}\right] \qquad (23)$$

is free of finite zeros, which may not be the case. Even though the Fourier transform might not contain any finite zeros, it drops quite rapidly at high frequencies. In practice, it is impossible to reconstruct the nondiffusive equivalent, d_n , in full detail, especially in the presence of noise. This is equivalent to being an ill-posed problem. By using exponential time sampling, which is similar to a log plot, we use more data points to represent early times. The intervals between data points are proportional to their time. Equation 22 is independent of the "*exponential time*", therefore, the effect of the deconvolution on the noise is also independent of the *exponential time*. A second consequence is the constant resolution on a logarithmic plot, results in the reduced resolution with travel time (Spies, 1989).

A third inversion method is based on the linearity of the transformation between the diffusive and the nondiffusive cases. It is possible to construct a transformation by expanding Xd(t) on a function base with a known transformation under equation (9). Particular pair of functions that are related by the transformation $s \rightarrow s^2$ is the exponential decay function, and its image the sinusoidal function. Expanding $x_d(t)$ yields

$$x_d(t) = \int_0^{\infty} g(a)\cdot e^{-at}da \qquad (24)$$

$g(a)$ is determined by the inverse Laplace transform. Performing the Laplace transform and after certain manipulations yields

$$X_d(s^2) = \int_0^\infty \frac{1}{s^2+a} g(a) da = \int_0^\infty \frac{2a}{s^2+a^2} g(a^2) da = X_n(s). \tag{25}$$

Finally inverting this expression for $X_n(s)$ results in the expression

$$x_n(t) = \int_0^\infty 2 \cdot \sin(a \cdot t) \cdot g(a^2) da \tag{26}$$

which is the definition of the sine transform with

$$g(\alpha) = L^{-1}\{x_{d(t)}\} \tag{27}$$

The sine transform can be approximated using the Discrete Sine Transform (DST). The method will be called *the Laplace-sine transform.* This result is quite interesting because it involves the inverse Laplace transform rather than the direct transform as an intermediate result. This method illustrates the analogy between the sine excitation in wave propagation and the decay exponential excitation in a diffusive propagation. A consequence of this is the analogy to the Doppler effect in diffusive media (seawater), which will be addressed in the discussion section.

APPLICATIONS

Most applications in seismic or electromagnetics can be employed for EM propagation in diffusive media using the preceding transformation. A few reservations have to be considered before applying the transformation. The most important distinctions are the reduced resolution and increased sensitivity to noise. A practical limitation is the large, slow rate of decay to the response of the media under excitation. Such a slow decay can dominate any returns and impose the need for a receiver with large dynamic range. This slow decay can cause false interpretations when the receiver is unstable relative to the media such as in a marine environment with a towed sensor.

Contrary to conventional sounding where the relative time difference between two sensors might be sufficient to determine direction, because of the use of logarithmic time scale, it is necessary to know the exact time with respect to the source excitation time. This is easily done in active sounding. Two such cases of active sounding will be presented following the discussion. When the time behavior of the source is not available, alternative approaches, which include generalized impedance and cross correlation, are possible and will be discussed.

The first alternative is the use of a generalized impedance (or admittance), which is the ratio between two measurable frequency-dependent responses. The magnetotelluric tensor is such a case; other possibilities include the S matrix and the ratio between the electric field at two different locations. Lee et al. (1989) show the utility of a similar transformation applied to the magnetotelluric impedance.The other alternative, when time information on the source is not available, is to use cross correlation. The cross correlation between two channels is defined as

$$r_{xy} = \int_{-\infty}^{\infty} x(t) \cdot \bar{y}(t-\tau) dt, \tag{28}$$

where the bar stands for complex conjugate. Cross correlation between two sensors is widely used in sonar to obtain time-delay estimates when dealing with an unknown signal source (Clifford, 1987). The Fourier transform of the cross correlation is given by

$$R_{xy} = X(\omega) \cdot \overline{Y}(\omega) \tag{29}$$

where a capital letter stands for the Fourier transformed function. Under the assumption of far field source in nondispersive media, $x_l(t)$ can be written as

$$X_l(\omega) = F(\omega) \cdot e^{-i\frac{\omega}{c}r_l}, \tag{30}$$

where r_l and r_m are the distances to the source. Hence,

$$R_{l,m}(\omega) = |F(\omega)|^2 \cdot e^{-i\frac{\omega}{c}(r_l - r_m)}. \tag{31}$$

Time delay information is available through the phase of cross correlation. In a diffusive media, the cross correlation of the two channels is given by

$$R_{l,m}(\omega) = |F(\omega)|^2 \cdot e^{\sqrt{\frac{-i\omega}{\sigma\mu}}(r_l - r_m)}. \tag{32}$$

Transformation of the time domain cross correlation into the q domain will transform equation (43) into a form analogous to equation (41). This allows extraction of a time-like parameter that is directly related to the bearing of the source. Contrary to the impedance case, time behavior of the source enters equation (41) and in principle it affects the choice of the order of Hermite polynomial. In practice, $F(\omega)$ is usually normalized in amplitude according to the noise estimate, so the order of the Hermite polynomial can be fixed arbitrarily as 0 or 1 for convenience. Even though the assumption of a far-field source cannot always be satisfied, time-delay information can still be useful in locating a target. Transforming the cross correlation into the q domain opens the possibility of adopting methods commonly used in sonar and radar; in particular, the use of an array of sensors (Van Veen and Buckley, 1988), which allows for localizing sources with high resolution.

DISCUSSION AND CONCLUSIONS

By using similarities between EM sounding in dielectric and conductive media, one can transform between solutions in one type to the other. The foremost important conclusion of this paper is that such a transformation can be constructed using alternate implementations. The implementations presented in this work are by no means an exhaustive search, only examples. The optimal transformation depends on the particular case. The important properties of the two types of propagation which enable those transformations are 1) that the transformations are local and require only the knowledge of the time history at the particular point of interest and 2) that the transformations are linear. It is important to remember that the above relations (which are derived directly from Maxwell equations) are independent of the particular frequency used.

The fact that only low frequencies can be obtained with sufficient signal to noise determines the possible resolution and implies that we cannot make the assumption of far field behavior. Far field behavior would allow the use of geometrical optics and simplify the interpretation of high frequency wave field sounding. Because $(\sigma\ \mu)^2 >> \varepsilon\ \mu$, the equivalent speed of propagation in the transformed media is much slower than that of the dielectric media. Therefore, in-spite of the low frequency of the sounding, we still can obtain a reasonable depth resolution. Three new methods were derived to perform the transformations. The first method is based on a direct relation between the Laplace transform of the two solutions. The second method is also based on the Laplace transform but avoids the intermediate calculation of the transform and derives direct time-domain relations between the diffusive and the wave solutions. Further simplification is achieved by using exponential time sampling. The third method is based on expanding the diffusive solution into exponential functions as a function base that is transformed into a sinusoidal base. The three methods were demonstrated on numerical examples. The direct time-domain transform appears to be the best in terms of numerical efficiency and ability to handle different levels of signal to noise. However for the direct Laplace transform the resolution of the GS algorithm is not sufficient. It should be remembered that this conclusion is partially based on the choice of GS for the inversion of the Laplace transform. Use of an alternative Laplace inversion routine may solve those shortfalls (Brian et al. 1979). A Laplace inversion routine based on Appendix B was tried successfully. Results will be published elsewhere. The significance of Laplace sine transform lies in the analogy to the Doppler effect it indicates. This analogy is useful where relative motion between the media, the target, the transmitter and the receiver, is expected. Underwater sounding from a moving platform is a typical case. While in wave propagation, the consequence of relative movement is an apparent shift in frequency; the equivalent in diffusive propagation is the apparent change in the decay rate of an exponential excitation.

REFERENCES

Abramowitz and Stegun, 1964, Handbook of mathematical functions: Dover publication, 775.

Carter, Clifford G., 1987, Coherence and time delay estimation Proceeding of IEEE, **75.** 236-255.

Cohen, L., 1992, Scale, frequency, time, and the scale time transform: IEEE. SSAP. workshop.

Craig, I. J. D., and Thompson, A. M., 1994, Why Laplace transforms are difficult to invert numerically: Computers in Physics, **8**, 648-654.

Davis B., and Martin B., 1979, Numerical inversion of the Laplace transform: a survey and comparison of methods. J. Compu. Phys., **33**, 1-32

Filatov, V. V., 1984 Construction of focusing transformations of transient electromagnet fields: Geologiya i Geofizika, **25,** 89-85.

Gershensom, M., Simple interpretation of time-domain electromagnetic sounding using similarities btween wave and diffusion propagation: Geophysics, **62**, 763-774.

Isaev, G. A., Filatov, V. V., 1981 Physicomathematical principles of visualization of nonstationary electromagnetic fields: Geologiya i Geofizika, **22,** 89-95.

Knight, J. H., 1982 and Raich A P., Transient electromagnetic calculations using the Gaver-Stehfest inverse Laplace transform method: Geophysics, **47**, 47-50.

Kunetz G., 1972, Processing and interpretation of magnetotelluric sounding: Geophysics, 37, 1005-1021.

Lee, K. H., Liu, G., and Morrison, H. F., 1989, A new approach to modeling the electromagnetic response of conductive media: Geophysics, **54** 1180-1192.

Lee, K. H. and Xie, G. O., 1993, A new approach to imaging with low-frequency electromagnetic field: Geophysics, **58** 780-796.

Lee, S. McMehan G. A., and Aiken L. V., 1987, Phase-field imaging the electromagnetic equivalent of seismic migration: Geophysics, 52, 678-693.
Levy S., Oldenburg D. and Wang J., 1988 Subsurface imaging using magnetotelluric data: Geophysics, **53**, 104-117.
Roberts, G. E. and Kaufman, H., 1966, Tables of Laplace transforms: W.B. Saunders Company, 171,172.
Spies, B. R., 1989, Depth of investigation in electromagnetic sounding methods: Geophysics, **54** 872-888.
Van Veen B.D. and Buckley, K.M., 1988, Beam forming: A versatile approach to special filtering: IEEE ASSP Magazine, 4-24

AUTHOR INDEX

GPSR Compliance

The European Union's (EU) General Product Safety Regulation (GPSR) is a set of rules that requires consumer products to be safe and our obligations to ensure this.

If you have any concerns about our products, you can contact us on ProductSafety@springernature.com

In case Publisher is established outside the EU, the EU authorized representative is:

Springer Nature Customer Service Center GmbH
Europaplatz 3
69115 Heidelberg, Germany

Batch number: 10755284

Printed by Printforce, the Netherlands